Topics in
REAL ANALYSIS

[For B.C.A. IInd Year Students of K.U., Kurukshetra]

By

Dr. KULBHUSHAN PARKASH

Head, Deptt. of Mathematics
D.A.V. College (Lahore),
Ambala City
Haryana

FIREWALL MEDIA

(An Imprint of Laxmi Publications Pvt. Ltd.)

BANGALORE • CHENNAI • COCHIN • GUWAHATI • HYDERABAD
JALANDHAR • KOLKATA • LUCKNOW • MUMBAI • RANCHI
NEW DELHI • BOSTON, USA

Published by :

FIREWALL MEDIA

(An Imprint of Laxmi Publications Pvt. Ltd.)
113, Golden House, Daryaganj,
New Delhi-110002

Phone : 011-43 53 25 00
Fax : 011-43 53 25 28

www.laxmipublications.com
info@laxmipublications.com

Price : ₹ **295.00** *Only.*

First Edition : 2009
Reprint : 2013

OFFICES

✆ **Bangalore**	080-26 75 69 30	✆ **Chennai**	044-24 34 47 26
✆ **Cochin**	0484-237 70 04, 405 13 03	✆ **Guwahati**	0361-251 36 69, 251 38 81
✆ **Hyderabad**	040-24 65 23 33	✆ **Jalandhar**	0181-222 12 72
✆ **Kolkata**	033-22 27 43 84	✆ **Lucknow**	0522-220 99 16
✆ **Mumbai**	022-24 91 54 15, 24 92 78 69	✆ **Ranchi**	0651-220 44 64

FRE-3010-295-TOPICS IN REAL ANALYSIS-PAR
Typeset at : Goswami Associates, Delhi.

C—6801/013/04
Printed at : Ajit Printers, Delhi.

Topics in
REAL ANALYSIS

CONTENTS

SYLLABUS

BCA-206 : REAL ANALYSIS

Max. Marks : 100
Min. Pass Marks : 35
Time : 3 Hours

Note: Ten questions are to be set : Four from section A, three each from Section B and C. The candidates are required to attempt five questions in all selecting at least one questions from each section.

SECTION-A **(4 Questions)**

Denumerable and non-denumerable sets, Denumerability of integers and rationals and non-denumerability of real numbers. Properties of the real number system as a complete ordered field. The concepts of bounds, neighbourhoods,interior points, isolate points and limit points in R open and closed sets with their properties in R Bolzano Weierstrass Theorem. *Heine-Borel Theorem*. Real sequences and their convergence. Cauchy sequences. Cauchy's general principle of convergence. Suprema and infinima of bounded sequences.

Infinite series of Real numbers of their convergence as well as divergence. Comparison, ratio, root integral Leibnitz's Gauss and Cauchy's Condensation tests.

SECTION-B **(3 Questions)**

Absolute convergence and re-arrangement of series. Product of them two absolutely convergent series, Cauchy product of two series. Convergence of infinite product.

Properties of continuous functions, definition of uniform continuity. Statement of theorem : "A continuous function in a closed bounded interval in uniformly continuous" Types of discontinuities with examples. Rigorous proofs of Rolle's Lagrange's mean value and Taylor's theorems.

SECTION-C **(3 Questions)**

Definition and existence of Riemann integral of a bounded function, Darboius condition of integrability. Riemann integrability of continuous functions, and monotonic function. Riemann integral of functions with finite number of discontinuities and of function with discontinuity points having a finite number of limit points. Riemann integral as the limit of a sum. The fundamental theorem of integral calculus.

Sequence and series of functions and their point-wise convergence. Uniform convergence of sequences and series of functions, Weierstrass M-test Uniform convergence and continuity, uniform convergence and integration, uniform convergence and differentiation. Improper integral ; Convergence of an improper integral comparison test Dirichlet's test. Beta and Gamma functions, their properties and relationships, Differentiation under integral sign.

PREFACE

It is indeed a matter of immense pleasure for me that the book 'A Textbook on Real Analysis' has been appreciated and liked by both students and teachers of Kurukshetra University, Kurukshetra and M.D. University, Rohtak. The present book is under the title 'Topics in Real Analysis' but according to the latest syllabus as prescribed by Kurukshetra University, Kurukshetra for BCA II. However, the book can be found to be useful by the students for undergraduate, B.Tech. and other competitive examinations. The subject has been treated on the lines as I feel it to be developed in the classroom. The text of the book has been prepared with the following salient features :

(*i*) The language of the book is simple and easy to understand.

(*ii*) Starting with a brief introduction, each topic has been presented in a *systematic, simple, lucid* and *exhaustive* manner.

(*iii*) Keeping in view the nature of the subject, a large number of typical and important solved examples properly selected from various university question papers have been provided to enable the student to have a clear grasp of the subject and to equip him for attempting problems in the university examination without any difficulty.

Apart from providing a large number of solved examples, different type of questions in ample quantity without necessary repetition, have also been provided for a thorough practice to the students. Typical and difficult questions have been provided with sufficient hints.

(*iv*) A large number of '*notes*' and '*remarks*' have been added for a *better* understanding of the subject.

In fact, no pains have been spared to make the book interesting and useful. Now I look forward to the valuable suggestions and comments of the readers in order to make the book more purposeful and responsive in the proper perspective.

I am grateful to all those whose works have been consulted in the preparation of this book. Some references have been provided in the Bibliography at the end of the book.

—AUTHOR

1

Preliminaries on Sets and Functions

In this chapter, we describe some fundamental concepts on Set Theory and Functions which are essential for the study of the rest of the book.

1.1. SET

1.1.1. Set. *A set is a well-defined collection of distinct objects.*

By a 'well-defined collection' we mean we must be given a rule which enables us to say without doubt or ambiguity whether a particular object belongs to the collection or not. For example, the collection of all vowels of the English alphabet is a well defined collection of distinct objects and is thus a set. Collection of numbers 1, 2, 3, 4, 5 is a well-defined collection of distinct objects and is therefore a set. But collection of rich people in India is not well-defined and is thus not a set. Likewise the collection of interesting books in a library is not a set.

The words 'set', 'collection', 'aggregate' and 'family' are used synonymously. Each object (whatsoever) which belongs to a set under consideration is called a member or an element of the set.

Sets are usually denoted by capital letters *e.g.* A, B, C, X, Y, Z and the elements by the lower case letters *e.g.* a, b, c, x, y.

If an object x is a member of a set A, we write '$x \in$ A' and read as 'x belongs to A' or 'x is in A' or 'x is an element of A' or 'A contains x'. If an object is not a member of A, we write '$x \notin$ A' and say that 'x does not belong to A' or 'x is not in A' or 'x is not an element of A' or 'A does not contain x'. For instance, if N denotes the set of all natural numbers, then $5 \in$ N but $-5 \notin$ N.

Some typical sets used in this book are :

N : Set of all natural numbers
Z : Set of all integers
Q : Set of all rational numbers
Q^+ : Set of all positive rational numbers
I : Set of all irrational numbers
R : Set of all real numbers
R^+ : Set of all positive real numbers
R^- : Set of all negative real numbers
C : Set of all complex numbers.

We shall use the symbols :

$\exists$ for 'there exists'

$\forall$ for 'for all'

iff for 'if and only if'.

If the truth of the proposition P implies the truth of proposition Q, we write $P \Rightarrow Q$ and read as 'P implies Q'.

If the propositions P and Q are such that the truth of either implies the truth of other, we write '$P \Leftrightarrow Q$' and read as 'P implies Q and P is implied by Q'.

1.1.2. Representation of Sets. A set may be written by the following two methods :

(*i*) **Roster Method** (*Tabulation Method or Listing Method*). In this method a set is described by listing all its elements, separating them by commas and enclosing them by curly brackets.

For example,

(*i*) the set of all vowels of the English language is written as $\{a, e, i, o, u\}$.

(*ii*) the set of natural numbers less than 5 is written as $\{1, 2, 3, 4\}$.

If the listing of a few elements of a set enables us to know about its remaining elements, then we write only those elements followed by dots, *e.g.*

(*i*) the set N of all numbers can be written as $N = \{1, 2, 3, ...\}$.

(*ii*) the set Z of all integers can be written as $Z = \{..., -3, -2, -1, 0, 1, 2, 3, ...\}$.

(The dots in both the cases indicate that there are other objects also in the set).

(*ii*) **Property Method** (*Rule Method or Set Builder Form*). Generally, we have to deal with sets, the elements of which are not even explicitly, known to us, though we know a certain property which distinguishes the objects of our set. For instance, we do not know immediately, the elements of the set which consists of all the roots of the equation

$x^{10} + 9x^5 + 1 = 0$. We write this set A, as $A = \{x \mid x^{10} + 9x^5 + 1 = 0\}$.

More generally, if P is a property and A is the set of all those elements for which the property P is true, then we write

$$A = \{x \mid x \text{ satisfies P}\}$$

or

$$A = \{x : x \text{ satisfies P}\}$$

and we read as A is the set of all those elements x for which the property P is true.

1.1.3. Null Set. *A set having no element is called a null set or an empty set or a void set.* It is denoted by ϕ. For example,

$$A = \{x \mid x \in N \text{ and } x^2 + 5 = 0\}$$

$$B = \{x \mid x \in R \text{ and } x^2 < 0\}$$

are null sets.

1.1.4. Singleton. *A set having only one element is called a singleton.* For example :

The sets $\{a\}$, $\{b\}$, $\{w\}$, $\{1\}$ and $A = \{x \mid x \geq 1 \text{ and } x \leq 1\}$ are all singletons.

1.1.5. Finite Set. *A set having a finite number of elements is called a finite set.*

For example,

(*i*) $V = \{a, e, i, o, u\}$

(*ii*) $A = \{1, 2, 5, 7\}$

are finite sets.

ϕ is also a finite set.

1.1.6. Infinite Sets. *A set having infinitely many elements is called an infinite set.*

For example, the sets N, Z, Q, R and C are all infinite sets.

1.1.7. Equal Sets (Identical Sets). *Two sets A and B are said to be equal or identical when they have same elements, i.e.,* every element of A is an element of B and every element of B is an element of A.

If A is equal to B, we write A = B.

If A is not equal to B, we write A ≠ B.

For example, if A = {2, 5, 7}, B = {5, 7, 2} and C = {5, 9, 2},

then A = B but A ≠ C and B ≠ C.

Note. *(i)* $\{\phi\} \neq \phi$. *(ii)* $\phi \in \{\phi\}$.

1.1.8. Subset and Super-set. *If A and B are sets, then A is called a subset of B if every element of A is an element of B.* In this case we also say that A is contained in B (or B contains A) and we write

A ⊆ B (or B ⊇ A). We also say that B is a super-set of A.

If A is not a subset of B, we say that A ⊈ B or B ⊉ A.

(This is the case when there is at least one x such that $x \in$ A and $x \notin$ B.)

For example, (i) If A = {2, 7, 9}, B = {1, 2, 7, 9, 11, 25}, then A ⊆ B.

(ii) N ⊆ Z, Z ⊆ Q, Q ⊆ R, R ⊆ C.

Note. A = B iff A ⊆ B and B ⊆ A.

1.1.9. Proper Subset. *A non-empty set A is called a proper subset of a set B* (A is properly contained in B or B properly contains A).

iff *(i)* A ⊆ B and *(ii)* A ≠ B.

If A is a proper subset of B, then we write A ⊂ B.

In the above example,

(i) A ⊂ B *(ii)* N ⊂ Z, Z ⊂ Q, Q ⊂ R, R ⊂ C.

Note. *(i) Empty set is a subset of every set.*

(ii) A ⊆ A, for every set A.

(iii) If A, B and C are three sets such that A ⊆ B and B ⊆ C, then A ⊆ C.

(iv) ϕ and A are improper subsets of A.

1.1.10. Power Set of a Set. *The set of all subsets of a given set A is called the* **power set of A** *and is denoted by P(A).*

For example, (i) P(ϕ) = {ϕ}

(ii) If A = {1, 2}, then P(A) = {ϕ, {1}, {2}, A}.

(iii) If A = {a, b, c}, then P(A) = {ϕ, {a}, {b}, {c}, {a, b}, {b, c}, {c, a}, A}.

1.1.11. Theorem. *If a set A has n elements (n finite), then power set P(A) has 2^n elements.*

1.1.12. Universal Set. *If in a particular problem all the sets we deal with become subsets of a fixed set, then this fixed set is called the* **universal set.** *It is generally denoted by X or U.* For example, when we discuss only integers, then the set Z of all integers is the universal set.

1.1.13. Venn Diagrams. In order to clarify ideas of sets with the help of our geometric intution, we represent relations among sets by means of certain diagrams known as Venn diagrams or Venn-Euler diagrams*. In Venn diagrams, the universal set X is generally indicated by a rectangular region and any subset A (say) of X is indicated by the region enclosed within a closed curve.

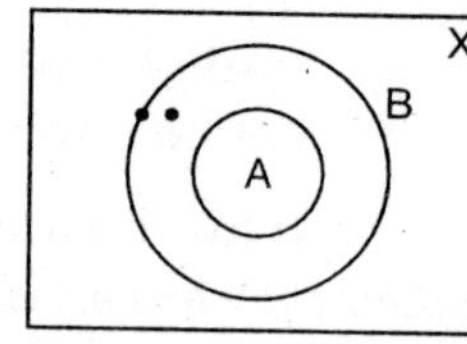

A, B $\subseteq$ X and A $\subseteq$ B

1.1.14. Indexed Family of sets. Let Λ be any non-empty set. Let for each $\lambda \in \Lambda$, there be a set A_λ. Then the set $\{A_\lambda \mid \lambda \in \Lambda\}$ is called a family of sets indexed by the set Λ. Λ is called the *indexing set.* We also denote the indexed family by $\{A_\lambda\}_{\lambda \in \Lambda}$.

For example, if $\Lambda = \{1, 2, 3, 4, 5\}$, *then the set*

$\{A_1, A_2, A_3, A_4, A_5\}$ or $\{A_\lambda\}_{\lambda \in \Lambda}$ *is the indexed family and* Λ *is the indexing set.*

SET OPERATIONS

1.1.15. Union of two sets. *The union (sum or join) of two sets A and B, written as A $\cup$ B, is the set of all those elements which belong to A or to B (or to both A and B).*

For example,

(*i*) If A = N, B = Z, then A $\cup$ B = Z

(*ii*) If A = Q and B = I (set of irrationals), then A $\cup$ B = R

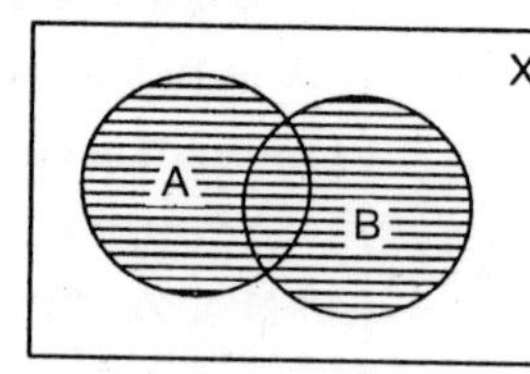

A $\cup$ B

1.1.16. Properties of Union. *If X is the universal set and A, B, C $\subseteq$ X, then the operation of union satisfies the following properties :*

(*i*) A $\cup$ B = B $\cup$ A (Commutativity)

(*ii*) A $\cup$ ϕ = ϕ $\cup$ A = A (Identity), ϕ being the null set

(*iii*) A $\cup$ A = A (Idempotent)

(*iv*) A $\cup$ X = X

(*v*) A $\subseteq$ A $\cup$ B and B $\subseteq$ A $\cup$ B

(*vi*) (A $\cup$ B) $\cup$ C = A $\cup$ (B $\cup$ C) (Associativity)

1.1.17. Union of a finite number of sets. The union of n sets $A_1, A_2, \ldots, A_n$ is written as $\bigcup_{i=1}^{n} A_i$ and is a set defined by

$$\bigcup_{i=1}^{n} A_i = \{x \mid x \in A_i \text{ for at least one } i,\ 1 \le i \le n\}.$$

1.1.18. Union of an arbitrary family of sets. Let $\{A_\lambda\}_{\lambda \in \Lambda}$ be an arbitrary family of sets. Their union is denoted by $\bigcup_{\lambda \in \Lambda} A$ and is a set defined by

$$\bigcup_{\lambda \in \Lambda} A_\lambda = \{x \mid x \in A_\lambda \text{ for at least one } \lambda \in \Lambda\}.$$

---

*After the names of an English logician John Venn (1839—1923) and the famous Swiss mathematician Leonhard Euler (1707—1787).

1.1.19. Intersection of two sets. *The intersection (product or meet) of two sets A and B, written as $A \cap B$, is the set of elements common to both A and B, e.g.*

$$A \cap B = \{x \mid x \in A \text{ and } x \in B\}$$

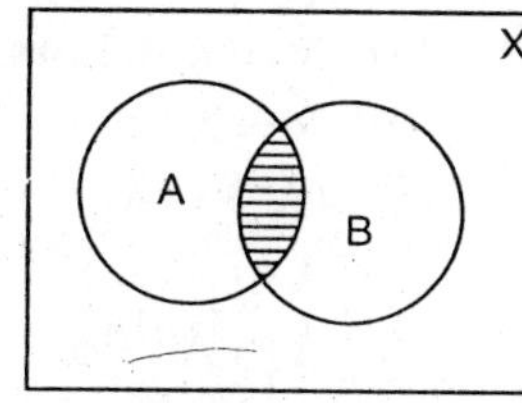

$A \cap B$

For example,

(*i*) if A = N, B = Z, then

$$A \cap B = N$$

(*ii*) if A = Q and B = I, then

$$A \cap B = \phi.$$

1.1.20. Disjoint sets. Two sets A and B are said to be disjoint sets if $A \cap B = \phi$.

1.1.21. Properties of intersection. If X is the universal set and A, B, C are subsets of X, then the operation of intersection satisfies the following properties :

(*i*) $A \cap B = B \cap A$ (Commutativity)

(*ii*) $A \cap A = A$ (Idempotent)

(*iii*) $A \cap X = A$

(*iv*) $A \cap \phi = \phi$, ϕ being the empty set

(*v*) $(A \cap B) \cap C = A \cap (B \cap C)$

(*vi*) $A \cap B \subseteq A$ and $A \cap B \subseteq B$.

1.1.22. Intersection of a finite number of sets. The intersection of n sets $A_1, A_2, A_3, \ldots, A_n$ is written as $\bigcap_{i=1}^{n} A_i$ and is a set defined by $\bigcap_{i=1}^{n} A_i = \{x \mid x \in A_i,$ for each $i, 1 \le i \le n\}$.

1.1.23. Intersection of an arbitrary family of sets. Let $\{A_\lambda\}_{\lambda \in \Lambda}$ be an arbitrary family of sets. Their intersection is denoted by $\bigcap_{\lambda \in \Lambda} A_\lambda$ and is a set defined by

$$\bigcap_{\lambda \in \Lambda} A_\lambda = \{x \mid x \in A_\lambda \text{ for every } \lambda \in \Lambda\}.$$

1.1.24. Disjoint family of sets. A family of sets is called a disjoint family of sets if no two sets have any common element.

1.1.25. Difference of two sets. If A and B are two given sets, then the set of all those elements which are in A but not in B, is called 'the difference of A and B' (more accurately difference of B from A) and is denoted by A – B, *i.e.*,

$$A - B = \{x \mid x \in A \text{ but } x \notin B\}.$$

For example, (*i*) if $A = \{a, b, c, d\}$, $B = \{c, d, e, f\}$, then

$$A - B = \{a, b\}.$$

(*ii*) if A = Q, B = I, then A – B = Q.

Note. (*i*) $A - B \neq B - A$ in general

(*ii*) $A - B \subseteq A$ and $B - A \subseteq B$.

1.1.26. Complement of a set. Let X be the universal set and $A \subseteq X$. Then the set of those elements of X which are not in A, is called the complement of A and is denoted by A^c (or X – A), *i.e.*,

$$A^c = \{x \mid x \in X \text{ and } x \notin A\}.$$

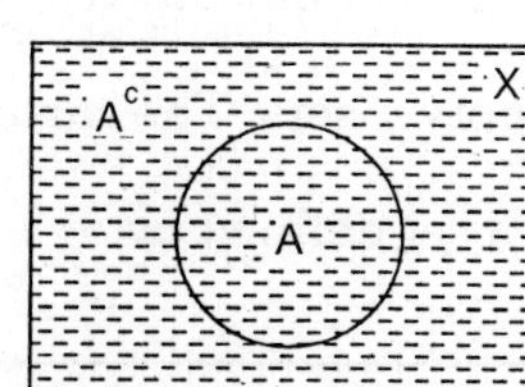

For example, if X = R

and A = Q

then $A^c = I$.

1.1.27. Properties of Complement. If X is the universal set and A, B, C are its subsets, then the operation of complementation satisfies the properties :

(*i*) $X^c = \phi$, where ϕ is the empty set

(*ii*) $\phi^c = X$

(*iii*) $(A^c)^c = A$

(*iv*) $A \cup A^c = X$

(*v*) $A \cap A^c = \phi$

(*vi*) If $A \subseteq B$, then $B^c \subseteq A^c$

(*vii*) $A = B \Leftrightarrow A^c = B^c$

(*viii*) De Morgan's laws :*

(*a*) $(A \cup B)^c = A^c \cap B^c$

[Complement of union is intersection of complements]

(*b*) $(A \cap B)^c = A^c \cup B^c$

[Complement of intersection is union of complements]

(*ix*) Generalized De Morgan's laws

(*a*) $[\bigcap_{\lambda \in \Lambda} A_\lambda]^c = \bigcup_{\lambda \in \Lambda} A_\lambda{}^c$

(*b*) $[\bigcup_{\lambda \in \Lambda} A_\lambda]^c = \bigcap_{\lambda \in \Lambda} A_\lambda{}^c$.

1.1.28. Symmetric difference of two sets. If A and B are sets then their symmetric difference is denoted by $A \Delta B$ and is defined by

$$A \Delta B = (A - B) \cup (B - A).$$

For example, if $A = \{a, b, c, d\}$, $B = \{b, c, e, f\}$, then

$$A - B = \{a, d\},\ B - A = \{e, f\} \text{ and } A \Delta B = \{a, d, e, f\}.$$

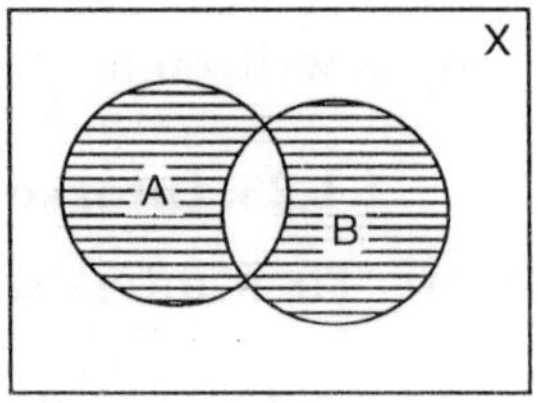

$A \Delta B$

CARTESIAN PRODUCT

1.1.29. Cartesian product of two sets. Let A and B be two non-empty sets. Then, the set of all ordered pairs (a, b), where $a \in A$ and $b \in B$, is called the Cartesian product of A and B, and is denoted by $A \times B$. *i.e.*

$$A \times B = \{(a, b) \mid a \in A \text{ and } b \in B\}.$$

For example, (*i*) if $A = \{0, 1\}$, $B = \{2, 5\}$, then

$$A \times B = \{(0, 2), (0, 5), (1, 2), (1, 5)\}.$$

(*ii*) if $A = B = R$, then

$$A \times B = R \times R \text{ (also denoted by } R^2)$$

represents the Euclidean plane.

1.1.30. Cartesian product of a finite number of sets. Let $A_1, A_2, ..., A_n$ be any non empty n sets. The cartesian product of the sets A_i is denoted by $A_1 \times A_2 \times ... \times A_n$ or $\prod_{i=1}^{n} A_i$ and is a set defined by

*De Morgan was an Indian Born British Mathematician, born in Madurai.

$$\prod_{i=1}^{n} A_i = \{(a_1, a_2, ..., a_n) \mid a_i \in A_i, 1 \leq i \leq n\}$$

$(a_1, a_2, ..., a_n)$ is called an ordered n-tuple.

For example, if $n = 3$, then R × R × R (or written as R^3) is the set of all points of three dimensional Euclidean space.

1.2. FUNCTIONS

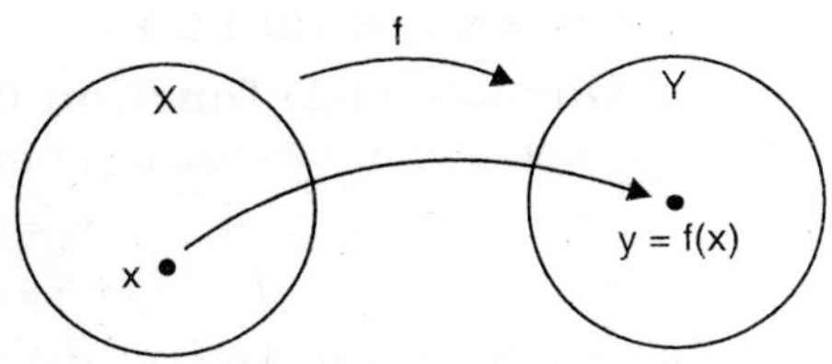

1.2.1. Function. Let X and Y be any two non-empty sets. Then a rule which associates *to each* x in X, *a unique* y in Y, is called a function (or a map or mapping or correspondence) from the set X to the set Y. It is denoted by f, g, ϕ, ψ, etc. If f is a function from X to Y, we write $f : X \to Y$ or $X \to Y$.

The unique element $y \in Y$ associated by f to an element $x \in X$, is called the value of the function f at x or image of x under f or f-image of x and we denote it by $f(x)$, *i.e.*, $y = f(x)$. The element $x \in X$, is called the pre-image of y.

X is called the domain of the function f and Y is called the codomain or range space or image space or just range set of the function f. The set of all f-images is called the range of f. If $f : X \to Y$ is a function and $A \subseteq X$, then $f(A) = \{f(a) \mid a \in A\}$ is called the image of A under f. $f(X)$ is the range of the function f and $f(X) \subseteq Y$.

Examples :

(*i*) Let X = Y = Z (set of integers). The rule which associates to each $n \in Z$, its square n^2, is a function from X to Y.

(*ii*) Let $X = \{n^2 \mid n \in Z\}$ and Y = Z. The rule which associates to each $n \in X$, its square root, is not a function because this rule associates to the element 1, two elements + 1 and – 1, and not the unique element.

(*iii*) If $X = \{n^2 \mid n \in Z\}$ and $Y = n \mid n \geq 0, n \in Z\}$, then the rule defined in (*ii*) is a function.

(*iv*) If X = Z, $Y = \{n \mid n \in Z, n \geq 0\}$, then the rule defined in (*ii*) is not a function because no element of Y is associated to – 1 in X.

Note 1. Sometimes $f(x)$ is also used for a function. But this notation should be avoided as it indicates the value of the function at x.

2. The definition of the function $f : X \to Y$ asserts that each element of X has a unique image in Y but each element of Y may not have a unique pre-image.

3. Since by definition 'to each $x \in X$', there corresponds a unique $y \in Y$,

$$\therefore \qquad f(X) \neq \phi.$$

4. We define $f(\phi) = \phi$.

1.2.2. Types of Functions

(*i*) **Constant function.** *A function $f : X \to Y$ is said to be a constant function if $f(x) = a$ (some $a \in Y$), $\forall\, x \in X$, i.e., if range of the function is a singleton.*

For example, the function $f : R \to R$ such that $f(x) = 2, \forall\, x \in R$ is a constant function with range {2}.

(*ii*) **Identity function**. *Let X be a non-empty set. Then a function $I_X : X \to X$ defined by $I_X(x) = x, \forall\, x \in X$ is called the identity function of X to itself.*

For example, let X = {1, 5, 7, 9, 20}.

Then the function $I_X : X \to X$ such that $I_X(1) = 1$, $I_X(5) = 5$. $I_X(7) = 7$, $I_X(9) = 9$, $I_X(20) = 20$ is the identity function on X.

(*iii*) **Into function.** *A function $f : X \to Y$ is said to be an into function if there exists at least one element in Y which has no pre-image* under *f*, *i.e.*, $f(X) \neq Y$. Functions in examples (*i*) and (*v*) are into functions.

(*iv*) **Onto function (surjection).** *A function $f : X \to Y$ is said to be an onto function if given $y \in Y$, there exists at least one $x \in X$ such that $f(x) = y$, i.e., $f(X) = Y$.*

See example (*iii*) of 1.2.1.

(*v*) **One-one (1-1) function (injection).** *A function $f : X \to Y$ is said to be one-one if distinct elements of X have distinct images under f.*

i.e., $$x_1 \neq x_2,\ x_1, x_2 \in X$$

$$\Rightarrow \quad f(x_1) \neq f(x_2)$$

See examples (*iii*) and (*v*) of 1.2.1.

(*vi*) **Many one function.** *A function which is not one-one is said to be many one function. i.e.*, a function $f : X \to Y$ is many one if $\exists\, x_1, x_2 \in X$ such that $x_1 \neq x_2$ and $f(x_1) = f(x_2)$.

See example (*i*) of 1.2.1.

(*vii*) **One-one onto function** (*one to one correspondence or bijection*). A function which is both one-one and onto is called a one-one onto function.

For example, the function $f : N \to Z$ defined by

$$f(n) = \begin{cases} \frac{1}{2}(n-1), & \text{if } n = 1, 3, 5, \ldots.. \\ -\dfrac{n}{2}, & \text{if } n = 2, 4, 6, \ldots.. \end{cases}$$

is both one-one and onto.

(*viii*) **Inclusion function.** *Let $X \subseteq Y$. Then the function $i : X \to Y$ defined by $i(x) = x\ \forall\, x \in X$, is called the inclusion function of X into Y. It is one-one.*

(*ix*) **Restriction of a function f.** *Let $f : X \to Y$ be a function. Let $A \subseteq X$. Then the function f_A defined by $f_A(x) = f(x)\ \forall\, x \in A$, is called the restriction of f to A.* It is one-one if *f* is one-one.

But f_A may be one-one even when *f* is not.

f_A may not be onto even if *f* is onto.

f is called the extension of f_A.

(*x*) **Equality of two functions.** *Two functions f and g from a set X to a set Y are said to be equal if $f(x) = g(x)\ \forall\, x \in X$.*

1.2.3. Direct Images. Let $f : X \to Y$ be a function. If $A \subseteq X$, then its direct image $f(A)$ is a subset of Y defined by

$$f(A) = \{f(x) \mid x \in A\}.$$

Examples : (*i*) Let $f : R \to R$ be a function defined by

$$f(x) = 2x + 5$$

Let $$A = \{2, 5, 7, 9\}.$$

Then $$f(A) = \{9, 15, 19, 23\}$$

(*ii*) Let $f : R \to R$ be a function defined by

$$f(x) = x^2$$

Let $A = [-1, 0]$, then $f(A) = [0, 1]$.

Note. $x \in A \Rightarrow f(x) \in f(A)$, but $f(x) \in f(A) \nRightarrow x \in A$.

For example, in example *(ii)*, $f(\frac{1}{2}) = \frac{1}{4} \in [0, 1]$ but $\frac{1}{2} \notin A$.

1.2.4. Theorem. If $f : X \to Y$ is a function and $A \subseteq X$, $B \subseteq X$.

Then, *(i)* $A \subseteq B \Rightarrow f(A) \subseteq f(B)$ *(ii)* $f(A \cup B) = f(A) \cup f(B)$

(iii) $f(A \cap B) \subseteq f(A) \cap f(B)$.

Following example shows that inclusion in *(iii)* may be proper.

Let $f : R \to R^+$ be a function defined by $f(x) = x^2$

Let $A = \{-2, 5\}$, $B = \{2, 5\}$. Then $A \cap B = \{5\}$

$$f(A) = \{4, 25\}, f(B) = \{4, 25\}$$

$$f(A) \cap f(B) = \{4, 25\}.$$

$$f(A \cap B) = \{25\}$$

Thus, $$f(A \cap B) \subset f(A) \cap f(B)$$

However, equality exists in *(iii)* iff f is one-one.

Generalization. If $f : X \to Y$ is a function and $\{A_i\}$, $1 \le i \le n$ is a family of subsets of X, then

(i) $f(\bigcup_{i=1}^{n} A_i) = \bigcup_{i=1}^{n} f(A_i)$ *(ii)* $f(\bigcap_{i=1}^{n} A_i) \subseteq \bigcap_{i=1}^{n} f(A_i)$.

1.2.5. Inverse Images. Let $f : X \to Y$ be a function and B a subset of Y. Then inverse image of B is denoted by $f^{-1}(B)$ and is a subset of X defined by

$$f^{-1}(B) = \{x \mid x \in X \text{ and } f(x) \in B\}.$$

Example : Let $X = \{-2, -1, 0, 1, 2\}$ and $Y = \{0, 1, 2, 3, 4, 5, 6, 7\}$

Let $f : X \to Y$ such that $f(x) = x^2 \ \forall \ x \in X$

(i) If $B = \{0, 1, 2\}$, then $f^{-1}(B) = \{-1, 0, 1\}$

(ii) If $B = \{2, 3, 5, 7\}$, then $f^{-1}(B) = \phi$

(iii) If $B = \{0, 1, 4\}$, then $f^{-1}(B) = X$.

Note. *(i)* $f^{-1}(\phi) = \phi$

(ii) $f^{-1}(Y) = X$

(iii) $f^{-1}(B) = \phi \nRightarrow B = \phi$ as is clear from example *(ii)* above

(iv) $f^{-1}(y) \neq f^{-1}(\{y\})$.

In fact $f^{-1}(y)$ is defined iff f is one-one and onto whereas $f^{-1}\{y\}$ is always defined.

1.2.6. Theorem. If $f : X \to Y$ is a function and $A \subseteq Y$, $B \subseteq Y$, then

(i) $A \subseteq B \Rightarrow f^{-1}(A) \subseteq f^{-1}(B)$ *(ii)* $f^{-1}(A \cup B) = f^{-1}(A) \cup f^{-1}(B)$

(iii) $f^{-1}(A \cap B) = f^{-1}(A) \cap f^{-1}(B)$.

Generalization. If $f : X \to Y$ is a function and $\{A_i\}$, $1 \le i \le n$ is a family of subsets of Y, then

(i) $f^{-1}(\bigcup_{i=1}^{n} A_i) = \bigcup_{i=1}^{n} f^{-1}(A_i)$ *(ii)* $f^{-1}(\bigcap_{i=1}^{n} A_i) = \bigcap_{i=1}^{n} f^{-1}(A_i)$.

1.2.7. Composite (or Resultant) of functions. Let $f : X \to Y_1$ and $g : Y_2 \to Z$ be two functions. (Here Z is not necessarily the set of integers.)

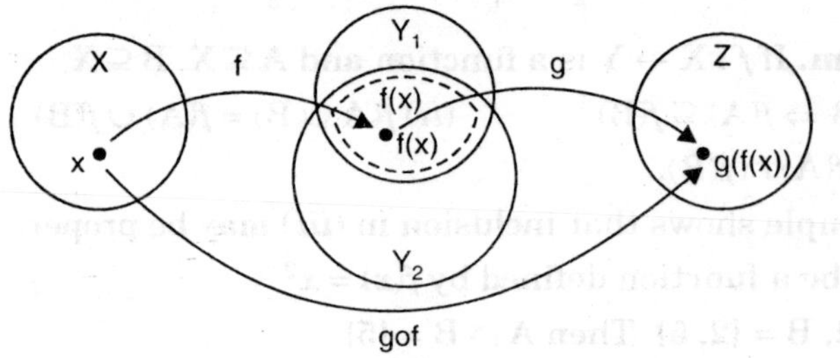

Let $f(X) \subseteq Y_2$ (*i.e.*, range of f is a subset of domain of g). Then the function $h : X \to Z$ defined by $h(x) = g(f(x)) \ \forall \ x \in X$

is called the composite (or resultant) of two functions f and g. It is denoted by gof, *i.e.*, $gof : X \to Z$. It is also called function of a function.

Examples : (*i*) Let $X = \{1, 2, 3, 4, 5\}$, $Y_1 = \{7, 8, 9, 10\}$,

$Y_2 = \{8, 9, 10, 12\}$, $Z = \{15, 16, 17, 18, 19\}$.

Let $f : X \to Y_1$ and $g : Y_2 \to Z$.

Association of elements is indicated by arrows.

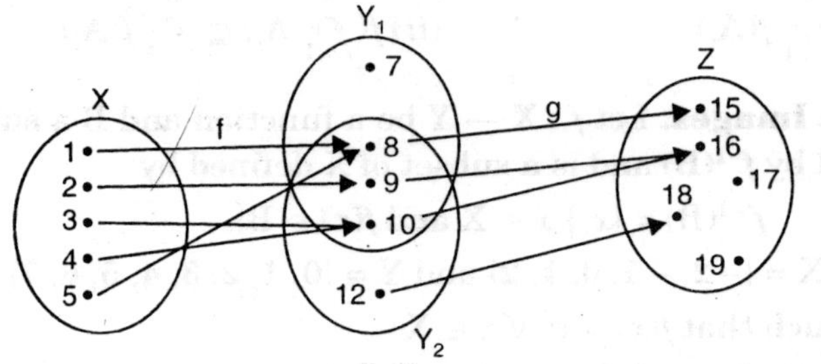

The composite of f and g is a function $gof : X \to Z$ defined by

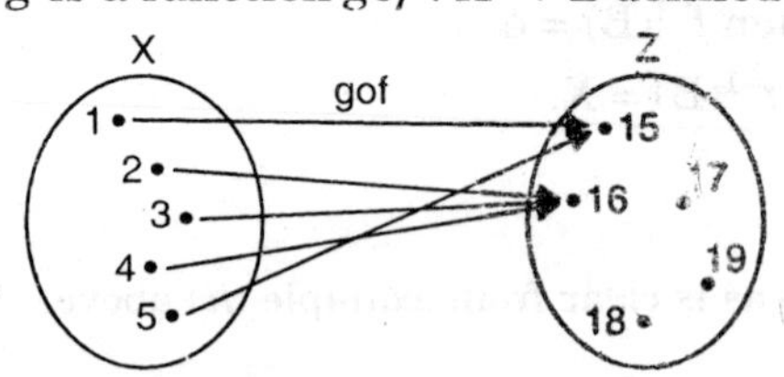

(*ii*) Let X = Y = Z = R (set of reals)

Let $f : X \to Z$ defined by $f(x) = \sin x$

and $\quad g : Y \to Z$ defined by $g(x) = x^2$.

Then, the function $h : X \to Z$ defined by

$h(x) = \sin^2 x$ is the composite gof of f and g.

Note 1. It may happen that for certain pairs of functions f and g

(*i*) none of gof and fog is defined

(*ii*) only one of gof and fog is defined.

(*iii*) gof and fog are both defined but are unequal (*i.e.*, the composition of functions is not commutative).

For example,

(*i*) Let $f : R^+ \to R^+$ defined by $f(x) = x$ and $g : R—R^+ \to R—R^+$ defined by $g(x) = x$.

Then neither *gof* nor *fog* is defined.

(*ii*) Let $f : R \to R$ defined by $f(x) = x$ and $g : R^+ \to R^+$ defined by $g(x) = x$.

Then *gof* is defined but *fog* is not defined.

(*iii*) Let $f : R \to R$ defined by $f(x) = x^2$ and

$g : R \to R$ defined by $g(x) = x + 5$.

Then $gof : R \to R$ defined by

$$(gof)(x) = g(f(x)) = g(x^2) = x^2 + 5 \text{ and } fog : R \to R \text{ defined by}$$

$$f(g(x)) = f(x+5) = (x+5)^2$$

$$\therefore \quad gof \neq fog.$$

2. Composite *gof* of two functions *f* and *g* is defined only if the domain of *g* (second function) is contained in the range of *f* (first function).

1.2.8. Theorem. ***(Composition of functions is associative).*** **If $f : X \to Y$, $g : Y \to Z$ and $h : Z \to T$ are three functions, then**

$$(hog)of = ho(gof).$$

1.2.9. Inverse Functions. Let $f : X \to Y$ be a function.

A function $g : Y \to X$ is said to be a an inverse of the function if $gof = I_X$ and $fog = I_Y$.

For example, let $f : R^+ \to R^+$ defined by $f(x) = x^2$

and $\quad g : R^+ \to R^+$ defined by $g(x) = \sqrt{x}$.

Then $gof = I$ and $fog = I$ (identity function)

Therefore, *g* is the inverse of *f*.

If a function has an inverse, then it is said to be invertible.

1.2.10. Theorem. (*i*) **A function *f* has an inverse if and only if *f* is one-one and onto.**

(*ii*) **Inverse of a function, if exists is unique.**

(*iii*) **Composite of two one-one and onto functions is a one-one and onto function. (That is, the composite of two bijections is a bijection).**

(*iv*) **If $f : X \to Y$ and $g : Y \to Z$ are two one-one and onto functions, then $(gof)^{-1} = f^{-1}og^{-1}$.**

2

Countability of Sets and the Real Number System

2.1. COUNTABILITY OF SETS

The act of counting is perhaps one of the oldest activities of man. The earliest men having domesticated animals might be keeping a record of the number of goats by means of a pile of stones. If the herd was counted in each night by removing one stone from the pile for each goat accounted for, then the stones left over would have indicated strays and herdsmen would have gone out to search for them. This simple idea is the idea of one-to-one correspondence between the stones and the goats.

Today, we use the infinite set N of natural numbers as 'a pile of stones' for counting. Whenever, we want to count a set, say, the set of all alphabets in English, we start through the set N and tally off one alphabet against each natural number as we come to it. Since the last number we reach, corresponding to the last alphabet is 26, so we say that there are 26 alphabets in English.

If X is a non-empty finite set, say, $X = \{x_1, x_2, x_3, ..., x_n\}$, then there is a one-one onto function $f : X \rightarrow \{1, 2, 3, ..., n\}$ defined by $f(x_i) = i$, $1 \leq i \leq n$. The same idea of one to one correspondence is used in counting infinite sets.

To compare the sizes of two sets, we use the concept of equivalence.

2.1.1. Equivalent Sets. Two sets A and B are said to be equivalent if there exists a one to one correspondence between A and B, *i.e.*, if there exists a one-one and onto function from A to B. If this is the case, we write '$A \sim B$' and read as 'A is equivalent to B'.

2.1.2. Theorem. *The relation '~' is an equivalence relation**.

Proof. (*i*) '~' is reflexive

$A \sim A$ for each set A.

($\because$ the identity function $I_A : A \rightarrow A$ is one-one and onto)

(*ii*) '~' is symmetric.

If $A \sim B$ then $B \sim A$ for any two sets A and B.

($\because$ if $f : A \rightarrow B$ is one-one and onto, then its inverse function $f^{-1} : B \rightarrow A$ is also one-one and onto.)

(*iii*) '~' is transitive.

If $A \sim B$ and $B \sim C$, then $A \sim C$ for any three sets A, B and C.

*A relation is said to be an equivalence relation if it is (*i*) reflexive (*ii*) symmetric and (*iii*) transitive.

($\because$ If $f : A \to B$ and $g : B \to C$ are two one-one and onto functions, then $gof : A \to C$ is also one-one and onto.)

Finite Set. *A set A is said to be finite if A is either* ϕ *or* $A \sim N_n$ *for some fixed natural number n, where* $N_n = \{1, 2, 3, ..., n\}$ *is the subset of natural numbers truncated at n.*

(K.U. BCA II 2008)

A set which is not finite, is called an infinite set.

For example,

(*i*) The set {1, 2, 3, 5, 20, 1000} is a finite set.

(*ii*) Set of all vowels in English is a finite set.

(*iii*) N, Z, Q, R are all infinite sets.

Two non-empty finite sets are equivalent, of course, means that they have the same number of elements. We are surprised when we consider equivalence between infinite sets.

It appears that the set N of all natural numbers has 'more elements than the set {2, 4, 6, ...} of all positive even integers. But, we must remember that our criterion in these matters is whether there exists a one to one correspondence between the sets or not.

On tallying $\quad 1, 2, 3, ..., n, ...$

$\quad\quad 2, 4, 6, ..., 2n, ...$

We see that the function $f : N \to \{2, 4, 6, ...\}$ defined by $f(n) = 2n$ establishes a one to one correspondence between these two sets. So these sets are '~' and we regard them as having the same number of elements.

Also, N ~ Z (set of all integers), because on tallying

$$\begin{matrix} 1, 2, & 3, 4, & 5, 6, ... \\ 0, -1, & 1, -2, & 2, -3, ... \end{matrix}$$

We see that the function $f : N \to Z$ defined by

$$f(n) = \begin{cases} \frac{1}{2}(n-1), & \text{if } n = 1, 3, 5, ... \\ -\frac{n}{2}, & \text{if } n = 2, 4, 6, ... \end{cases}$$

is both one-one and onto. Thus, all that is really necessary in showing that an infinite set X is equivalent to N is to list the elements of X into a first, a second, a third, and so on, in such a way that it is completely exhausted by tallying off its elements with N. That is why an infinite set which is equivalent to N is sometimes said to be countably infinite. Every infinite set is not equivalent to N. For example, the set R of all reals is one such set. (See Cor. to Theorem 2.1.11.). Such sets are known as uncountably infinite. This simply means that the set of all real numbers represents a 'higher type of infinity' than that of N.

2.1.3. Enumerable (or denumerable or countably infinite) Set. A set X is said to be *enumerable* if there exists a one-one and onto function from the set N to the set X.

(K.U. BCA (II) 2008)

2.1.4. Countable set. A set X is said to be countable if it is either finite or enumerable.

2.1.5. Uncountable set. A set which is not countable is said to be uncountable.

Examples : (*i*) ϕ is countable. *(K.U. BCA (II) 2006)*

(*ii*) The set {1, 5, 9, 20} is countable.

(*iii*) The sets N and Z are countable.

(*iv*) The set R is uncountable.

Some basic results about countable and uncountable sets.

2.1.6. Theorem. *Every subset of a finite set is finite.*

Proof. Let A be a non empty finite set.

($\because$ If A = ϕ, we have nothing to prove)

Let $$B \subseteq A.$$

If B = ϕ or B = A, then B is obviously a finite set. So, let B $\neq \phi$ and B $\neq$ A.

Since A is finite,

$\therefore$ A ~ {1, 2, 3, ..., n} for some natural number n.

Since $$B \subset A,$$

$\therefore$ number of elements in B is, say, k less than n.

$\therefore$ $$B \sim \{1, 2, 3, ..., k\}$$

$\therefore$ B is finite.

Cor. *Every super-set of an infinite set is infinite.*

Proof. Let A be an infinite set and B a super-set of A.

Let, if possible, B be finite.

Since A $\subseteq$ B and every subset of a finite set is finite.

$\therefore$ A is finite which is a contradiction to the given hypothesis that A is infinite.

Hence B is infinite.

2.1.7. Theorem. *A set A is finite if and only if it is not equivalent to any proper subset B of A.* (K.U. BCA (II) 2003, 2004)

Proof. (i) Let A be finite.

If A = ϕ, then there is no proper subset of A and we have nothing to prove.

If A $\neq \phi$, then A ~ {1, 2, 3, ..., n} for some n, *i.e.*,

A has just n elements.

Since B is a proper subset of A.

$\therefore$ B has just k elements, $k < n$.

$\therefore$ A cannot be equivalent to B.

(ii) Suppose A is not equivalent to any of its proper subset. If possible, let A be infinite.

Let $$a_1 \in A.$$

Since A is infinite,

$\therefore$ $$A \neq \{a_1\}$$

$\therefore$ $\exists$ an element $a_2 \neq a_1$ and $a_2 \in$ A.

Again, A $\neq \{a_1, a_2\}$ and hence must contain an element a_3. Continuing like this, we see that A contains elements $a_1, a_2, a_3, ..., a_i \neq a_j$ for $i \neq j$. The process continues indefinitely, because otherwise A will be finite.

Let $$B = A - \{a_1\}$$

Define a function f : A $\to$ B such that

$$f(a_i) = a_{i+1} \quad i = 1, 2, 3, ...$$

and $$f(a) = a \text{ if } a \neq a_1, a_2, a_3, ...$$

Then, f is one-one and onto.

Hence A ~ B which is a contradiction.

Hence A must be finite.

2.1.8. Theorem. *Every subset of a countable set is countable.*

Proof. Let A be a countable set and B a subset of A.

Case (i) If A is a finite, then B being a subset of a finite set is finite and hence countable.

Case (ii) If A is enumerable, then A can be written as A = $\{a_1, a_2, a_3, ...\}$

If B = ϕ, then B is countable.

If B $\neq \phi$, then in the sequence of elements of A, there is first element a_{k_1} which belongs to B. Let $b_1 = a_{k_1}$. This is followed again by an element a_{k_2} in A such that $a_{k_2} \in$ B. Let $b_2 = a_{k_2}$. This process does or does not terminate according as B is finite or not. Since A contains all the elements of B, the possibly terminating sequence $b_1, b_2, b_3, ...$ comprizes precisely the elements of B.

$\therefore$ B is either finite or enumerable.

Hence B is countable.

Cor. 1. If A and B are countable, then A $\cap$ B is also countable.

$(\because \quad A \cap B \subseteq A)$.

2. Every super-set of an uncountable set is uncountable.

2.1.9. Theorem. *Union of two countable sets is countable.*

Proof. Let A and B be two countable sets.

Following cases arise :

(i) A and B are both finite.

Since union of two finite sets is a finite set, so A $\cup$ B is a finite set.

(ii) One of A or B is a finite set and the other is enumerable.

Let A be finite and B enumerable.

Then, we can write elements of A as $a_1, a_2, a_3, ..., a_m$ for some fixed m and arrange the elements of B as a sequence $b_1, b_2, b_3, ...$ *i.e.*, B = $\{b_1, b_2, b_3, ...\}$.

Also, B – A is enumerable and hence its elements can be arranged as a sequence $c_1, c_2, c_3, ...$

$$\therefore \quad A \cup B = A \cup (B - A)$$
$$= \{a_1, a_2, ..., a_m, c_1, c_2, ...\}.$$

Define a function $f : N \to A \cup B$ such that

$$f(n) = \begin{cases} a_n, & \text{if } n = 1, 2, ..., m \\ c_{n-m}, & \text{if } n \geq m + 1. \end{cases}$$

Clearly f is one-one and onto.

So, A $\cup$ B is enumerable and hence countable.

(iii) A and B are both enumerable.

Then, elements of A and B can be written as infinite sequences $a_1, a_2, a_3, ...$ and $b_1, b_2, b_3, ...$ respectively.

Also, B – A is a subset of B and A $\cup$ B = A $\cup$ (B – A).

Subcase *(i)* B – A is a finite set $\{c_1, c_2, ..., c_m\}$ **(say)**

Then, $\quad A \cup B = \{c_1, c_2, ..., c_m, a_1, a_2, ...\}$

Define a function $f : \mathrm{N} \to \mathrm{A} \cup \mathrm{B}$ such that

$$f(n) = \begin{cases} c_n, & \text{if } n = 1, 2, 3, \ldots\ldots, m \\ a_{n-m}, & \text{if } n \geq m + 1. \end{cases}$$

f is one-one and onto. So $\mathrm{A} \cup \mathrm{B}$ is enumerable and hence countable.

Subcase (*ii*) $\mathrm{B} - \mathrm{A}$ is an infinite subset of the enumerable set B and hence is enumerable. Therefore, elements of $\mathrm{B} - \mathrm{A}$ can be written as an infinite sequence $\{c_1, c_2, c_3, \ldots\ldots\}$.

$$\therefore \quad \mathrm{A} \cup \mathrm{B} = \mathrm{A} \cup (\mathrm{B} - \mathrm{A})$$
$$= \{a_1, c_1, a_2, c_2, a_3, c_3, \ldots\ldots\}$$

Define a mapping $f : \mathrm{N} \to \mathrm{A} \cup \mathrm{B}$ such that

$$f(n) = \begin{cases} a_{\frac{n+1}{2}} & \text{if } n \text{ is odd} \\ c_{\frac{n}{2}} & \text{if } n \text{ is even} \end{cases}$$

Then, f is one-one and onto. So, $\mathrm{A} \cup \mathrm{B}$ is enumerable and hence countable.

Cor. *Finite union of countable sets is countable.*

2.1.10. Theorem. *Prove that enumerable union of enumerable sets is enumerable.*

(K.U. BCA (II) 2001)

Or

Prove that denumerable union of denumerable sets is denumerable.

Proof. Let $\{\mathrm{A}_i\}_{i \in \mathrm{N}}$ be an enumerable collection of enumerable sets.

Since each A_i is enumerable.

$\therefore$ We can arrange the elements of A_i as a sequence $a_{i_1}, a_{i_2}, a_{i_3} \ldots$

i.e., $\mathrm{A}_i = \{a_{i_1}, a_{i_2}, \ldots, a_{i_n} \ldots\}$

$$\mathrm{A}_1 = \{a_{11}, a_{12}, a_{13}, a_{14}, \ldots\}$$

$$\mathrm{A}_2 = \{a_{21}, a_{22}, a_{23}, a_{24}, \ldots\}$$

$$\therefore \quad \mathrm{A}_3 = \{a_{31}, a_{32}, a_{33}, a_{34}, \ldots\}$$

$$\mathrm{A}_4 = \{a_{41}, a_{42}, a_{43}, a_{44}, \ldots\}$$

$$\vdots \quad \vdots \quad \vdots \quad \vdots \quad \vdots$$

Let us write the elements of $\bigcup_{i=1}^{\infty} \mathrm{A}_i$ in the order of the sum of the suffixes of elements. That is, we first write those elements in which sum of suffixes is 2 (namely a_{11}), then those elements in which sum of suffixes is 3 (namely a_{12}, a_{21}) and so on (this is also indicated by the arrows) leaving out the elements which have once occurred. In this way, we can arrange all the elements of $\bigcup_{i=1}^{\infty} \mathrm{A}_i$ as a sequence $a_{11}, a_{12}, a_{21}, a_{31}, a_{22}, a_{13}, \ldots$

Hence $\bigcup_{i=1}^{\infty} \mathrm{A}_i$ is enumerable.

2.1.11. Theorem. *The set of all rational numbers is countable.*

Proof. Let us first deal with positive rational numbers. Every positive rational number can be written as $\frac{p}{q}$, where p and q are positive integers. First write all whole numbers, *i.e.*, numbers with denominator 1, then all fractions with denominator 2, then fractions with denominator 3, ... and so on.

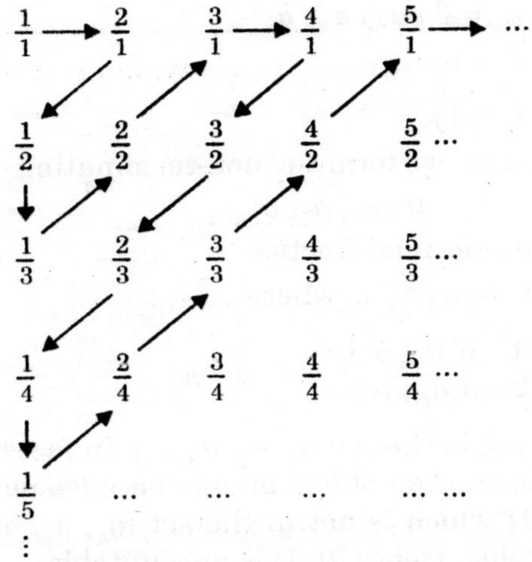

If we write the numbers in order of succession indicated by the arrows, leaving out the elements which have already occurred, then, every positive rational number certainly appears and also only once. Thus all the positive rationals can be written as a sequence

$$1, 2, \frac{1}{2}, \frac{1}{3}, 3, 4, \frac{3}{2}, \frac{2}{3}, \frac{1}{4}, \ldots$$

If we denote this sequence by $r_1, r_2, r_3, \ldots$, then clearly, the set of all rational numbers can be written as a sequence

$$0, -r_1, r_1, -r_2, r_2, -r_3, r_3, \ldots$$

Hence the set of all rational numbers is countable.

Cor. Set of all rational numbers in the interval [0, 1] is countable.

[For, since the set of all rational numbers is countable and subset of a countable set is countable, hence the set of all rational numbers in [0, 1] is countable.]

2.1.12. Theorem. *Set of all real numbers in the interval [0, 1] is uncountable.*

Proof. Let, if possible, the set of all real numbers in [0, 1] be countable. Since [0, 1] is an infinite set, therefore, [0, 1] is enumerable. Thus, the set [0, 1] can be written as

$$\{a_1, a_2, a_3, a_4, \ldots\}$$

Since every real number in [0, 1] can be written as a non-terminating decimal fraction $0 . x_1 x_2 x_3 \ldots$, $0 \le x_i \le 9$ and in fact in a unique way. [Here, of course, we suppose that for a real number of the type $\frac{1}{2}$, we consider the infinite decimal fraction, .49999 ... and not .5].

Therefore, expressing every real number a_i in the decimal form, we have

$$a_1 = 0\,.\,a_{11}\,a_{12}\,a_{13}\,\ldots$$
$$a_2 = 0\,.\,a_{21}\,a_{22}\,a_{23}\,\ldots$$
$$a_3 = 0\,.\,a_{31}\,a_{32}\,a_{33}\,\ldots$$
$$a_4 = 0\,.\,a_{41}\,a_{42}\,a_{43}\,\ldots$$
$$\vdots \qquad \vdots \quad \vdots \quad \vdots$$
$$a_n = 0\,.\,a_{n_1}\,a_{n_2}\,a_{n_3}\,\ldots$$
$$\vdots \quad \ldots \quad \ldots \quad \ldots \quad \ldots$$

where $a_{ij} \in \{0, 1, 2, 3, ..., 9\}$, $\forall$ i and j.

From the principal diagonal, we form the non-terminating decimal fraction :

$$0\,.\,a_{11}\,a_{22}\,a_{33}\,\ldots$$

From this construct a new decimal fraction

$$b = 0\,.\,b_1\,b_2\,b_3\,\ldots, \text{ where}$$

$$b_n = \begin{cases} 1 & \text{if } a_{nn} \neq 1 \\ 2 & \text{if } a_{nn} = 1, \end{cases} \qquad \forall \quad n$$

Clearly $b \in [0, 1]$ and is not in the set $\{a_1, a_2, a_3, ...\}$. In fact by our choice, b differs from a_n in the nth place. Also, the expansion of b is unique because no b_n is 0 or 9. Thus we have found a real number $b \in [0, 1]$ which is not in the set $\{a_1, a_2, a_3, ...\}$. This contradicts the assumption that $[0, 1]$ is countable. Hence $[0, 1]$ is uncountable.

Cor. *The set R of all real numbers is uncountable.*

Proof. Since $[0, 1] \subset$ R and $[0, 1]$ is uncountable.

$\therefore$ R is uncountable

($\because$ Super-set of an uncountable set is uncountable).

2.1.13. Theorem. *The set of all irrationals is uncountable.*

Proof. Let I denote the set of all irrationals and Q denote the set of all rationals, then

$$\text{R} = \text{I} \cup \text{Q}.$$

Let, if possible, I be countable.

Since Q is countable and union of two countable sets is countable.

$\therefore$ $\text{I} \cup \text{Q}$ is countable.

or R is countable.

which contradicts the fact that R is uncountable.

Hence I is uncountable.

***2.1.14. Algebraic and Transcendental numbers.** Consider a polynomial of the type $f(x) \equiv a_0x^n + a_1x^{n-1} + a_2x^{n-2} + \ldots + a_n$, where $a_0\,(\neq 0), a_1, a_2, \ldots\ldots, a_n$ are integers and n is a natural number. Such a polynomial has exactly n zeros real or imaginary [If α is a root of a polynomial equation $f(x) = 0$, then α is called a zero of the polynomial $f(x)$].

An *algebraic number* is a complex number which is a zero of such a polynomial. The set of algebraic numbers contains the set Q (*e.g.*, $\frac{2}{3}$ is a zero of $3x - 2$) and many other numbers (*e.g.*, $\sqrt{5}$ is a zero of $x^2 - 5$ and $2 + 3i$ is a zero of $x^2 - 4x + 13$).

---

*§2.1.14 to 2.1.16 may be omitted on first reading.

The complex numbers which are not algebraic are called transcendental numbers. e and π are well known examples of real transcendental numbers.

2.1.15. Theorem. *The set of all real algebraic numbers is countable.*

Proof. By definition, a real algebraic number is a zero of polynomial

$$f(x) \equiv a_0 x^n + a_1 x^{n-1} + \dots + a_n,$$

where $a_0 \neq 0$ and all a_i's are integers.

Without loss of generality, suppose that $a_0 > 0$. Define the height of the polynomial $f(x)$ as the positive integer $h = n + a_0 + |a_1| + |a_2| + \dots + |a_n|$. Clearly, $h \geq 1$. The same height is possessed only by a finite number of polynomials ($\because$ $n \leq h$ and each $|a_i| \leq h$). This makes us possible to write the set of all real algebraic numbers as a sequence in the following way.

First write all the real algebraic numbers given by polynomials of height 2. These polynomials are x and 2 ; we obtain the only algebraic number zero. The polynomials of height 3 are x^2, $2x$, $x + 1$, $x - 1$ and 3. These gives new real algebraic numbers -1, $+1$. The real algebraic numbers given by polynomials of height 4 are $-2, -\frac{1}{2}, \frac{1}{2}$ and 2, and so on. Thus, by allowing the height to run through the natural numbers and writing the new algebraic numbers corresponding to each value of the height, we get a sequence of distinct real algebraic numbers. Since every polynomial has a height, therefore, all real algebraic numbers appear in the sequence. Hence the set of all real algebraic numbers is countable.

2.1.16. Theorem. *The set of all real transcendental numbers is uncountable.*

Proof. Let A be the set of all real algebraic numbers and T be the set of all real transcendental numbers.

Then $\quad A \cup T = R$ (set of all reals)

Let if possible, T be countable.

Since A is countable and union of two countable sets is countable.

$\therefore$ R is countable.

But, this is a contradiction to the fact that R is uncountable.

Hence, T is uncountable.

Example 1. *Prove that N × N is countable.*

Proof. $\quad N \times N = \{(m, n) \mid m, n \in N\}$.

Let

$$A_1 = \{(1, 1), (1, 2), (1, 3), \dots\dots\}$$
$$A_2 = \{(2, 1), (2, 2), (2, 3), \dots\dots\}$$
$$A_3 = \{(3, 1), (3, 2), (3, 3), \dots\dots\}$$
$$\dots\dots\dots\dots\dots\dots\dots\dots\dots\dots$$
$$A_n = \{(n, 1), (n, 2), (n, 3), \dots\dots\}$$
$$\dots\dots\dots\dots\dots\dots\dots\dots\dots\dots$$

Then, $\quad \bigcup_{n \in N} A_n = N \times N$

Define a function $f : A_n \to N$ such that $f(n, k) = k$, $k \in N$. f is one-one and onto. Therefore, each A_n is enumerable. Since enumerable union of enumerable sets is enumerable, N × N is enumerable and hence countable.

Example 2. ***Prove that the set of all rational points in the coordinate plane (i.e., all the points both of whose coordinates are rationals) is enumerable.***

Proof. Since the set Q of all rational numbers is enumerable, we can write members of Q as a sequence $a_1, a_2, a_3, \ldots$

i.e.,
$$Q = \{a_1, a_2, a_3, \ldots\}$$

Now, set of all points whose coordinates are rationals

$$= Q \times Q = \{(x, y) \mid x, y \in Q\}$$

Let
$$A_1 = \{(a_1, a_1), (a_1, a_2), (a_1, a_3), \ldots\}$$
$$A_2 = \{(a_2, a_1), (a_2, a_2), (a_2, a_3), \ldots\}$$
$$\ldots\ldots\ldots\ldots\ldots\ldots\ldots\ldots\ldots$$
$$A_n = \{(a_n, a_1), (a_n, a_2), (a_n, a_3), \ldots\}$$
$$\ldots\ldots\ldots\ldots\ldots\ldots\ldots\ldots\ldots$$

so that
$$\bigcup_{n \in N} A_n = Q \times Q$$

Define a function $f : A_n \to N$ such that $f(a_n, a_k) = k$ for each fixed n. The function f so defined is one-one and onto. Therefore, each A_n is enumerable. Since enumerable union of enumerable sets is enumerable, $\bigcup_{n \in N} A_n$, *i.e.*, $Q \times Q$ is enumerable.

EXERCISE 2(*a*)

1. Prove that union of two finite sets is a finite set.

   [**Proof.** Let A and B be two finite sets.

   Case (*i*) If $A = \phi$, $B = \phi$, then $A \cup B = \phi$. So, $A \cup B$ is a finite set.

   Case (*ii*) If one of A or B is empty and the other is non-empty, then $A \cup B$ is equa[illegible] [illegible]ne of A and B which is non-empty finite set. Therefore, $A \cup B$ is finite.

   Case (*iii*) $A \neq \phi$, $B \neq \phi$.

   Then, $\exists$ positive integers m and n such that $N_m \sim A$ and A and $N_n \sim B$.

   Subcase (*i*) If $A \cap B = \phi$, then $A \cup B$ has $n + m$ elements and hence is finite.

   Subcase (*ii*) If $A \cap B \neq \phi$, then $A \cup B$ has $m + n - r$ elements where $A \cap B$ has r elements such that $r \leq \min\{m, n\}$.

   Hence, $A \cup B$ is a finite set.
2. Show that intersection of two infinite sets may not be infinite.
3. Prove that the set of all squares of integers is countable (Galileo's conjecture).
4. Prove that every subset of an enumerable set is at the most enumerable.
5. Prove that the set of all prime numbers is countable.
6. Prove that the set of all positive real numbers is uncountable.
7. Prove that the set of all positive irrational numbers is uncountable.
8. Prove that the set of all polynomials with integral co-efficients is countable.
9. If A is denumerable set, then prove that $A \times A$ is denumerable.
10. Prove that every infinite set has a countable subset.
11. Prove that $Z \times Z$ is countable where Z is a set of integers.

2.2. THE REAL NUMBER SYSTEM

2.2.1. Binary Composition. Let A be a given non-empty set. An operation denoted by '$*$' is called a binary composition (or just a composition) in A if to each ordered pair (a, b) of elements of A, it associates a unique element denoted by $a * b$ of A. In other words, the binary composition '$*$' is a function from A × A to A denoted by $(a, b) \to a * b \in A$.

If $a * b \in A, \forall\, a, b \in A$, then we also say that A is *closed w.r.t. the binary composition* '$*$' or we say that the binary composition '$*$' *is defined* on A.

For example, ordinary addition (+) is a binary composition on the set N of all natural numbers because sum of any two natural numbers is always a natural number. Thus, N is closed w.r.t.(+). Similarly, ordinary addition (+) and multiplication (.) are binary compositions in N, Z, Q and R.

Subtraction (–) is not a binary composition on N as $6 \in N$, $10 \in N$ but $6 - 10 = -4 \notin N$. Thus, N is not closed w.r.t. subtraction.

But subtraction is a binary composition in Z, Q and R.

2.2.2. Algebraic Structure. A non-empty set with one or more binary compositions defined on it, is called an algebraic structure. For example, the set R along with the two binary compositions—addition (+) and multiplication (.), is an algebraic structure. It is denoted by (R, +, .).

We assume that the reader is familiar with real numbers, the two fundamental compositions (operations) of addition (+) and multiplication (.) on them and order relations—greater than (>) and less than (<), between them. We shall describe some basic properties about real numbers. These properties will be regarded as axioms and all other properties of real numbers will be deduced from them. These axioms will be (*i*) Field axioms (*ii*) Order axioms (*iii*) Completeness axioms.

2.2.3. The Field Axioms. The set R of all real numbers with the two usual compositions—called addition and multiplication defined on it, and denoted by '+' and ' . ' respectively satisfies the following axioms :

The Addition Axioms. The addition composition '+' has the following properties :

A_1. **Commutativity**

$$a + b = b + a \ \forall\, a, b \in R$$

A_2. **Associativity**

$$(a + b) + c = a + (b + c) \ \forall\, a, b, c \in R$$

A_3. **Existence of additive identity**

There exists a real number zero '0' such that

$$a + 0 = a \ \forall\, a \in R$$

'0' is called an additive identity.

A_4. **Existence of additive inverse**

For each real number zero '0', $\exists$ a real number $-a$ (called a negative of a) such that

$$a + (-a) = 0.$$

The Multiplication Axioms

The multiplication composition ' . ' has the following properties :

M_1. **Commutativity**

$$a \,.\, b = b \,.\, a \ \forall\, a, b \in R$$

M_2. **Associativity**

$$(a \,.\, b) \,.\, c = a \,.\, (b \,.\, c) \ \forall\ a, b, c \in R$$

M_3. **Existence of multiplicative identity**

There exists a real number unity '1' such that

$$a \,.\, 1 = a \ \forall\ a \in R$$

M_4. **Existence of multiplicative inverse**

For each real number a ($\neq 0$), $\exists$ another real number $\frac{1}{a}$ (or a^{-1}) such that

$$a \,.\, \frac{1}{a} = 1.$$

$\frac{1}{a}$ is called a multiplicative inverse (or reciprocal) of a.

D. **Distributivity of multiplication over addition**

$$a \,.\, (b + c) = a \,.\, b + a \,.\, c \ \forall\ a, b, c \in R.$$

Since the set R satisfies the above 9 properties w.r.t. the compositions '+' and ' . ' defined on it, we say that R is a 'field'* or R has field structure.

Since the set Q of all rational numbers and the set C of all complex numbers also satisfy all the above 9 properties w.r.t. the compositions '+' and ' . ', so (Q, +, .) and (C, +, .) are also fields.

The set N of natural numbers satisfies only A_1 – A_2 w.r.t. addition, M_1 – M_3 w.r.t. multiplication and D.

$\therefore$ (N, +, .) is not a field.

The set Z of integers satisfies only A_1 – A_4 w.r.t. addition, M_1 – M_3 w.r.t. multiplication and D.

$\therefore$ (Z, +, .) is not a field.

2.2.4. Theorem. *Prove that in the set R.*

(*i*) *additive identity is unique*

(*ii*) *additive inverse in unique*

(*iii*) *multiplicative identity is unique*

(*iv*) *multiplicative inverse is unique.*

Proof. (*i*) If possible, let there be two additive identities 0 and 0′ in R.

Since 0 $\in$ R and 0′ is identity,

$$\therefore \qquad 0 + 0' = 0 \qquad \ldots(i)$$

Since 0′ $\in$ R and 0 is identity,

$$\therefore \qquad 0' + 0 = 0' \qquad \ldots(ii)$$

But $\qquad 0 + 0' = 0' + 0$

$\therefore$ from (*i*) and (*ii*),

$$0 = 0'$$

---

*A non-empty set A with at least two elements in it and with two compositions defined on it, two compositions being denoted by '+' and '.' and satisfying the above 9 properties, is called a field.

(*ii*) Let, if possible, b and c be two additive inverses of the real number a.

$\therefore \quad a + b = 0$

and $\quad a + c = 0$

Now, $\quad b = 0 + b$ (By A_3)

$= (a + c) + b$

$= (c + a) + b$ (By A_1)

$= c + (a + b)$ (By A_2)

$= c + 0$ ($\because \ a + b = 0$)

$= c$ (By A_3)

(*iii*) Let, if possible, 1 and 1′ be two multiplicative identities in R.

Since $1 \in$ R and 1′ is multiplicative identity

$\therefore \quad 1.1' = 1$ ...(*i*)

Since $1' \in$ R and 1 is multiplicative identity

$\therefore \quad 1'.1 = 1'$ ...(*ii*)

But $\quad 1.1' = 1'.1$ (By M_1)

$\therefore$ from (*i*) and (*ii*),

$$1 = 1'$$

(*iv*) Let $a\ (\neq 0) \in$ R.

Let, if possible, b and c be two multiplicative inverses of a.

$\therefore \quad a.b = b.a = 1$

$a.c = c.a = 1$

Now $\quad b = b.1$ (By M_3)

$= b.(a.c)$ ($\because \ a.c = 1$)

$= (b.a).c$ (By M_2)

$= 1.c$ ($\because \ b.a = 1$)

$= c$

In view of the above theorem, we shall be refering in future :

(*i*) 'the additive identity' instead of 'an additive identity' in A_3.

(*ii*) 'the additive inverse' instead of 'an additive inverse' in A_4.

(*iii*) 'the multiplicative identity' instead of 'a multiplicative identity' in M_3.

(*iv*) 'the multiplicative inverse' instead of 'a multiplicative inverse' in M_4.

2.2.5. Theorem. (*i*) (*Cancellation law for addition*)

If a, b, c ∈ R such that $a + b = a + c$, *then* $b = c$

(*ii*) (*Cancellation law for multiplication*)

If a, b, c ∈ R such that $a.c = b.c$ *and* $c \neq 0$ *then* $a = b$.

Proof. (*i*) Since $\quad a + b = a + c$

$\Rightarrow \quad (-a) + (a + b) = (-a) + (a + c)$

$\Rightarrow \quad (-a + a) + b = (-a + a) + c$ (By A_2)

$\Rightarrow \quad 0 + b = 0 + c$ (By A_4)

$\Rightarrow \quad b = c$ (By A_3)

(*ii*) Proof is on same lines and is left as an exercise for readers.

2.2.6. Theorem. *Prove*

(*i*) $a.0 = 0 \ \forall \ a \in R$

(*ii*) $a.b = 0 \Rightarrow$ *either* $a = 0$ *or* $b = 0 \ \forall \ a, b \in R$.

Proof. (*i*)

$$a.0 + a.0 = a.(0 + 0) \quad \text{(By D)}$$
$$= a.0 \quad \text{(By } A_3\text{)}$$
$$= a.0 + 0 \quad \text{(By cancellation law for addition)}$$
$$\Rightarrow \quad a.0 = 0$$

(*ii*) If $a = 0$, then there is nothing to prove.

So assume that $a \neq 0$, then a^{-1} exists and $a.a^{-1} = 1$

Now
$$a.b = 0$$
$$\Rightarrow \quad a^{-1}.(a.b) = a^{-1}.0$$
$$\Rightarrow \quad (a^{-1}.a).b = 0 \quad \text{(By } M_2 \text{ and by (i))}$$
$$\Rightarrow \quad 1.b = 0 \quad \text{(By } M_4\text{)}$$
$$\Rightarrow \quad b = 0. \quad \text{(By } M_3\text{)}$$

2.2.7. Subtraction and Division in R

If $a, b \in$ R, then the difference of a and b is written as $a - b$ and is defined by $a + (-b)$. The process of finding $a - b$ is called subtraction.

If $b \neq 0$, then the quotient of a by b is written as $\frac{a}{b}$ or $a \div b$ and is denoted by $a.b^{-1}$. The process of finding $\frac{a}{b}$ is called division.

EXERCISE 2(*b*)

1. If $a, b, c, d \in$ R, prove the following :

   (*i*) $a + [b + (c + d)] = (a + b) + (c + d)$ (*ii*) $a.(b.c).d = (a.b).(c.d)$

   (*iii*) $a.(-b) = -(a.b) = (-a).b$ (*iv*) $(-a).(-b) = a.b$

   (*v*) $(-1).a = -a$ (*vi*) $a^{-1} = b^{-1} \Rightarrow a = b$

   (*vii*) $(ab)^{-1} = a^{-1}.b^{-1}$ (*viii*) $(a - b) - c = a - (b + c)$

   (*ix*) $\frac{a}{b} \cdot \frac{c}{d} = \frac{a \cdot c}{b \cdot d}$. $(b \neq 0, d \neq 0)$

2. Solve the equations for x in R ($x, a, b \in$ R)

   (*i*) $x + a = b$ (*ii*) $ax = b$

   (*iii*) $ax + b = 0$.

3. Prove that 0 has no reciprocal.

2.2.8. The Order Axioms. R satisfies the following axioms w.r.t. the relation 'greater than' (denoted by >) :

O_1. Law of trichotomy

For any two real numbers a, b exactly one of the following holds :

$$a > b \text{ or } a = b \text{ or } b > a.$$

O_2. Transitivity

If $a > b$ and $b > c$, then $a > c \ \forall \ a, b, c \in R$

O_3. Monotonicity for addition

$$a > b \Rightarrow a + c > b + c \ \forall \ a, b, c \in R$$

O_4. Monotonicity for multiplication

If $a, b, c \in$ R such that

$$a > b \text{ and } c > 0, \text{ then } a.c > b.c$$

In view of the above 4 axioms, we say that (R, +, .) is an ordered field.

Remark (Q, +, .) satisfies $O_1 - O_4$ and is therefore an ordered field. Since the order relation '>' does not exist in C, set of all complex numbers, so, (C +, .) is not an ordered field.

2.2.9. Order relations. <, ≥, ≤. We often use the relations :

(*i*) less than '<' (*ii*) greater than or equal to '≥', (*iii*) less than or equal to '≤'.

For all $a, b \in$ R, we say

(*i*) $a < b$ iff $b > a$

(*ii*) $a \geq b$ iff either $a > b$ or $a = b$

(*iii*) $a \leq b$ iff either $a < b$ or $a = b$.

2.2.10. Positive and negative numbers. *A real number a is said to be positive or negative according as $a > 0$ or $a < 0$.*

EXERCISE 2(*c*)

If $a, b, c \in$ R, prove the following :

(*i*) $a > 0, b > 0 \ \Rightarrow \ a + b > 0$

(*ii*) $a > 0, b > 0 \ \Rightarrow \ a \, . \, b > 0$

(*iii*) $a < b \ \Leftrightarrow \ a + c < b + c$

(*iv*) $a < b$ and $c > 0 \ \Leftrightarrow \ a.c > b.c$

(*v*) $a \leq b$ and $b \leq a \ \Leftrightarrow \ a = b$

(*vi*) $a \leq b$ and $b \leq c \ \Rightarrow \ a \leq c$

(*vii*) If $a > 0$ and $b > 0$, then

$$a^2 = b^2 \ \Leftrightarrow \ a = b$$
$$a^2 > b^2 \ \Leftrightarrow \ a > b$$
$$a^2 < b^2 \ \Leftrightarrow \ a < b.$$

2.2.11. Interval. *A subset I of R having at least two elements is called an interval if whenever $a, b \in I$ and $a \leq x \leq b$, then $x \in I$.*

2.2.12. Finite Interval. Let $a, b \in$ R and $a < b$. Then

(*i*) the set $\{x \mid a < x < b \text{ and } x \in R\}$ is called an open interval and is denoted by (a, b) or $]a, b[$.

(*ii*) the set $\{x \mid a \leq x \leq b \text{ and } x \in R\}$ is called the closed interval and is denoted by $[a, b]$.

(*iii*) the sets $\{x \mid a \leq x < b \text{ and } x \in R\}$ and $\{x \mid a < x \leq b \text{ and } x \in R\}$ are called semi-open (or semi-closed) intervals and are respectively denoted by $[a, b)$ and $(a, b]$. The interval $[a, b)$ is left closed and right open.

The above four intervals are called finite intervals, each of length $b - a$. The word finite means the length of interval is finite, though each interval contains infinitely many elements.

2.2.13. Infinite Interval. Let $a \in$ R. Then

(*i*) the set $\{x \mid x > a \text{ and } x \in \text{R}\}$ is called an open right ray emanating from a and is denoted by (a, ∞).

(*ii*) the set $\{x \mid x < a \text{ and } x \in \text{R}\}$ is called an open left ray and is denoted by $(-\infty, a)$.

(*iii*) the sets $\{x \mid x \geq a \text{ and } x \in \text{R}\}$ and $\{x \mid x \leq a \text{ and } x \in \text{R}\}$ are called right closed ray and left closed ray respectively.

(*iv*) $\text{R} = (-\infty, \infty)$.

The intervals in (*i*)—(*iv*) above are infinite intervals.

Remark. Every interval is an infinite set but an infinite set may not be an interval. For example, N and Z are infinite sets but are not intervals.

BOUNDEDNESS OF A SET

2.2.14. Set bounded above. *A subset S of R is said to be bounded above if there exists a real number K such that*

$$x \leq K, \quad \forall \quad x \in S.$$

The real number K is called an upper bound of S.

For example,

(*i*) the sets (1, 2), [1, 2], [1, 2), (1, 2] are bounded above.

(*ii*) the set of negative integers is bounded above by 0.

2.2.15. Set bounded below. *A subset S of R is said to be bounded below if there exists a real number k such that*

$$x \geq k \quad \forall \quad x \in S.$$

The real number k is called a lower bound of S.

For example, the sets (1, 2), [1, 2], [1, 2), (1, 2] and N are all bounded below.

2.2.16. Bounded set. *A subset S of R is said to be a bounded set if it is bounded below as well as above, i.e., if there exist two real numbers k and K such that*

$$k \leq x \leq K \quad \forall \quad x \in S. \qquad \text{(K.U. BCA II 2008)}$$

Equivalently, if there exists an interval I (= [k, K]) such that

$$\text{S} \subseteq \text{I}.$$

Examples. (*i*) The sets (0, 1) [0, 1], [0, 1), (0, 1] are bounded.

(*i*) Every finite set is bounded

(*ii*) The set $\left\{\frac{1}{2}, \frac{1}{3}, \frac{1}{4}, \ldots\right\}$ is bounded.

2.2.17. Unbounded set. *A subset S of R, which is not bounded is called an unbounded set.*

For example,

(*i*) N, Z, Q, R are unbounded sets.

(*ii*) Set of all prime numbers is an unbounded set.

Remarks. (*i*) If a set is bounded above, then it has infinitely many upper bounds in as much as every number greater than an upper bound is also an upper bound.

(*ii*) If a set is bounded below, then it has infinitely many lower bounds in as much as every number smaller than a lower bound is also a lower bound.

One natural question arises—'Does the set of all upper bounds of a set S bounded above has the smallest element' ? If the answer is 'Yes', then that smallest element is called a least upper bound (*l.u.b.*) or supremum (sup) of the set S and we write l.u.b.S. or sup S.

2.2.18. Least upper bound (or Supremum). Let S be a non-empty subset of R.

Definition. (1) *A real number u is said to be a least upper bound of S if*

(*i*) *u is an upper bound of S.*

and (*ii*) *if u′ is another upper bound of S, then* $u \le u'$.

i.e., no real number less than *u* can be an upper bound of S.

Definition. (2) *A real number u is a l.u.b. of S if*

(*i*) $u \ge x \ \forall \ x \in S$

and (*ii*) *given any* $\varepsilon > 0$ *(however small), there exists at least one* $x \in S$ *such that* $x > u - \varepsilon$.

2.2.19. Theorem. *Least upper bound of a set, if exists, is unique.*

Proof. Let S be a non-empty subset of R.

Let, if possible, l_1 and l_2 be two least upper bounds of S.

Since l_1 is *l.u.b.* of S and l_2 is an upper bound of S

$\therefore$ $$l_1 \le l_2 \qquad ...(1)$$

Since l_2 is *l.u.b.* of S and l_1 is an upper bound of S

$\therefore$ $$l_2 \le l_1 \qquad ...(2)$$

$\therefore$ from (1) and (2),

$$l_1 = l_2.$$

Remark. In view of the above theorem, we use the word 'the *l.u.b.*' rather than 'a *l.u.b.*'

2.2.20. Theorem. *Prove that definitions (1) and (2) are equivalent.*

Proof. (*i*) Let *u* be the *l.u.b.* of S according to def. (1)

$\therefore$ *u* is an upper bound of S

$\therefore$ $$u \ge x \quad \forall \quad x \in S$$

Also, for each $\varepsilon > 0$, (however small)

$u - \varepsilon$ is not an upper bound of S

$\therefore$ $\exists$ at least one $x \in$ S such that

$$x > u - \varepsilon$$

$\therefore$ *u* is the *l.u.b.* of S according to def. (2).

(*ii*) Let *u* be the *l.u.b.* of S according to def. (2)

$\therefore$ $$u \ge x \ \forall \ x \in S$$

$\therefore$ *u* is an upper bound of S

Let *u′* be a real number $< u$

Let $$\varepsilon = u - u' > 0$$

$\therefore$ by condition (*ii*) of def. (2), $\exists$ at least one $x \in$ S such that

$$x > u - \varepsilon$$

$\therefore$ $$x > u' \qquad (\because \ \varepsilon = u - u')$$

$\therefore$ u' is not an upper bound of S.

$\therefore$ no real number less than u is an upper bound of S.

Hence u is the *l.u.b.* of S according to def. (1).

Remarks. 1. The least upper bound of a set may or may not belong to the set.
For example,

(*i*) If $S = \left\{\frac{1}{n} \mid n \in N\right\}$, then *l.u.b.* S is 1 and $1 \in S$

(*ii*) If S = (0, 1), then *l.u.b.* S is 1 and $1 \notin S$

(*iii*) If S = (0, 1] or [0, 1], then *l.u.b.* S is 1 and $1 \in S$.

(*iv*) If S is a finite set, then *l.u.b.* S is the largest element and it belongs to the set.

2. If the *l.u.b.* of a set, belongs to the set, then *l.u.b.* is the largest element of the set.

Now we state the completeness axiom which answers the above question in the affermative.

2.2.21. C. Completeness Axiom. *The set of all upper bounds of a non-empty and bounded above subset of real numbers has the smallest element.*

Equivalently, we say—*Every non-empty subset of reals which is bounded above, has the least upper bound.*

Since the set R of all real numbers satisfies the completeness axiom and is also an ordered field w.r.t. + and . compositions, so we say that (R, +, .) is a complete ordered field.

We are, now, in a position to distinguish between the ordered fields (Q, + , .) and (R, +, .) . (Q, +, .) is not a complete ordered field as will be seen in the following theorems.

2.2.22. Theorem. $\sqrt{2}$ *is not a rational number*.*

Proof. Let, if possible, $\sqrt{2}$ be a rational number

$$\Rightarrow \qquad \sqrt{2} = \frac{p}{q} \qquad \text{...(1)}$$

where $p, q \in Z$ and $q \neq 0$.

We further suppose that p and q do not have any factor other than 1 in common.

Squaring (1),

$$2 = \frac{p^2}{q^2}$$

or

$$2q^2 = p^2 \qquad \text{...(2)}$$

$\Rightarrow$ p^2 is even

$\Rightarrow$ p is even

Let $p = 2m$, where $m \in Z$

$$\Rightarrow \qquad p^2 = 4m^2 \qquad \text{...(3)}$$

$\therefore$ from (2) and (3),

$$2q^2 = 4m^2$$

$$\Rightarrow \qquad q^2 = 2m^2$$

***Rational number.** A number of the type $\frac{p}{q}$ $(p, q \in Z$ and $q \neq 0)$ is called a rational number.

Irrational number. A real number which is not a rational number, is called an irrational number.

$\Rightarrow$ q^2 is even

$\Rightarrow$ q is even

$\therefore$ p and q have 2 as a common factor.

This contradicts our supposition that p and q do not have any factor other than 1 in common.

Hence $\sqrt{2}$ is not a rational number.

2.2.23. Theorem. *(Q, +, .) is not a complete ordered field.*

Proof. To prove the theorem, we shall show that there exists a non-empty subset of Q which is bounded above but has no rational number as its *l.u.b.*

Consider a subset S of Q defined by

$$S = \{\alpha \mid \alpha \in Q^+, 0 < \alpha^2 < 2\}$$

$$S \neq \phi \qquad (\because \ 1 \in S)$$

Also, S is bounded above. ($\because$ each element of S is less than 2)

We shall show that no rational number is the *l.u.b.* of S

Let x be *any* rational number.

Following cases arise :

Case (*i*) $x \leq 0$

Since every element of S is > 0,

$\therefore$ x is not an upper bound of S

Hence x is not the *l.u.b.* of S.

Case (*ii*) $x > 0$ and $0 < x^2 < 2$.

Let $$y = \frac{3x+4}{2x+3} \qquad ...(1)$$

$$y \in Q^+ \qquad (\because \ x \in Q^+)$$

Also, on simplification,

$$y^2 - 2 = \frac{x^2 - 2}{(2x+3)^2} \qquad ...(2)$$

and $$y - x = \frac{2(2 - x^2)}{2x+3} \qquad ...(3)$$

Since $x^2 < 2$,

$\therefore$ from (2), $y^2 < 2$

$\therefore$ $y \in S$

But, from (3), $y - x > 0$

$\Rightarrow$ $y > x$

$\Rightarrow$ y is an element of S which is larger than x

$\Rightarrow$ x is not the *l.u.b.* of S.

Case (*iii*) $x > 0$ and $x^2 = 2$

This case is not possible, as by the above theorem $\sqrt{2} \notin Q$.

Case (*iv*) $x > 0$ and $x^2 > 2$

Define y as in case (*ii*)

From (2), $y^2 > 2$ $(\because \; x^2 > 2)$

From (3), $y - x < 0$

i.e., $y < x$

Let z be any element of S

$\Rightarrow$ $0 < z^2 < 2$

$\Rightarrow$ $0 < z^2 < y^2$ and $y < x$

$\Rightarrow$ $0 < z < y < x$

$\Rightarrow$ x and y are both upper bounds of S

$\Rightarrow$ x is not the *l.u.b.* of S $(\because \; y < x)$

Since the above four cases are mutually exclusive and exhaustive.

$\therefore$ x cannot be the *l.u.b.* of S.

Since x is any rational number.

$\therefore$ no rational number is the *l.u.b.* of S.

Hence Q is not a complete ordered field.

We now introduce the concept of 'greatest lower bound' (or infimum) of a set on the lines we introduced least upper bound (or supremum) of a set.

If the set of all lower bounds of a non-empty set S ($\subseteq$ R) bounded below has the greatest element, then that element (is called the greatest lower bound of S (*g.l.b.*) or infimum of S (inf.) and we write *g.l.b.* S or inf. S.

2.2.24. Greatest lower bound (or Infimum). Let S be a non-empty subset of R.

Definition 1. *A real number l is called a greatest lower bound of S if*

(*i*) *l is a lower bound of S*

and (*ii*) *if l' is another lower bound of S, then $l' \leq l$*

i.e., no real number greater than l can be a lower bound of S.

Definition 2. *A real number l is a greatest lower bound of S if*

(*i*) $l \leq x \; \forall \; x \in S$

and (*ii*) *given any $\varepsilon > 0$ (however small), there exists at least one $x \in S$ such that $x < l + \varepsilon$.*

2.2.25. Theorem. *Greatest lower bound of a set if exists, is unique.*

(K.U. B.C.A. II 2007)

Proof. Proof is on the same lines as of theorem 2.2.19 and is left as an exercise for the reader.

Remark. In view of the above theorem, we use the word 'the *g.l.b.*' rather than 'a *g.l.b.*'

2.2.26. Theorem. *Prove that definitions (1) and (2) are equivalent.*

Proof. Proof is on the same lines as of theorem 2.2.20 and is left as an exercise for the reader.

Remark 1. The *g.l.b.* of a set may or may not belong to the set.

For example,

(*i*) In N, 1 is the *g.l.b.* and it belongs to N.

(*ii*) If $S = \left\{ \frac{1}{n} \mid n \in N \right\}$, then *g.l.b.* of S is 0 and $0 \notin S$.

(*iii*) If S is a finite non-empty set, then its smallest element is *g.l.b.* of S and it belongs to S.

(*iv*) If S = [0, 1) or [0, 1], then *g.l.b* of S is 0 and $0 \in$ S.

2. If the *g.l.b.* of a set, belongs to the set, then it is the smallest member of the set.

The following theorem asserts the existence of the *g.l.b.* of any non-empty subset of R and is in fact, the statement of 'Completeness Axiom' in terms of *g.l.b.*

2.2.27. Theorem. *Prove that every non-empty subset of real numbers, which is bounded below has the greatest lower bound.* (K.U. BCA (II) 2006)

Proof. Let S be a non-empty subset of R and S be bounded below.

To prove : S has the *g.l.b.*

Define a subset T of R such that

$$T = \{-x \mid x \in S\}$$

$$T \neq \phi \qquad (\because S \neq \phi)$$

Let k be a lower bound of S

$\Rightarrow \quad k \le x \quad \forall \quad x \in S$

$\Rightarrow \quad -k \ge -x \quad \forall \quad x \in S$

$\Rightarrow \quad -x \le -k \quad \forall \quad x \in S$

$\Rightarrow$ T is bounded above and $-k$ is an upper bound of T.

$\therefore$ by completeness axiom of real numbers, T has the *l.u.b.* ξ (say).

We shall show that $-\xi$ is the *g.l.b.* of S.

$t \le \xi \quad \forall \quad t \in T \qquad (\because \xi = l.u.b.\ T)$

$\Rightarrow \quad -t \ge -\xi \quad \forall \quad t \in T$

$\Rightarrow \quad x \ge -\xi \quad \forall \quad x \in S$

$\Rightarrow \quad -\xi \le x, \quad \forall \quad x \in S$

$\Rightarrow$ $-\xi$ is a lower bound of S

Let η be another lower bound of S

$\therefore \quad \eta \le x \quad \forall \quad x \in S$

$\Rightarrow \quad -\eta \ge -x \quad \forall \quad x \in S$

$\Rightarrow \quad -\eta \ge t \quad \forall \quad t \in T$

$\Rightarrow$ $-\eta$ is an upper bound of T

$\Rightarrow \quad -\eta \ge \xi \qquad (\because \xi \text{ is } l.u.b.\ T)$

$\Rightarrow \quad \eta \le -\xi$

$\therefore$ every lower bound of S other than $-\xi$ is less than or equal to $-\xi$.

Hence, $-\xi$ is the *g.l.b.* of S.

Now, we give some applications of 'completeness axiom' to prove certain important properties of real numbers.

2.2.28. Theorem. *(Archimedean property of real numbers) Given any two real numbers x and y with x > 0, there exists a natural number n such that nx > y.* (K.U. BCA (II) 2001, 08)

Proof. If $y \le 0$, the result is obvious.

So, let $\quad y > 0$

Suppose that the theorem is false.

$\therefore \quad nx \le y \quad \forall \quad n \in N.$

Consider the set $S = \{nx \mid n \in N\}$

Now, S is a non-empty set bounded above by y.

$\therefore$ By completeness axiom of reals, S has the *l.u.b.* ξ (say).

$$\therefore \qquad nx \le \xi \quad \forall \quad n \in N$$

$$\Rightarrow \qquad (n+1)\,x \le \xi \quad \forall \quad n \in N$$

$$\Rightarrow \qquad nx \le \xi - x \quad \forall \quad n \in N$$

$\Rightarrow$ $\xi - x$, which is strictly less than ξ, is an upper bound of S.

This contradicts the fact that ξ is the *l.u.b.* of S.

Hence our supposition that $nx \le y, \ \forall \ n \in N$ is wrong.

$\therefore$ $\exists\, n \in N$ such $nx > y$.

Cor. 1. *Given any real number x, there exists a natural number n such that $n > x$.*

Proof. Taking $y = x$ and $x = 1$ in the above theorem, we see that $\exists$ a natural number n such that

$$n \,.\, 1 > x$$

i.e.

$$n > x.$$

Cor. 2. *For each positive real number x, there exists $n \in N$ such that $x > \dfrac{1}{n}$.*

Proof. Taking $y = 1$, by Archimedean property of reals, $\exists$ a natural number n such that $n \,.\, x > 1$.

i.e.

$$x > \frac{1}{n}.$$

Cor. 3. *Given any $x \in R$, there exists integers p and q such that $p < x < q$.*

Proof. By Cor. 1, for $x \in$ R, $\exists\, q \in$ N such that $q > x$

i.e.

$$x < q \qquad \text{...(1)}$$

Again, for $-x \in$ R, $\exists\, n \in$ N such that

$$n > -x$$

$$\Rightarrow \qquad -n < x \qquad \text{...(2)}$$

$\therefore$ from (1) and (2), $\exists$ integers p and q (where $p = -n$) such that $p < x < q$.

2.2.29. Theorem. *Prove that N is not bounded above.*

Proof. Let, if possible, N be bounded above.

$$N \ne \phi \qquad (\because 1 \in N)$$

Since N is non-empty and bounded above subset of R.

$\therefore$ By completeness axiom of reals, N has the *l.u.b.* ξ (say).

$$\Rightarrow \qquad n \le \xi \quad \forall \quad n \in N$$

$$\Rightarrow \qquad n + 1 \le \xi \quad \forall \quad n \in N$$

$$\Rightarrow \qquad n \le \xi - 1 \quad \forall \quad n \in N$$

$\Rightarrow$ $\xi - 1$ is an upper bound of N.

But this contradicts the fact that ξ is the *l.u.b.* of N.

Hence N is not bounded above.

2.2.30. Theorem. *Given any $x \in R$, there exists a unique integer m such that*

$$m \le x < m + 1.$$

[Such an integer m is called the integral part of x and is denoted by $[x]$].

Proof. Let $S = \{p \mid p \in Z \text{ and } p \leq x\}$

$S \neq \phi$ (By Cor. 3 above)

and S is bounded above by x.

$\therefore$ by completeness axiom of reals,

S has the *l.u.b.* m (say). Obviously $m \in Z$ and m is unique.

Since m is *l.u.b.* of S and x is an upper bound of S.

$\therefore \quad m \leq x$ ...(1)

To prove : $x < m + 1$

Let, if possible, $x \geq m + 1$

$\Rightarrow \quad m + 1 \leq x$

$\Rightarrow \quad m + 1 \in S.$

This contradicts the fact that m is the *l.u.b.* of S

Hence $x < m + 1$ ...(2)

$\therefore$ from (1) and (2),

$$m \leq x < m + 1.$$

Cor. *Given any $x \in R$, there exists a unique integer k such that*

$$x - 1 \leq k < x.$$

Proof. By above theorem, $\exists$ integers m and n such that $m < x < n$. Choose the largest integer k from the finite collection $m, m + 1, m + 2, ..., n$ such that $k < x$. Then $k + 1 \geq x$.

$\Rightarrow \quad k \geq x - 1$

$\Rightarrow \quad x - 1 \leq k$

Hence $x - 1 \leq k < x.$

2.2.31. Theorem. *Between any two different real numbers, there lie infinitely many rational numbers.* (K.U. BCA (II) 2003, 2004)

Proof. Let x and y be any two different real numbers. Without loss of generality, let

$$y > x.$$

$\Rightarrow \quad y - x > 0$

$\Rightarrow \quad \exists\, n \in N$ such that $n\,.\,(y - x) > 1$ (By Archimedean Property)

$\Rightarrow \quad ny > nx + 1$ ...(1)

Since $nx \in R,$

$\therefore \quad \exists$ a unique integer m such that $m - 1 \leq nx < m$

$\Rightarrow \quad m - 1 \leq nx,\ nx < m$

$\Rightarrow \quad m \leq nx + 1,\ nx < m$

$\Rightarrow \quad nx + 1 \geq m > nx$ ...(2)

$\therefore$ from (1) and (2),

$$ny > nx + 1 \geq m > nx$$

$\Rightarrow \quad ny > m > nx$

$\Rightarrow \quad y > \dfrac{m}{n} > x$

$\Rightarrow \quad y > r > x$ where $r = \dfrac{m}{n} \in Q.$

Repeating the above arguments, taking y and r or r and x as real numbers, we can prove that $\exists$ another rational number s between x and y and so on. Hence, between any two real numbers there lie infinitely many rational numbers.

Cor. (*Rationals are dense*).

Between any two different rational numbers, there lie infinitely many rational numbers.

2.2.32. Theorem. *Between any two different real numbers, there lie infinitely many irrational numbers.* (K.U. BCA (II) 2006)

Proof. Let x, y be any two different real numbers. Without loss of generally, let $x < y$.

$\sqrt{2}$ is an irrational number.

$$\Rightarrow \qquad x - \sqrt{2} < y - \sqrt{2}$$

Now, $\qquad x, y, \sqrt{2} \in R \quad \Rightarrow \quad x - \sqrt{2}, y - \sqrt{2} \in R$

$\Rightarrow$ $\exists$ a rational number r (say), between $x - \sqrt{2}$ and $y - \sqrt{2}$ (in fact there lie infinitely many rationals between $x - \sqrt{2}$ and $y - \sqrt{2}$)

$$\Rightarrow \qquad x - \sqrt{2} < r < y - \sqrt{2}$$

$$x < r + \sqrt{2} < y$$

i.e. $r + \sqrt{2}$ is an irrational number lying between x and y.

($\because$ Sum of a rational number and an irrational number is an irrational number)

Repeating the above arguments, we can prove that infinitely many irrational numbers lie between x and y.

In the light of above two theorems, we can say that R is dense.

Cor. Between any two different irrational numbers, there lie infinitely many irrational numbers.

Example. *Let A and B be two non-empty bounded subsets of R and $\lambda \in R$. If*

$$A + B = \{x + y \mid x \in A \text{ and } y \in B\}$$

and $\qquad \lambda A = \{\lambda x \mid x \in A\}$, *then* A + B *and* λA *are bounded and*

(*i*) *l.u.b.* (A + B) = *l.u.b.* A + *l.u.b.* B

(*ii*) *l.u.b.* $\lambda A = \lambda$ *l.u.b.* A, *if* $\lambda \geq 0$

$\qquad = \lambda$ *g.l.b.* A, *if* $\lambda \leq 0$.

Sol. (*i*) Since A and B are bounded, there exist positive real numbers k_1 and k_2 such that

$$|x| \leq k_1, \; |y| \leq k_2 \; \forall \; x \in A, y \in B$$

$$\therefore \qquad |x + y| \leq |x| + |y| \leq k_1 + k_2 \; \forall \; x \in A, y \in B$$

Hence A + B is bounded and so it has the *l.u.b.*

Let $\qquad \xi_1 = l.u.b.\ A, \; \xi_2 = l.u.b.$ of B

and $\qquad \xi = l.u.b.\ (A + B)$.

Then, $\qquad \xi_1 \geq x, \; \xi_2 \geq y, \; \forall \; x \in A, y \in B$

$$\Rightarrow \qquad \xi_1 + \xi_2 \geq x + y \; \forall \; x \in A, y \in B$$

$$\Rightarrow \qquad \xi_1 + \xi_2 \geq \xi \qquad \ldots(1)$$

Also, for any $\varepsilon > 0$, $\exists\, x_1 \in A$, $y_1 \in B$ such that

$$x_1 > \xi_1 - \frac{\varepsilon}{2}, \; y_1 > \xi_2 - \frac{\varepsilon}{2}$$

$\Rightarrow$ $$x_1 + y_1 > \xi_1 + \xi_2 - \varepsilon$$

$\Rightarrow$ $$\xi \geq x_1 + y_1 > \xi_1 + \xi_2 - \varepsilon$$

$\Rightarrow$ $$\xi > \xi_1 + \xi_2 - \varepsilon, \quad \forall \quad \varepsilon > 0$$

$\Rightarrow$ $$\xi \geq \xi_1 + \xi_2 \qquad ...(2)$$

$\therefore$ from (1) and (2), $$\xi = \xi_1 + \xi_2,$$

i.e. $$l.u.b.\ (A + B) = l.u.b.\ A + l.u.b.\ B.$$

(*ii*) If $\lambda = 0$, $l.u.b.\ \lambda A = l.u.b.\ \{0\} = 0$

$$= \lambda\, l.u.b.\ A = \lambda . g.l.b.\ A \qquad ...(I)$$

If $\lambda > 0$, let $$\xi = l.u.b.\ A$$

$\Rightarrow$ $$\xi \geq x \quad \forall \quad x \in A$$

$\Rightarrow$ $$\lambda\xi \geq \lambda x \quad \forall \quad \lambda x \in \lambda A$$

$\Rightarrow$ $$\lambda\xi \geq l.u.b.\ \lambda A \qquad ...(1)$$

Also, for any $\varepsilon > 0$, $\exists$ some $x_1 \in A$ such that

$$x_1 > \xi - \frac{1}{\lambda}\varepsilon$$

$\Rightarrow$ $$\lambda x_1 > \lambda\xi - \varepsilon$$

$\Rightarrow$ $$l.u.b.\ \lambda A \geq \lambda x_1 > \lambda\xi - \varepsilon$$

$\Rightarrow$ $$l.u.b.\ \lambda A > \lambda\xi - \varepsilon \ \forall\ \varepsilon > 0$$

$\Rightarrow$ $$l.u.b.\ \lambda A \geq \lambda\xi \qquad ...(2)$$

From (1) and (2), we have

$$l.u.b.\ \lambda A = \lambda\xi = \lambda\ l.u.b.\ A, \text{ when } \lambda > 0 \qquad ...(II)$$

If $\lambda < 0$, let $$\eta = g.l.b.\ A.$$

$\Rightarrow$ $$x \geq \eta \quad \forall \quad x \in A$$

$\Rightarrow$ $$\lambda\eta \geq \lambda x \quad \forall \quad \lambda x \in \lambda A$$

$\Rightarrow$ $$\lambda\eta \geq l.u.b.\ \lambda A \qquad ...(3)$$

($\because$ $\lambda\eta$ is an upper bound of λA)

Also, for any $\varepsilon > 0$, $\exists$ some $x_1 \in A$ such that

$$x_1 < \eta - \frac{1}{\lambda}\varepsilon$$

$\Rightarrow$ $$\lambda x_1 > \lambda\eta - \varepsilon \qquad (\because \ \lambda < 0)$$

$\Rightarrow$ $$l.u.b.\ \lambda A \geq \lambda x_1 > \lambda\eta - \varepsilon$$

$\Rightarrow$ $$l.u.b.\ \lambda A > \lambda\eta - \varepsilon, \quad \forall \quad \varepsilon > 0$$

$\Rightarrow$ $$l.u.b.\ \lambda A \geq \lambda\eta \qquad ...(4)$$

From (3) and (4), $$l.u.b.\ \lambda A = \lambda\eta = \lambda . g.l.b\ A, \text{ when } \lambda < 0. \qquad ...(III)$$

2.2.33. Modulus of a real number. *The modulus (or absolute value or numerical value) of a real number x is denoted by* $|\,x\,|$ *and is defined by*

$$|\,x\,| = \begin{cases} x & \text{if } x \geq 0 \\ -x & \text{if } x < 0 \end{cases}$$

$|\,x\,|$ is read as mod. x and is defined for each $x \in R$.

Clearly, (*i*) $|\,x\,| \geq 0 \quad \forall \quad x$

(*ii*) $| x | \geq x$ and $| x | \geq -x$.

Infact, $| x | = \max. \{x, -x\}$, where $\max. \{x, -x\}$ denotes greater of the numbers x and $-x$.

2.2.34. Some results involving moduli :

Let $x, y \in R$, then

(*i*) $| x | = 0 \Leftrightarrow x = 0$

(*ii*) $| x | > 0$ for $x \neq 0$

(*iii*) $| -x | = | x |$

(*iv*) $| x |^2 = x^2$

(*v*) $| xy | = | x | . | y |$

(*vi*) **Triangle inequality**

$$| x + y | \leq | x | + | y |$$

(*vii*) $| x - y | \geq | x | - | y |$

(*viii*) $| x - y | < k, k > 0 \Leftrightarrow x - k < y < x + k$

(*ix*) $| x - y | \leq | x | + | y |$

For, $| x - y | = | x + (-y) | \leq | x | + | -y |$

$= | x | + | y |$ $\qquad (\because \ | -y | = | y |)$

(*x*) $| x | < k \Leftrightarrow -k < x < k$

(*xi*) A subset S of R is bounded iff $\exists$ a real number $k > 0$ such that $| x | < k \ \forall \ x \in S$.

For, (*i*) Let S be bounded.

$\Rightarrow$ S is bounded below as well as bounded above.

$\Rightarrow \exists\, l$ and $m \in R$ such that

$$l \leq x \leq m, \quad \forall \quad x \in S \qquad ...(1)$$

Let $k = \max. \{ | l |, | m | \}$

$\Rightarrow$ $| l | \leq k, | m | \leq k$

$\Rightarrow$ $-k \leq l \leq k, -k \leq m \leq k$ $\qquad ...(2)$

$\therefore$ from (1) and (2), $-k \leq x \leq k \ \forall \ x \in S$

$\Rightarrow$ $| x | \leq k \ \forall \ x \in S$.

(*ii*) Let $| x | \leq k \ \forall \ x \in S$, where $k > 0$

$\Rightarrow$ $-k \leq x \leq k \ \forall \ x \in S$

$\Rightarrow$ S is bounded.

EXERCISE 2 (*d*)

1. (*a*) Find the greatest element, least element, the *l.u.b.* and the *g.l.b.*, whichever exists, of the following sets :

(*i*) $\{1, 5, 9, 20, \pi\}$ (*ii*) $\left\{\frac{1}{n} \mid n \in N\right\}$ (*iii*) $\{3n - 5 \mid n \in N\}$

(*iv*) $\left\{\frac{(-1)^n}{n} \mid n \in N\right\}$ (*v*) $\left\{1 - \frac{1}{n} \mid n \in N\right\}$ (*vi*) $\left\{\left(\frac{1}{3}\right)^n \mid n \in N \cup \{0\}\right\}$.

(*b*) Find the l.u.b. and g.l.b. of the following sets :

(*i*) $\left\{\frac{1}{5} \middle| n \in Z, n \neq 0\right\}$ (*ii*) $\left\{\frac{3n+2}{2n+1} \middle| \in N\right\}$.

2. Give one example of a set which is
(*i*) bounded above but not below (*ii*) bounded below but not above
(*iii*) neither bounded above not bounded below
(*iv*) both bounded above and below.

3. Prove that every subset of a set bounded above (below or both) is bounded above (below or both). Is the converse true ?

4. If A and B are bounded subsets of R, then prove that the set $A + B = \{x + y \mid x \in A \text{ and } y \in B\}$ is also bounded and $g.l.b.\ (A + B) = g.l.b.\ A + g.l.b.\ B$.

5. For a real number λ and a subset A of R, let λA be the set defined by $\lambda A = \{\lambda x \mid x \in A\}$. Prove that if A is bounded, then λA is also bounded and

$$g.l.b.\ \lambda A = \lambda\ g.l.b.\ A \text{ if } \lambda \geq 0$$
$$= \lambda\ l.u.b.\ A \text{ if } \lambda \leq 0.$$

6. (*i*) Prove that $l.u.b.\ (-A) = g.l.b.\ A$. $A \subseteq R$.
(*ii*) If X is bounded subset of r, then prove $g.l.b.\ (-X) = l.u.b.\ X$.

7. (*i*) If $0 \leq a < \frac{1}{n}\ \forall\ n$, then $a = 0$
(*ii*) If $0 \leq a < \varepsilon\quad \forall \quad \varepsilon > 0$, then $a = 0$.

8. If A and B are two bounded subsets of R, then $A \cup B$ and $A \cap B\ (\neq \phi)$ are also bounded and
(*i*) Sup $(A \cup B)$ = max. (sup A, sup B) (*ii*) Inf $(A \cup B)$ = min. (inf A, inf B)
(*iii*) Sup $(A \cap B)$ = min. (sup A, sup B) (*iv*) Inf $(A \cap B)$ = max. (inf A, inf B).

9. If u is an upper bound of a set A ($\subseteq$ R) and $u \in A$, then $u = l.u.b.\ A$.

10. Give an example of a set in which $l.u.b. = g.l.b.$

11. (*a*) Prove that the set $\{x \mid x \in Q, x > 0 \text{ and } x^2 < 3\}$ does not have any $l.u.b.$ in Q.

[**Hint.** Proceed exactly on the lines of proof in § 2.2.23 and noting that $\sqrt{3}$ is not a rational number.

In case (*ii*), define $y = \dfrac{3(x+1)}{x+3}$

Then, $y \in Q^+$

Also, on simplification, $y^2 - 3 = \dfrac{6(x^2-3)}{(x+3)^2}$ and $y - x = \dfrac{3 - x^2}{x+3}$]

(*b*) Prove : $\{x \mid x \in Q, x > 0 \text{ and } x < 5\}$ does not have any l.u.b. in Q. (*K.U. BCA (II) 2007*)

12. If $B \supseteq A\ (\neq \phi)$ and B is bounded, then sup B $\geq$ sup A $\geq$ inf A $\geq$ inf B.

13. If A and B are two non-empty bounded subsets of R, and $A - B = \{a - b \mid a \in A, b \in B\}$, $AB = \{a.b \mid a \in A, b \in B\}$, then
(*i*) Sup (A – B) = sup A – inf B (*ii*) Inf (A – B) = inf A – sup B
(*iii*) Sup AB $\leq$ sup A . sup B (*iv*) Inf AB $\geq$ inf A . inf B.

14. What are supermum and infimum of a set ? Prove that the supermum or infimum if they exist are unique. Do they always belong to the set ?

15. Define a complete ordered field. (*K.U. BCA (II) 2001*)

3

Topology of Real Numbers

In the present chapter, we are going to introduce the concept of neighbourhood of a point, interior point of a set, open set, closed set, limit point of a set, isolated point, derived set, closure of a set, perfect set, dense set and compact set. We shall also prove the Bolzano-Weierstrass theorem which gives the sufficient condition for the existence of a limit point of a set. Our universal set in this chapter will be the set R, of all real numbers.

3.1.1. Neighbourhood of a point. *A set A ($\subseteq$ R) is called a neighbourhood of a point ξ, if there exists an open interval (a, b) containing ξ and is contained in A, i.e., $\xi \in (a, b) \subseteq A$.*

If $\xi \in (a, b)$, then for $\varepsilon = \min. \{\xi - a, b - \xi\} > 0$, $(\xi - \varepsilon, \xi + \varepsilon) \subseteq (a, b)$. Therefore, if A is a neighbourhood of ξ, then it contains all the points at a distance ε from ξ.

Equivalently, a set A is a neighbourhood of ξ if there exists an $\varepsilon > 0$ such that $(\xi - \varepsilon, \xi + \varepsilon) \subseteq A$.

The choice of ε depends upon ξ and A. For instance, if $\xi = \frac{1}{2}$ and A = R, then ε can have any value but if A = (0, 1), then max. value of ε can be $\frac{1}{2}$. For an $\varepsilon > 0$, $(\xi - \varepsilon, \xi + \varepsilon)$ is itself taken as a neighbourhood of ξ and is sometimes called an ε-neighbourhood of ξ.

3.1.2. Deleted neighbourhood of ξ. *Any neighbourhood of ξ from which ξ itself has been deleted (removed) is called a deleted neighbourhood of ξ.*

If A is a neighbourhood of ξ, then A – $\{\xi\}$ is a deleted neighbourhood of ξ.

Then set
$$(\xi - \varepsilon, \xi + \varepsilon) - \{\xi\} = (\xi - \varepsilon, \xi) \cup (\xi, \xi + \varepsilon)$$
$$= (x \mid 0 < | x - \xi | < \varepsilon)$$
is a deleted ε-neighbourhood of ξ.

Examples. (*i*) Every open interval (*a*, *b*) is a neighbourhood of each of its points.
[$\because \quad \xi \in (a, b) \subseteq (a, b)$]

(*ii*) The interval [*a*, *b*) is a neighbourhood of each point of (*a*, *b*) but is not a neighbourhood of *a*, because for a neighbourhood of *a*, it has to contain points on the left of *a* also. Similarly, the interval (*a*, *b*] is a neighbourhood of each point of (*a*, *b*) but is not a neighbourhood of *b*.

(*iii*) The interval [*a*, *b*] is a neighbourhood of each point of (*a*, *b*) but is not a neighbourhood of *a* and *b*.

(*iv*) R is a neighbourhood of each of its points.
[$\because \quad (\xi - 1, \xi + 1) \subseteq R \quad \forall \quad \xi \in R$]

(*v*) A finite non-empty set is not a neighbourhood of any of its points.

[For, A neighbourhood of a point has to contain an open interval containing the point and so must contain infinitely many points].

(*vi*) $(0, 1) \cup (1, 2)$ is a neighbourhood of each of its points.

(*vii*) N, Z and Q are not neighbourhoods of any real number.

[For, If $x \in R$, then $(x - \frac{1}{3}, x + \frac{1}{3})$ is a neighbourhood of x and $(x - \frac{1}{3}, x + \frac{1}{3}) \not\subseteq Q$, as every interval contains infinitely many irrationals also. So Q is not a neighbourhood of any real number].

(*viii*) ϕ is a neighbourhood of each of its points.

($\because$ there is no point in ϕ of which it is not a neighbourhood).

3.1.3. Theorem. (*i*) *If A and B are two neighbourhoods of a point ξ, then $A \cap B$ is a neighbourhood of ξ.*

(*ii*) *If A is a neighbourhood of ξ and $B \supseteq A$, then B is a neighbourhood of ξ.*

Proof. (*i*) Since A is a neighbourhood of ξ,

$\therefore\ \exists\ \varepsilon_1 > 0$ such that $(\xi - \varepsilon_1, \xi + \varepsilon_1) \subseteq A$

Since B is a neighbourhood of ξ,

$\therefore\ \exists\ \varepsilon_2 > 0$ such that $(\xi - \varepsilon_2, \xi + \varepsilon_2) \subseteq B$

Let $\varepsilon = \min \{\varepsilon_1, \varepsilon_2\} > 0$

Then, $(\xi - \varepsilon, \xi + \varepsilon) \subseteq (\xi - \varepsilon_1, \xi + \varepsilon_1) \subseteq A$

and $(\xi - \varepsilon, \xi + \varepsilon) \subseteq (\xi - \varepsilon_2, \xi + \varepsilon_2) \subseteq B$

$\Rightarrow$ $(\xi - \varepsilon, \xi + \varepsilon) \subseteq A \cap B$

$\Rightarrow$ $A \cap B$ is a neighbourhood of ξ.

(*ii*) Since A is a neighbourhood of ξ,

$\therefore\ \exists\ \varepsilon > 0$ such that $(\xi - \varepsilon, \xi + \varepsilon) \subseteq A$

$\Rightarrow$ $(\xi - \varepsilon, \xi + \varepsilon) \subseteq B$ ($\because$ $B \supseteq A$)

$\Rightarrow$ B is a neighbourhood of ξ.

3.2.1. Interior point. *If A is a neighbourhood of a point ξ, then ξ is called an interior point of A.* (K.U. BCA (II) 2006)

3.2.2. Interior of a set. *The set of all interior points of a set A is called the interior of A and is denoted by A°.*

Remarks. (*i*) $A^\circ \subseteq A$

(*ii*) If $A \subseteq B$, then $A^\circ \subseteq B^\circ$.

Examples. (*i*) 2 is an interior point of the interval (1, 3) but 1 is not an interior point of (1, 3).

(*ii*) $N^\circ = \phi$, $Z^\circ = \phi$, $Q^\circ = \phi$

(*iii*) $R^\circ = R$

(*iv*) $[a, b]^\circ = (a, b]^\circ = [a, b)^\circ = (a, b)$

(*v*) If A is a finite set, then $A^\circ = \phi$.

3.3.1. Open set. *A set A ($\subseteq$ R) is said to be an open set if it is a neighbourhood of each of its points.* (K.U. BCA (II) 2007)

Equivalently, A is open if for each $x \in A$, $\exists$ an $\varepsilon > 0$ such that $(x - \varepsilon, x + \varepsilon) \subseteq A$.

Remarks. (*i*) If A is an open set, then each point of A is an interior point, *i.e.*, $A = A^\circ$.

(*ii*) A set A is not open, means $\exists$ at least one point of A of which A is not a neighbourhood, *i.e.*, $\exists$ some $x \in A$ such that for each $\varepsilon > 0$ (however small) $(x - \varepsilon, x + \varepsilon) \not\subseteq A$.

Examples. (*i*) Open intervals are open sets.

(*ii*) None of $[a, b)$, $(a, b]$, $[a, b]$, $[a, \infty)$ and $(-\infty, a]$ is an open set.

(*iii*) R and ϕ are open sets.

(*iv*) Non-empty finite sets are not open sets.

(*v*) N, Z and Q are not open sets.

(*vi*) $(0, 1) \cup (1, 2)$ is an open set.

(*vii*) R – N and R – Z are open sets.

[For, $x \in R - N$

$\Rightarrow \quad x \notin N$

Case (*i*) $x > 1$

Let $n \in N$ such that $n < x < n + 1$

Let $\varepsilon = \min\left[\frac{1}{2}(x - n), \frac{1}{2}(n + 1 - x)\right] > 0$

Then $\qquad (x - \varepsilon, x + \varepsilon) \cap N = \phi$

$\Rightarrow \qquad (x - \varepsilon, x + \varepsilon) \subseteq R - N$

$\Rightarrow$ R – N is a neighbourhood of x

$\Rightarrow$ R – N is an open set

Case (*ii*) $x < 1$

Then $(-\infty, 1)$ is a neighbourhood of x and $(-\infty, 1) \subseteq R - N$.

$\therefore$ R – N is an open set.

Similarly, R – Z is an open set.]

3.3.2. Theorem. *The union of an arbitrary family of open sets is open.*

Proof. Let $\{A_\lambda\}_{\lambda \in \Lambda}$ be an arbitrary family of open sets.

Let $x \in \bigcup_{\lambda \in \Lambda} A_\lambda$

$\Rightarrow \quad x \in A_\lambda$ for at least one $\lambda \in \Lambda$

$\Rightarrow \quad A_\lambda$ is a neighbourhood of x $\qquad (\because A_\lambda$ is open)

$\Rightarrow \quad \bigcup_{\lambda \in \Lambda} A_\lambda$ is a neighbourhood of x $\qquad$ (By Theorem 3.1.3)

$\Rightarrow \quad \bigcup_{\lambda \in \Lambda} A_\lambda$ is a neighbourhood of each of its points $\qquad (\because x$ is arbitrary)

$\Rightarrow \quad \bigcup_{\lambda \in \Lambda} A_\lambda$ is open.

3.3.3. Theorem. *The intersection of a finite number of open sets is open.*

Proof. Let $\{A_i\}_{1 \le i \le n}$ be a family of n open sets.

If $\bigcap_{i=1}^{n} A_i = \phi$, then the result is proved $\qquad (\because \phi$ is open)

So, let $\bigcap_{i=1}^{n} A_i \neq \phi$,

Let $x \in \bigcap_{i=1}^{n} A_i$

$\Rightarrow$ $x \in A_i\ \forall\, i,\ 1 \le i \le n$

$\Rightarrow$ A_i is a neighbourhood of x ($\because$ A_i is open), $1 \le i \le n$

$\Rightarrow$ $\exists$ an $\varepsilon_i > 0$ such that $(x - \varepsilon_1, x + \varepsilon_i) \subseteq A_i$, $1 \le i \le n$

Let $\varepsilon = \min. \{\varepsilon_1, \varepsilon_2, \ldots, \varepsilon_n\} > 0$

Then, $x \in (x - \varepsilon, x + \varepsilon) \subseteq A_i,\ \forall_i\ 1 \le i \le n$

$\Rightarrow$ $x \in (x - \varepsilon, x + \varepsilon) \subseteq \bigcap_{i=1}^{n} A_i$

$\Rightarrow$ $\bigcap_{i=1}^{n} A_i$ is a neighbourhood of x.

$\Rightarrow$ $\bigcap_{i=1}^{n} A_i$ is an open set ($\because$ x in arbitrary)

Remark. The intersection of an infinite family of open sets may or may not be an open set as is clear from the following examples :

Examples. (*i*) Let $A_n = \left(\dfrac{-1}{n}, \dfrac{1}{n}\right)$, $n \in N$. Then, for each $n \in N$, A_n is an open set.

Let $A = \bigcap_{n=1}^{\infty} A_n$. We shall show that $A = \{0\}$. Clearly, $0 \in A$. If $x \in R$ and $x \ne 0$, then by Archimedean property of reals, there exists a natural number m such that $m \mid x \mid > 1$ or $\mid x \mid > \dfrac{1}{m}$. Thus, $x \notin A_m$. This shows that $0 \ne x \notin A$. Hence, $A = \{0\}$, which is not an open set.

(*ii*) Let $A_n = (0, n)$, $n \in N$. Then, A_n is an open set for each $n \in N$ and $\bigcap_{n=1}^{\infty} A_n = (0, 1)$, which is an open set.

Note. Subset and super-set of an open set may or may not be open.

For example,

(*i*) $(0, 1) \subseteq [0, 1]$

$(0, 1)$ is open but $[0, 1]$ is not open

(*ii*) $(0, \frac{1}{2}) \subseteq (0, 1)$

$(0, \frac{1}{2})$ and $(0, 1)$ are both open

(*iii*) $(0, \frac{1}{2}] \subseteq (0, 1)$

$(0, \frac{1}{2}]$ is not open but $(0, 1)$ is open.

3.3.4. Theorem. *Interior of a set is an open set.*

Proof. Let A be a given set

If $A^\circ = \phi$, then A° is open ($\because$ ϕ is an open set)

If $A^\circ \ne \phi$, let $x \in A^\circ$

$\Rightarrow$ A is neighbourhood of x

$\Rightarrow$ $\exists$ some $\varepsilon > 0$ such that $(x - \varepsilon, x + \varepsilon) \subseteq A$

$\Rightarrow$ A is a neighbourhood of each point of $(x - \varepsilon, x + \varepsilon)$

$\Rightarrow$ $(x - \varepsilon, x + \varepsilon) \subseteq A^\circ$

$\Rightarrow$ A° is a neighbourhood of x

$\Rightarrow$ A° is open.

Cor. For any set A, $(A^\circ)^\circ = A^\circ$.

3.3.5. Theorem. *A° is the largest open set contained in A.*

Proof. Let $B \subseteq A$ and B be open. If $B = \phi$, we have nothing to prove.

Let $B \neq \phi$. Let $x \in B$.

$\Rightarrow$ B is a neighbourhood of x ($\because$ B is open)

$\Rightarrow$ A is a neighbourhood of x ($\because$ $A \supseteq B$)

$\Rightarrow x \in A^\circ$.

Since $B \subseteq A$ $\therefore$ $B^\circ \subseteq A^\circ$

$\therefore$ $B \subseteq A^\circ$ ($\because$ B is open $\Rightarrow B = B^\circ$)

$\therefore$ every open set contained in A, is a subset of A°.

Also A° is open.

Hence A° is the largest open set contained in A.

Cor. *A° is equal to the union of all open subsets of A.*

For, Let $\mathfrak{I} = \{S \mid S \text{ is an open set and } S \subseteq A\}$

Let $B = \cup S$, $S \in \mathfrak{I}$. Clearly, $B \subseteq A$. Since A° is an open set contained in A, $A^\circ \in \mathfrak{I}$ and hence $A^\circ \subseteq B$. However, B being the union of open subsets of A is an open set contained in A. Therefore, $B \subseteq A^\circ$.

Hence, $A^\circ = B$.

3.4.1. Closed Set. *A set $A(\subseteq R)$ is said to be a closed set if its complement A^c $(= R - A)$ is open.* (B.C.A. II 2007)

Examples. (*i*) ϕ and R are closed sets. ($\because$ $\phi^c = R$ and $R^c = \phi$ are open sets)

(*ii*) Closed intervals are closed sets.

[For, $[a, b]^c = (-\infty, a) \cup (b, \infty)$ is open]

(*iii*) None of $[a, b)$, $(b, a]$, (a, b) (a, ∞), $(-\infty, a)$ is a closed set.

(*iv*) Every finite set is a closed set.

(*v*) N and Z are closed sets. [For N^c and Z^c are open sets]

Note. (*i*) ϕ and R are the only sets which are both open and closed.

(*ii*) Sets $[a, b)$ and $(a, b]$ are neither open nor closed.

3.4.2. Theorem. *The intersection of an arbitrary family of closed sets is closed.*

Proof. Let $\{A_\lambda\}_{\lambda \in \Lambda}$ be an arbitrary family of closed sets.

Since A_λ is a closed set $\forall\ \lambda \in \Lambda$

$\therefore$ $A_\lambda^{\,c}$ is an open set $\forall\ \lambda \in \Lambda$

$\Rightarrow$ $\bigcup_{\lambda \in \Lambda} A_\lambda^{\,c}$ is an open set [Theorem 3.3.2]

$\Rightarrow$ $[\bigcap_{\lambda \in \Lambda} A_\lambda]^c$ is an open set [By De Morgan's Law]

$\Rightarrow \bigcap_{\lambda \in \Lambda} A_\lambda$ is a closed set.

3.4.3. Theorem. *The union of a finite number of closed sets is a closed set.*

Proof. Let $\{A_i\}_{1 \le i \le n}$ be a finite family of n closed sets.

To prove : $\bigcup_{i=1}^{n} A_i$ is a closed set.

Since A_i is a closed set, $1 \le i \le n$

$\therefore$ A_i^c is an open set, $1 \le i \le n$

$\Rightarrow \bigcap_{i=1}^{n} A_i^c$ is an open set [Theorem 3.3.3]

$\Rightarrow [\bigcup_{i=1}^{n} A_i]^c$ is an open set [By De Morgan's Law]

$\Rightarrow \bigcup_{i=1}^{n} A_i$ is a closed set.

Remark. The union of an infinite family of closed sets need not be a closed set. For example, if $A_n = \left[-1+\frac{1}{n}, 1-\frac{1}{n}\right]$, $n \in N$, then $\bigcup_{n=1}^{\infty} A_n = (-1, 1)$ which is not a closed set.

EXERCISE 3(*a*)

1. Give one example of each of the following :
   (*i*) a set which is a neighbourhood of each of its points
   (*ii*) a set which is not a neighbourhood of any of its points.
   (*iii*) a set which is a neighbourhood of all its points except n given points ($n \ge 1$). Can this set be an interval.
   (*iv*) a set which is neither open nor closed
   (*v*) a set which is both open and closed
   (*vi*) an interval which is not an open set
   (*vii*) an open set which is not an interval.
2. Show that the set of all natural numbers N is not a neighbourhood of any of its points.
3. Prove that a set A is a neighbourhood of ξ iff there exists $n \in N$ such that $\left(\xi - \frac{1}{n}, \xi + \frac{1}{n}\right) \subseteq A$.
4. Show that the intersection of the family of all neighbourhoods of an arbitrary point $x \in R$ is $\{x\}$.
5. Prove or disapprove the following :
   (*i*) Q is open
   (*ii*) R – Q is closed (*K.U. BCA (II) 2007*)
   (*iii*) $(0, 1) \cup (1, 2)$ is open
   (*iv*) a finite set is open
   (*v*) every infinite set is a closed set
   (*vi*) $[1, 2] \cup [2, 5]$ is a closed set
   (*vii*) $\{x \mid 0 \le x < 1\}$ is open
   (*viii*) Singleton is a closed set.

6. Prove that every non-empty open set is a union of open intervals.
7. Prove that union of two open sets is an open set. *(K.U. BCA (II) 2008)*
8. (*a*) Prove that intersection of two closed sets is a closed set.
   (*b*) Prove that union of two closed sets is a closed set. *(K.U. BCA (II) 2001)*
9. Find the interiors of the following sets :
   (*i*) {1, 2, 3, 4, 5} (*ii*) [0, 1] (*iii*) $[0, 1] \cup [3, 5]$
   (*iv*) $\left\{\frac{1}{n} \mid n \in N\right\}$ (*v*) Z (*vi*) Q
   (*vii*) R
10. Let A and B be two subsets of R. Prove the following :
   (*i*) If $A \subseteq B$, then $A^\circ \subseteq B^\circ$ (*ii*) $A^\circ \cup B^\circ \subseteq (A \cup B)^\circ$
   (*iii*) $(A \cap B)^\circ = A^\circ \cap B^\circ$
   Give an example to show that $A^\circ \cup B^\circ \neq (A \cup B)^\circ$.
11. Which of the following sets are closed ? Give arguments in support of your answer.
   (*i*) The set of all integers.
   (*ii*) The set of all rational numbers. *(K.U. BCA (II) 2006)*

3.5.1. Limit Point of a Set

Definition 1. *A real number ξ is called a limit point of a set A ($\subseteq$ R) if every neighbourhood of ξ contains at least one point of A other than ξ, i.e. if for every neighbourhood B of ξ, $B \cap (A - \{\xi\}) \neq \phi$.*

Definition 2. *A real number ξ is called a limit point of a set A ($\subseteq$ R) if every neighbourhood of ξ contains infinitely many elements of A.*

A limit point of a set is also called a *limiting point* or an *accumulation point* or a *condensation point* or *cluster point* of the set.

Remark. *Definitions (1) and (2) are equivalent.*

Proof. Let $A \subseteq R$.

(*i*) Let ξ be a limit point of A according to definition (1). Let, if possible, ξ be not a limit point of A according to definition (2). Then, there exists a neighbourhood B (say) of ξ which contains only finitely many elements of A.

Let $$B \cap A - \{\xi\} = \{a_1, a_2, ..., a_n\}.$$
$$\varepsilon = \min_{1 \le i \le n} \{ \mid \xi - a_i \mid \} > 0$$

Then, $(\xi - \varepsilon, \xi + \varepsilon)$ is a neighbourhood of ξ and $(\xi - \varepsilon, \xi + \varepsilon) \cap A - \{\xi\} = \phi$.

$\Rightarrow$ ξ is not a limit point of A according to definition (1), a contradiction.

(*ii*) Let ξ be a limit point of A according to definition (2).

$\Rightarrow$ every neighbourhood of ξ contains infinitely many elements of A.

$\Rightarrow$ every neighbourhood of ξ contains an element of A different from ξ.

$\Rightarrow$ ξ is a limit point of A according to definition (1) also.

Note. A finite set has no limit point.

Examples : (*i*) Every real number is a limit point of Q and R – Q.

[For, if $x \in R$, then for each $\varepsilon > 0$ (however small), $(x - \varepsilon, x + \varepsilon)$ is a neighbourhood of x and it contains infinitely many rational numbers

$\Rightarrow (x - \varepsilon, x + \varepsilon) \cap Q - \{x\} \neq \phi.$

$\Rightarrow x$ is a limit point of Q.

Similar is the case with R – Q]

(*ii*) Every element of $[a, b]$ is a limit point of (a, b), $a, b \in R$.

[For, if $x \in [a, b]$, then for any $\varepsilon > 0$, $(x - \varepsilon, x + \varepsilon)$ is a neighbourhood of x. Let $c = \max. \{x - \varepsilon, a\}$ and $d = \min. \{x + \varepsilon, b\}$.

Then $(x - \varepsilon, x + \varepsilon) \cap (a, b) = (c, d)$ and the interval (c, d) contains infinitely many elements. Hence x is a limit point of (a, b).]

(*iii*) Every element of $[a, b]$ is a limit point of each of the set $[a, b]$, $[a, b)$ and $(a, b]$.

(*iv*) N and Z have no limit point.

[For, if $x \in R$, then $(x - \frac{1}{2}, x + \frac{1}{2})$ is a neighbourhood of x and it contains at the most one element of N. Hence x is not a limit point of N.

Similar is the case with Z.]

(*v*) If ξ is a limit point of A and $A \subseteq B$, then ξ is also a limit point of B.

(*vi*) 0 is the only limit point of the set $\left\{\frac{1}{n} \mid n \in N\right\}$

[For, consider any neighbourhood $(-\varepsilon, \varepsilon)$. $(\varepsilon > 0)$ of 0.

By Archimedean property of reals, $\exists\, n \in N$ such that $n > \frac{1}{\varepsilon}$

$$\Rightarrow \quad \frac{1}{n} < \varepsilon \qquad \Rightarrow \quad 0 \neq \frac{1}{n} \in (-\varepsilon, \varepsilon)$$

$\Rightarrow$ any neighbourhood $(-\varepsilon, \varepsilon)$ of 0, contains a point $\frac{1}{n}$ other than 0 of A

$\Rightarrow$ 0 is a limit point of A.

Uniqueness. $A = \left\{\frac{1}{n} \mid n \in N\right\} \subseteq (0, 1]$

We shall show that no real number other than 0 is a limit point of A.

Let $x (\neq 0)$ be any real number.

Following cases arise :

Case (*i*) If $x < 0$, then $(-\infty, 0)$ is neighbourhood of x and $(-\infty, 0) \cap A = \phi$.

$\therefore$ x is not a limit point of A.

Case (*ii*) If $x > 1$, then $(1, \infty)$ is a neighbourhood of x and $(1, \infty) \cap A = \phi$.

$\therefore$ x is not a limit point of A.

Case (*iii*) If $x \in (0, 1]$, then $\exists\, n \in N$ such that

$$n \leq \frac{1}{x} < n + 1 \qquad \left(\because \; 0 \neq \frac{1}{x} \in R\right)$$

$$\Rightarrow \quad \frac{1}{n} \geq x > \frac{1}{n+1}$$

$\Rightarrow \quad \dfrac{1}{n+1} < x \le \dfrac{1}{n} \qquad \Rightarrow \quad \dfrac{1}{n+1} < x < \dfrac{1}{n-\frac{1}{2}}$

$\Rightarrow \left(\dfrac{1}{n+1}, \dfrac{1}{n-\frac{1}{2}}\right)$ is a neighbourhood of x and contains at the most $\dfrac{1}{n} \in$ A.

$\Rightarrow x$ is not a limit point of A.

$\Rightarrow$ no element of (0, 1] is a limit point of A.

Hence 0 is the only limit point of A.]

3.5.2. Isolated point. *If $\xi \in A$ and ξ is not a limit point of A, then ξ is called an isolated point of A.*

Examples. (*i*) Every element of a finite set is an isolated point of the set.

(*ii*) Each element of N is an isolated point of N.

(*iii*) Each element of Z is an isolated point of Z.

(*iv*) 2 is an isolated point of each of $(0, 1) \cup \{2\}$, $(0, 1] \cup \{2\}$, $[0, 1) \cup \{2\}$ and $[0, 1] \cup \{2\}$.

We have seen in the examples after § 3.5.1 that a limit point of a set may or may not belong to the set. Further, a set may have no limit point, one limit point or any number of limit points.

We, now state and prove Bolzano-Weierstrass theorem which gives a sufficient condition for the existence of a limit point of a set.

3.5.3. Theorem. *(Bolzano-Weierstrass Theorem)*. Every infinite and bounded subset of R has a limit point.* *(K.U. BCA (II) 2003, 2006, 2007)*

Proof. Let A be an infinite and bounded subset of R.

Let k = g.l.b. A and K = l.u.b. A

Define a set B = $\{x \mid x \in$ R and x exceeds at the most only a finite number of elements of A$\}$.

$$B \neq \phi \qquad (\because \ k \in B)$$

Also, B is bounded above $\qquad (\because \ x < K \ \forall \ x \in B)$

$\therefore$ by completeness property of real numbers, B has the l.u.b. ξ (say).

We shall show that ξ is a limit point of A.

Let $\varepsilon > 0$ be given.

Since ξ is the l.u.b. of B

$\therefore \quad \exists$ at least one $x \in$ B such that $x > \xi - \varepsilon$.

Since $x \in$ B,

$\therefore \quad x$ exceeds at the most a finite number of elements of A

$\Rightarrow \quad$ at the most a finite number of elements of A are $< \xi - \varepsilon$ ...(1)

Since $\qquad \xi$ = l.u.b. B

$\therefore \qquad \xi + \varepsilon \notin$ B

$\Rightarrow \quad \xi + \varepsilon$ exceeds infinitely many elements of A ...(2)

$\therefore \quad$ from (1) and (2),

infinitely many elements of A lie in $(\xi - \varepsilon, \xi + \varepsilon)$

$\therefore \quad \xi$ is a limit point of A.

Remark. Since a finite set has no limit point and N has no limit point, therefore, the conditions 'infinite and bounded' in the statement of the theorem are necessary.

**After the names of two German mathematicians Bolzano and Weierstrass.*

3.5.4. Theorem. (*i*) *If A is an infinite bounded above subset of R and l = l.u.b. A such that $l \notin$ A, then l is a limit point of A.*

(*ii*) *If A is an infinite bounded below subset of R and g = g.l.b. A such that g $\notin$ A, then g is a limit point of A.*

Proof. (*i*) Let $\varepsilon > 0$ be given

Since l = l.u.b. A

$\therefore$ $\exists$ some $x \in$ A such that $x > l - \varepsilon$

Also, $x \le l < l + \varepsilon$ $(\because\ l = \text{l.u.b. A})$

$\Rightarrow$ $l - \varepsilon < x < l + \varepsilon$

$\Rightarrow$ $(l - \varepsilon, l + \varepsilon) \cap \text{A}—\{l\}$ contains at least x $(\because\ l \notin \text{A and } x \in \text{A})$

$\Rightarrow$ l is a limit point of A.

(*ii*) Proof is on same lines and is left as an exercise for the readers.

3.5.5. Derived Set. *The set of all limit points of a set A $\subseteq$ R, is called the derived set of A and is denoted by A′.*

Examples : (*i*) $(0, 1)' = [0, 1]$, $[0, 1]' = [0, 1]$

(*ii*) Q′ = R, (R – Q)′ = R

(*iii*) N′ = ϕ, Z′ = ϕ

[**To show** Z′ = ϕ.

Let x be any real number.

If $x \in$ Z, then $(x - \frac{1}{2}, x + \frac{1}{2})$ is a neighbourhood of x and contains no member of Z different from x. Therefore, x is not a limit point of Z.

If $x \notin$ Z, then there exists an integer n such that $n < x < n + 1$. Therefore, $(n, n + 1)$ is a neighbourhood of x and contains no member of Z. Hence, x cannot be a limit point of Z. Hence, Z′ = ϕ]

(*iv*) $\phi' = \phi$

(*v*) If $S = \left\{\frac{1}{n} \mid n \in N\right\}$, then $S' = \{0\}$

(*vi*) If $S = \left(\frac{1}{m} + \frac{1}{n} \mid m, n \in N\right)$, then $S' = \{0\} \cup \left\{\frac{1}{n} \mid n \in N\right\}$.

[**Sol***. If $x \in$ R, following cases arise :

(*i*) $x < 0$. Then, $(2x, 0)$ is a neighbourhood of x and does not contain any positive number. So, it does not contain any point of S.

$\therefore$ $x \notin S'$.

(*ii*) $x = 0$. For any $\varepsilon > 0$, $\exists\ m \in$ N such that

$$\tfrac{1}{2}\varepsilon > \frac{1}{n}\ \forall\ n \ge m$$

$$\Rightarrow \quad \varepsilon > \frac{1}{n} + \frac{1}{m} > 0\ \forall\ n \ge m$$

$\Rightarrow (-\varepsilon, \varepsilon) \cap S$ is infinite $\Rightarrow$ $0 \in S'$.

*May be omitted on first reading.

(iii) $x = \frac{1}{n}, n \in \mathbf{N}$

For any $\varepsilon > 0$, $\exists\, m_0 \in \mathbf{N}$ such that $\varepsilon > \frac{1}{m}$ $\forall\, m \geq m_0$

$\Rightarrow$ every neighbourhood $\left(\frac{1}{n} - \varepsilon, \frac{1}{n} + \varepsilon\right)$ of $\frac{1}{n}$ contains infinitely many points $\frac{1}{n} + \frac{1}{m}$, $(\forall\, n, m \geq m_0)$ of S

$$\Rightarrow \qquad \frac{1}{n} \in S' \quad \forall\, n \in \mathbf{N}.$$

(iv) $0 < x < 2$ and $x \neq \frac{1}{n}, n \in \mathbf{N}$.

Since $\frac{1}{x} > 0$, there exists unique $n_0 \in \mathbf{N}$ such that $n_0 + 1 > \frac{1}{x} > n_0$

$$\Rightarrow \quad \frac{1}{n_0} > x > \frac{1}{n_0 + 1} \quad \Rightarrow \quad \left(x - \frac{1}{n_0 + 1}\right)^{-1} > 0$$

$\therefore$ there exists $m_0 \in \mathbf{N}$ such that

$$m_0 + 1 > \left(x - \frac{1}{n_0 + 1}\right)^{-1} \geq m_0$$

$$\Rightarrow \qquad \frac{1}{m_0 + 1} < \left(x - \frac{1}{n_0 + 1}\right) \leq \frac{1}{m_0}$$

$$\Rightarrow \qquad \frac{1}{m_0} \geq x - \frac{1}{n_0 + 1} > \frac{1}{m_0 + 1}$$

$$\Rightarrow \qquad \frac{1}{m_0} + \frac{1}{n_0 + 1} \geq x > \frac{1}{m_0 + 1} + \frac{1}{n_0 + 1}$$

$\Rightarrow$ $\left(\frac{1}{m_0 + 1} + \frac{1}{n_0 + 1}, 2\right)$ is a neighbourhood of x and contains only a finite number of points $m_0 n_0 - 1$,
(for $n = 1, 2, \ldots, n_0$, $m = 1, 2, \ldots, m_0$ except $n = n_0$, $m = m_0$ simultaneously) of S.

$$\therefore \qquad x \notin S'.$$

(v) $x \geq 2$

Then, $\left(x - \frac{1}{4}, x + \frac{1}{4}\right)$ is a neighbourhood of x and contains at the most $2 \in S$ $\therefore$ $x \notin S'$.

Hence, $$S' = \{0\} \cup \left\{\frac{1}{n} \,\middle|\, n \in \mathbf{N}\right\}.]$$

3.5.6. Perfect Set. *A set A ($\subseteq$ R) is said to be a perfect set if*

$$A = A'.$$

For example, the set ϕ, $[a, b]$ and R are perfect sets. *(BCA (II) 2007)*

3.5.7. Theorem. *If A, B $\subseteq$ R, then*

(*i*) $A \subseteq B \Rightarrow A' \subseteq B'$ (*ii*) $(A \cup B)' = A' \cup B'$

(*iii*) $(A \cap B)' \subseteq A' \cap B'$ (*iv*) $A'' \subseteq A'$

(*v*) $\phi' = \phi$.

Proof. (*i*) Let $x \in A'$

$\Rightarrow$ for each $\varepsilon > 0$, $(x - \varepsilon, x + \varepsilon) \cap A - \{x\} \neq \phi$

$\Rightarrow$ for each $\varepsilon > 0$, $(x - \varepsilon, x + \varepsilon) \cap B - \{x\} \neq \phi$ $(\because A \subseteq B)$

$\Rightarrow$ $x \in B'$

$\therefore$ $x \in A' \Rightarrow x \in B'$

Hence $A' \subseteq B'$.

(*ii*) Since $A \subseteq A \cup B$ and $B \subseteq A \cup B$

$\therefore$ $A' \subseteq (A \cup B)'$ and $B' \subseteq (A \cup B)'$ [By (*i*)]

$\therefore$ $A' \cup B' \subseteq (A \cup B)'$ ...(1)

To prove that $(A \cup B)' \subseteq A' \cup B'$, we shall show that if

$$x \notin A' \cup B', \text{ then } x \notin (A \cup B)'$$

Now, $x \notin A' \cup B'$

$\Rightarrow$ $x \notin A'$ and $x \notin B'$

$\Rightarrow$ $\exists$ a neighbourhood N_1 of x such that

$$N_1 \cap A—\{x\} = \phi$$

and $\exists$ a neighbourhood N_2 of x such that

$$N_2 \cap B—\{x\} = \phi$$

$\Rightarrow$ $(N_1 \cap N_2) \cap A—\{x\} = \phi$

and $(N_1 \cap N_2) \cap B—\{x\} = \phi$

$\Rightarrow$ the neighbourhood $N_1 \cap N_2$ of x contains at the most one element x of $A \cup B$

$\Rightarrow$ $x \notin (A \cup B)'$

$\therefore$ $x \notin A' \cup B' \Rightarrow x \notin (A \cup B)'$ ...(2)

$\therefore$ from (1) and (2), $(A \cup B)' = A' \cup B'$.

(*iii*) Since $A \cap B \subseteq A$ and $A \cap B \subseteq B$

$\therefore$ $(A \cap B)' \subseteq A'$ and $(A \cap B)' \subseteq B'$

$\Rightarrow$ $(A \cap B)' \subseteq A' \cap B'$.

(*iv*) If $A'' = \phi$, we have nothing to prove.

If $A'' \neq \phi$, let $x \in A''$. Then, for any

$$\varepsilon > 0, (x - \varepsilon, x + \varepsilon) \cap A' - \{x\} \neq \phi.$$

Let $y \in (x - \varepsilon, x + \varepsilon) \cap A' - \{x\}$

$\Rightarrow$ $y \in (x - \varepsilon, x + \varepsilon)$, $y \in A'$ and $y \neq x$.

Since $(x - \varepsilon, x + \varepsilon)$ is a neighbourhood of y and $y \in A'$, therefore $(x - \varepsilon, x + \varepsilon) \cap A$ is an infinite set.

So, $(x - \varepsilon, x + \varepsilon) \cap A - \{x\}$ is infinite.

$\Rightarrow$ $x \in A'$

Thus, $x \in A'' \Rightarrow x \in A'$

Hence, $A'' \subseteq A'$.

(*v*) Let $x \in R$. Then, for any $\varepsilon > 0$, $(x - \varepsilon, x + \varepsilon) \cap \phi = \phi$

$\Rightarrow \quad x \notin \phi' \Rightarrow \phi' = \phi.$

3.5.8. Theorem. (*i*) *If an infinite set A ($\subseteq$ R) is bounded then* (*i*) *A′ is bounded* (*ii*) *A′ has the smallest and greatest numbers.*

Proof. (*i*) Since A is an infinite and bounded subset of R, $A' \neq \phi$ (Bolzano-Weierstrass Theorem) and there exist real numbers h, k such that $A \subseteq [h, k]$.

Let $x \in [h, k]$, then either $x < h$ or $x > k$.

If $x < h$, $(x - 1, h)$ is a neighbourhood of x containing no element of $[h, k]$ and hence of A.

$\Rightarrow \quad x \notin A'$

If $x > k$, $(k, x + 1)$ is a neighbourhood of x containing no element of $[h, k]$ and hence of A.

$\Rightarrow \quad x \notin A'$

Thus, $x \notin [h, k] \Rightarrow x \notin A'$.

Hence, $A' \subseteq [h, k]$, *i.e.*, A′ is bounded and bounds of A are bounds of A′.

(*ii*) Since A′ is non-empty and bounded, it has the g.l.b. ξ (say).

Then, for any $\varepsilon > 0$, there exists $x \in A'$ such that $\xi \le x < \xi + \varepsilon$ and so $x \in (\xi - \varepsilon, \xi + \varepsilon)$.

$\Rightarrow$ $(\xi - \varepsilon, \xi + \varepsilon)$ is a neighbourhood of $x \in A'$

$\Rightarrow$ $(\xi - \varepsilon, \xi + \varepsilon)$ contains infinitely many elements of A

$\Rightarrow$ $\xi \in A'$. Hence A′ contains the smallest number (the g.l.b. ξ).

Similarly, A′ contains the greatest number (the l.u.b).

Remark. *An unbounded set may or may not have a bounded derived set. For example,*

(*i*) *Q′ = R is unbounded and*

(*ii*) $\left[N \cup \left\{\frac{1}{n} \mid n \in N\right\}\right]' = \{0\}$ *is bounded.*

The smallest and greatest numbers ξ and η (say), of the derived set A′ of an infinite and bounded set A, as they do exist are usually denoted by $\underline{lim}$ A and $\overline{lim}$ A respectively and are called the *inferior* (or *lower*) limit of A and the *superior* (or upper) limit of A. Clearly, $\xi \le \eta$ and for each $\varepsilon > 0$, $(\xi - \varepsilon, \eta + \varepsilon)$ contains infinitely many elements of A and at the most only a finitely many elements of A can lie outside the interval $(\xi - \varepsilon, \eta + \varepsilon)$.

3.5.9. Adherent Point. *A real number x is called an adherent point of a set A $\subseteq$ R if every neighbourhood of x contains at least one point of A.*

Clearly, a real number x is an adherent point of A iff either $x \in A$ or $x \in A'$.

3.5.10. Closure of a set. *The set of all adherent points of a set A is called the closure of the set A and is denoted by $\overline{A}$ i.e.,* $\overline{A} = A \cup A'$.

Examples. (*i*) $\overline{(0, 1)} = [0, 1]$, $\overline{(0, 1]} = [0, 1]$, $\overline{[0, 1]} = [0, 1]$

(*ii*) $\overline{Q} = R$, $\overline{R - Q} = R$

(*iii*) $\overline{N} = N$, $\overline{Z} = Z$.

(*iv*) If $A = \left\{1 + \frac{1}{n} \mid n \in N\right\} \cup \left\{\frac{1}{n} \mid n \in N\right\}$, then $\overline{A} = A \cup \{0\}$.

3.6.1. Some relations between closed sets, derived sets and closures :

(*i*) For a set $A \subseteq R$, A' is a closed set (*K.U. BCA 2007*)

For, A' is a closed set iff $(A')^c$ is an open set.

Let $x \in (A')^c$

$\Rightarrow$ $x \notin A'$

$\Rightarrow$ $\exists$ some $\varepsilon > 0$ such that $(x - \varepsilon, x + \varepsilon) \cap A - \{x\} = \phi$

We shall show that $(x - \varepsilon, x + \varepsilon) \subseteq (A')^c$.

Let $y \in (x - \varepsilon, x + \varepsilon)$

$\Rightarrow$ $(x - \varepsilon, x + \varepsilon)$ is a neighbourhood of y and

$$(x - \varepsilon, x + \varepsilon) \cap A - \{x\} = \phi$$

$\Rightarrow$ $y \notin A'$

$\Rightarrow$ $y \in (A')^c$

$\therefore$ $y \in (x - \varepsilon, x + \varepsilon) \Rightarrow y \in (A')^c$

$\therefore$ $(x - \varepsilon, x + \varepsilon) \subseteq (A')^c$

$\Rightarrow$ $(A')^c$ is a neighbourhood of x

$\Rightarrow$ $(A')^c$ is an open set ($\because$ x is arbitrary)

$\Rightarrow$ A' is a closed set.

(*ii*) $\overline{A}$ is a closed set.

For, $\overline{A}$ is a closed set iff $(\overline{A})^c$ is an open set.

Let $x \in (\overline{A})^c$

$\Rightarrow$ $x \notin \overline{A}$

$\Rightarrow$ $\exists\, \varepsilon > 0$ such that $(x - \varepsilon, x + \varepsilon) \cap A = \phi$,

$\Rightarrow$ no element of $(x - \varepsilon, x + \varepsilon)$ is in $\overline{A}$

[$\because$ if $y \in (x - \varepsilon, x + \varepsilon)$, then $(x - \varepsilon, x + \varepsilon)$ is a neighbourhood of y and $(x - \varepsilon, x + \varepsilon) \cap A = \phi$.

$\therefore$ $y \notin \overline{A}$]

$\Rightarrow (x - \varepsilon, x + \varepsilon) \subseteq (\overline{A})^c$

$\Rightarrow (\overline{A})^c$ is a neighbourhood of x

$\Rightarrow (\overline{A})^c$ is an open set. ($\because$ x is arbitrary)

(*iii*) If $A \subseteq B$, then $\overline{A} \subseteq \overline{B}$

For, let $x \in \overline{A}$ $\Rightarrow$ $x \in A \cup A'$

$\Rightarrow$ $x \in A$ or $x \in A'$ $\Rightarrow$ $x \in B$ or $x \in B'$ ($\because$ $A \subseteq B \Rightarrow A' \subseteq B'$)

$\Rightarrow$ $x \in B \cup B'$ $\Rightarrow$ $x \in \overline{B}$

$\therefore$ $\overline{A} \subseteq \overline{B}$.

(*iv*) A is a closed set iff $A = \overline{A}$ [*K.U. BCA (II) 2003*]

i.e., *A is a closed set iff A contains all its limit points.*

For, if $A = \overline{A}$, then A is a closed set [by ((*iv*)]

So, let A be a closed set.

Let $x \in A'$. We shall show that $x \in A$

Since A is closed

$\therefore$ A^c is open.

If $x \notin A$, then $x \in A^c$

$\Rightarrow$. A^c is a neighbourhood of x

Since $x \in A'$

$\therefore$ $A^c \cap A \neq \phi$ which is not true.

$\therefore$ $x \in A$

$\therefore$ $x \in A' \Rightarrow x \in A$

$\therefore$ $A' \subseteq A$.

(*v*) A non-empty bounded closed set contains its l.u.b. and g.l.b.

(*vi*) For any set A, $\overline{A}$ is the smallest closed set containing A.

For, $\overline{A}$ is a closed set and $\overline{A} \supseteq A$

Let $B \supseteq A$ and B be a closed set

$\Rightarrow$ $\overline{B} \supseteq \overline{A}$

$\Rightarrow$ $B \supseteq \overline{A}$ $(\because \; B = \overline{B}$ as B is closed)

Hence the result.

(*vii*) For any set A, $\overline{(\overline{A})} = \overline{A}$.

(*viii*) **A set A is closed if A contains all its limit points.** Using this as definition of a closed set, prove : **A set is closed if and only if its complement is open.**

Proof. Let A be closed. We shall show that A^c is open.

Let $x \in A^c$

$\Rightarrow$ $x \notin A$.

$\Rightarrow x$ is not a limit point of A ($\because$ A is closed)

$\Rightarrow \exists$ some $\varepsilon > 0$ such that $(x - \varepsilon, x + \varepsilon) \cap A = \phi$

$\Rightarrow$ $(x - \varepsilon, x + \varepsilon) \subseteq A^c$

$\Rightarrow A^c$ is a neighbourhood of x, any point of A^c

$\Rightarrow A^c$ is open

Conversely, let A^c be open. We shall show that A is closed.

Let $\xi \in A'$.

Let if possible, $\xi \in A^c$.

$\Rightarrow A^c$ is a neighbourhood of ξ ($\because$ A^c is open)

$\Rightarrow \exists\, \varepsilon > 0$ such that $\xi \in (\xi - \varepsilon, \xi + \varepsilon) \subseteq A^c$

$\Rightarrow \exists$ a neighbourhood $(\xi - \varepsilon, \xi + \varepsilon)$ of ξ such that

$$(\xi - \varepsilon, \xi + \varepsilon) \cap A = \phi$$

$\Rightarrow \xi \notin A'$, which is a contradiction.

$\Rightarrow \xi \notin A^c \Rightarrow \xi \in A$

Thus, $A' \subseteq A$ and hence A is closed.

3.6.2. Theorem. *The g.l.b. and the l.u.b. of a set A are also the g.l.b. and the l.u.b. of $\bar{A}$ and are contained in $\bar{A}$ according as A is bounded below or above.*

Proof. Let A be a bounded below subset of R. Then, it has the g.l.b. ξ (say). If $\xi \in A$, then $\xi \in A \cup A' = \bar{A}$. If $\xi \notin A$, then for any $\varepsilon > 0$ there exists an $x \in A$ such that $\xi < x < \xi + \varepsilon$. This implies that the neighbourhood $(\xi - \varepsilon, \xi + \varepsilon)$ of ξ contains at least one element x of A other than ξ and so $\xi \in A'$. Thus, $\xi \in A \cup A' = \bar{A}$ in both the cases.

Now, A is bounded below $\Rightarrow$ $\bar{A}$ is bounded below. [For $A \subseteq [k, \infty)$ $\Rightarrow$ $\bar{A}$ is bounded below]. Therefore, $\bar{A}$ has the g.l.b. ξ_1 (say). $\xi_1 \not> \xi$, because $A \subseteq \bar{A}$. Let, if possible, $\xi_1 < \xi$. Then, $\xi_1 \notin A$.

Taking $\varepsilon = \dfrac{\xi - \xi_1}{2}$, $(\xi_1 - \varepsilon, \xi_1 + \varepsilon)$ is a neighbourhood of ξ_1 and contains no element of A.

So, $\xi_1 \notin A'$. Hence, $\xi_1 \notin A$ and $\xi_1 \notin A'$ imply that $\xi_1 \notin A \cup A' = \bar{A}$. But, $\xi_1 \in \overline{(\bar{A})} = \bar{A}$ (by the first part proved above). This contradiction proves that if $\xi_1 =$ g.l.b. $\bar{A}$, then $\xi_1 \not> \xi$ and $\xi_1 \not< \xi$. Thus, $\xi_1 = \xi \in \bar{A}$.

i.e.,

$$\text{g.l.b. } A = \text{g.l.b. } \bar{A} \in \bar{A}.$$

When $\bar{A}$ is bounded above, one can similarly show that

$$\text{l.u.b. } A = \text{l.u.b. } \bar{A} \in \bar{A}.$$

3.6.3. Dense set. *Let A, B $\subseteq$ R and A $\subseteq$ B. The set A is said to be dense (or everywhere dense) in B if $B \subseteq \bar{A}$.*

For example, if A = (0, 1] and B = [0, 1], then A is dense in B.

But, if B = [0, 1] $\cup$ {5}, then A is not dense in B ($\because$ $5 \notin \bar{A}$)

A set A is said to be dense in R if $\bar{A}$ = R.

Since $\bar{Q}$ = R and $\overline{R - Q}$ = R

$\therefore$ Q and R – Q are dense in R.

3.7.1. Compact set. *A set A ($\subseteq$ R) is said to be a compact set if it is closed and bounded.*

Examples. (*i*) Every finite set being bounded and closed, is a compact set.

(*ii*) The closed interval [*a*, *b*], where *a* and *b* are both finite, is compact. No other type of interval is compact.

(*iii*) The set $A = \{-1, 1\} \cup \left\{-1 - \dfrac{1}{n} \mid n \in N\right\} \cup \left\{1 + \dfrac{1}{n} \mid n \in N\right\}$ is compact.

[For, A is bounded, – 2 and 2 being the bounds of A. Also, the only, limit points – 1 and 1 of A belong to A. So, A is closed.]

(*iv*) Union of a finite number of compact sets is compact.

An infinite union of compact sets may or may not be compact.

(*v*) Intersection of any family of compact sets is compact.

3.7.2. Open Cover. *A family* $\{A_\lambda\}_{\lambda \in \Lambda}$ of open sets is said to *be an open cover of a set A if* $A \subseteq \bigcup_{\lambda \in \Lambda} A_\lambda$, *i.e.*, each element of A belongs to some A_λ.

3.7.3. Sub-cover. *If* $\{A_\lambda\}_{\lambda \in \Lambda}$ *is an open cover of a set A and there exists a subset* $\Lambda' \subseteq \Lambda$ *such that* $\{A_\lambda\}_{\lambda \in \Lambda'}$ *also covers A, then the family* $\{A_\lambda\}_{\lambda \in \Lambda'}$ *is called a sub-cover of the open cover* $\{A_\lambda\}_{\lambda \in \Lambda}$ or just a sub-cover in case there is no confusion to which cover we are referring to at a particular time.

Examples. (*i*) The family $\mathfrak{I}$ of open intervals.

$\left(\frac{1}{n}, \frac{2}{n}\right)$ for $n = 2, 3, 4, ...$, is an open cover of (0, 1). Note that there are infinitely many open sets in this cover of (0, 1) and that no finite number of these sets form a cover.

(*ii*) The family $\mathfrak{I}$ of all open intervals (a, b) is an open cover of R. Further, the family of all intervals of the type $(n, n + 2)$ is a sub-cover of $\mathfrak{I}$.

3.7.4. Heine-Borel Property. *A set A* ($\subseteq$ *R*) *is said to have the Heine-Borel Property if every open cover of A has a finite subcover.*

Next is an important theorem known as Heine-Borel Theorem concerning open covers of compact sets.

3.7.5. Heine-Borel Theorem. *Every open cover of a compact set has a finite sub-cover.*

Proof. To prove the above theorem, we first prove the following Lemma.

Lemma. *If a set A* ($\subseteq$ *R*) *satisfies the Heine-Borel property, then any closed subset of A also satisfies the Heine-Borel property.*

Proof of Lemma. Let A satisfy the Heine-Borel property. Let B be a closed subset of A. If $\{A_\lambda\}_{\lambda \in \Lambda}$ is an open cover of B, then the family $\mathfrak{I}$ containing all the sets $A_{\lambda(\lambda \in \Lambda)}$ and the set B^c is an open cover of A. (B closed $\Rightarrow B^c$ is open). By the given hypothesis, $\{A_\lambda\}_{\lambda \in \Lambda}$ has a finite subcover of A. Therefore, there exist sets $A_{\lambda_1}, A_{\lambda_2}, ..., A_{\lambda_n}$ (say) such that

$$A_{\lambda_1} \cup A_{\lambda_2} \cup ...... A_{\lambda_n} \cup B^c \supseteq A \supseteq B.$$

This implies, $A_{\lambda_1} \cup A_{\lambda_2} \cup ...... \cup A_{\lambda_n} \supseteq B.$ $(\because \quad B \cap B^c = \phi)$

Hence, the Lemma.

Proof of the theorem. Let A be a compact subset of R. Since A is bounded, its g.l.b. as well as the l.u.b. exist.

Let a = g.l.b. A and b = l.u.b. A.

Then, $A \subseteq [a, b]$.

By the above Lemma, it is enough to show that the interval $I = [a, b]$ satisfies the Heine-Borel property.

If $a = b$, *i.e.*, A is a singleton, we have nothing to prove.

So, let $a \neq b$. Let, if possible, the interval I do not satisfy the Heine-Borel property. Then, there exists a family $\mathfrak{I}$ of open sets which covers I but no finite sub-family of which covers I. Divide I into two equal closed intervals I′ and I″ but not both of them can be covered by finitely many members of $\mathfrak{I}$. Call the interval I_1 that of I′ and I″, which is not covered by finitely many members of $\mathfrak{I}$. Again, divide I_1 into two equal closed intervals and continuing as above, we get a sequence of closed intervals.

$$I = I_0 \supset I_1 \supset I_2 \supset \ldots\ldots \text{ such that}$$

(*i*) length of $I_n = l\,(I_n) = \frac{1}{2}\, l\,(I_{n-1})$

$$= \frac{1}{2^n}\, l(I) \to 0 \text{ as } n \to \infty.$$

(*ii*) No I_n can be covered by finitely many members of $\mathfrak{I}$.

By nested interval property of sequences, (See § 4.3.5)

$\bigcap_{n=1}^{\infty} I_n$ is a singleton. Let $\bigcap_{n=1}^{\infty} I_n = \{\xi\}$.

Since $\xi \in I$ and $\mathfrak{I}$ is an open cover of I,

$\therefore$ there exists an open set $B \in \mathfrak{I}$ such that $\xi \in B$.

$\therefore$ there exists an $\varepsilon > 0$ such that $(\xi - \varepsilon, \xi + \varepsilon) \subseteq B$.

Since $l\,(I_n) \to 0$ as $n \to \infty$,

$\therefore$ We can choose a natural number m so large that $l\,(I_m) < \varepsilon$.

Then, $\qquad I_m \subseteq (\xi - \varepsilon, \xi + \varepsilon) \subseteq B.$

Thus, the single member B of $\mathfrak{I}$ covers I_m which contradicts the fact that no I_n can be covered by finitely many members of $\mathfrak{I}$. Hence, I satisfies the Heine-Borel property.

3.7.6. Converse of Heine-Borel Theorem. The above theorem shows that every compact set possesses the Heine-Borel Property. The converse is also true, *i.e.*, *every set having the Heine-Borel Property is compact*.

Proof. Let A ($\subseteq$ R) satisfy Heine-Borel Property. We shall show that A is closed and bounded.

(*i*) **To show that A is closed,** we shall show that A^c is open, *i.e.*, A^c is a neighbourhood of each of its points. To show this, let $x \in A^c$ be any element. Then, for any $y \in A$, x is never equal to y. Choose $\varepsilon = \frac{1}{3} \mid x - y \mid > 0$. Then, $A_{x,\,y} = (x - \varepsilon, x + \varepsilon)$ and $B_y = (y - \varepsilon, y + \varepsilon)$ are neighbourhoods of x and y respectively, such that $A_{x,\,y} \cap B_y = \phi$*.

The family $\{B_y\}_{y \in A}$ (for each $y \in A$) is an open cover of A. Since A satisfies Heine-Borel Property, the family $\{B_y\}_{y \in A}$ has a finite subcover consisting of elements $B_{y_1}, B_{y_2}, \ldots\ldots, B_{y_n}$ (say) such that $A \subseteq \bigcup_{i=1}^{n} B_{y_i}$.

Let $V = \bigcap_{i=1}^{n} A_{x,\,y_i}$ be the finite intersection of corresponding A_x, y_i

$$V \neq \phi \text{ as } x \in V.$$

Then, V is a neighbourhood of x and $V \cap A = \phi$. This implies $V \subseteq A^c$. Hence, A^c is open, *i.e.* A is closed.

---

*__Hausdorff Property.__ *R has Hausdorff Property, namely : if $x, y \in R$ and $x \neq y$, then x and y have disjoint neighbourhoods.*

For, $x \neq y \Rightarrow \mid x - y \mid > 0$. Then, for $\varepsilon = \frac{1}{3} \mid x - y \mid > 0$, $(x - \varepsilon, x + \varepsilon)$ and $(y - \varepsilon, y + \varepsilon)$ are neighbourboods of x and y, and $(x - \varepsilon, x + \varepsilon) \cap (y - \varepsilon, y + \varepsilon) = \phi$.

(*ii*) **To show that A is bounded.** Clearly, $\{(-n, n)\}_{n \in N}$ is an open cover of R and hence an open cover of A also. Since A has Heine-Borel Property, this open cover has a finite subcover $\{(-n_i, n_i)\}$, $i = 1, 2, 3, \ldots\ldots k$ (say). If $K = \max. \{n_1, n_2, \ldots\ldots, n_k\}$, then $A \subseteq (-K, K)$ and hence A is bounded.

EXERCISE 3 (*b*)

1. Give one example of each of the following :
   (*i*) an infinite set having no limit point.
   (*ii*) an infinite set having one limit point.
   (*iii*) a set having two limit points.
   (*iv*) a set having an infinite number of limit points.
   (*v*) a set every point of which is a limit point.
   (*vi*) a set with only $\sqrt{2}$ as a limit point.
   (*vii*) $(A \cap B)' \neq A' \cap B'$.
2. Find the derived set of each of the following :
   (*i*) $\left\{\frac{1}{n} \mid n \in N, n \neq 0\right\}$
   (*ii*) $\left\{\frac{1}{p} + \frac{1}{q} \mid p, q \in N\right\}$
   (*iii*) $\left\{1 + \frac{1}{5^n} \mid n \in N\right\}$
   (*iv*) $(1, \infty)$
   (*v*) $(-\infty, -1)$
   (*vi*) $\left\{1 + \frac{(-1)^n}{n} \mid n \in N\right\}$
   (*vii*) $\left\{(-1)^n + \frac{1}{n} \mid n \in N\right\}$
   (*viii*) $\left\{\frac{1 + (-1)^n}{n} \mid n \in N\right\}$
   (*ix*) $\{r\sqrt{2} \mid r \in Q\}$.
3. Prove that a real number x is a limit point of a set $S \subseteq R$ iff for each $n \in N$,
$$\left\{x - \frac{1}{n}, x + \frac{1}{n}\right\} \cap S - \{x\} \neq \phi.$$
4. Prove : (*i*) Union of a finite family of compact sets is compact.
   (*ii*) Intersection of an arbitrary family of compact sets, having at least one point in common, is compact.
5. (*a*) Prove that the l.u.b. of the set B is also the l.u.b. of $\overline{B}$ and also prove that it is contained in $\overline{B}$.
   (*b*) Prove that the set of limit points of an infinite and bounded set B has the greatest number.

4

Sequences

In this chapter, we will study a special class of functions—known as sequences. We shall also introduce the concept of boundedness, monotonicity, cluster point, convergence of a sequence and of subsequence of a sequence.

4.1.1. Sequence. *A sequence is a function whose domain is the set N of all natural numbers and range can be any set.*

A sequence whose range is a subset of R, the set of all real numbers, is called *a real sequence*. Since we are going to study only real sequences, therefore, we shall be writing just 'sequences' instead of 'real sequences'.

If $a : N \to R$ is a sequence, then the real number associated to $n \in N$ by the sequence 'a' is denoted by a_n and is called the nth term (element or member) of the sequence a. A sequence 'a' can thus be regarded as an infinite tuple $\{a_1, a_2, a_3, \ldots\ldots, a_n \ldots\ldots\}$ and $a_1, a_2, a_3, \ldots\ldots, a_n, \ldots\ldots$ are the first, second, third, ..., nth, ... terms of the sequence 'a'. If $n \neq m$, then nth and mth terms a_n and a_m of the sequence a are treated as different even if a_m and a_n have the same value. Since the domain of all sequences is N, the set of all natural numbers, therefore, a sequence is completely known if we know a_n for each n. Thus a sequence 'a' whose nth term is a_n is denoted by $\{a_n\}$ or $< a_n >$.

4.1.2. Methods to Specify a Sequence. Some common methods used to describe sequences are :

(*i*) A sequence may be described by writing a first few terms of a sequence till the rule for writing down the other terms of the sequence becomes evident. For example,

(*a*) the sequence $< 1, 4, 9, \ldots\ldots, >$ is the sequence whose nth term is n^2 ;

(*b*) the sequence $\{1, \frac{1}{2}, \frac{1}{3}, \ldots\ldots\}$ is the sequence whose nth term is $\dfrac{1}{n}$.

(*ii*) A sequence may be described by giving a formula for its nth term. For example, the sequence $< 1, 4, 9, \ldots\ldots >$ can be written as $<n^2>, n \in N$ or just $<n^2>$ and the sequence $< 1, \frac{1}{2}, \frac{1}{3}, \ldots\ldots >$ can be written as $<\dfrac{1}{n}>$.

(*iii*) **Recursive Formula.** A sequence may be described by specifying its first few terms and a rule (formula) to determine the other terms of the sequence in terms of the preceeding terms. For example,

if $\qquad a_1 = 1$ and $a_{n+1} = 5a_n \ \forall \ n \in N,$

then it defines a sequence whose nth term is 5^{n-1}. As another example,

let $a_1 = \sqrt{2}$ and $a_{n+1} = \sqrt{2 . a_n} \ \ \forall \ n \in N$. Then, we can determine $a_2, a_3, \ldots\ldots$, in order.

In fact $a_2 = \sqrt{2 . a_1} = \sqrt{2 . \sqrt{2}}$

$$a_3 = \sqrt{2 . a_2} = \sqrt{2 . \sqrt{2 . \sqrt{2}}}$$

$$\vdots$$

Thus the sequence $< a_n >$ is completely determined.

(*iv*) Sometimes a sequence is described by more than one relation. For example, the sequence $< a_n >$ where a_n is given by

$$a_{2n-1} = 2n - 1, n \in \mathrm{N}$$
$$a_{2n} = 2^n, n \in \mathrm{N}$$

is the sequence $< 1, 2, 3, 2^2, 5, 2^3, \ldots\ldots >$.

4.1.3. Range of a sequence. *The set of all distinct terms of a sequence* $< a_n >$ *is called the range of the sequence, i.e., the set* $\{a_n : n \in N\}$ *is called the range of the sequence.* For example, the range of the sequence $< (-1)^n >$ is the set $\{-1, 1\}$.

It may be noted that a sequence $< a_n >$ always has infinite terms whereas the range of a sequence may have a finite number of elements.

Remarks. (*i*) Sometimes we start a sequence with a zero-th term also and write $< a_n >, n \geq 0$. Similarly, if we start a sequence from any non-negative integer k, then we write $< a_n >, n \geq k$.

(*ii*) Two sequences $< a_n >$ and $< b_n >$ are equal if, and only if $a_n = b_n \ \forall \ n$ as is clear from the definition of equality of two functions.

(*iii*) *To distinguish between a sequence and its range set, we use* $< >$ *to denote a sequence and* $\{ \}$ *to describe a set.*

4.1.4. Bounded sequences. *A sequence is said to be bounded above, bounded below or bounded according as the range of the sequence is bounded above, bounded below or bounded.* Thus, a sequence $< a_n >$ is

(*i*) bounded above if there exists a real number K such that $a_n \leq \mathrm{K} \ \forall \ n \in \mathrm{N}$.

(*ii*) bounded below if there exists a real number k such that $a_n \geq k_n \ \forall \ n \in \mathrm{N}$.

(*iii*) bounded if there exist real numbers k and K such that

$$k \leq a_n \leq \mathrm{K} \quad \forall \, n \in \mathrm{N}.$$

The l.u.b. and the g.l.b. of the range set of a sequence $< a_n >$ are respectively called the l.u.b. and the g.l.b. of the sequence $< a_n >$.

Since in a sequence, terms with equal values can occur, therefore, a sequence may have more than one term with smallest value. In such a case, any of these can be taken as the smallest value. Similar is the case with greatest value. Further, like sets of real numbers, a sequence bounded below or above may or may not have a smallest or a greatest member. Moreover, an unbounded sequence cannot have a smallest or a greatest member.

Remark. It can be seen that a sequence $< a_n >$ is bounded if and only if, there exists a positive real number K such that $| a_n | \leq \mathrm{K} \ \forall \ n \in \mathrm{N}$.

Examples : (*i*) The sequence $< -n >$ is bounded above but unbounded below :

(*ii*) The sequence $< n >$ is unbounded above but bounded below.

(*iii*) The sequence $< (-1)^n >$ is bounded.

(*iv*) The sequence $< (-1)^n . n >$ is neither bounded below nor bounded above.

(*v*) Every constant sequence is bounded.

4.1.5. Convergent sequences. *A sequence* $< a_n >$ *is said to* **converge** *to (or tend to) a real number* l *if given any* $\varepsilon > 0$ *(however small), there exists a positive integer* m *(depending upon* ε*) such that*

$$| a_n - l | < \varepsilon \qquad \forall\, n \geq m.$$

The real number l is called a *limit* of the sequence $< a_n >$.

The definition asserts that given any positive number ε (however small), then, after a few terms of the sequence (*i.e.*, after first $m - 1$ terms of the sequence), the absolute value of the difference between a term of the sequence and the number l can be made less than ε. That is, all the terms of the sequence $< a_n >$, except the first $m - 1$ terms, lie in the interval $(l - \varepsilon, l + \varepsilon)$. The first $m - 1$ terms of the sequence may be scattered anywhere. The number $m - 1$ of the terms left out from the interval $(l - \varepsilon, l + \varepsilon)$ depends upon the size of ε. The smaller the size of ε, larger will be the number of terms left out of the interval $(l - \varepsilon, l + \varepsilon)$.

Since $\varepsilon > 0$ can assume any value, $(l - \varepsilon, l + \varepsilon)$ is an arbitrary neighbourhood of l. So, we can say that if a sequence $< a_n >$ converges to l, then all except a first few terms of the sequence lie in every neighbourhood of l. Or that the sequence is eventually in every neighbourhood of l.

If a sequence $< a_n >$ converges to l, we write $\lim\limits_{n \to \infty} a_n = l$ or $a_n \to l$ as $n \to \infty$ or simply $a_n \to l$.

Remarks. (*i*) A sequence $< a_n >$ is said to be a convergent sequence if $< a_n >$ converges to some limit, otherwise $< a_n >$ is said to be a non-convergent sequence.

(*ii*) If we can find one $\varepsilon > 0$ for which an infinite number of terms of a sequence lie outside $(l - \varepsilon, l + \varepsilon)$, the sequence cannot converge to l.

4.1.6. Theorem. *A sequence, if convergent converges to a unique limit.*

Proof. Let, if possible, the sequence $< a_n >$ converge to two different real numbers l and l'.

Let $\qquad \varepsilon = \frac{1}{2} | l - l' | > 0.$

Then, there exist positive integers m_1 and m_2 such that

$$| a_n - l | < \frac{\varepsilon}{2} \qquad \forall\, n \geq m_1$$

and
$$| a_n - l' | < \frac{\varepsilon}{2} \qquad \forall\, n \geq m_2$$

Let $\qquad m = \max. \{m_1, m_2\}$

Then,
$$| a_m - l | < \frac{\varepsilon}{2}$$

and
$$| a_m - l' | < \frac{\varepsilon}{2}$$

Now
$$\begin{aligned} | l - l' | &= | l - a_m + a_m - l' | \\ &\leq | a_m - l | + | a_m - l' | \\ &< \frac{\varepsilon}{2} + \frac{\varepsilon}{2} = \varepsilon \end{aligned}$$

$\therefore \qquad | l - l' | < \varepsilon,$

which is a contradiction to the fact that $\varepsilon = \frac{1}{2} | l - l' |$.

Hence, $\qquad l = l'.$

Remark. In the light of the above theorem, we replace the words 'a limit' by 'the limit' in the above definition.

Example 1. *The sequence* $< \frac{1}{n} >$ *converges to 0.*

Sol. Let $\varepsilon > 0$ be given. Choose a natural number m such that $m > \frac{1}{\varepsilon}$. Then for $n \geq m$.

$$\left| \frac{1}{n} - 0 \right| < \varepsilon$$

Hence, $$\lim_{n \to \infty} \frac{1}{n} = 0.$$

Example 2. *The sequence* $< \frac{1}{3^n} >$ *converges to 0.* (K.U. BCA (II) 2008)

Sol. Let $\varepsilon > 0$ be given.

Now $$\left| \frac{1}{3^n} - 0 \right| = \left| \frac{1}{3^n} \right| = \frac{1}{3^n} < \varepsilon$$

if $$\frac{1}{3^n} < \varepsilon$$

i.e., if $$3^n > \frac{1}{\varepsilon}$$

i.e., if $$n \log 3 > \log \left(\frac{1}{\varepsilon} \right)$$

i.e., if $$n > \frac{\log \left(\frac{1}{\varepsilon} \right)}{\log 3}$$

If m is any positive integer $> \frac{\log \left(\frac{1}{\varepsilon} \right)}{\log 3}$, then

$$\left| \frac{1}{3^n} - 0 \right| < \varepsilon \quad \forall\ n \geq m$$

Hence, $$\lim_{n \to \infty} \frac{1}{3^n} = 0.$$

Example 3. *The sequence* $< \frac{n^2 + 3n + 5}{2n^2 + 5n + 7} >$ *converges to* $\frac{1}{2}$.

Sol. Let $\varepsilon > 0$ be given

Now $$\left| \frac{n^2 + 3n + 5}{2n^2 + 5n + 7} - \frac{1}{2} \right| = \frac{1}{2} \left| \frac{n + 3}{2n^2 + 5n + 7} \right| \leq \frac{1}{2} \left| \frac{n + 3n}{2n^2 + 5n + 7} \right| \qquad (\forall\ n \geq 1)$$

$$= \frac{2n}{2n^2 + 5n + 7}$$

$$< \frac{2n}{2n^2} = \frac{1}{n}$$

$$< \varepsilon \ \text{ for } \ \frac{1}{n} < \varepsilon, \quad i.e., \quad \text{for } n > \frac{1}{\varepsilon}.$$

$\therefore$ If m is a positive integer $> \frac{1}{\varepsilon}$, then

$$\left| \frac{n^2 + 3n + 5}{2n^2 + 5n + 7} - \frac{1}{2} \right| < \varepsilon \quad \forall\ n \geq m$$

Hence, $$\lim_{n \to \infty} \frac{n^2 + 3n + 5}{2n^2 + 5n + 7} = \frac{1}{2}.$$

Example 4. *The sequence $< (-1)^n >$ does not converge.*

Sol. Let, if possible, the sequence $< (-1)^n >$ converge to l. Let $0 < \varepsilon < 1$. Then, there exists a positive integer m such that

$$| (-1)^n - l | < \frac{\varepsilon}{2} \quad \forall\, n \geq m$$

Taking some $n \geq m$,

$$\begin{aligned} | (-1)^{n+1} - (-1)^n | &= | (-1)^{n+1} - l + l - (-1)^n | \\ &\leq | (-1)^{n+1} - l | + | (-1)^n - l | \\ &< \frac{\varepsilon}{2} + \frac{\varepsilon}{2} = \varepsilon \end{aligned}$$

i.e., $$2 = | (-1)^{n+1} - (-1)^n | < \varepsilon < 1$$

i.e., $2 < 1$, which is impossible.

Hence, the sequence $< (-1)^n >$ does not converge.

4.1.7. Theorem. *If the sequence $< a_n >$ converges to l, then the sequence $< | a_n | >$ converges to $| l |$.*

Proof. Proof follows from the inequality.

$$| \, | a_n | - | l | \, | \leq | a_n - l | .$$

Remark. Converse of the above theorem is false.

For example, if $a_n = (-1)^{n-1}$, then $| a_n | = 1 \quad \forall \quad n$.

Obviously $< | a_n | >$ converges to 1, but $< a_n >$ is not convergent.

4.1.8. Theorem. *Every convergent sequence is bounded.* (K.U. BCA (II) 2001)

Proof. Let $< a_n >$ be a convergent sequence, converging to l.

Taking $\varepsilon = 1$, there exists a positive integer m such that

$$| a_n - l | < 1 \qquad \forall\, n \geq m$$

$\Rightarrow$ $$l - 1 < a_n < l + 1 \;\forall\, n \geq m$$

Let $$k = \min. \{l - 1, a_1, a_2, \ldots\ldots, a_{m-1}\},$$
$$K = \max. \{l + 1, a_1, a_2, \ldots\ldots, a_{m-1}\}$$

Then, $k \leq a_n \leq K \;\forall\, n$.

$\Rightarrow$ The sequence $< a_n >$ is a bounded sequence.

Remark. The converse of the above theorem is not true, *i.e.*, a bounded sequence need not be convergent. For example, the sequence $< (-1)^n >$ is bounded but is not convergent (see example 4 above).

Examples (*i*) The sequence $< n^2 >$ is not convergent because it is unbounded.

(*ii*) The sequence $< a_n >$, where $a_n = \begin{cases} n \text{ when } n \text{ is odd} \\ \dfrac{1}{n} \text{ when } n \text{ is even} \end{cases}$

is not convergent because it is not bounded.

4.1.9. Divergent Sequences

(*i*) *A sequence $< a_n >$ is said to diverge to $+\infty$ if given any positive real number K (however large), there exists a positive integer m (depending upon K) such that*

$$a_n > K \quad \forall\, n \geq m.$$

We write $\lim\limits_{n \to \infty} a_n = \infty$ or simply $a_n \to \infty$ as $n \to \infty$.

(ii) A sequence $< a_n >$ is said to diverge to $-\infty$ if given any positive real number K (however large) there exists a positive integer m (depending upon K) such that $a_n < -K \quad \forall\, n \geq m$.

We write $\lim\limits_{n \to \infty} a_n = -\infty$ or $a_n \to -\infty$ as $n \to \infty$.

(iii) A sequence $< a_n >$ is said to be a divergent sequence if

$$a_n \to \infty \text{ or } -\infty.$$

Examples (*i*) The sequences $< n >$, $< n^2 >$, $< 2^n >$ and $< x^n >$ for $x > 1$ are all divergent to ∞.

(*ii*) The sequences $< -n >$, $< -n^2 >$, $< -2n^n >$ and $< -x^n >$ for $x > 1$ are all divergent to $-\infty$.

4.1.10. Theorem. *(i) If a sequence $< a_n >$ diverges to ∞, then $< a_n >$ is bounded below but unbounded above.*

(ii) If a sequence $< a_n >$ diverges to $-\infty$, then $< a_n >$ is unbounded below but bounded above.

Proof. (*i*) Since the sequence $< a_n >$ diverges to ∞.

$\therefore$ for any positive real number K, all but first finitely many terms of the sequence $< a_n >$ exceed K. This implies $< a_n >$ is not bounded above. Taking K = 1, there exists a positive integer m such that

$$a_n > 1 \quad \forall\, n \geq m$$

Let $$k = \min. \{a_1, a_2, \ldots\ldots, a_{m-1}, 1\}, \text{ then}$$

$$a_n \geq k \quad \forall\, n.$$

$\Rightarrow$ the sequence $< a_n >$ is bounded below.

(*ii*) Proof is left as an exercise for the reader.

Remark. Converse of the above theorem need not be true.

For example,

(*i*) The sequence $< a_n >$, where

$$a_n = \begin{cases} n & \text{when } n \text{ is even} \\ \dfrac{1}{n} & \text{when } n \text{ is odd} \end{cases}$$

is bounded below ($a_n \geq 0$) and unbounded above but does not diverge to ∞;

(*ii*) The sequence $< a_n >$, where

$$a_n = \begin{cases} -n & \text{when } n \text{ is even} \\ \dfrac{-1}{n} & \text{when } n \text{ is odd} \end{cases}$$

is bounded above ($a_n \leq 0$) and unbounded below but does not diverge to $-\infty$.

4.1.11. Oscillating Sequences

(i) A sequence is said to be oscillating (or oscillatory) if it is neither convergent nor divergent to $+\infty$ or $-\infty$.

(ii) A sequence is said to be finitely oscillating if it is bounded as well as oscillating.

(iii) A sequence is said to be infinitely oscillating if it is unbounded as well as oscillating.

For example, (*i*) the sequence $< (-1)^n >$ is finitely oscillating

(*ii*) the sequence $< (-1)^n \,.\, n >$ is in infinitely oscillating.

4.1.12. Null sequence. *A sequence $< a_n >$ is said to be a null sequence if $< a_n >$ converges to zero.*

For example, the sequences $< \frac{1}{n} >$, $< (-1)^{n-1} . \frac{1}{n} >$, $< \frac{1}{n^2} >$ and $< \frac{1}{2^n} >$ are null sequences.

Remarks. (*i*) A sequence $< a_n >$ is a null sequence if, and only if $< | a_n | >$ is a null sequence.

For, $a_n \to 0$. Then, for any $\varepsilon > 0$, there exists a positive integer m such that

$$| a_n - 0 | < \varepsilon \ \forall \ n \geq m$$

$$\Rightarrow \quad | a_n | < \varepsilon \ \forall \ n \geq m$$

$$\Rightarrow \quad | \, | a_n | - 0 \, | < \varepsilon \ \forall \ n \geq m$$

$$\Rightarrow \quad | a_n | \to 0 \text{ as } n \to \infty$$

Converse is obvious.

(*ii*) A sequence $< a_n >$ converges to l if, and only if the sequence $< a_n - l >$ converges to 0.

(*iii*) If $b_n \to 0$ and $| a_n | \leq | b_n | \ \forall \ n$, then $a_n \to 0$.

The result however is false, if $b_n \to l \ (\neq 0)$.

For example, let $b_n = 1$ and $a_n = (-1)^n \ \forall \ n$. Then $< b_n >$ converges but $< a_n >$ does not converge.

EXERCISE 4 (*a*)

1. Discuss the boundedness of the following sequences $< a_n >$ where a_n is given by

(*i*) $a_n = 5$ (*ii*) $a_n = (-1)^n . 3$ (*iii*) $a_n = \frac{2n+3}{3n+4}$

(*iv*) $a_n = \left(1 + \frac{1}{n}\right)^n$ (*v*) $a_n = \frac{1}{n^2} + \frac{1}{(n+1)^2} + \ldots + \frac{1}{(2n)^2}$

(*vi*) $a_n = n$th prime (*vii*) $a_n = \frac{1}{3^n}$ (*viii*) $a_n = n^3$

(*ix*) $a_n = -n^2$ (*x*) $a_n = 1 + (-1)^n$.

2. Discuss the convergence or divergence of sequences in question No. 1.

3. (*a*) Establish the truth or falsify the following statements :

(*i*) If $< a_{n+1} - a_n >$ oscillates finitely, then $< a_n >$ oscillates

(*ii*) If $< a_{n+1} - a_n >$ oscillates infinitely, then $< a_n >$ oscillates.

(*iii*) If given k (however large), we can find m for which $a_m > k$, then $a_n \to \infty$.

(*iv*) Every bounded sequence is convergent.

(*b*) Give an example of a sequence $< a_n >$ which is not bounded but for which

$$\lim_{n \to \infty} \frac{a_n}{n} = 0.$$

4. By definition show that

(*i*) $\lim\limits_{n \to \infty} \frac{2n+1}{n+3} = 2$

(*ii*) $\lim\limits_{n \to \infty} \frac{n^2+1}{2n^2+3} = \frac{1}{2}$

(*iii*) $\lim\limits_{n \to \infty} \frac{1}{n^2} = 0.$

5. If $< a_n >$ and $< b_n >$ both are null sequences, show that $< a_n + b_n >$ is a null sequence.

6. If $< a_n >$ is a null sequence and $< b_n >$ is a bounded sequence, then $< a_n . b_n >$ is a null sequence. (K.U. BCA (II) 2006)
7. If $< a_n >$ is a null sequence and c is a constant, then $< ca_n >$ is a null sequence.
8. Give examples of sequences $< a_n >$ for which

$$\lim_{n \to \infty} \frac{a_{n+1}}{a_n} = 1$$

and (*i*) $a_n \to \infty$, (*ii*) $a_n \to 2$ and (*iii*) $a_n \to 0$.

9. If $a_n = \sqrt{n+1} - \sqrt{n}$, show that $a_n \to 0$.
10. Prove that $< \frac{n!}{n^n} >$ is a null sequence.

$$\left[\textbf{Hint. } \frac{n.(n-1).(n-2)......2.1}{n.n.n......n.n} = \left(1-\frac{1}{n}\right)\left(1-\frac{2}{n}\right)......\left(1-\frac{n-2}{n}\right).\frac{1}{n} = \frac{k}{n}, \text{ where } k < 1.\right]$$

4.2. SOME BASIC THEOREMS ON LIMITS

4.2.1. Theorem. *If $< a_n >$ converges to a and $a \neq 0$, then there exists a positive number k and a positive integer m such that $| a_n | > k$, for each $n \geq m$.*

Proof. Take $\varepsilon = \frac{1}{2} | a | > 0 \quad (\because \quad a \neq 0)$ ...(1)

Since $a_n \to a$,

$\therefore$ $\exists$ a positive integer m such that

$$| a_n - a | < \varepsilon \quad \forall \, n \geq m \qquad ...(2)$$

Now $| a | = | a - a_n + a_n | = | (a - a_n) + a_n |$

$$\leq | a - a_n | + | a_n |$$

$$< \varepsilon + | a_n | \quad \forall \, n \geq m \qquad \text{[By (2)]}$$

$\Rightarrow$ $| a | - \varepsilon < | a_n | \quad \forall \, n \geq m$

$\Rightarrow$ $| a_n | > | a | - \varepsilon \quad \forall \, n \geq m$

$\Rightarrow$ $| a_n | > \frac{1}{2} | a | \quad \forall \, n \geq m$ [By (1)]

Therefore, we have found a positive real number k $(= \frac{1}{2} | a |)$ and a positive integer m such that $| a_n | > k \quad \forall \, n \geq m$.

If $< a_n >$ and $< b_n >$ are two given sequences, then the sequences $< a_n + b_n >$, $< a_n - b_n >$ and $< a_n . b_n >$ are called the **sum, difference** *and* **product** *of the sequences $< a_n >$ and $< b_n >$. If $a_n \neq 0 \; \forall \, n$, then sequence $< \frac{1}{a_n} >$ is called the reciprocal of the sequence $< a_n >$. If $b_n \neq 0 \; \forall \, n$, then the sequence $< \frac{a_n}{b_n} >$ is called the* **quotient** *of the sequence $< a_n >$ by $< b_n >$. If k is a real number, then the sequence $< ka_n >$ is called the scalar multiple of the sequence $< a_n >$.*

We shall now discuss the relation of two convergent sequences w.r.t. their sum, difference, product, quotient, etc.

4.2.2. Theorem. *If $< a_n >$ converges to a and $< b_n >$ converges to b, then*

(*i*) *$< ka_n >$ converges to ka, where k is a fixed real number.*

(*ii*) *$< a_n + b_n >$ converges to $a + b$.*

(*iii*) *$< a_n - b_n >$ converges to $a - b$.*

(iv) $< a_n \,.\, b_n >$ *converges to* $a \,.\, b$.

(v) $< \frac{1}{a_n} >$ *converges to* $\frac{1}{a}$, *if* $a_n \neq 0 \; \forall \; n$ *and* $a \neq 0$.

(vi) $< \frac{a_n}{b_n} >$ *converges to* $\frac{a}{b}$, *if* $b_n \neq 0 \quad \forall \, n$ *and* $b \neq 0$.

Proof. *(i)* Let $\varepsilon > 0$ be given $| ka_n - ka | = | k | \; | a_n - a |$

Since $a_n \to a$, there exists a positive integer m such that

$$| a_n - a | < \frac{\varepsilon}{|k|+1} \qquad \forall \; n \geq m$$

Thus,

$$| k | \; | a_n - a | < \frac{|k|\varepsilon}{|k|+1} < \varepsilon \quad \forall \; n \geq m$$

$\Rightarrow$ $$| ka_n - ka | < \varepsilon \qquad \forall \; n \geq m$$

$\Rightarrow$ $< ka_n >$ converges to ka.

(ii) Let $\varepsilon > 0$ be given.

Since $a_n \to a$ and $b_n \to b$

$\therefore$ $\exists$ positive integers m_1 and m_2 such that

$$| a_n - a | < \frac{\varepsilon}{2} \quad \forall \; n \geq m_1$$

and $$| b_n - b | < \frac{\varepsilon}{2} \quad \forall \; n \geq m_2$$

Let $m = \text{max.} \, \{m_1, m_2\}$. Then

$$\begin{aligned} | a_n + b_n - (a + b) | &= | a_n - a + b_n - b | \\ &\leq | a_n - a | + | b_n - b | \\ &< \frac{\varepsilon}{2} + \frac{\varepsilon}{2} = \varepsilon \quad \forall \; n \geq m \end{aligned}$$

i.e., $$| a_n + b_n - (a + b) | < \varepsilon \quad \forall \; n \geq m$$

$\Rightarrow$ $< a_n + b_n >$ converges to $a + b$.

(iii) $$\begin{aligned} | a_n - b_n - (a - b) | &= | a_n - a + b - b_n | \\ &\leq | a_n - a | + | b_n - b | \end{aligned}$$

The result now follows as in *(ii)* above.

(iv) We have, $$\begin{aligned} | a_nb_n - ab | &= | a_nb_n - ab_n + ab_n - ab | = | b_n(a_n - a) + a(b_n - b) | \\ &\leq | b_n | \; | a_n - a | + | a | \; | b_n - b | \end{aligned}$$

The sequence $< b_n >$ being convergent, is bounded.

$\therefore$ $\exists$ a positive real number k such that

$$| b_n | < k \quad \forall \; n$$

$\therefore$ $$| a_nb_n - ab | < k \; | a_n - a | + | a | \; | b_n - b | \; \forall \; n \qquad ...(1)$$

Let $\varepsilon > 0$ be given

Since $a_n \to a$ and $b_n \to b$

$\therefore$ $\exists$ integers m_1 and m_2 such that

$$| a_n - a | < \frac{\varepsilon}{2|k|} \qquad \forall \; n \geq m_1 \qquad ...(2)$$

and $$| b_n - b | < \frac{\varepsilon}{2(|a|+1)} \quad \forall\, n \ge m_2 \qquad ...(3)$$

Let $m = \max.\{m_1, m_2\}$

Then, from (1), (2) and (3),

$$| a_n b_n - ab | < k \cdot \frac{\varepsilon}{2|k|} + | a | \frac{\varepsilon}{2(|a|+1)} \quad \forall\, n \ge m$$

$$< \varepsilon \quad \forall\, n \ge m$$

Hence $< a_n b_n >$ converges to ab.

(*v*) $$\left| \frac{1}{a_n} - \frac{1}{a} \right| = \frac{|a - a_n|}{|a|\,|a_n|} \qquad ...(1)$$

Since $a_n \to a\ (\ne 0)$,

$\therefore$ $\exists$ a positive number k and a positive integer m such that

$$| a_n | > k \quad \forall\, n \ge m \quad \text{[Theorem 4.2.1]} \qquad ...(2)$$

Let $\varepsilon > 0$ be given.

Since $a_n \to a$,

$\therefore$ $\exists$ a positive integer m_1 such that

$$| a_n - a | < k\, | a | \,.\, \varepsilon \qquad \forall\, n \ge m_1 \qquad ...(3)$$

Let $M = \max.\{m_1, m\}$

From (1) to (3), we have

$$\left| \frac{1}{a_n} - \frac{1}{a} \right| < \frac{1}{|a|} \cdot \frac{k|a|.\varepsilon}{k} \qquad \forall\, n \ge M$$

$$\Rightarrow \quad \left| \frac{1}{a_n} - \frac{1}{a} \right| < \varepsilon \quad \forall\, n \ge M$$

$$\Rightarrow \quad < \frac{1}{a_n} > \text{ converges to } \frac{1}{a}.$$

(*vi*) Since $b_n \to b$, $b_n \ne 0$, $\forall\, n$ and $b \ne 0$

$\therefore$ by (*v*) above, $\frac{1}{b_n} \to \frac{1}{b}$

Since $a_n \to a$ also,

$\therefore$ by (*iv*) above, $< a_n \cdot \frac{1}{b_n} >$ converges to $a \cdot \frac{1}{b}$

i.e., $$< \frac{a_n}{b_n} > \text{ converges to } \frac{a}{b}.$$

Independent Proof of (*vi*)

We have $$\left| \frac{a_n}{b_n} - \frac{a}{b} \right| = \left| \frac{a_n b - a b_n}{b b_n} \right| = \left| \frac{a_n b - ab + ab - a b_n}{b b_n} \right|$$

$$= \left| \frac{b(a_n - a) + a(b - b_n)}{b b_n} \right|$$

$$\le \frac{|b|\,|a_n - a| + |a|\,|b - b_n|}{|b|\,|b_n|} \qquad ...(1)$$

Since $\quad b_n \to b \ (\neq 0),$

$\therefore\ \exists$ a positive number k and a positive integer m such that

$$| b_n | > k \quad \forall\, n \geq m \text{ [By Theorem 4.3.1]} \qquad ...(2)$$

Let $\quad \varepsilon > 0$ be given

Since $\quad a_n \to a$ and $b_n \to b,$

$\therefore\ \exists$ positive integers m_1 and m_2 such that

$$| a_n - a | < \frac{k\varepsilon}{2} \quad \forall\, n \geq m_1 \qquad ...(3)$$

and
$$| b_n - b | < \frac{k|b|\varepsilon}{2(|a|+1)} \quad \forall\, n \geq m_2 \qquad ...(4)$$

Let $\quad \text{M} = \max. \{m, m_1, m_2\}$

From (1) to (4),

$$\left| \frac{a_n}{b_n} - \frac{a}{b} \right| < \frac{|b|k.\varepsilon}{2|b|.k} + \frac{|a|.k|b|.\varepsilon}{2(|a|+1)|b|k}$$

$$< \varepsilon \quad \forall\, n \geq \text{M}.$$

$$\Rightarrow \quad < \frac{a_n}{b_n} > \text{ converges to } \frac{a}{b}.$$

Remarks. (*i*) The above theorem is also sometimes referred to by saying that the operation of limit is compatible with 'scalar multiplication', 'addition', 'subtraction', 'multiplication' and 'division' compositions in the set of all real convergent sequences.

(*ii*) The convergence, divergence or oscillation behaviour of a sequence does not change if a finite number of terms of the sequence are altered. Therefore, if $< a_n >$ and $< b_n >$ are two sequences and p, q are positive integers such that $a_{p+m} = b_{q+m}\ \forall\, m$, then the sequences $< a_n >$ and $< b_n >$ converge, diverge or oscillate, together, *i.e.*, they behave alike.

(*iii*) In the results (*v*) and (*vi*), b_n can be zero for finitely many n.

(*iv*) The converse of the results (*ii*), (*iii*), (*iv*) and (*vi*) of the above theorem need not be true as the following examples show :

Let $\quad a_n = (-1)^n$ and $b_n = (-1)^{n-1}, \forall\, n$

Then, neither $< a_n >$ nor $< b_n >$ is convergent,

but $\quad a_n + b_n = 0 \quad \forall\, n$

$\therefore\ < a_n + b_n >$ is convergent.

Similarly, one can see that each of the sequences $< a_n - b_n >$, $< a_n . b_n >$ and $< \frac{a_n}{b_n} >$ is also convergent.

4.2.3. Theorem. *If* $< a_n >$ *converges to* a *and* $a_n \geq 0\ \ \forall\, n$, *then* $a \geq 0$. *i.e., if a sequence of non-negative terms converges to a, then prove that a is also non-negative.*

Proof. Suppose that $\quad a < 0$

Let $\quad \varepsilon = -\frac{1}{2} a > 0$

Since $\quad a_n \to a$

$\therefore\ \exists$ a positive integer m such that

$$| a_n - a | < \varepsilon \qquad \forall\, n \geq m$$

$$\Rightarrow \quad a - \varepsilon < a_n < a + \varepsilon \qquad \forall\, n \geq m$$

$$\Rightarrow \quad \frac{3a}{2} < a_n < \frac{a}{2} \quad \forall\, n \geq m$$

$$\Rightarrow \quad a_n < 0 \quad \forall\, n \geq m, \text{ which is not true.}$$

$\therefore$ our supposition that $a < 0$ is false.

Hence $a \geq 0$.

Cor. 1. If $< a_n >$ converges to a, $< b_n >$ converges to b

and $a_n \leq b_n \quad \forall\, n,$

then $a \leq b$

[For, $a_n \to a$ and $b_n \to b \;\Rightarrow\; (a_n - b_n) \to (a - b)$

Since $a_n \leq b_n \quad \forall\, n$

$\therefore \quad b_n - a_n \geq 0 \quad \forall\, n$

$\Rightarrow \quad b - a \geq 0 \;\Rightarrow\; a \leq b$.]

Cor. 2. If $a_n \leq k \;\forall\, n$ and $a_n \to a$, then $a \leq k$.

Remark. If in the above theorem, $a_n > 0 \;\forall\, n$, even then we cannot say that $a > 0$. For example, if $a_n = \frac{1}{n}$, $n \in \mathrm{N}$, then $a_n \to 0$.

4.2.4. Theorem (*Squeeze Principle*). *If $< a_n >$, $< b_n >$ and $< c_n >$ are three sequences such that*

(i) $< a_n >$ converges to l and $< b_n >$ converges to l,

(ii) for some positive integer m,

$$a_n \leq c_n \leq b_n \quad \forall\, n \geq m$$

then, $< c_n >$ converges to l.

Proof. Let $\varepsilon > 0$ be given

Since $a_n \to l$ and $b_n \to l$,

$\therefore$ $\exists$ positive integers m_1 and m_2 such that

$$|\, a_n - l \,| < \varepsilon \quad \forall\, n \geq m_1 \quad \text{and} \quad |\, b_n - l \,| < \varepsilon \quad \forall\, n \geq m_2$$

$$\Rightarrow \quad l - \varepsilon < a_n < l + \varepsilon \quad \forall\, n \geq m_1 \quad \text{and} \quad l - \varepsilon < b_n < l + \varepsilon \quad \forall\, n \geq m_2$$

Let $\mathrm{M} = \max.\ \{m_1, m_2, m\}$. Then

$$l - \varepsilon < a_n \leq c_n \leq b_n < l + \varepsilon \quad \forall\, n \geq \mathrm{M}$$

$$\Rightarrow \quad l - \varepsilon < c_n < l + \varepsilon \quad \forall\, n \geq \mathrm{M}$$

$$\Rightarrow \quad |\, c_n - l \,| < \varepsilon \quad \forall\, n \geq \mathrm{M}$$

$\Rightarrow \quad < c_n >$ converges to l.

4.2.5. Theorem. (*Cauchy's first theorem on limits*)

If a sequence $< a_n >$ converges to l and $b_n = \dfrac{a_1 + a_2 + \ldots\ldots + a_n}{n}$, then the sequence $< b_n >$ also converges to l. (KU BCA II 2003)

Proof. *Case (i)* $l = 0$.

Let $\varepsilon > 0$ be given

Since $a_n \to 0$

$\therefore$ $\exists$ a positive integer m such that

$$|\, a_n \,| < \frac{\varepsilon}{2} \;\forall\, n \geq m \qquad \ldots(1)$$

Since every convergent sequence is bounded,

$\therefore$ $\exists$ a positive number k such that $|a_n| < k, \forall n$ ...(2)

Now for $n \geq m$,

$$\frac{|a_1 + a_2 + \ldots\ldots + a_n|}{n} = \frac{|a_1 + a_2 + \ldots\ldots + a_{m-1} + a_m + a_{m+1} + \ldots\ldots + a_n|}{n}$$

$$\leq \frac{|a_1| + |a_2| + \ldots\ldots + |a_{m-1}|}{n} + \frac{|a_m| + |a_{m+1}| + \ldots\ldots + |a_n|}{n}$$

$$< \frac{k(m-1)}{n} + \frac{\varepsilon(n-m+1)}{2n} \qquad \text{[By (1) and (2)]}$$

$$< \frac{k(m-1)}{n} + \frac{\varepsilon}{2} < \frac{\varepsilon}{2} + \frac{\varepsilon}{2} = \varepsilon$$

if $\dfrac{k(m-1)}{n} < \dfrac{\varepsilon}{2}$ *i.e.*, if $\dfrac{2k(m-1)}{\varepsilon} < n$

i.e., if $n > \dfrac{2k(m-1)}{\varepsilon}$

Let M be a positive integer $> \dfrac{2k(m-1)}{\varepsilon}$. Then,

$$\left|\frac{a_1 + a_2 + \ldots\ldots + a_n}{n}\right| < \varepsilon \quad \forall\, n \geq \text{M}.$$

$\Rightarrow$ the sequence $< \dfrac{a_1 + a_2 + \ldots\ldots + a_n}{n} >$ converges to 0.

Case (ii) $l \neq 0$

Let $a_n = l + c_n$, then $c_n \to 0$ and $b_n = l + \dfrac{c_1 + c_2 + \ldots\ldots + c_n}{n}$

By case (*i*), $\dfrac{c_1 + c_2 + \ldots\ldots + c_n}{n} \to 0$

Hence $< b_n >$ converges to l.

Remark. The converse of the theorem need not be true. For example, if $a_n = (-1)^n$, then $b_n = \dfrac{a_1 + a_2 + \ldots\ldots + a_n}{n} = \dfrac{-1}{n}$ or 0 according as n is odd or even. $< b_n >$ converges but $< a_n >$ does not converge.

4.2.6. Theorem*. *If $< a_n >$ is a sequence of positive terms converging to $l > 0$ and $b_n = (a_1 a_2 \ldots\ldots a_n)^{\frac{1}{n}}$, then $< b_n >$ converges to l.*

Proof. Since $a_n \to l$

$\therefore$ $\log a_n \to \log l$

$\Rightarrow$ $\dfrac{\log a_1 + \log a_2 + \ldots\ldots + \log a_n}{n} \to \log l$ (Cauchy's first theorem on limits)

$\Rightarrow$ $\dfrac{1}{n} \log (a_1 a_2 \ldots\ldots a_n) \to \log l$ $\Rightarrow$ $\log (a_1 a_2 \ldots\ldots a_n)^{\frac{1}{n}} \to \log l$

$\Rightarrow$ $(a_1 a_2 \ldots\ldots a_n)^{\frac{1}{n}} \to l$ $\Rightarrow$ $< b_n >$ converges to l.

---

*$a_n \to l \Leftrightarrow \log a_n \to \log l$, provided $a_n > 0 \quad \forall\, n$ and $l > 0$.

4.2.7. Theorem (*Cauchy's second theorem on limits*). *If* $< a_n >$ *is a sequence of positive terms, then*

$$\lim_{n \to \infty} a_n^{\frac{1}{n}} = \lim_{n \to \infty} \frac{a_{n+1}}{a_n}$$

provided the limit on the right hand side exists, whether finite or infinite. (K.U. BCA (II) 2008)

Proof. Let $\displaystyle\lim_{n \to \infty} \frac{a_{n+1}}{a_n} = l$

Case (*i*) l is finite.

Let $\varepsilon > 0$ be given. Then, there exists a positive integer m such that

$$l - \frac{\varepsilon}{2} < \frac{a_{n+1}}{a_n} < l + \frac{\varepsilon}{2}, \quad \forall\, n \geq m \qquad ...(1)$$

Putting $n = m, m + 1, ..., n - 1$, in (1), we have

$$l - \frac{\varepsilon}{2} < \frac{a_{m+1}}{a_m} < l + \frac{\varepsilon}{2}$$

$$l - \frac{\varepsilon}{2} < \frac{a_{m+2}}{a_{m+1}} < l + \frac{\varepsilon}{2}$$

....

$$l - \frac{\varepsilon}{2} < \frac{a_n}{a_{n-1}} < l + \frac{\varepsilon}{2}$$

Multiplying the above $n - m$ inequalities, we have

$$\left(l - \frac{\varepsilon}{2}\right)^{n-m} < \frac{a_n}{a_m} < \left(l + \frac{\varepsilon}{2}\right)^{n-m}$$

$$\Rightarrow \quad a_m \left(l - \frac{\varepsilon}{2}\right)^{n-m} < a_n < \left(l + \frac{\varepsilon}{2}\right)^{n-m} a_m \qquad (\because \ a_m > 0)$$

$$\Rightarrow \quad a_m^{\frac{1}{n}} \left(l - \frac{\varepsilon}{2}\right)^{1-\frac{m}{n}} < a_n^{\frac{1}{n}} < \left(l + \frac{\varepsilon}{2}\right)^{1-\frac{m}{n}} a_m^{\frac{1}{n}} \qquad ...(2)$$

Now, if $n \to \infty$, then

$$a_m^{\frac{1}{n}} \left(l - \frac{\varepsilon}{2}\right)^{1-\frac{m}{n}} \to l - \frac{\varepsilon}{2} \qquad (\because \ a_m^{\frac{1}{n}} \to 1 \text{ as } n \to \infty)$$

and

$$a_m^{\frac{1}{n}} \left(l + \frac{\varepsilon}{2}\right)^{1-\frac{m}{n}} \to l + \frac{\varepsilon}{2}$$

$\therefore$ by definition of convergence, for above $\varepsilon > 0$, there exist positive integers m_1 and m_2 such that

$$\left| a_m^{\frac{1}{n}} \left(l - \frac{\varepsilon}{2}\right)^{1-\frac{m}{n}} - \left(l - \frac{\varepsilon}{2}\right) \right| < \frac{\varepsilon}{2}, \quad \forall\, n \geq m_1$$

and

$$\left| a_m^{\frac{1}{n}} \left(l + \frac{\varepsilon}{2}\right)^{1-\frac{m}{n}} - \left(l + \frac{\varepsilon}{2}\right) \right| < \frac{\varepsilon}{2}, \quad \forall\, n \geq m_2$$

$$\Rightarrow \quad \left(l-\frac{\varepsilon}{2}\right)-\frac{\varepsilon}{2}<a_m^{\frac{1}{n}}\left(l-\frac{\varepsilon}{2}\right)^{1-\frac{m}{n}}<\left(l-\frac{\varepsilon}{2}\right)+\frac{\varepsilon}{2}, \quad \forall\, n \geq m_1 \qquad ...(3)$$

and

$$\left(l+\frac{\varepsilon}{2}\right)-\frac{\varepsilon}{2}<a_m^{\frac{1}{n}}\left(l+\frac{\varepsilon}{2}\right)^{1-\frac{m}{n}}<\left(l+\frac{\varepsilon}{2}\right)+\frac{\varepsilon}{2} \quad \forall\, n \geq m_2 \qquad ...(4)$$

Let $\mathrm{M} = \max.\ \{m_1, m_2\}$.

Then, from (2), (3) and (4), we have

$$l-\varepsilon < a_n^{\frac{1}{n}} < l+\varepsilon \qquad \forall\, n \geq \mathrm{M}$$

$$\Rightarrow \quad \left|a_n^{\frac{1}{n}} - l\right| < \varepsilon \qquad \forall\, n \geq \mathrm{M}$$

$$\Rightarrow \quad a_n^{\frac{1}{n}} \to l \quad \text{as } n \to \infty.$$

Case (*ii*) l is infinite

Then, $\displaystyle\lim_{n\to\infty} \frac{\frac{1}{a_{n+1}}}{\frac{1}{a_n}} = 0$ (finite)

Hence by case (*i*), $\displaystyle\lim_{n\to\infty}\left(\frac{1}{a_n}\right)^{\frac{1}{n}} = 0 \Rightarrow \lim_{n\to\infty} a_n^{\frac{1}{n}} = \infty.$

Remark. The sequence $< a_n^{\frac{1}{n}} >$ can converge without $< \frac{a_{n+1}}{a_n} >$ being convergent. For example, if $a_n = 2^{-n+(-1)^n}$, then $< a_n^{\frac{1}{n}} >$ converges to $\frac{1}{2}$ but $< \frac{a_{n+1}}{a_n} >$ oscillates between $\frac{1}{2^3}$ and 2 and hence does not converge.

4.2.8. An important Sequence $< a^n >$. *The behaviour of the sequence $< a_n >$ where $a_n = a^n$, depends upon the value of 'a'. Following cases arise.*

Case (*i*) $a = 1$

Then $a_n = 1 \quad \forall\, n$

$\therefore$ $a_n \to 1$ as $n \to \infty$.

Case (*ii*) $a = 0$. Then $a_n \to 0$ as $n \to \infty$.

Case (*iii*) $a > 1$

Let $a = 1 + h, h > 0 \Rightarrow a_n = (1 + h)^n > 1 + nh$

(by taking only first two terms of the binomial expansion)

As $n \to \infty, 1 + nh \to \infty \Rightarrow a_n \to \infty$ as $n \to \infty$.

Case (*iv*) $0 < a < 1 \Rightarrow \frac{1}{a} > 1$

$$\Rightarrow \quad \frac{1}{a} = 1 + h \text{ for some } h > 0$$

$$\Rightarrow \qquad \frac{1}{a^n} = (1+h)^n$$
$$\geq 1 + nh \quad \text{for } n \geq 1$$
$$\Rightarrow \qquad a^n \leq \frac{1}{1+nh} < \frac{1}{nh}$$

Since h is fixed, $\frac{1}{nh} \to 0$ as $n \to \infty$

Hence $a^n \to 0$ as $n \to \infty$.

Case (v) $-1 < a < 0$

Let $a = -b$. Then $0 < b < 1$.

By case (iv), $b^n \to 0$ as $n \to \infty$ and hence

$$a_n = (-b)^n \to 0 \quad \text{as} \quad n \to \infty.$$

Case (vi) $a = -1$. Then $a_n = (-1)^n$, which is an oscillating sequence.

Case (vii) $a < -1$.

Let $b = -a$, $b > 1$

Then by case (iii), $b^n \to \infty$ as $n \to \infty$.

So, $a_n = (-b)^n$ takes values, alternatively negative and positive, numerically greater than any assigned number, *i.e.*, $< a_n >$ oscillates infinitely.

EXERCISE 4 (*b*)

1. Verify the following :

(i) $\lim_{n \to \infty} \frac{3n-5}{4-2n} = -\frac{3}{2}$ (ii) $\lim_{n \to \infty} [(n^2+1)^{1/8} - (n+1)^{1/4}] = 0$

(iii) $\lim_{n \to \infty} \left[\frac{1}{n^2} + \frac{1}{(n+1)^2} + \ldots\ldots + \frac{1}{(2n)^2}\right] = 0$ (iv) $\lim_{n \to \infty} \left(1 - \frac{1}{n}\right)^{-n} = e$

(v) $\lim_{n \to \infty} \left[\frac{1}{n^2} + \frac{1}{(n+2)^2} + \ldots\ldots + \frac{1}{(2n)^2}\right] = \infty$

(vi) $\lim_{n \to \infty} \frac{2n^4 - 7n^3 + 3n^2 - 4n + 5}{3n^4 + 5n^3 + 9n^2 + 2n + 7} = \frac{2}{3}$ (vii) $\lim_{n \to \infty} \frac{1 + \frac{1}{2} + \frac{1}{3} + \ldots\ldots + \frac{1}{n}}{n} = 0$

[**Hint.** Apply Cauchy's first theorem on limits to the sequence $< \frac{1}{n} >$.]

($viii$) $\lim_{n \to \infty} \frac{1 + \frac{1}{3} + \frac{1}{5} + \ldots\ldots + \frac{1}{2n-1}}{n} = 0.$

2. If $a_n < b_n \;\forall\, n$, then $\lim_{n \to \infty} a_n \leq \lim_{n \to \infty} b_n$, provided both the limits exist.
   Give an example to show that the equality may exist between the limits.
3. If $k > 0$ is a fixed real number and $< b_n >$ is divergent, show that $< kb_n >$ is divergent.
4. Let $< a_n >$ and $< b_n >$ be two convergent sequences.
   Let $u_n = \max. \{a_n, b_n\}$ and $v_n = \min. \{a_n, b_n\} \;\forall\, n$

Show that $< u_n >$ and $< v_n >$ are also convergent and that

$$\lim_{n\to\infty} u_n = \max. \{ \lim_{n\to\infty} a_n, \lim_{n\to\infty} b_n \},$$

$$\lim_{n\to\infty} v_n = \min. \{ \lim_{n\to\infty} a_n, \lim_{n\to\infty} b_n \}.$$

5. Show that $\lim_{n\to\infty} . \frac{(n!)^{1/n}}{n} = \frac{1}{e}$.

6. Prove that $< a_n >$ converges to a implies $< | a_n | >$ converges to $| a |$. What can you say about the converse ?

7. If $< a_n >$ and $< b_n >$ are two convergent sequences, prove that $< a_n + b_n >$, $< a_n - b_n >$, $< a_n b_n >$ and $< \frac{a_n}{b_n} >$, (where $b_n \neq 0 \ \forall \ n$ in the last case) are also convergent. Give suitable examples to show that converse may not be true.

8. If $< a_n >$ converges and $< b_n >$ diverges to $+\infty$ (or $-\infty$), prove that $< a_n + b_n >$ also diverges to $+\infty$ (or $-\infty$).

9. If $< a_n >$ diverges to $+\infty$ and $< b_n >$ diverges to $-\infty$, then show by examples that $< a_n + b_n >$ may (*i*) converge, (*ii*) diverge to ∞, (*iii*) diverge to $-\infty$ and (*iv*) oscillate.

10. If $< a_n >$ diverges to ∞, then $< \frac{a_1 + a_2 + a_3 + ... + a_n}{n} >$ diverges to ∞. Give an example to show that

$$< \frac{a_1 + a_2 + a_3 + ... + a_n}{n} >$$

diverges to ∞ but $< a_n >$ does not.
[See for example, the sequence $< a_n >$ where

$$a_n = \left[\begin{array}{l} 1 \text{ if } n \text{ is odd} \\ n \text{ if } n \text{ is even} \end{array} \right].]$$

11. If $< a_n >$ diverges to ∞ and $b_n \geq a_n \ \ \forall \ n$, show that $< b_n >$ diverges to ∞.

12. If $< a_n >$ diverges to $-\infty$ and $b_n \leq a_n \ \ \forall \ n$, show that $< b_n >$ diverges to $-\infty$.

13. Using $\lim_{n\to\infty} \frac{n}{n-1} = 1$, show that $\lim_{n\to\infty} \left(\frac{2}{1} \frac{3}{2} \frac{4}{3} \cdots\cdots \frac{n}{n-1} \right)^{\frac{1}{n}} = 1$.

14. Using Cauchy's first theorem on limits, show that

$$\lim_{n\to\infty} \left[\frac{1}{\sqrt{n^2+1}} + \frac{1}{\sqrt{n^2+2}} + \ldots + \frac{1}{\sqrt{n^2+n}} \right] = 1.$$

4.3.1. Monotonic Sequences. (*i*) *A sequence $< a_n >$ is said to be monotonically increasing if $a_{n+1} \geq a_n$ for each n.*

(*ii*) *A sequence $< a_n >$ is said to be monotonically decreasing if $a_{n+1} \leq a_n$ for each n.*

(*iii*) *A sequence $< a_n >$ is said to be strictly monotonically increasing if $a_{n+1} > a_n$ for each n.*

(*iv*) *A sequence $< a_n >$ is said to be strictly monotonically decreasing if $a_{n+1} < a_n$ for each n.*

(*v*) *A sequence which is either monotonically increasing or decreasing is called a monotonic sequence.*

Examples. (*i*) The sequences $<n>$, $<n^2>$, $<\frac{n+1}{n}>$ and $<2^n>$ are strictly monotonically increasing.

(*ii*) The sequences $<\frac{1}{n}>$, $<\frac{1}{n^2}>$, $<\frac{n}{n+1}>$ and $<\frac{1}{2^n}>$ are strictly monotonically decreasing.

(*iii*) The sequence $<(-1)^n \cdot \frac{1}{n}>$ is neither increasing nor decreasing.

4.3.2. Theorem (*Monotone Convergence Theorem*). *(i) Every monotonically increasing sequence bounded above converges to the least upper bound.*

(ii) Every monotonically decreasing sequence bounded below converges to the greatest lower bound.

Proof. (*i*) Let $<a_n>$ be a monotonically increasing sequence bounded above.

Let ξ be the l.u.b. of the sequence $<a_n>$. We shall show that $<a_n>$ converges to ξ.

Let $\varepsilon > 0$ be given.

Since $\xi - \varepsilon < \xi$,

$\therefore$ $\xi - \varepsilon$ is not an upper bound of $<a_n>$.

$\Rightarrow$ $\exists\, a_m > \xi - \varepsilon$, for some $m \in \mathrm{N}$

Since $<a_n>$ is monotonically increasing,

$$\therefore \quad a_n \ge a_m > \xi - \varepsilon, \quad \forall\, n \ge m \qquad \text{...(1)}$$

Since ξ is the l.u.b. of $<a_n>$

$$\therefore \quad a_n \le \xi < \xi + \varepsilon \quad \forall\, n \qquad \text{...(2)}$$

From (1) and (2),

$$\xi - \varepsilon < a_n < \xi + \varepsilon, \quad \forall\, n \ge m$$

$$\Rightarrow \quad |\, a_n - \xi \,| < \varepsilon \quad \forall\, n \ge m$$

$\Rightarrow$ $<a_n>$ converges to ξ.

(*ii*) Let $<a_n>$ be a monotonically decreasing sequence bounded below.

Let $b_n = -a_n \quad \forall\, n$

$\Rightarrow$ $<b_n>$ is a monotonically increasing sequence bounded above.

$\Rightarrow$ $<b_n>$ converges to ξ

where ξ = l.u.b. of the sequence $<b_n>$.

$\Rightarrow$ $<-a_n>$ converges to ξ.

$\Rightarrow$ $<a_n>$ converges to $-\xi$,

where $-\xi$ = g.l.b. of the sequence $<a_n>$.

Cor. Every bounded monotonic sequence is convergent.

4.3.3. *(i) Every monotonically increasing sequence unbounded above diverges to ∞.*

(ii) Every monotonically decreasing sequence unbounded below diverges to $-\infty$.

Proof. (*i*) Let $<a_n>$ be a monotonically increasing sequence not bounded above.

Let k be a positive number (however large).

$\Rightarrow$ $\exists$ a positive integer m such that $a_m > k$

$\Rightarrow$ $a_n \ge a_m \quad \forall\, n \ge m$ $\qquad (\because\ <a_n>$ is monotonically increasing)

$\Rightarrow \quad a_n > k \quad \forall\, n \geq m$

$\Rightarrow \quad < a_n >$ diverges to ∞.

(*ii*) Proof is exactly on same lines and is left as an exercise for the readers.

Note. For a monotonic sequence, there are only two types of behaviours—(*i*) a monotonically increasing sequence is either convergent or divergent to ∞ ; (*ii*) a monotonically decreasing sequence is either convergent or divergent to $-\infty$.

4.3.4. Nested Sequence. *A sequence of intervals* $< I_n >$ *is said to be nested if either*

$$I_1 \subseteq I_2 \subseteq I_3 \subseteq \ldots\ldots \quad or \quad I_1 \supseteq I_2 \supseteq I_3 \supseteq \ldots\ldots$$

4.3.5. Nested Interval Property *(or Cantor* Intersection Theorem).*

Let $< I_n = [a_n, b_n] >$ *be a sequence of closed intervals such that*

(*i*) $I_{n+1} \subseteq I_n \quad \forall\, n \geq 1$ (*ii*) $l(I_n) = b_n - a_n \to 0$ *as* $n \to \infty$

Then, $\bigcap_{n=1}^{\infty} I_n$ *is a singleton.*

Proof. Since $I_{n+1} \subseteq I_n$,

$\therefore \quad a_n \leq a_{n+1} < b_{n+1} \leq b_n \quad \forall\, n$

$\Rightarrow \quad < a_n >$ is monotonically increasing and $< b_n >$ is monotonically decreasing.

Now $\quad I_n \subseteq I_1\ \forall\, n$

$\Rightarrow \quad a_n \leq b_1$ and $b_n \geq a_1$

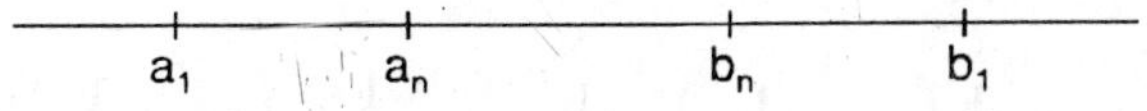

$\Rightarrow$ the sequence $< a_n >$ is bounded above by b_1 and the sequence $< b_n >$ is bounded below by a_1.

$\Rightarrow$ by monotone convergence theorem, the sequence $< a_n >$ converges to the l.u.b. ξ_1 (say) and the sequence $< b_n >$ converges to the g.l.b. ξ_2 (say).

$$b_n = b_n - a_n + a_n \quad \forall\, n$$

$\Rightarrow \quad \lim_{n \to \infty} b_n = \lim_{n \to \infty} (b_n - a_n) + \lim_{n \to \infty} a_n$

$\Rightarrow \quad \lim_{n \to \infty} b_n = \lim_{n \to \infty} a_n \qquad (\because \lim_{n \to \infty} (b_n - a_n) = 0)$

$\Rightarrow \quad \xi_1 = \xi_2 = \xi$ (say)

Since ξ is the l.u.b. of the sequence $< a_n >$ as well as g.l.b. of the sequence $< b_n >$.

$\Rightarrow \quad a_n \leq \xi \leq b_n \quad \forall\, n \quad \Rightarrow \quad \xi \in I_n \quad \forall\, n \quad \Rightarrow \quad \xi \in \bigcap_{n=1}^{\infty} I_n$

Let $\quad \eta \neq \xi$

Since $\quad b_n - a_n \to 0$ as $n \to \infty$

$\therefore\ \exists$ some $n \in N$ such that

$$| b_n - a_n | < | \xi - \eta |$$

$\Rightarrow \quad \eta \notin I_n \quad \Rightarrow \quad \eta \notin \bigcap_{n=1}^{\infty} I_n$

Hence $\quad \bigcap_{n=1}^{\infty} I_n = \{\xi\}$.

*After the name of G. Cantor (1845–1918), a famous German mathematician and one of the creators of modern Set Theory.

Remark. The word 'closed' in the statement of the above theorem cannot be dropped *i.e.*, the intersection of a decreasing sequence of open intervals may be empty. For example, let $I_n = \left(0, \frac{1}{n}\right)$, $n \in N$, then $\bigcap\limits_{n=1}^{\infty} I_n = \phi$.

For, if $x \le 0$, then $x \notin I_n$ for any n and if $x > 0$, then by Archimedean property of reals, $\exists$ a positive integer m such that $m > \frac{1}{x} \Rightarrow \frac{1}{m} < x \Rightarrow x \notin I_m \Rightarrow x \notin \bigcap\limits_{n=1}^{\infty} I_n$.

Example 5. *Prove that the sequence $< a_n >$ where $a_n = \left(1 + \frac{1}{n}\right)^n$ is convergent and its limit lies between 2 and 3.*

Sol. $a_n = \left(1 + \frac{1}{n}\right)^n$ (By binomial theorem for positive integral index)

$$= 1 + n \cdot \frac{1}{n} + \frac{n(n-1)}{2!}\frac{1}{n^2} + \ldots\ldots + \frac{n(n-1)\ldots 1}{n!}\frac{1}{n^n}$$

$$= 1 + 1 + \frac{1}{2!}\left(1 - \frac{1}{n}\right) + \ldots\ldots + \frac{1}{n!}\left(1 - \frac{1}{n}\right)\left(1 - \frac{2}{n}\right)\ldots\left(1 - \frac{n-1}{n}\right) \quad \ldots(1)$$

Similarly,

$$a_{n+1} = 1 + 1 + \frac{1}{2!}\left(1 - \frac{1}{n+1}\right) + \ldots\ldots + \frac{1}{(n+1)!}\left(1 - \frac{1}{n+1}\right)\left(1 - \frac{2}{n+1}\right)\ldots\left(1 - \frac{n}{n+1}\right) \quad \ldots(2)$$

Comparing (1) and (2), we see that $a_{n+1} \ge a_n$

$\Rightarrow$ the sequence $< a_n >$ is monotonically increasing.

From (1),

$$2 < a_n < 1 + 1 + \frac{1}{2!} + \frac{1}{3!} + \ldots\ldots + \frac{1}{n!}$$

$$\le 1 + 1 + \frac{1}{2} + \frac{1}{2^2} + \ldots\ldots + \frac{1}{2^{n-1}}$$

$$= 1 + \frac{1 - \frac{1}{2^n}}{1 - \frac{1}{2}} = 3 - \frac{1}{2^{n-1}} < 3 \quad \forall\, n$$

$\Rightarrow$ The sequence $< a_n >$ is bounded.

Thus, the sequence $< a_n >$, being a monotonically increasing sequence bounded above by 3, is convergent.

Since $\quad 2 < a_n < 3 \quad \forall\, n$

$\Rightarrow \quad 2 \le \lim\limits_{n \to \infty} a_n \le 3 \quad \forall\, n$

$\Rightarrow$ Limit of the sequence $< a_n >$ lies between 2 and 3.

Example 6. *Show that the sequence $< a_n >$ defined by*

$$a_1 = \sqrt{2} \text{ and } a_{n+1} = \sqrt{2a_n} \text{ converges to 2.}$$

Sol. The given sequence is $\sqrt{2}, \sqrt{2\sqrt{2}}, \sqrt{2\sqrt{2\sqrt{2}}}, ...$

Now $\quad 2\sqrt{2} > 2$

$\Rightarrow \quad \sqrt{2\sqrt{2}} > \sqrt{2} \quad \Rightarrow \quad a_2 > a_1$

Suppose that $\quad a_n > a_{n-1} \quad \Rightarrow \quad 2a_n > 2a_{n-1}$

$\Rightarrow \quad \sqrt{2a_n} > \sqrt{2a_{n-1}} \quad \Rightarrow \quad a_{n+1} > a_n$

$\therefore$ by the principle of mathematical induction,

$$a_{n+1} > a_n \ \forall\, n.$$

$\therefore$ $< a_n >$ is a monotonically increasing sequence.

Also, $\quad a_1 = \sqrt{2} < 2$

Suppose that $\quad a_n < 2 \quad \Rightarrow \quad 2a_n < 4$

$\Rightarrow \quad \sqrt{2a_n} < 2 \quad \Rightarrow \quad a_{n+1} < 2$

$\therefore$ again by the principle of mathematical induction,

$$a_n < 2 \quad \forall\, n.$$

Hence $< a_n >$, being a monotonically increasing bounded above sequence, is convergent.

To find $\lim\limits_{n \to \infty} a_n$

Let $\quad \lim\limits_{n \to \infty} a_n = l \quad \Rightarrow \quad \lim\limits_{n \to \infty} a_{n+1} = l$

Now $\quad a_{n+1} = \sqrt{2a_n} \quad \Rightarrow \quad a^2_{n+1} = 2a_n$

$\Rightarrow \quad \lim\limits_{n \to \infty} a^2_{n+1} = 2 \lim\limits_{n \to \infty} a_n \quad \Rightarrow \quad \left(\lim\limits_{n \to \infty} a_{n+1}\right)^2 = 2 \lim\limits_{n \to \infty} a_n$

$\Rightarrow \quad l^2 = 2l \quad \Rightarrow \quad l^2 - 2l = 0$

$\Rightarrow \quad l = 0, 2.$

But $\quad l \neq 0 \qquad (\because \ a_n \geq \sqrt{2} \ \ \forall\, n)$

$\therefore \quad \lim\limits_{n \to \infty} a_n = 2.$

Example 7. *Prove that the sequence* $< a_n >$, *where*

$$a_n = \frac{1}{n+1} + \frac{1}{n+2} + ... + \frac{1}{2n},$$

is convergent and its limit lies between $\frac{1}{2}$ *and 1.*

Sol. $\quad a_{n+1} - a_n$

$$= \frac{1}{n+2} + \frac{1}{n+3} + ... + \frac{1}{(n+1)+(n+1)} - \left(\frac{1}{n+1} + \frac{1}{n+2} + ... + \frac{1}{n+n}\right)$$

$$= \frac{1}{2n+1} + \frac{1}{2n+2} - \frac{1}{n+1}$$

$$> \frac{1}{2n+2} + \frac{1}{2n+2} - \frac{1}{n+1} = \frac{1}{n+1} - \frac{1}{n+1} = 0$$

$\Rightarrow \quad a_{n+1} - a_n > 0 \quad \forall\, n$

$\Rightarrow \quad < a_n >$ is a monotonically increasing sequence.

Also, $$a_n \le \frac{1}{n+1} + \frac{1}{n+1} + \ldots + \frac{1}{n+1}$$

$$= \frac{n}{n+1} < 1 \quad \forall\, n$$

$\Rightarrow$ the sequence $< a_n >$ is bounded above

$\Rightarrow$ $< a_n >$ is convergent.

Clearly, $$a_n > \frac{1}{2n} + \frac{1}{2n} + \ldots + \frac{1}{2n} = \frac{n}{2n} = \frac{1}{2}$$

$\therefore$ $$\tfrac{1}{2} < a_n < 1 \quad \Rightarrow \quad \tfrac{1}{2} \le \lim_{n \to \infty} a_n \le 1.$$

Example 8. *If $a_1 > 0, b_1 > 0$ and $\forall\, n > 1$, $a_{n+1} = \frac{1}{2}(a_n + b_n)$, $b_{n+1} = \sqrt{a_n\, b_n}$, then prove that $< a_n >$ and $< b_n >$ are monotonic sequences converging to the same limit.*

Sol. Since the arithmetic mean of two positive numbers is always $\ge$ the geometric mean between them.

$\therefore$ $$a_{n+1} \ge b_{n+1} \quad \forall\, n$$

Now $$a_{n+1} = \tfrac{1}{2}(a_n + b_n)$$
$$\le \tfrac{1}{2}(a_n + a_n) = a_n$$

and $$b_{n+1} = \sqrt{a_n\, b_n}$$
$$\ge \sqrt{b_n \,.\, b_n} = b_n \ \forall\, n$$

$\therefore$ $< a_n >$ is a monotonically decreasing sequence and $< b_n >$ is a monotonically increasing sequence.

Further, $a_n \ge b_1$ and $b_n \le a_1 \quad \forall\, n$

$\therefore$ both the sequences $< a_n >$ and $< b_n >$ are convergent.

To show : $$\lim_{n \to \infty} a_n = \lim_{n \to \infty} b_n$$

Let $$a_n \to a \quad \text{and} \quad b_n \to b$$

Since $$2a_{n+1} = a_n + b_n$$

$\therefore$ $$\lim_{n \to \infty} 2a_{n+1} = \lim_{n \to \infty} (a_n + b_n)$$

$\Rightarrow$ $$2 \lim_{n \to \infty} a_{n+1} = \lim_{n \to \infty} a_n + \lim_{n \to \infty} b_n$$

$\Rightarrow$ $$2a = a + b \quad \Rightarrow \quad a = b.$$

Example 9. *Show that the sequence $< a_n >$ defined by $a_1 = 1$ and $a_{n+1} = \sqrt{2 + a_n}$, $\forall\, n \ge 1$ converges to the positive root of the equation $x^2 - x - 2 = 0$.* (K.U. BCA(II) 2003)

Sol. $$a_1 = 1$$
$$a_2 = \sqrt{2+1} = \sqrt{3} > a_1$$

Suppose that $a_n > a_{n-1} \quad \Rightarrow \quad 2 + a_n > 2 + a_{n-1}$

$\Rightarrow$ $\sqrt{2 + a_n} > \sqrt{2 + a_{n-1}} \quad \Rightarrow \quad a_{n+1} > a_n$

$\Rightarrow$ by the principle of mathematical induction

$$a_{n+1} > a_n \quad \forall\, n$$

$\Rightarrow$ $< a_n >$ is monotonically increasing.

Again, $a_1 < 2$

Suppose that $a_n < 2$ $\Rightarrow$ $2 + a_n < 2 + 2$

$\Rightarrow$ $\sqrt{2 + a_n} < 2$ $\Rightarrow$ $a_{n+1} < 2$

$\Rightarrow$ by principle of mathematical induction,

$$a_n < 2 \ \forall \ n.$$

$\therefore$ $< a_n >$, being a monotonically increasing, bounded above sequence, is convergent.

Let $\lim\limits_{n \to \infty} a_n = a$

Since $a_n > 0 \quad \forall \ n$

$\therefore$ $\lim\limits_{n \to \infty} a_n \geq 0, \quad i.e., \quad a \geq 0$

Now $a_n = \sqrt{2 + a_{n-1}}$ $\Rightarrow$ $a_n^2 = 2 + a_{n-1}$

$\Rightarrow$ $\lim\limits_{n \to \infty} a_n^2 = \lim\limits_{n \to \infty} (2 + a_{n-1})$ $\Rightarrow$ $(\lim\limits_{n \to \infty} a_n)^2 = 2 + \lim\limits_{n \to \infty} a_{n-1}$

$\Rightarrow$ $a^2 = 2 + a$ $\Rightarrow$ $a^2 - a - 2 = 0$

$\Rightarrow$ a is a root of the equation $x^2 - x - 2 = 0$.

Example 10. *A sequence* $< a_n >$ *is defined by* $a_1 = 1$, $a_{n+1} = \dfrac{4 + 3a_n}{3 + 2a_n}$, $n \geq 1$*; show that* $< a_n >$ *converges and find its limit.*

Sol. $$a_1 = 1, \quad a_2 = \frac{4 + 3a_1}{3 + 2a_1} = \frac{4 + 3}{3 + 2} = \frac{7}{5}.$$

$$a_3 = \frac{4 + 3a_2}{3 + 2a_2} = \frac{4 + \frac{21}{5}}{3 + \frac{14}{5}} = \frac{41}{29}, \text{ etc.}$$

$\therefore$ The sequence is $< 1, \frac{7}{5}, \frac{41}{29}, \ldots\ldots >$ and $a_n \geq 1, \forall \ n$.

Now, $a_{n+1} = \dfrac{4 + 3a_n}{3 + 2a_n} = \dfrac{3}{2} - \dfrac{1}{2(3 + 2a_n)} = \dfrac{3}{2}$ – (a positive quantity less than 1)

$\Rightarrow$ $a_{n+1} < \frac{3}{2}, \forall n.$

$\therefore$ $< a_n >$ is bounded above by $\frac{3}{2}$.

Now, $a_1 = 1, \quad a_2 = \frac{7}{5} > 1$

$\therefore$ $a_2 > a_1$

Suppose that $a_{n+1} > a_n$ for some n

$\therefore$ $$a_{n+2} - a_{n+1} = \frac{4 + 3a_{n+1}}{3 + 2a_{n+1}} - \frac{4 + 3a_n}{3 + 2a_n} = \frac{a_{n+1} - a_n}{(3 + 2a_{n+1})(3 + 2a_n)} > 0$$

$(\because \ a_{n+1} > a_n \text{ and } a_n \geq 1)$

$\therefore$ $a_{n+2} > a_{n+1}$.

$\therefore$ by principle of mathematical induction, $< a_n >$ is monotonically increasing.

Since $< a_n >$ is monotonically increasing and bounded above, it is convergent.

Let $<a_n>$ be convergent to l

i.e., $$\lim_{n \to \infty} a_n = l.$$

Since $$a_{n+1} = \frac{4+3a_n}{3+2a_n}$$

$$\lim_{n \to \infty} a_{n+1} = \frac{4+3\lim_{n \to \infty} a_n}{3+2\lim_{n \to \infty} a_n} \quad \Rightarrow \quad l = \frac{4+3l}{3+2l}$$

$$\Rightarrow \quad l^2 = 2 \quad \Rightarrow \quad l = \sqrt{2} \qquad \text{(Rejecting } l = -\sqrt{2}, \text{ as } a_n > 0)$$

EXERCISE 4 (c)

1. Prove that the following sequences whose nth terms are given, are monotonic. Which of the following are increasing or decreasing ?

(*i*) $\frac{1}{2^1} + \frac{1}{2^2} + \frac{1}{2^3} + \ldots + \frac{1}{2^n}$ (*ii*) $\frac{1}{n} + \frac{1}{n+1} + \frac{1}{n+2} + \ldots + \frac{1}{2n-1}$

(*iii*) $\frac{3n+7}{4n+8}$ (*iv*) $1 + \frac{1}{1!} + \frac{1}{2!} + \ldots\ldots + \frac{1}{(n-1)!}$

(*v*) $\frac{1}{2n+5}$ (*vi*) $2n+7$

(*vii*) $\frac{1}{n}$ (*viii*) $(-1)^n \frac{1}{n}$.

2. Discuss the convergence or divergence of the above sequences.

3. Show that the sequence $<a_n>$ defined by $a_1 = \sqrt{7}$, $a_{n+1} = \sqrt{7+a_n}$ is

(*i*) bounded above, (*ii*) monotonically increasing and (*iii*) converges to a positive root of the equation $x^2 - x - 7 = 0$. *(K.U. BCA (II) 2001)*

4. Show that the sequence $<a_n>$ defined by $a_{n+1} = \sqrt{3a_n}$, $a_1 = 1$ is convergent and converges to 3.

5. (*a*) Show that the sequence $< \frac{n}{n+1} >$ is bounded and is monotonically increasing.

(K.U. BCA (II) 2007)

(*b*) Prove that the sequence $< \frac{n+1}{n} >$ is monotonically decreasing and bounded below.

(K.U. BCA (II) 2008)

4.4.1. Cluster point of a sequence. Def. 1. *A real number ξ is said to be a cluster point of a sequence $<a_n>$ if given any $\varepsilon > 0$ (however small) and a positive integer m, there exists a positive integer $k > m$ such that*

$$| a_k - \xi | < \varepsilon, \quad i.e., \quad \xi - \varepsilon < a_k < \xi + \varepsilon.$$

The definition asserts that for a real number ξ to be a cluster point, one may take any $\varepsilon > 0$, however small and one may reach any stage in the sequence (*i.e.,* after first m terms of the sequence), a term a_k of the sequence can always be found such that $| a_k - \xi | < \varepsilon$. That is, after any number of terms of the sequence, there is always at least one term of the sequence which lies in the interval $(\xi - \varepsilon, \xi + \varepsilon)$. Since $\varepsilon > 0$ can assume any value, $(\xi - \varepsilon, \xi + \varepsilon)$ is an arbitrary neighbourhood of ξ. So, we can say that the sequence is frequently in every neighbourhood of ξ.

Def. 2. *A real number ξ is said to be a cluster point of a sequence $< a_n >$ if every neighbourhood of ξ contains infinitely many elements of $< a_n >$ i.e., for any $\varepsilon > 0$, $a_n \in (\xi - \varepsilon, \xi + \varepsilon)$ for infinitely many values of n.*

Clearly, the above two definitions are equivalent (see 3.5.1). Further note that if for an $\varepsilon > 0$, $a_n \in (\xi - \varepsilon, \xi + \varepsilon)$ for finitely many n, then ξ cannot be a cluster point of $< a_n >$.

A cluster point of a sequence is also called '*a limit point*', *a condensation point*' or '*an accumulation point*' or '*a subsequential limit*' of the sequence. One must carefully understand the distinction between '*the limit of a sequence*' and '*a limit point*' (cluster point) *of a sequence.* We have preferred to use the term a cluster point rather than a limit point of a sequence.

A sequence whose range is finite, for example the sequence $< 11, 13, 17, 1, 1, 1, \ldots >$ is convergent and has the limit 1, but its range set has no limit point. On the other hand a sequence may not converge and its range set may still have a limit point. For example, $< 1, \frac{1}{2}, 1, \frac{1}{3}, 1, \frac{1}{4}, \ldots >$ does not converge but its range set has 0 as its limit point.

Examples. (*i*) Each one of the sequences $< \frac{1}{n} >, < \frac{1}{n^2} >, < \frac{-1}{n} >$ and $< \frac{1}{2^n} >$ has 0 as the cluster point.

(*ii*) Each of the sequences $< (-1)^n >$ and $< (-1)^{n-1} >$ has -1 and 1 as its cluster points.

(*iii*) None of the sequences $< n >, < -n >, < n^2 >, < -n^2 >$ has a cluster point.

(*iv*) The sequence $< 0, 1, 2, 0, 1, 2, 0, 1, 2, \ldots\ldots >$ has 0, 1 and 2 as its cluster points.

(*v*) The sequence $< a_n >$ defined by

$$a_n = \begin{cases} n & \text{if } n \text{ is odd} \\ \dfrac{1}{n} & \text{if } n \text{ is even} \end{cases}$$

has 0 as its only cluster point.

Remark. Every cluster point ξ of the range set of a sequence $< a_n >$ is also a cluster point of the sequence $< a_n >$ because every neighbourhood of ξ contains infinitely many elements of the range set $\{a_n \mid n \in \mathbf{N}\}$ and so of the sequence $< a_n >$. On the other hand, a cluster point of $< a_n >$ may or may not be a cluster point of the range set. For example, the sequence $< (-1)^n >$ has two cluster points -1 and 1 but the range set $\{-1, 1\}$ has no cluster point.

If the values of only a finite number of terms of a sequence $< a_n >$ are not different, then the cluster points of $< a_n >$ and the range set $\{a_n \mid n \in \mathbf{N}\}$ are same. Thus, it follows that the cluster points of a sequence $< a_n >$ are either the points or the cluster points of the set $\{a_n \mid n \in \mathbf{N}\}$.

4.4.2. Theorem. *If a sequence $< a_n >$ converges to ξ, then ξ is the only cluster point of the sequence.*

Proof. (*i*) Let $\varepsilon > 0$ and a positive integer m be given. Since $< a_n >$ converges to ξ.

$\therefore$ $\exists$ a positive integer m_1, such that

$$| a_n - \xi | < \varepsilon \quad \forall\, n \geq m_1$$

Let k be a positive integer such that $k > \max. \{m_1, m\}$

Then, $\qquad | a_k - \xi | < \varepsilon$

Thus, given any $\varepsilon > 0$ and a positive integer m, we have found a positive integer $k > m$ such that $| a_k - \xi | < \varepsilon$.

$\Rightarrow$ ξ is a cluster point of $< a_n >$.

(*ii*) To show that ξ is the only cluster point of $< a_n >$, we shall show that if η is any cluster point of $< a_n >$, then $\xi = \eta$.

Let $\varepsilon > 0$ be given.

Since the sequence $< a_n >$ converges to ξ,

$\therefore$ $\exists$ a positive integer m such that

$$| a_n - \xi | < \frac{\varepsilon}{2} \quad \forall\, n \geq m \qquad ...(1)$$

Since η is a cluster point of $< a_n >$,

$\therefore$ $\exists$ a positive integer $m_1 > m$ such that

$$| a_{m_1} - \eta | < \frac{\varepsilon}{2} \qquad ...(2)$$

Since $m_1 > m$, therefore, from (1),

$$| a_{m_1} - \xi | < \frac{\varepsilon}{2} \qquad ...(3)$$

Now
$$| \xi - \eta | = | a_{m_1} - \eta - (a_{m_1} - \xi) |$$
$$\leq | a_{m_1} - \eta | + | a_{m_1} - \xi |$$
$$< \frac{\varepsilon}{2} + \frac{\varepsilon}{2} = \varepsilon$$

i.e.,
$$| \xi - \eta | < \varepsilon$$

Since $\varepsilon > 0$ is arbitrarily small, it follows that $\xi = \eta$.

Remark. The converse of the above theorem need not be true. For example, the sequence $< a_n >$, where

$$a_n = \begin{cases} \dfrac{1}{n} & \text{if } n \text{ is odd} \\ n & \text{if } n \text{ is even} \end{cases}$$

has '0' as the only cluster point, but the sequence is not convergent.

A sequence may or may not have a cluster point, but the following theorem gives the sufficient condition for the existence of a cluster point of a sequence.

4.4.3. Theorem *(Bolzano-Weierstrass Theorem for sequences).*

Every bounded sequence has a cluster point.

Proof. Let $< a_n >$ be a bounded sequence.

Let S be the range set of the sequence, *i.e.*,

$$S = \{a_n : n \in N\}$$

S is a bounded set $\qquad (\because \; < a_n > \text{ is bounded})$

Case (*i*) S is finite

Then, for infinitely many suffixes n, $a_n = \xi$, where ξ is some real number. Thus ξ is a cluster point of $< a_n >$.

Case (*ii*) S is infinite

Now, S being an infinite and bounded set has a limit point ξ (say). (Bolzano-Weierstrass Theorem).

$\Rightarrow$ every neighbourhood of ξ contains infinitely many elements of S

$\Rightarrow$ every neighbourhood of ξ contains infinitely many terms of $< a_n >$.

$\Rightarrow$ ξ is a cluster point of the sequence $< a_n >$.

Cor. 1. If S is a compact set (*i.e.*, closed and bounded), then every sequence in S* has a cluster point in S.

For, let $< a_n >$ be a sequence in S.

Now, $< a_n >$ is a bounded sequence (∵ S is bounded).

∴ $< a_n >$ has a cluster point ξ (say).

(Bolzano-Weierstrass Theorem). We shall show that $\xi \in S$.

If $\xi \in S^c$, then S^c being an open set is a neighbourhood of ξ and S^c contains no term of $< a_n >$. This contradicts the fact that ξ is a cluster point of $< a_n >$.

∴ $\xi \notin S^c \Rightarrow \xi \in S$.

Cor. 2. If I is a closed interval, then every sequence in I has a cluster point in I.

4.4.4. Theorem. *The set of cluster points of a bounded sequence is bounded.*

Proof. Let $< a_n >$ be a bounded sequence.

There exists real number k_1 and k_2 such that

$$k_1 \le a_n \le k_2 \quad \forall \quad n \in N$$

If $x \in (-\infty, k_1)$, then $(-\infty, k_1)$ is a neighbourhood of x and contains no element of $< a_n >$.

Similarly, if $x \in (k_2, \infty)$, then (k_2, ∞) is a neighbourhood of x and contains no element of $< a_n >$. Thus, no point outside $[k_1, k_2]$ is a cluster point of the bounded sequence $< a_n >$. So, the cluster points of the bounded sequence $< a_n >$ lie in $[k_1, k_2]$. Hence, the set of all cluster points of a bounded sequence $< a_n >$ is bounded (with the same bounds as that of the sequence).

Remark. The set of cluster points of an unbounded sequence may or may not be bounded.

For example, (*i*) the sequence $< 2, 1 + \frac{1}{2}, 2 + \frac{1}{2}, 1 + \frac{1}{3}, 2 + \frac{1}{3}, 3 + \frac{1}{3}, ... >$ is unbounded and the set of its cluster points is N.

(*ii*) the sequence $< 1, \frac{1}{2}, 2, \frac{1}{3}, 3, ... >$ is unbounded but its set of cluster points is {0}, which is bounded.

4.4.5. Theorem. *The set of cluster points of a sequence is a closed set.*

Proof. Let S be the set of all cluster points of a sequence $< a_n >$.

If $S' = \phi$, then clearly $S' \subseteq S$ and S is closed.

So, let $S' \ne \phi$.

Let $x \in S' \Rightarrow x$ is a cluster point of S.

$\Rightarrow$ every neighbourhood of x contains infinitely many elements of S

$\Rightarrow$ given any $\varepsilon > 0$, $(x - \varepsilon, x + \varepsilon) \cap S$ is infinite.

Let $y \in (x - \varepsilon, x + \varepsilon) \cap S \Rightarrow y \in (x - \varepsilon, x + \varepsilon)$ and $y \in S$

$\Rightarrow (x - \varepsilon, x + \varepsilon)$ is a neighbourhood of y and y is a cluster point of $< a_n >$

$\Rightarrow a_n \in (x - \varepsilon, x + \varepsilon)$ for infinitely many n.

$\Rightarrow x$ is a cluster point of $< a_n >$

$\Rightarrow x \in S$

∴ $x \in S' \Rightarrow x \in S$

$\Rightarrow S' \subseteq S$ and hence S is closed.

Cor. *Every bounded set of cluster points of a sequence $< a_n >$ contains the smallest and greatest members.*

Proof. Since the bounded set S of all cluster points of a sequence $< a_n >$ is closed, therefore, S has the smallest and greatest members.

---

*A sequence $< a_n >$ is in S if $a_n \in S \quad \forall n$.

4.5.1. Limit Superior and Limit Inferior. Def. 1. *The greatest and smallest cluster points of a bounded sequence (as they do exist) are respectively called the limit superior (or superior limit or upper limit) and limit inferior (or inferior limit or lower limit) of the sequence.*

Def. 2. Let $< a_n >$ be a bounded sequence. Let $< a_n >$ be bounded above by K. Then, for each $n \in$ N, the set $S_n = \{a_n, a_{n+1}, \ldots\ldots\}$ is bounded above by K. By completeness axiom, S_n has the l.u.b. M_n (say). Clearly, $M_n \geq M_{n+1}$ $\forall\, n \in$ N. Thus, the sequence $< M_n >$, being a decreasing sequence is either convergent or divergent to $-\infty$. If $< M_n >$ is convergent, then $\lim\limits_{n \to \infty} M_n$ is called the limit superior of $< a_n >$. If $< M_n >$ is divergent, then the limit superior of $< a_n >$ is taken as $-\infty$.

If $< a_n >$ is bounded below by k, then $\forall\, n \in$ N, the set $S_n = \{a_n, a_{n+1}, \ldots\ldots\}$ is bounded below by k. Hence, S_n has the g.l.b. m_n (say). Since $m_n \leq m_{n+1}$ $\forall\, n \in$ N, the sequence $< m_n >$ is monotonically increasing. Thus, $< m_n >$ is either convergent or divergent to ∞. If $< m_n >$ is convergent, then $\lim\limits_{n \to \infty} m_n$ is called the limit inferior of $< a_n >$. If $< m_n >$ is divergent, then the limit inferior of $< a_n >$ is taken as ∞.

The limit superior of a bounded sequence $< a_n >$ is denoted by $\lim\limits_{n \to \infty} \sup a_n$, $\lim \sup a_n$, $\overline{\lim\limits_{n \to \infty}}\, a_n$ or $\overline{\lim}\, a_n$.

The limit inferior of a bounded sequence $< a_n >$ is denoted by $\lim\limits_{n \to \infty} \inf a_n$, $\lim \inf a_n$, $\underline{\lim\limits_{n \to \infty}}\, a_n$ or $\underline{\lim}\, a_n$.

If $< a_n >$ is unbounded above, we write $\overline{\lim\limits_{n \to \infty}}\, a_n = \infty$ and if $< a_n >$ is unbounded below, we write

$$\underline{\lim_{n \to \infty}}\, a_n = -\infty.$$

Clearly,
$$\overline{\lim_{n \to \infty}}\, a_n \geq \underline{\lim_{n \to \infty}}\, a_n.$$

The above two definitions are equivalent.

Examples. (*i*) If
$$a_n = \begin{cases} 2, & \text{if } n \text{ is odd} \\ -n, & \text{if } n \text{ is even} \end{cases}$$

then
$$\overline{\lim_{n \to \infty}}\, a_n = 2 \text{ and } \underline{\lim_{n \to \infty}}\, a_n = -\infty.$$

(*ii*) If $a_n = (-1)^n n^2$, then $\overline{\lim\limits_{n \to \infty}}\, a_n = \infty$ and $\underline{\lim\limits_{n \to \infty}}\, a_n = -\infty$

(*iii*) If $a_n = (-1)^n$, then $\overline{\lim\limits_{n \to \infty}}\, a_n = 1$ and $\underline{\lim\limits_{n \to \infty}}\, a_n = -1$

(*iv*) If $a_n = (-1)\, n$, then $\overline{\lim\limits_{n \to \infty}}\, a_n = -\infty$ and $\underline{\lim\limits_{n \to \infty}}\, a_n = -\infty$

(*v*) If $a_n = n^2$, then $\overline{\lim\limits_{n \to \infty}}\, a_n = \infty$ and $\underline{\lim\limits_{n \to \infty}}\, a_n = \infty$.

4.5.2. Theorem. *A sequence $< a_n >$ is convergent to a if and only if* $\overline{\lim\limits_{n \to \infty}}\, a_n = \underline{\lim\limits_{n \to \infty}}\, a_n = a$.

Proof. (*i*) Let $< a_n >$ be convergent to a.

Let $\varepsilon > 0$ be given. Then, there exists a positive integer m such that $| a_n - a | < \varepsilon$, $\forall\, n \geq m$

$\Rightarrow\quad a-\varepsilon<a_n<a+\varepsilon\quad \forall\, n\geq m$

$\Rightarrow\quad \forall\, n\geq m,\ a+\varepsilon$ is an upper bound of the set

$S_n=\{a_n, a_{n+1}, \ldots\}$ and $a-\varepsilon$ is not an upper bound of S_n.

$\therefore$ if M_n = l.u.b. S_n, then

$$a-\varepsilon<M_n\leq a+\varepsilon,\quad \forall\, n\geq m$$

$$\Rightarrow\quad |\,M_n-a\,|\leq\varepsilon\qquad \forall\, n\geq m$$

$$\Rightarrow\quad \lim_{n\to\infty} M_n=a\quad\Rightarrow\quad \overline{\lim_{n\to\infty}}\, a_n=a$$

Similarly, $\underline{\lim}_{n\to\infty}\, a_n=a$.

(*ii*) Conversely, let $\overline{\lim}_{n\to\infty}\, a_n=\underline{\lim}_{n\to\infty}\, a_n=a$

Let $\varepsilon>0$ be given.

Since $\overline{\lim}_{n\to\infty}\, a_n=a$, there exists a positive integer m_1 such that for all $n\geq m_1$, $|\,M_n-a\,|<\varepsilon$, where M_n = l.u.b. $\{a_n, a_{n+1}, \ldots\}$.

$$\Rightarrow\quad a-\varepsilon<M_n<a+\varepsilon$$

$$\Rightarrow\quad a_n<a+\varepsilon\qquad \forall\, n\geq m_1\qquad \ldots(1)$$

Since $\underline{\lim}_{n\to\infty}\, a_n=a$, proceeding as above, we can show that there exists a positive integer m_2 such that

$$a_n>a-\varepsilon\qquad \forall\, n\geq m_2\qquad \ldots(2)$$

Let $m_0=\max.\{m_1, m_2\}$

Then, from (1) and (2),

$$a-\varepsilon<a_n<a+\varepsilon\quad \forall\, n\geq m_0$$

$$\Rightarrow\quad |\,a_n-a\,|<\varepsilon\qquad \forall\, n\geq m_0$$

$$\Rightarrow\quad \lim_{n\to\infty} a_n=a$$

A similar result for diverging sequences is :

4.5.3. Theorem. *If a sequence $<a_n>$ is such that*

$\overline{\lim}_{n\to\infty}\, a_n=\underline{\lim}_{n\to\infty}\, a_n=\infty$, then $<a_n>$ is divergent to ∞.

Proof. Let m_n = g.l.b. $\{a_n, a_{n+1}, \ldots\}$, $\forall\, n\in N$

Since $\underline{\lim}_{n\to\infty}\, a_n=\infty$

$$\therefore\quad \lim_{n\to\infty} m_n=\infty$$

$\therefore$ given any positive real number k, (however large), there exists a positive integer m such that

$$m_n>k,\quad \forall\, n\geq m$$

But, $a_n\geq m_n\quad \forall\, n\in N$

$$\Rightarrow\quad a_n>k\quad \forall\, n\geq m$$

$$\Rightarrow\quad \lim_{n\to\infty} a_n=\infty.$$

4.5.4. Theorem. *For a sequence $<a_n>$,*

$$\underline{\lim_{n\to\infty}}\, a_n\leq \overline{\lim_{n\to\infty}}\, a_n.$$

Proof. If $< a_n >$ is not bounded, then either

$$\underline{\lim}_{n \to \infty} a_n = \infty \quad \text{or} \quad \underline{\lim}_{n \to \infty} a_n = -\infty$$

and hence there is nothing to prove.

So, let $< a_n >$ be a bounded sequence.

Let $\quad m_n = \text{g.l.b. } \{a_n, a_{n+1}, \ldots\ldots\}$

and $\quad M_n = \text{l.u.b. } \{a_n, a_{n+1}, \ldots\ldots\} \quad \forall\, n \in N$

Then, $\quad m_n \leq M_n$

$$\Rightarrow \quad \lim_{n \to \infty} m_n \leq \lim_{n \to \infty} M_n \quad \Rightarrow \quad \underline{\lim}_{n \to \infty} a_n \leq \overline{\lim}_{n \to \infty} a_n.$$

4.5.5. Theorem. *If $< a_n >$ and $< b_n >$ are bounded sequences such that for all $n \in N$, $a_n \leq b_n$, then*

(*i*) $\quad \underline{\mathit{lim}}_{n \to \infty} a_n \leq \overline{\mathit{lim}}_{n \to \infty} b_n$

and (*ii*) $\quad \underline{\mathit{lim}}_{n \to \infty} a_n \leq \underline{\mathit{lim}}_{n \to \infty} b_n.$

Proof. Let $\quad M_n = \text{l.u.b. } \{a_n, a_{n+1}, \ldots\ldots\}.$

$M_n' = \text{l.u.b. } \{b_n, b_{n+1}, \ldots\ldots\},$

$m_n = \text{g.l.b. } \{a_n, a_{n+1}, \ldots\ldots\}$

$m_n' = \text{g.l.b. } \{b_n, b_{n+1}, \ldots\ldots\}$

Since $\quad a_n \leq b_n \quad \forall\, n$

$\therefore \quad M_n \leq M_n'$ and $m_n \leq m_n' \;\forall\, n$

$$\Rightarrow \quad \lim_{n \to \infty} M_n \leq \lim_{n \to \infty} M_n' \quad \text{and} \quad \lim_{n \to \infty} m_n \leq \lim_{n \to \infty} m_n'$$

$$\Rightarrow \quad \overline{\lim}_{n \to \infty} a_n \leq \overline{\lim}_{n \to \infty} b_n \quad \text{and} \quad \underline{\lim}_{n \to \infty} a_n \leq \underline{\lim}_{n \to \infty} b_n.$$

4.6.1. Cauchy Sequences*. Def. 1. *A sequence $< a_n >$ is said to be a Cauchy sequence if given any $\varepsilon > 0$ (however small), there exists a positive integer m such that*

$$| a_n - a_m | < \varepsilon \quad \forall\, n \geq m.$$

Def. 2. A sequence $< a_n >$ is said to be a Cauchy sequence if given any $\varepsilon > 0$, there exists a positive integer m such that

$$| a_{m+p} - a_m | < \varepsilon \quad \forall\, p > 0 \qquad (p \text{ being an integer})$$

A Cauchy sequence is also called a *Fundamental sequence*.

Remarks. (*i*) Definitions (1) and (2) are equivalent.

(*ii*) The inequality '$n \geq m$' in definition (1) may be replaced by '$n > m$'.

(*iii*) The inequality in definition (2) is obviously true if $p = 0$ and we may therefore, rewrite it as $| a_{m+p} - a_m | < \varepsilon \quad \forall\, p \geq 0.$

(*iv*) Some times, we also write the definition as follows :

A sequence $< a_n >$ is said to be a Cauchy sequence if given any $\varepsilon > 0$, there exists a natural number k such that

$$| a_n - a_m | < \varepsilon, \quad \forall\, n \geq m > k.$$

Example. (*i*) *The sequence* $< \dfrac{1}{2^n} >$ *is a Cauchy sequence.* (K.U. BCA (II) 2007)

For, let $\varepsilon > 0$ be given.

---

*After the name of a distinguished French Mathematician A.L. Cauchy (1789-1857).

Now, $\left|\dfrac{1}{2^n}-\dfrac{1}{2^m}\right|=\dfrac{1}{2^m}\left|\dfrac{1}{2^{n-m}}-1\right|, n-m \geq 0$

$$< \frac{1}{2^m}$$

$$< \varepsilon \text{ if } m > \frac{\log\left(\frac{1}{\varepsilon}\right)}{\log 2}$$

$\therefore$ $\exists$ a positive integer $m\left(> \dfrac{\log\left(\frac{1}{\varepsilon}\right)}{\log 2}\right)$

such that $\left|\dfrac{1}{2^n}-\dfrac{1}{2^m}\right| < \varepsilon \quad \forall\, n \geq m$

$\therefore$ $<\dfrac{1}{2^n}>$ is a Cauchy sequence.

(ii) The sequence $< n^2 >$ is not a Cauchy sequence.

For, $|n^2 - m^2| = n^2 - m^2 = (n-m)(n+m) \geq n+m > 1$ for $n > m$.

$\therefore$ taking $\varepsilon = \frac{1}{2}$, we cannot find any positive integer m such that for $n > m$,

$$| n^2 - m^2 | < \tfrac{1}{2}$$

$\therefore$ $< n^2 >$ is not a Cauchy sequence.

(iii) The sequence $< (-1)^n >$ is not a Cauchy sequence

For, take $\varepsilon = \frac{1}{2}$

$$| a_{n+1} - a_n | = | (-1)^{n+1} - (-1)^n | = 2$$

$\therefore$ $$| a_{n+1} - a_n | \nless \tfrac{1}{2}$$

$\Rightarrow$ $< a_n >$ is not a Cauchy sequence.

4.6.2. Theorem. *Every Cauchy sequence is bounded.*

Proof. Let $< a_n >$ be a Cauchy sequence.

Taking $\varepsilon = 1$, there exists a positive integer m such that

$$| a_n - a_m | < 1 \qquad \forall\, n \geq m$$

$\Rightarrow$ $$a_m - 1 < a_n < a_m + 1 \quad \forall\, n \geq m$$

Let $k = \min. \{a_m - 1, a_1, a_2, \ldots\ldots, a_{m-1}\}$

$\mathrm{K} = \max. \{a_m + 1, a_1, a_2, \ldots\ldots, a_{m-1}\}$

Then, $k \leq a_n \leq \mathrm{K}\ \forall\, n$

$\Rightarrow$ the sequence $< a_n >$ is bounded.

Remark. Converse of the above theorem is not true, *i.e.*, a bounded sequence need not be a Cauchy sequence. For example, the sequence $< (-1)^n >$ is bounded but is not a Cauchy sequence.

The following theorem gives a convergence criterion—known as Cauchy's General Principle of Convergence. The importance of the criterion lies in the fact that it decides with precision the convergence or non-convergence of a sequence without any idea of the limit of the sequence and involves only the terms of the sequence.

4.6.3. Theorem. (*Cauchy's General Principle of Convergence*). ***A sequence is convergent if and only if it is a Cauchy sequence.***

Proof. (*i*) Let $< a_n >$ be a convergent sequence. Let this sequence converge to l.

$\therefore$ given any $\varepsilon > 0$ there exists a positive integer m such that

$$|a_n - l| < \frac{\varepsilon}{2} \quad \forall\, n \geq m \qquad \text{...(1)}$$

In particular, for $n = m$,

$$|a_m - l| < \frac{\varepsilon}{2} \qquad \text{...(2)}$$

Now
$$\begin{aligned} |a_n - a_m| &= |a_n - l + l - a_m| \\ &\leq |a_n - l| + |a_m - l| \\ &< \frac{\varepsilon}{2} + \frac{\varepsilon}{2} \quad \forall\, n \geq m \end{aligned}$$

i.e.,
$$|a_n - a_m| < \varepsilon \qquad \forall\, n \geq m$$

$\Rightarrow$ $< a_n >$ is a Cauchy sequence.

(*ii*) Conversely, let $< a_n >$ be a Cauchy sequence.

$\Rightarrow$ $< a_n >$ is a bounded sequence (By Theorem 4.6.2)

$\Rightarrow$ $< a_n >$ has a cluster point ξ (say) (By Theorem 4.4.3)

We shall show that $< a_n >$ converges to ξ.

Let $\varepsilon > 0$ be given.

Since $< a_n >$ is a Cauchy sequence,

$\therefore$ $\exists$ a positive integer m such that

$$|a_n - a_m| < \frac{\varepsilon}{3} \qquad \forall\, n \geq m \qquad \text{...(1)}$$

Since ξ is a cluster point of $< a_n >$.

$\therefore$ for above choice of ε and m, $\exists$ a positive integer $k > m$ such that

$$|a_k - \xi| < \frac{\varepsilon}{3} \qquad \text{...(2)}$$

Since $k > m$, therefore, from (1),

$$|a_k - a_m| < \frac{\varepsilon}{3} \qquad \text{...(3)}$$

Now
$$\begin{aligned} |a_n - \xi| &= |a_n - a_m + a_m - a_k + a_k - \xi| \\ &\leq |a_n - a_m| + |a_m - a_k| + |a_k - \xi| \\ &< \frac{\varepsilon}{3} + \frac{\varepsilon}{3} + \frac{\varepsilon}{3} \quad \forall\, n \geq m \qquad \text{[By (1) to (3)]} \end{aligned}$$

i.e.,
$$|a_n - \xi| < \varepsilon \quad \forall\, n \geq m$$

$\Rightarrow$ $< a_n >$ is convergent to ξ.

Example. *Using Cauchy's general principle of convergence, show that the sequence* $< a_n >$ *where* $a_n = 1 + \frac{1}{2} + \frac{1}{3} + \ldots + \frac{1}{n}$, *is not convergent.*

Sol. Suppose that $< a_n >$ is convergent.

$\therefore$ by Cauchy's general principle of convergence, $< a_n >$ is a Cauchy sequence.

Taking $\varepsilon = \frac{1}{2}$, there exists a positive integer m such that

$$| a_n - a_m | < \tfrac{1}{2} \quad \forall\, n \geq m.$$

In particular, $$| a_{2m} - a_m | < \tfrac{1}{2}$$

i.e., $$\frac{1}{m+1} + \frac{1}{m+2} + \ldots + \frac{1}{m+m} < \frac{1}{2}$$

which is not true, because

$$\frac{1}{m+1} + \frac{1}{m+2} + \ldots + \frac{1}{2m}$$

$$> \frac{1}{2m} + \frac{1}{2m} + \ldots + \frac{1}{2m} = \frac{m}{2m} = \frac{1}{2}.$$

Hence, $< a_n >$ is not a convergent sequence.

Remark. Since $< a_n >$ is monotonically increasing and is not a convergent sequence, therefore, $< a_n >$ is divergent to ∞.

EXERCISE 4 (*d*)

1. Give examples of sequences having (*i*) no cluster point, (*ii*) one cluster point, (*iii*) two cluster points and (*iv*) infinitely many cluster points.

2. Examine the sequences whose nth terms are given below, for cluster points :

(*i*) $(-1)^n$ (*ii*) 5

(*iii*) $\dfrac{(-1)^n}{n}$ (*iv*) n

(*v*) $(-1)^n \left(1 + \dfrac{1}{n}\right)$ (*vi*) $1 + \dfrac{1}{1!} + \dfrac{1}{2!} + \dfrac{1}{3!} + \ldots + \dfrac{1}{n!}$.

3. Find the limit superior and limit inferior of each of the following sequences :

(*i*) $< 1, 5, 17, 19, 1, 5, 17, 19, 1, 5, 17, 19, \ldots >$

(*ii*) $< a_n >$ where $a_n = (-10)^n \left(1 + \dfrac{1}{n}\right)^2$ (*iii*) $< a_n >$ where $a_n = \sin\left(\dfrac{n\pi}{3}\right)$

(*iv*) $< a_n >$ where $a_n = (-1)^n (2^n + 3^n)$.

4. Prove by definition that the following sequences whose nth terms are given below, are Cauchy sequences :

(*i*) $\dfrac{1}{n}$ (*ii*) $\dfrac{1}{n^2}$

(*iii*) $(-1)^n \dfrac{1}{n}$ (*iv*) $\dfrac{n}{n+1}$.

5. Show with the help of Cauchy's general principle of convergence that the sequence $< a_n >$ where $a_n = 1 + \dfrac{1}{3} + \dfrac{1}{5} + \ldots + \dfrac{1}{2n-1}$ is not convergent.

6. Show that the sequences whose nth terms are given below are not Cauchy sequences :

(i) $(-1)^n$ (ii) n

(iii) $(-1)^n \cdot n$ (iv) $<(-1)^n \cdot 2>$. *(K.U. BCA (II) 2006)*

7. If $<a_n>$ and $<b_n>$ are two Cauchy sequences, then show that

(i) $<a_n + b_n>$ (ii) $<a_n - b_n>$

(iii) $<a_n b_n>$ and (iv) $<\dfrac{a_n}{b_n}>$, where $b_n \neq 0$, are Cauchy sequences.

8. If a set A of limit points (*i.e.*, cluster points) of a sequence, prove that A contains all limit points of A.

[Hint. See 4.4.5.]

4.7. SOME MORE EXAMPLES ON SEQUENCES

Example 1. *Prove that* $\lim_{n \to \infty} a_n = 1$,

where $a_n = n^{1/n}$.

Sol. For $n = 1, \quad a_n = 1$

For $n \geq 2, \quad a_n > 1$

Let $a_n = 1 + \alpha_n, \alpha_n > 0 \quad \forall\, n \geq 2$

$$n = a_n^{\ n}$$

$$= (1 + \alpha_n)^n \qquad \text{(By binomial theorem)}$$

$$= 1 + n\,\alpha_n + \frac{n(n-1)}{2!}\alpha_n^{\ 2} + \ldots + \alpha_n^{\ n}$$

$$\Rightarrow \qquad n \geq \frac{n(n-1)}{2!}\alpha_n^{\ 2}$$

$$\Rightarrow \qquad 0 < \alpha_n^{\ 2} \leq \frac{2}{n-1}$$

$$\Rightarrow \qquad 0 < \alpha_n \leq \sqrt{\frac{2}{n-1}}$$

Since $\sqrt{\dfrac{2}{n-1}} \to 0$ as $n \to \infty$

$\therefore \quad \alpha_n \to 0$ as $n \to \infty$ (By Squeeze Principle)

Hence $a_n \to 1$ as $n \to \infty$.

Example 2. *If* $<a_n>$ *is a sequence of positive real numbers such that*

$a_n = \frac{1}{2}(a_{n-1} + a_{n-2}) \quad \forall\, n \geq 3$, *then show that* $a_n \to \frac{1}{3}(a_1 + 2a_2)$.

Sol. To show that $<a_n>$ is convergent, we shall show that it is a Cauchy sequence.

Now,
$$|\,a_n - a_{n-1}\,| = \left|\frac{a_{n-1} + a_{n-2}}{2} - a_{n-1}\right| = \tfrac{1}{2}\,|\,a_{n-1} - a_{n-2}\,|$$

Changing n to $n - 1$,

$$|\,a_{n-1} - a_{n-2}\,| = \frac{1}{2}\,|\,a_{n-2} - a_{n-3}\,|$$

$$\Rightarrow \qquad |\,a_n - a_{n-1}\,| = \frac{1}{2^2}\,|\,a_{n-2} - a_{n-3}\,|$$

Continuing like this, we see that

$$|a_n - a_{n-1}| = \frac{1}{2^{n-2}} |a_2 - a_1|$$

Therefore, for $n \geq m$,

$$\begin{aligned} |a_n - a_m| &= |(a_n - a_{n-1}) + (a_{n-1} - a_{n-2}) + \dots + (a_{m+1} - a_m)| \\ &\leq |a_n - a_{n-1}| + |a_{n-1} - a_{n-2}| + \dots + |a_{m+1} - a_m| \\ &= \left(\frac{1}{2^{n-2}} + \frac{1}{2^{n-3}} + \dots + \frac{1}{2^{m-1}}\right) |a_2 - a_1| \\ &= \frac{1}{2^{m-1}} \left(1 + \frac{1}{2} + \dots + \frac{1}{2^{n-m-1}}\right) |a_2 - a_1| \\ &= \frac{1}{2^{m-1}} \frac{\left(1 - \frac{1}{2^{n-m}}\right)}{1 - \frac{1}{2}} |a_2 - a_1| \\ &< \frac{1}{2^{m-2}} |a_2 - a_1| \end{aligned}$$

Let $\varepsilon > 0$ be given. Choose a positive integer m such that

$$\frac{|a_2 - a_1|}{2^{m-2}} < \varepsilon. \text{ Then, for } n \geq m, \ |a_n - a_m| < \varepsilon.$$

$\Rightarrow$ $\langle a_n \rangle$ is a Cauchy sequence.

Hence $\langle a_n \rangle$ is a convergent sequence.

To find $\lim\limits_{n \to \infty} a_n$

Let $a_n \to l$

Putting $n = 3, 4, 5, \dots\dots, n$ in the relation

$a_n = \frac{1}{2}(a_{n-1} + a_{n-2})$, we have

$$a_3 = \tfrac{1}{2}(a_2 + a_1)$$
$$a_4 = \tfrac{1}{2}(a_3 + a_2)$$
$$\vdots \qquad \vdots$$
$$a_n = \tfrac{1}{2}(a_{n-1} + a_{n-2})$$

Adding these equations, we have

$$a_n + a_{n-1} = \tfrac{1}{2} a_1 + a_2 + \tfrac{1}{2} a_{n-1}$$

$$\Rightarrow \qquad a_n + \tfrac{1}{2} a_{n-1} = \tfrac{1}{2} a_1 + a_2$$

Taking limits as $n \to \infty$,

$$\lim_{n \to \infty} a_n + \tfrac{1}{2} \lim_{n \to \infty} a_{n-1} = \tfrac{1}{2} a_1 + a_2$$

$$\Rightarrow \qquad l + \tfrac{1}{2} l = \tfrac{1}{2} a_1 + a_2 \ \Rightarrow \ l = \tfrac{1}{3}(a_1 + 2a_2).$$

Example 3. *If* $$a_n = \left[\left(\frac{2}{1}\right)^1 \left(\frac{3}{2}\right)^2 \left(\frac{4}{3}\right)^3 \ldots\ldots \left(\frac{n+1}{n}\right)^n\right]^{\frac{1}{n}},$$

then $a_n \to e$. *Hence or otherwise show that* $\lim\limits_{n\to\infty}\left(\dfrac{n^n}{n!}\right)^{\frac{1}{n}} = e$.

Sol. Let $$x_n = \left(\frac{2}{1}\right)^1 \left(\frac{3}{2}\right)^2 \left(\frac{4}{3}\right)^3 \ldots\ldots \left(\frac{n+1}{n}\right)^n$$

so that $$a_n = x_n^{\frac{1}{n}}$$

Also, $$\frac{x_{n+1}}{x_n} = \left(\frac{n+2}{n+1}\right)^{n+1} = \left(1+\frac{1}{n+1}\right)^{n+1}$$

Now $\lim\limits_{n\to\infty} \dfrac{x_{n+1}}{x_n} = e.$

Therefore, by Cauchy's second theorem on limits,

$$\lim_{n\to\infty} a_n = \lim_{n\to\infty} x_n^{\frac{1}{n}} = e.$$

Also, $$a_n = \left[2.\left(\frac{3}{2}\right)^2 . \left(\frac{4}{3}\right)^3 \ldots\ldots \left(\frac{n+1}{n}\right)^n\right]^{\frac{1}{n}}$$

$$= \left[\frac{(n+1)^n}{n!}\right]^{\frac{1}{n}} = \left[\frac{(n+1)^n}{n^n} . \frac{n^n}{n!}\right]^{\frac{1}{n}} = \frac{n+1}{n}\left(\frac{n^n}{n!}\right)^{\frac{1}{n}}$$

$$\therefore \quad \lim_{n\to\infty} a_n = \lim_{n\to\infty}\left[\left(\frac{n+1}{n}\right).\left(\frac{n^n}{n!}\right)^{\frac{1}{n}}\right] = \lim_{n\to\infty} \frac{n+1}{n} \lim_{n\to\infty}\left(\frac{n^n}{n!}\right)^{\frac{1}{n}}$$

$$e = 1 . \lim_{n\to\infty}\left(\frac{n^n}{n!}\right)^{\frac{1}{n}} \Rightarrow \lim_{n\to\infty}\left(\frac{n^n}{n!}\right)^{\frac{1}{n}} = e.$$

4.8.1. Subsequences. *Let* $< a_n >$ *be a given sequence. If* $< n_k >$ *is a strictly increasing sequence of natural numbers (i.e.,* $n_1 < n_2 < n_3 < \ldots\ldots$*), then* $< a_{n_k} >$ *is called a subsequence of* $< a_n >$.

Examples. (*i*) $< a_{2n-1} >$, $< a_{2n} >$, $< a_{n^2} >$ and $< a_{n^3} >$ are all subsequences of $< a_n >$.

(*ii*) The sequences < 1, 3, 5, ... >, < 2, 4, 6, 8, ... >. < 1, 2, 4, 8, ...>, < 1, 4, 9, 16, 25, ...> are all subsequences of the sequence $< n >$.

Remarks. (*i*) The terms of the subsequence occur in the same order in which they occur in the original sequence, *i.e.*, if a_5 and a_7 occur in a subsequence of $< a_n >$, then a_7 must follow a_5 and not *vice versa*. For example, the sequence < 2, 6, 4, 8, ... > is not a subsequence of the sequence < 1, 2, 3, ...> as 6 and 4 do not occur in the same order in which they occur in the sequence < 1, 2, 3, ... >.

(*ii*) Given a term a_m of the sequence $< a_n >$, there is a term of the subsequence following it.

(*iii*) The interval in the various terms of a subsequence need not be regular as is clear from the above examples.

(*iv*) Every sequence is a subsequence of itself.

(*v*) Every subsequence of a subsequence $< a_{n_k} >$ of a sequence $< a_n >$, is a subsequence of $< a_n >$.

4.8.2. Theorem. *If a sequence $< a_n >$ converges to l or diverges to ∞ or $-\infty$, then every subsequence of $< a_n >$ also converges to l or diverges to ∞ or $-\infty$.*

Proof. Let $< a_n >$ converge to l.

Let $< a_{n_k}$ be a subsequence of $< a_n >$

Let $\varepsilon > 0$ be given. Then, there exists a positive integer m such that

$$| a_n - l | < \varepsilon \quad \forall\, n \geq m$$

If $a_{k_0} \geq m$, then for $k \geq k_0$, $n_k \geq n_{k_0} \geq m$ and hence,

$$| a_{n_k} - l | < \varepsilon$$

which implies $< a_{n_k} >$ converges to l.

The proof of the remaining two parts is on similar lines, and is left as an exercise for the reader.

Remark. The converse of the above theorem is not always true. Following examples illustrate this.

(*i*) Let $a_n = (-1)^n \;\; \forall\, n$. Then $< a_n >$ does not converge but its two subsequences $< a_{2n} >$ and $< a_{2n-1} >$ converge to 1 and -1 respectively.

(*ii*) Let $a_n = \begin{cases} n & \text{if } n \text{ is odd} \\ 0 & \text{if } n \text{ is even} \end{cases}$

Then, the subsequence $< a_{2n-1} >$ diverges to ∞ whereas the original sequence does not converge.

(*iii*) Let $a_n = \begin{cases} -n & \text{if } n \text{ is even} \\ 0 & \text{if } n \text{ is odd} \end{cases}$

Then, the subsequence $< -2n >$ diverges to $-\infty$ but the original sequence does not diverge to $-\infty$.

4.8.3. Subsequential limit. *Let $< a_n >$ be a sequence. A real number ξ is called a* **subsequential limit** *of the sequence $< a_n >$ if there exists a subsequence of $< a_n >$ converging to ξ.*

For example, -1 and 1 are two subsequential limits of $< (-1)^n >$.

Remark. Some authors also define a cluster point of a sequence as subsequential limit of the sequence. This definition is equivalent to those, we used in this book. The following theorem (which we state without proof) gives the equivalence between the cluster points of a sequence and its subsequential limits.

4.8.4. Theorem. *A real number ξ is a cluster point of a sequence $< a_n >$ if, and only if there exists a subsequence of $< a_n >$ converging to ξ.*

In the light of the above theorem, we can restate Bolzano-Weierstrass Theorem on sequences as '*Every bounded sequence has a convergent subsequence*'.

5

Infinite Series

Infinite series are essential in the calculation of values of many functions, and can frequently be used for the evaluation of definite integrals. They can also serve to define new and useful functions that are fundamental in many investigations in advanced mathematics and its applications. An understanding of infinite series is the key to many doors in scientific research.

In this chapter, we are going to discuss the convergence behaviour of infinite series of real numbers and shall obtain a few tests for ascertaining the convergence of the infinite series. The study of infinite series is based on sequences. So, for an easy understanding of the topic, the reader is advised to review the fundamental results on sequences in Chapter 4, before going through this chapter.

5.1. SERIES AND THE SEQUENCE OF ITS PARTIAL SUMS

If $< a_n >$ *is a sequence of real numbers, then the expression* $a_1 + a_2 + a_3 + \ldots\ldots + a_n + \ldots\ldots$ *is called an infinite series* and is written as $\sum_{n=1}^{\infty} a_n$ or just as Σa_n. *The numbers* $a_1, a_2, a_3, \ldots a_n, \ldots$ *are called the first, the second, the third, ..., nth term (or general term), ... of the series. If* $S_n = a_1 + a_2 + a_3 + \ldots\ldots + a_n$, *then* S_n *is the sum of first n terms of the series* $\sum_{n=1}^{\infty} a_n$ *and is called the nth partial sum of the series.* Thus, $S_1 = a_1$, $S_2 = a_1 + a_2$, $S_3 = a_1 + a_2 + a_3$, and so on. *The sequence* $< S_n >$ *is called the sequence of partial sums of the series* $\sum_{n=1}^{\infty} a_n$.

For example, in the series $\frac{1}{2} + \frac{1}{4} + \frac{1}{8} + \ldots\ldots$,

the sum S_n of the first n terms is $\dfrac{\frac{1}{2}\left(1 - \frac{1}{2^n}\right)}{1 - \frac{1}{2}} = 1 - \frac{1}{2^n}$.

Thus, the sequence of partial sums of the series is the sequence $< S_n >$,

where $S_n = 1 - \frac{1}{2^n}$.

5.2. CONVERGENCE, DIVERGENCE OR OSCILLATION OF A SERIES

An infinite series $\sum_{n=1}^{\infty} a_n$ *is said to be*

(*i*) *convergent if the sequence* $< S_n >$ *of its partial sums converges to a real number S (say) and in that case S is called the sum of the series* $\sum_{n=1}^{\infty} a_n$ and we write $\sum_{n=1}^{\infty} a_n = S$. In this case, we also say that the series is convergent to S.

(*ii*) *converges absolutely if* $\sum_{n=1}^{\infty} | a_n |$ *converges.*

(*iii*) *converges conditionally if* $\sum_{n=1}^{\infty} a_n$ *converges but* $\sum_{n=1}^{\infty} | a_n |$ *does not converge.*

(*iv*) *diverges to* ∞ *(or* $-\infty$*) if the sequence* $< S_n >$ *diverges to* ∞ *(or* $-\infty$*) and in that case we write* $\sum_{n=1}^{\infty} a_n = \infty \left(\text{or } \sum_{n=1}^{\infty} a_n = -\infty \right)$.

(*v*) *oscillates finitely (or infinitely) if the sequence* $< S_n >$ *oscillates finitely (or infinitely).*

To illustrate, consider the examples :

Example (*i*) Consider the series $\frac{1}{2} + \frac{1}{2^2} + \frac{1}{2^3} + \ldots$..

Here,
$$S_n = \frac{1}{2} + \frac{1}{2^2} + \frac{1}{2^3} + \ldots.. + \frac{1}{2^n} = \frac{\frac{1}{2}\left(1 - \frac{1}{2^n}\right)}{1 - \frac{1}{2}} = 1 - \frac{1}{2^n}$$

$$\therefore \quad \lim_{n \to \infty} S_n = \lim_{n \to \infty} \left(1 - \frac{1}{2^n}\right) = 1$$

Thus, the sequence of partial sums is convergent with limit 1. Hence, the given series is convergent with sum 1.

Example (*ii*) Consider the series $\sum_{n=1}^{\infty} \frac{1}{n(n+1)}$.

Here, T_n (nth term) $= \frac{1}{n(n+1)} = \frac{1}{n} - \frac{1}{n+1}$

$\therefore$ the given series can be written as

$$\sum_{n=1}^{\infty}\left(\frac{1}{n}-\frac{1}{n+1}\right)$$

$$\therefore \quad S_n = \left(\frac{1}{1}-\frac{1}{2}\right)+\left(\frac{1}{2}-\frac{1}{3}\right)+\left(\frac{1}{3}-\frac{1}{4}\right)+\ldots+\left(\frac{1}{n}-\frac{1}{n+1}\right) = 1-\frac{1}{n+1}$$

$$\therefore \quad \lim_{n\to\infty} S_n = \lim_{n\to\infty}\left(1-\frac{1}{n+1}\right) = 1.$$

Thus, the sequence $< S_n >$ is convergent to 1. Hence, the series is convergent to 1.

Example (*iii*) Every infinite arithmetic series

$$a + (a + d) + (a + 2d) + \ldots\ldots$$

is divergent (except when both a and d are zero).

For, $\quad S_n = a + (a + d) + (a + 2d) + \ldots\ldots$ to n terms

$$= \frac{n}{2}\,[2a + (n - 1)d]$$

$\therefore$ as $n \to \infty$, $S_n \to \infty$ or $-\infty$ except when both a and d are zero.

Hence, every infinite arithmetic series is divergent.

Example (*iv*) Consider the series $-1-1-1-1-\ldots\ldots.$

Here, $\quad S_n = -n.$

$\therefore$ the sequence $< S_n >$ is divergent to $-\infty$. Hence, the series is divergent to $-\infty$.

Example (*v*) Consider the series $1-1+1-1+1-1+\ldots..$

$$\text{Here,} \quad S_n = \begin{cases} 0 & \text{if } n \text{ is even} \\ 1 & \text{if } n \text{ is odd} \end{cases}$$

$\therefore$ the sequence $< S_n >$ and hence the series oscillates finitely.

Example (*vi*) The series $\sum_{n=1}^{\infty} n(-1)^n$

i.e., $\;-1, 2, -3, 4, -5, 6, \ldots..$ oscillates infinitely.

Since the convergence, divergence or oscillation behaviour of a series $\sum_{n=1}^{\infty} a_n$ depends upon that of the associated sequence $< S_n >$ of its partial sums, we have the following theorems.

5.2.1. Theorem. *The convergence, divergence or oscillation behaviour of a series remains unchanged by the addition, omission or alteration of first finite number of terms of the series*. That is,

if $\sum_{n=1}^{\infty} a_n$ and $\sum_{n=1}^{\infty} b_n$ *be two series such that* $b_n = a_{n+p}$, $n \geq M$, *where* M *is some natural number and* $p \geq 0$ *is an integer, then the two series converge, diverge or oscillate together.*

Proof. Let $S_n = \sum_{k=1}^{n} a_k$, $S_n' = \sum_{k=1}^{n} b_k$, $n = 1, 2, 3, \ldots\ldots$

Then, for $n > M$,

$$\begin{aligned} S_n' &= b_1 + b_2 + \ldots\ldots + b_M + b_{M+1} + \ldots\ldots + b_n \\ &= \underbrace{b_1 + b_2 + \ldots\ldots + b_M}_{} + a_{M+p+1} + \ldots\ldots + a_{n+p} \\ &= S_M' + a_{M+p+1} + \ldots\ldots + a_{n+p} \\ &= S_M' + S_{n+p} - S_{M+p} = K + S_{n+p}, \end{aligned}$$

where $S_M' - S_{M+p} = K$

is a constant independent of n.

$\therefore$ the sequences $< S_n >$ and $< S_n' >$ of the partial sums of the series $\sum_{n=1}^{\infty} a_n$ and $\sum_{n=1}^{\infty} b_n$ have the same behaviour.

5.2.2. Theorem. *If $\sum_{n=1}^{\infty} a_n$ and $\sum_{n=1}^{\infty} b_n$ converge to a and b respectively, and c is any number, then*

(i) $\sum_{n=1}^{\infty} c\, a_n$, *(ii)* $\sum_{n=1}^{\infty} (a_n + b_n)$ *and* *(iii)* $\sum_{n=1}^{\infty} (a_n - b_n)$

all converge to ca, a + b and a – b respectively.

Proof. *(i)* If S_n denotes the nth partial sum of the series $\sum_{n=1}^{\infty} a_n$, then $c\, S_n$ is the nth partial sum of the series $\sum_{n=1}^{\infty} c\, a_n$.

Since $\sum_{n=1}^{\infty} a_n$ converges to a, $\lim_{n \to \infty} S_n = a$

Now, $\lim_{n \to \infty} c\, S_n = c \lim_{n \to \infty} S_n = ca$

Hence, $\sum_{n=1}^{\infty} ca_n$ converges to ca.

(ii) If S_n', S_n'' and S_n denote the nth partial sums of the series $\sum_{n=1}^{\infty} a_n$, $\sum_{n=1}^{\infty} b_n$ and $\sum_{n=1}^{\infty} (a_n + b_n)$ respectively, then

$$\lim_{n \to \infty} S_n' = a, \quad \lim_{n \to \infty} S_n'' = b.$$

Now,
$$\begin{aligned} S_n &= (a_1 + b_1) + (a_2 + b_2) + \ldots\ldots + (a_n + b_n) \\ &= (a_1 + a_2 + \ldots\ldots + a_n) + (b_1 + b_2 + \ldots\ldots + b_n) = S_n' + S_n''. \end{aligned}$$

$\therefore \quad \lim_{n\to\infty} S_n = \lim_{n\to\infty} (S_n' + S_n'') = \lim_{n\to\infty} S_n' + \lim_{n\to\infty} S_n = a + b.$

$\therefore \quad \sum_{n=1}^{\infty} (a_n + b_n)$ is convergent to $a + b$.

Proof of (*iii*) is simple and is left as an exercise to the readers.

Cor. If $\sum_{n=1}^{\infty} a_n$ and $\sum_{n=1}^{\infty} b_n$ are convergent to a and b respectively and s, t are real numbers, then

$$\sum_{n=1}^{\infty} (s\, a_n + t\, b_n) \text{ converges to } sa + tb.$$

Proofs to the following two theorems follow directly from the definition of covergence.

5.2.3. Theorem. *The convergence, divergence or oscillation behaviour of a series* $\sum_{n=1}^{\infty} a_n$ *remains unchanged by multiplication of each term of series by a non-zero real number (independent of n).*

5.2.4. Theorem. *If* $\sum_{n=1}^{\infty} a_n$ *converges to a and* $\sum_{n=1}^{\infty} b_n$ *diverges to* ∞ *(or* $-\infty$*), then* $\sum_{n=1}^{\infty} (a_n + b_n)$ *diverges to* ∞ *(or* $-\infty$*).*

An Important Remark. If a series $\sum_{n=1}^{\infty} a_n$ converges to a (or diverges to ∞ or $-\infty$), then every series obtained on grouping its terms in brackets, keeping the order of the terms in tact also converges to a (or diverges to ∞ or $-\infty$). So, we can insert brackets in a convergent (or divergent) series, but we cannot omit them.

For example, the series $1 - 1 + 1 - 1 + \ldots\ldots$ oscillates, but the series $(1 - 1) + (1 - 1) + \ldots\ldots$ converges to 0. However, if the terms in the brackets are only non-negative or non-positive, then the removal of the brackets in such a series does not affect its convergence or divergence.

EXERCISE 5 (*a*)

1. Show that the series $\sum_{n=1}^{\infty} \frac{1}{n(n+2)}$ converges to $\frac{3}{4}$.

2. Show that the series $\sum_{n=1}^{\infty} \frac{1}{n(n+1)(n+2)}$ converges to $\frac{1}{4}$.

3. Show that the series $1^2 + 2^2 + 3^2 + \ldots\ldots$ diverges to ∞.
4. Show that the series $2 + 2^2 + 2^3 + \ldots\ldots$ diverges to ∞.
5. Show that the series $-1 - 2 - 3 - 4 - 5 \ldots\ldots$ diverges to $-\infty$.

6. Show that the series 3 – 3 + 3 – 3 + ... oscillates.

7. Show that the series $\sum_{n=1}^{\infty}\left(\frac{3}{4}\right)^{n-1}$ converges to 4.

8. Assuming that the series $\sum_{n=1}^{\infty}\frac{1}{n^2}, \sum_{n=1}^{\infty}\frac{1}{n^3}, \sum_{n=1}^{\infty}\frac{1}{n^4}$ are convergent and using theorem 5.2.2, show that the following series are convergent :

(*i*) $\sum_{n=1}^{\infty}\frac{3n+2}{n^3}$ (*ii*) $\sum_{n=1}^{\infty}\frac{n-2}{n^3}$ (*iii*) $\sum_{n=1}^{\infty}\frac{3n^2+4}{n^4}$.

9. Show that the series $\sum_{n=1}^{\infty}\frac{1}{\sqrt{n}+\sqrt{n+1}}$ is divergent.

[**Hint.** On rationalizing $a_n = \sqrt{n+1} - \sqrt{n}$.]

In the above examples, we considered those series in which it was easy to calculate S_n, the sum of first n terms of the series. However, in most of the series, it is either difficult or impossible to calculate S_n. Now, the question is—'How to know the convergence behaviour of such series ?' For ascertaining the behaviour of such series, we describe some *tests*.

Since the convergence of a series is directly connected with the convergence of the sequence of its partial sums, we deduce a test for the series from Cauchy's general principle of convergence for sequences.

5.3.1. Theorem. *(Cauchy's General Principle of Convergence of Series).*

A necessary and sufficient condition for a series $\sum_{n=1}^{\infty} a_n$ *to be convergent is that to each* $\varepsilon > 0$*, there exists a positive integer m such that*

$$| a_{m+1} + a_{m+2} + \ldots\ldots + a_n | < \varepsilon \text{ whenever } n > m.$$

Proof. Let $S_n = a_1 + a_2 + \ldots\ldots + a_n$, so that $< S_n >$ is the sequence of partial sums of the series $\sum_{n=1}^{\infty} a_n$.

Now, the series $\sum_{n=1}^{\infty} a_n$ converges.

iff, the sequence $< S_n >$ converges

i.e., iff, by Cauchy's general principle of convergence, given any $\varepsilon > 0$, $\exists$ a positive integer m such that

$$| S_n - S_m | < \varepsilon, \text{ whenever } n > m.$$

i.e., iff $| a_{m+1} + a_{m+2} + \ldots\ldots + a_n | < \varepsilon$, whenever $n > m$

Hence the result.

Remark. *Sometimes we write the statement of the theorem as follows :*

A necessary and sufficient condition for a series $\sum_{n=1}^{\infty} a_n$ *to be convergent is that to each* $\varepsilon > 0$*, there exists a positive integer k such that*

$$| a_{m+1} + a_{m+2} + \ldots\ldots + a_n | < \varepsilon, \textit{ whenever } n > m \geq k.$$

Example 1. *Prove with the help of Cauchy's general principle of convergence that the series* $\sum_{n=1}^{\infty} \frac{1}{n} = 1 + \frac{1}{2} + \frac{1}{3} + \ldots\ldots$ *is not convergent.*

Sol. Let, if possible, the series be convergent.

Take $\varepsilon = \frac{1}{2}$. Then, by Cauchy's general principle of convergence, there exists a positive integer m such that

$$\frac{1}{m+1} + \frac{1}{m+2} + \ldots\ldots + \frac{1}{n} < \frac{1}{2} \quad \forall\, n > m.$$

Taking $n = 2m$, we have

$$\frac{1}{m+1} + \frac{1}{m+2} + \ldots\ldots + \frac{1}{n} = \frac{1}{m+1} + \frac{1}{m+2} + \ldots\ldots + \frac{1}{2m}$$

$$> \frac{1}{2m} + \frac{1}{2m} + \ldots\ldots + \frac{1}{2m} = \frac{m}{2m} = \frac{1}{2}$$

Thus, we have a contradiction.

Hence, the series is not convergent.

We give below a necessary condition for the convergence of a series.

5.3.2. Theorem. *If the series* $\sum_{n=1}^{\infty} a_n$ *converges, then*

$$\lim_{n \to \infty} a_n = 0.$$

Proof. Let $\quad S_n = a_1 + a_2 + a_3 + \ldots\ldots + a_n$

Then, $\quad S_{n-1} = a_1 + a_2 + a_3 + \ldots\ldots + a_{n-1}$

$\therefore \quad a_n = S_n - S_{n-1}$

If the series $\sum_{n=1}^{\infty} a_n$ is convergent to a (say), then

$$\lim_{n \to \infty} S_n = a.$$

Also, $\quad \lim_{n \to \infty} S_{n-1} = a.$

$$\therefore \quad \lim_{n \to \infty} a_n = \lim_{n \to \infty} (S_n - S_{n-1}) = \lim_{n \to \infty} S_n - \lim_{n \to \infty} S_{n-1} = a - a = 0.$$

Remark. The converse of the above theorem is not necessarily true. For example, in the series

$$\sum_{n=1}^{\infty} \frac{1}{n} = 1 + \frac{1}{2} + \frac{1}{3} + \ldots\ldots, a_n = \frac{1}{n} \to 0 \text{ as } n \to \infty$$

but the series is not convergent (See above solved example).

Thus, if in a series $\sum_{n=1}^{\infty} a_n$, $a_n \to 0$ as $n \to \infty$, we cannot say anything about the convergence of the series $\sum_{n=1}^{\infty} a_n$ but if a_n does not tend to 0, then the series is certainly not convergent.

Example 2. *Show that the series*

$$\frac{2}{3}+1+\frac{8}{7}+\ldots\ldots+\frac{3n-1}{2n+1}+\ldots\ldots$$

does not converge.

Sol. Here, $$a_n=\frac{3n-1}{2n+1}=\frac{3-\frac{1}{n}}{2+\frac{1}{n}}$$

$$\therefore \qquad \lim_{n\to\infty} a_n=\frac{3}{2}\neq 0.$$

Hence, the series does not converge.

EXERCISE 5 (*b*)

Show that none of the following series is convergent :

(*i*) $\sum_{n=1}^{\infty}\frac{n}{2n+1}$ (*ii*) $\sum_{n=1}^{\infty}\frac{n^2}{n+1}$

(*iii*) $\sum_{n=1}^{\infty}\frac{n^2-2n+3}{2n^2+n+1}$ (*iv*) $\sum_{n=1}^{\infty}(-1)^{n+1}\frac{e^n}{n^3}$

(*v*) $\frac{1}{2}+\frac{2}{3}+\frac{3}{4}+\ldots\ldots$

5.4.1. Geometric Series. The geometric series

$$1+r+r^2+\ldots\ldots+r^{n-1}+\ldots\ldots$$

(*i*) converges to $\frac{1}{1-r}$ if $|r|<1$,

(*ii*) diverges to ∞ if $r\geq 1$,

(*iii*) oscillates finitely if $r=-1$.

and (*iv*) oscillates infinitely if $r<-1$.

Sol. Here, $S_n=1+r+r^2+\ldots\ldots+r^{n-1}=\begin{cases}\frac{1-r^n}{1-r} & \text{if } r\neq 1\\ n & \text{if } r=1\end{cases}$

Case (*i*) $|r|<1$

Then, $\lim_{n\to\infty} r^n=0$ so that

$$\lim_{n\to\infty} S_n=\frac{1}{1-r}.$$ Hence, the series converges to $\frac{1}{1-r}$.

Case (*ii*) $r>1$.

Then, $\lim_{n\to\infty} r^n=\infty$ so that

$$S_n=\frac{1-r^n}{1-r}=\frac{1}{1-r}+\frac{r^n}{r-1}\to\infty \text{ as } n\to\infty.$$

Hence, the series is divergent to ∞.

If $r = 1$, then $\quad S_n = 1 + 1 + 1 + \ldots\ldots + 1 = n.$

Thus, the sequence $< S_n >$ diverges and hence the series diverges.

Case (*iii*) $r = -1$.

Thus, $S_n = 0$ or 1 according as n is even or odd. Therefore, the sequence $< S_n >$ oscillates between 0 and 1. Hence, the series oscillates finitely between 0 and 1.

Case (*iv*) $r < -1$.

Let $r = -a$, where $a > 1$.

Then, $$S_n = \frac{1}{1+a} - \frac{(-1)^n a^n}{1+a} \text{ so that}$$

$$S_{2n} \to -\infty \quad \text{and} \quad S_{2n+1} \to \infty$$

$\therefore$ the sequence $< S_n >$ oscillates infinitely between $-\infty$ and ∞.

Hence, the series oscillates infinitely.

5.5. SERIES OF POSITIVE TERMS

A series $\sum_{n=1}^{\infty} a_n$, *for which* $a_n > 0 \ \forall\ n$, *is called a series of positive terms or a positive term series.*

5.5.1. Theorem. *A positive term series* $\sum_{n=1}^{\infty} a_n$ *either converges to a finite limit or diverges to* ∞.

Proof. Let $$S_n = a_1 + a_2 + a_3 + \ldots\ldots + a_n$$
so that $$S_{n+1} = a_1 + a_2 + a_3 + \ldots\ldots + a_{n+1}$$
$\therefore$ $$S_{n+1} - S_n = a_{n+1} > 0$$
$\Rightarrow$ $$S_{n+1} > S_n \ \forall\ n$$
$\Rightarrow$ $< S_n >$ is a monotonically increasing sequence.

Since a monotonically increasing sequence is either convergent to a finite limit or divergent to ∞, the sequence $< S_n >$ of partial sums of the series $\sum_{n=1}^{\infty} a_n$ is either convergent or divergent to ∞. Hence, the series $\sum_{n=0}^{\infty} a_n$ either converges to a finite limit or diverges to ∞.

Note. (*i*) In view of the above theorem, a positive term series has only two possible behaviours—convergence or divergence while a general term series has got five behaviours.

(*ii*) If in a positive term series $\sum_{n=1}^{\infty} a_n$, a_n does not tend to 0 as $n \to \infty$, the series is divergent.

(*iii*) Similarly, it can be proved that a negative term series either converges to a finite number or diverges to $-\infty$.

Most of the time, in this chapter, we shall deal with only positive term series.

5.6. GENERAL TEST FOR THE CONVERGENCE OF POSITIVE TERM SERIES

5.6.1. Theorem. *A series of positive terms* $\sum_{n=1}^{\infty} a_n$ *is convergent if and only if the sequence* $< S_n >$ *(where* $S_n = a_1 + a_2 + a_3 + \ldots + a_n$*) of its partial sums is bounded above.*

Proof. Since $a_n > 0 \; \forall \; n$, the sequence $< S_n >$ of partial sums of the series is monotonically increasing.

Now, the series $\sum_{n=1}^{\infty} a_n$ is convergent iff the sequence $< S_n >$ is convergent.

i.e., iff the sequence $< S_n >$ is bounded above.

($\because$ a monotonically increasing sequence is convergent iff it is bounded above.)

Hence the result.

Remarks 1. To show that a series of positive terms is convergent, it is enough to show that the sequence of its partial sums is bounded above and to show that the series of positive terms is divergent, we have to show that the sequence of its partial sums is unbounded above.

2. A series $\sum_{n=1}^{\infty} a_n$, where the terms are not necessarily positive, may fail to be convergent even if the sequence of its partial sums is bounded above. For example, consider the series $\sum_{n=1}^{\infty} a_n$, where $a_n = (-1)^n$. Then,

$$S_n = \begin{cases} -1 & \text{if } n \text{ is odd} \\ 0 & \text{if } n \text{ is even} \end{cases}$$

The sequence $< S_n >$ is bounded above but is not convergent and as such, the series is not convergent. Hence, boundedness of the sequence of partial sums of a series $\sum_{n=1}^{\infty} a_n$ is only a necessary condition and not a sufficient one. However, it is a sufficient condition for a positive term series.

5.7. COMPARISON TESTS

The most important technique for deciding whether a series is convergent or not is to compare it with another suitably chosen series which is already known to be convergent or divergent.

There are two types of comparison tests—the first one pertains to the comparison of the general term of one series with the general term of the second series and the second to the comparison of the ratios of two consecutive terms of one series with the ratio of the corresponding consecutive terms of the second series.

5.7.1. First Form. *Let* $\sum_{n=1}^{\infty} a_n$ *and* $\sum_{n=1}^{\infty} b_n$ *be two series of positive terms such that* $a_n \le b_n \; \forall \; n$.

Then, (*i*) $\sum_{n=1}^{\infty} b_n$ *converges* $\Rightarrow$ $\sum_{n=1}^{\infty} a_n$ *converges*

(*ii*) $\sum_{n=1}^{\infty} a_n$ *diverges* $\Rightarrow$ $\sum_{n=1}^{\infty} b_n$ *diverges.*

Proof. Let S_n and σ_n denote the nth partial sums of the series $\sum_{n=1}^{\infty} a_n$ and $\sum_{n=1}^{\infty} b_n$ respectively.

Since $a_n \le b_n$, therefore $S_n \le \sigma_n \quad \forall n.$ ...(1)

Case (*i*) $\sum_{n=1}^{\infty} b_n$ is convergent.

$\Rightarrow$ the sequence $<\sigma_n>$ of its partial sums is bounded above by K (say).

i.e., $\sigma_n \le K \quad \forall\, n$ ...(2)

From (1) and (2),

$$S_n \le K \quad \forall\, n$$

$\Rightarrow$ the sequence $< S_n >$ of partial sums of the series $\sum_{n=1}^{\infty} a_n$ is bounded above.

$\Rightarrow$ the series $\sum_{n=1}^{\infty} a_n$ is convergent.

Case (*ii*) $\sum_{n=1}^{\infty} a_n$ is divergent.

$\Rightarrow$ the sequence $< S_n >$ of its partial sums is unbounded above.

$\Rightarrow$ given any positive number K, however large, $\exists$ a positive integer m such that

$S_n > K \quad \forall\, n \ge m$ ...(3)

From (1) and (3), we have

$$\sigma_n > K \ \forall\, n \ge m$$

$\Rightarrow$ the sequence $<\sigma_n>$ is unbounded above.

$\Rightarrow$ the sequence $<\sigma_n>$ and hence the series $\sum_{n=1}^{\infty} b_n$ is divergent.

Remark. If $0 < a_n \le b_n \ \forall n$, then the series $\sum_{n=1}^{\infty} b_n$ is said to *dominate* the series $\sum_{n=1}^{\infty} a_n$. Therefore, by the above theorem, every series dominated by a convergent series is convergent and every series dominating a divergent series is divergent.

Since the convergence or divergence behaviour of a series remains unchanged by the deletion of first finite number of terms of a series, the above comparison test may hold for the mth term onwards, where m is some positive integer and can be restated as :

5.7.2. Second Form. *Let* $\sum_{n=1}^{\infty} a_n$ *and* $\sum_{n=1}^{\infty} b_n$ *be two series of positive terms such that* $a_n \le b_n \ \forall \ n \ge m$, *where m is some positive integer.*

Then, (*i*) $\sum_{n=1}^{\infty} b_n$ *converges* $\Rightarrow \sum_{n=1}^{\infty} a_n$ *converges*

(*ii*) $\sum_{n=1}^{\infty} a_n$ *diverges* $\Rightarrow \sum_{n=1}^{\infty} b_n$ *diverges.*

5.7.3. Third Form. *Let* $\sum_{n=1}^{\infty} a_n$ *and* $\sum_{n=1}^{\infty} b_n$ *be two series of positive terms. Let k and K be two positive numbers such that*

$$k\,b_n \le a_n \le K\,b_n \quad \forall\, n.$$

Then, the series $\sum_{n=1}^{\infty} a_n$ *and* $\sum_{n=1}^{\infty} b_n$ *converge or diverge together.*

Proof. We have

$$k\,b_n \le a_n \le K\,b_n \quad \forall\, n \qquad \text{...(1)}$$

(*i*) If the series $\sum_{n=1}^{\infty} b_n$ is convergent, then $\sum_{n=1}^{\infty} K\,b_n$ is convergent and hence from second part of the inequality (1), the series $\sum_{n=1}^{\infty} a_n$ is convergent.

(*ii*) If the series $\sum_{n=1}^{\infty} a_n$ is convergent, then from first part of the inequality (1), $\sum_{n=1}^{\infty} k\,b_n$ is convergent and hence $\sum_{n-1}^{\infty} b_n \left(= \frac{1}{k} \sum_{n=1}^{\infty} k b_n \right)$ is convergent.

(*iii*) If the series $\sum_{n=1}^{\infty} a_n$ is divergent, then from second part of inequality (1), $\sum_{n=1}^{\infty} K\,b_n$ is divergent and hence $\sum_{n=1}^{\infty} b_n$ is divergent.

(*iv*) If the series $\sum_{n=1}^{\infty} b_n$ is divergent, then $\sum_{n=1}^{\infty} k\,b_n$ is divergent and hence from first part of the inequality (1), $\sum_{n=1}^{\infty} a_n$ is divergent.

Hence the result.

5.7.4. Fourth Form. *Let* $\sum_{n=1}^{\infty} a_n$ *and* $\sum_{n=1}^{\infty} b_n$ *be two series of positive terms. Let k and K be two positive numbers such that*

$k\, b_n \le a_n \le K\, b_n \;\forall\, n \ge m$, *for some positive integer m. Then the series* $\sum_{n=1}^{\infty} a_n$ *and* $\sum_{n=1}^{\infty} b_n$ *converge or diverge together.*

Proof. Proof follows from the above theorem and the fact that the convergence or divergence of a series is not changed by deleting a finite number of terms of the series.

Another powerful result that is most useful for studying series is contained in the *limiting form* of the comparison test given below.

5.7.5. Fifth Form. *Let* $\sum_{n=1}^{\infty} a_n$ *and* $\sum_{n=1}^{\infty} b_n$ *be two series of positive terms such that* $\lim_{n \to \infty} \frac{a_n}{b_n} = l$ *(finite and non-zero), then both the series converge or diverge together.*

Proof. Since $\frac{a_n}{b_n} > 0 \;\forall\, n$

$\therefore \quad l = \lim_{n \to \infty} \frac{a_n}{b_n} \ge 0$

But, $l \ne 0$ (given)

$\therefore \quad l > 0$

Choose $\varepsilon > 0$ such that $l - \varepsilon > 0$

Now, $\lim_{n \to \infty} \frac{a_n}{b_n} = l$

$\Rightarrow$ $\exists$ a positive integer m such that

$$\left| \frac{a_n}{b_n} - l \right| < \varepsilon \quad \forall\, n \ge m$$

$$\Rightarrow \quad l - \varepsilon < \frac{a_n}{b_n} < l + \varepsilon \quad \forall\, n \ge m$$

$$\Rightarrow \quad (l - \varepsilon) b_n < a_n < (l + \varepsilon)\, b_n \quad \forall\, n \ge m.$$

$\therefore$ From the *fourth form* of the comparison test, the series $\sum_{n=1}^{\infty} a_n$ and $\sum_{n=1}^{\infty} b_n$ converge or diverge together.

Remark. In the above form of the comparison test, the condition '$\lim_{n \to \infty} \frac{a_n}{b_n}$ must be finite and non-zero' cannot be dropped. For example,

(*i*) Take $a_n = \frac{1}{n}$ and $b_n = \frac{1}{n^2}$.

Then, $\sum_{n=1}^{\infty} a_n$ is divergent, $\sum_{n=1}^{\infty} b_n$ is convergent (to be proved later in p-series.)

and
$$\lim_{n \to \infty} \frac{a_n}{b_n} = \lim_{n \to \infty} n = \infty.$$

(*ii*) Take $a_n = \frac{1}{n^2}$ and $b_n = \frac{1}{n}$

Then, $\sum_{n=1}^{\infty} a_n$ is convergent, $\sum_{n=1}^{\infty} b_n$ is divergent and
$$\lim_{n \to \infty} \frac{a_n}{b_n} = \lim_{n \to \infty} \frac{1}{n} = 0.$$

However, if $\lim_{n \to \infty} \frac{a_n}{b_n} = 0$ or ∞, we have the following result.

5.7.6. Theorem. *Let* $\sum_{n=1}^{\infty} a_n$ *and* $\sum_{n=1}^{\infty} b_n$ *be two series of positive terms, then*

(i) if $\lim_{n \to \infty} \frac{a_n}{b_n} = 0$ *and* $\sum_{n=1}^{\infty} b_n$ *is convergent, then* $\sum_{n=1}^{\infty} a_n$ *is convergent*

(ii) if $\lim_{n \to \infty} \frac{a_n}{b_n} = \infty$ *and* $\sum_{n=1}^{\infty} b_n$ *is divergent, then* $\sum_{n=1}^{\infty} a_n$ *is divergent.*

Proof. (*i*) Take $\varepsilon = 1$

Since $\lim_{n \to \infty} \frac{a_n}{b_n} = 0$, there exists a positive integer m such that
$$\left| \frac{a_n}{b_n} - 0 \right| < 1 \quad \forall \, n \geq m$$

$$\Rightarrow \qquad -1 < \frac{a_n}{b_n} < 1 \quad \forall \, n \geq m$$

From second part of the inequality, we have
$$a_n < b_n \; \forall \, n \geq m.$$

$\therefore \quad \sum_{n=1}^{\infty} b_n$ is convergent

$\Rightarrow \quad \sum_{n=1}^{\infty} a_n$ is convergent.

(*ii*) Since $\lim\limits_{n \to \infty} \dfrac{a_n}{b_n} = \infty$,

$\therefore$ $\exists$ a positive integer m such that

$$\frac{a_n}{b_n} > 2^* \quad \forall\, n \geq m$$

$$\Rightarrow \qquad a_n > 2b_n \quad \forall\, n \geq m$$

$\therefore$ $\sum\limits_{n=1}^{\infty} b_n$ is divergent

$\Rightarrow$ $\sum\limits_{n=1}^{\infty} a_n$ is divergent.

5.7.7. Sixth Form. (*Comparison by ratios.*) *Let* $\sum\limits_{n=1}^{\infty} a_n$ *and* $\sum\limits_{n=1}^{\infty} b_n$ *be two series of positive terms such that*

$$\frac{a_n}{a_{n+1}} \geq \frac{b_n}{b_{n+1}} \quad \forall\, n \geq m,$$

where m is some positive integer.

Then, (*i*) $\sum\limits_{n=1}^{\infty} b_n$ *converges* $\Rightarrow$ $\sum\limits_{n=1}^{\infty} a_n$ *converges*

(*ii*) $\sum\limits_{n=1}^{\infty} a_n$ *diverges* $\Rightarrow$ $\sum\limits_{n=1}^{\infty} b_n$ *diverges*.

Proof. We have $\dfrac{a_n}{a_{n+1}} \geq \dfrac{b_n}{b_{n+1}} \quad \forall\, n \geq m$ ...(1)

Putting $n = m, m + 1, \ldots\ldots, n - 1$, in (1), we have

$$\frac{a_m}{a_{m+1}} \geq \frac{b_m}{b_{m+1}}$$

$$\frac{a_{m+1}}{a_{m+2}} \geq \frac{b_{m+1}}{b_{m+2}}$$

$$\ldots\ldots\ldots\ldots\ldots\ldots\ldots$$

$$\frac{a_{n-1}}{a_n} \geq \frac{b_{n-1}}{b_n}$$

Multiplying the above inequalities, we have

$$\frac{a_m}{a_n} \geq \frac{b_m}{b_n} \quad \forall\, n \geq m$$

$$\Rightarrow \qquad \frac{a_n}{a_m} \leq \frac{b_n}{b_m} \quad \forall\, n \geq m$$

---

*In fact we can choose any number > 1 instead of 2.

$$\Rightarrow \qquad a_n \leq \left(\frac{a_m}{b_m}\right) . b_n \ \forall \ n \geq m$$

$$\Rightarrow \qquad a_n \leq \text{K} . b_n \quad \forall \ n \geq m,$$

where $\text{K} = \dfrac{a_m}{b_m}$ is a fixed number.

Hence, $\sum_{n=1}^{\infty} b_n$ converges $\Rightarrow$ $\sum_{n=1}^{\infty} a_n$ converges

and $\sum_{n=1}^{\infty} a_n$ diverges $\Rightarrow$ $\sum_{n=1}^{\infty} b_n$ diverges.

For *comparison tests* to be useful, we must have in our hand a large number of series (of positive terms) whose convergence or divergence is known to us. So that whenever we come across with a new series of positive terms, we should have available with us some series for comparison. So far we have a few such series—a geometric series of positive terms and those series whose nth term does not tend to zero. We now consider a series which gives us a large collection of series useful for comparison.

5.8. THE *p*-SERIES OR GENERALIZED HARMONIC SERIES

The series $\sum_{n=1}^{\infty} \dfrac{1}{n^p} = 1 + \dfrac{1}{2^p} + \dfrac{1}{3^p} + \ldots + \dfrac{1}{n^p} + \ldots$ *is*

(i) convergent if $p > 1$ *and*

(ii) divergent if $p \leq 1$.

(Here, p is not necessarily an integer). (K.U. BCA (II) 2008)

Proof. Case *(i)* $p > 1$

We have $\dfrac{1}{1^p} = 1,$

$$\frac{1}{2^p} + \frac{1}{3^p} < \frac{1}{2^p} + \frac{1}{2^p} = \frac{2}{2^p} = 2^{1-p}$$

$$\frac{1}{4^p} + \frac{1}{5^p} + \frac{1}{6^p} + \frac{1}{7^p} < \frac{1}{4^p} + \frac{1}{4^p} + \frac{1}{4^p} + \frac{1}{4^p} = \frac{4}{4^p} = 4^{1-p} = (2^{1-p})^2$$

....

$$\frac{1}{(2^n)^p} + \frac{1}{(2^n + 1)^p} + \ldots + \frac{1}{(2^{n+1} - 1)^p} < (2^n)^{1-p} = (2^{1-p})^n$$

Adding the above inequalities, we have

$$S_{2^{n+1}-1} = 1 + \left(\frac{1}{2^p} + \frac{1}{3^p}\right) + \left(\frac{1}{4^p} + \frac{1}{5^p} + \frac{1}{6^p} + \frac{1}{7^p}\right) + \ldots + \left(\frac{1}{(2^n)^p} + \frac{1}{(2^n + 1)^p} + \ldots + \frac{1}{(2^{n+1} - 1)^p}\right)$$

$$< 1 + 2^{1-p} + (2^{1-p})^2 + \ldots + (2^{1-p})^n$$

(This is a geometric series of n terms with common ratio

$$2^{1-p} = \frac{1}{2^{p-1}} < 1 \text{ as } p > 1)$$

$$= \frac{1-(2^{1-p})^{n+1}}{1-2^{1-p}}$$

$$= \frac{1}{1-2^{1-p}} - \frac{(2^{1-p})^{n+1}}{1-2^{1-p}}$$

$$< \frac{1}{1-2^{1-p}} = \text{K (say)}, \quad \forall\, n.$$

Since the series is a series of positive terms and

$2^{n+1} - 1 > 2^n > n, \quad \forall\, n$, we have

$$S_n < S_{2^{n+1}-1} < K \quad \forall\, n$$

$\Rightarrow$ the sequence $< S_n >$ of partial sums of the series $\sum_{n=1}^{\infty} \frac{1}{n^p}$ is bounded above.

$\Rightarrow$ the given series is convergent.

Case (*ii*) $\quad p = 1$

$$1 + \tfrac{1}{2} = 1 + \tfrac{1}{2}$$

$$\tfrac{1}{3} + \tfrac{1}{4} > \tfrac{1}{4} + \tfrac{1}{4} = \tfrac{1}{2}$$

$$\frac{1}{5} + \frac{1}{6} + \frac{1}{7} + \frac{1}{8} > \frac{1}{8} + \frac{1}{8} + \frac{1}{8} + \frac{1}{8} = \frac{4}{8} = \frac{1}{2}$$

....

$$\frac{1}{2^{n-1}+1} + \frac{1}{2^{n-1}+2} + \ldots + \frac{1}{2^n} > \frac{1}{2^n}(2^n - 2^{n-1}) = 1 - \frac{1}{2} = \frac{1}{2}$$

Adding, we have $S_{2^n} > 1 + \frac{n}{2}$

Since $\frac{n}{2} \to \infty$ as $n \to \infty$, the sequence of partial sums of the series is unbounded above.

Hence the series is divergent.

Case (*iii*) $\quad p < 1$

Then, $\quad n^p < n \quad \forall\, n$

$$\Rightarrow \quad \frac{1}{n^p} > \frac{1}{n}$$

But, $\sum_{n=1}^{\infty} \frac{1}{n}$ is divergent [by case (*ii*)]

$\therefore$ by comparison test, $\sum_{n=1}^{\infty} \frac{1}{n^p}$ is divergent.

We, now give a few examples to illustrate the use of various comparison tests.

Example 3. *Test the series* $\sum_{n=1}^{\infty} \frac{1}{n!}$ *for convergence or divergence.*

Sol. Writing the first few terms, we have

$$\frac{1}{1!} = \frac{1}{1}, \frac{1}{2!} = \frac{1}{2}, \frac{1}{3!} = \frac{1}{1.2.3}, \ldots\ldots$$

Since each factor except 1 and 2 in $n\,!$ is larger than 2, we have the inequalities

$$n\,! \geq 2^{n-1} \quad \text{and} \quad \frac{1}{n!} \leq \frac{1}{2^{n-1}}$$

The series $\sum_{n=1}^{\infty} a_n$ with $a_n = \frac{1}{2^{n-1}}$, is a geometric series with common ratio = $\frac{1}{2}$ < 1 and is therefore, convergent.

Hence, by the comparison test $\sum_{n=1}^{\infty} \frac{1}{n!}$ is convergent.

Remark. The series of the above example, can also be used for comparison.

Example 4. *Show that the series* $\sum_{n=1}^{\infty} \frac{n+1}{n(2n-1)}$ *diverges.*

Sol. Here,
$$a_n = \frac{n+1}{n(2n-1)} = \frac{1}{n}\left(\frac{1+\frac{1}{n}}{2-\frac{1}{n}}\right)$$

$\therefore$ for large n, a_n behaves like $\frac{1}{2n}$

Take
$$b_n = \frac{1}{n}.$$

Then, $\sum_{n=1}^{\infty} b_n$ is divergent.

Now,
$$\lim_{n \to \infty} \frac{a_n}{b_n} = \lim_{n \to \infty} \frac{1+\frac{1}{n}}{2-\frac{1}{n}} = \frac{1}{2}$$

which is finite and non-zero.

$\therefore$ by comparison test (fifth form), the series $\sum_{n=1}^{\infty} a_n$ and $\sum_{n=1}^{\infty} b_n$ converge or diverge together.

Since $\sum_{n=1}^{\infty} b_n$ is divergent, the given series is divergent.

Example 5. *Examine the convergence of the series whose nth term is* $\sqrt{n^4+1}-\sqrt{n^4-1}$.

Sol. Here, $a_n = \sqrt{n^4+1}-\sqrt{n^4-1} > 0 \ \forall \ n$

$\therefore$ the given series is a series of positive terms

Also, $$a_n = \sqrt{n^4+1}-\sqrt{n^4-1} = \frac{(\sqrt{n^4+1}-\sqrt{n^4-1})(\sqrt{n^4+1}+\sqrt{n^4-1})}{\sqrt{n^4+1}+\sqrt{n^4-1}}$$

$$= \frac{n^4+1-(n^4-1)}{\sqrt{n^4+1}+\sqrt{n^4-1}} = \frac{2}{\sqrt{n^4+1}+\sqrt{n^4-1}} = \frac{2}{n^2\left(\sqrt{1+\frac{1}{n^4}}+\sqrt{1-\frac{1}{n^4}}\right)}$$

For sufficiently large n, a_n behaves like $\frac{1}{n^2}$.

Take $$b_n = \frac{1}{n^2}.$$

Then, $$\lim_{n\to\infty}\frac{a_n}{b_n} = \lim_{n\to\infty}\frac{2}{\sqrt{1+\frac{1}{n^4}}+\sqrt{1-\frac{1}{n^4}}} = \frac{2}{1+1} = 1$$

which is finite and non-zero.

Therefore, by comparison test, the series $\sum_{n=1}^{\infty} a_n$ and $\sum_{n=1}^{\infty} b_n$ converge or diverge together.

Since $\sum_{n=1}^{\infty}\frac{1}{n^2}$ is convergent , the given series is convergent.

Example 6. *Examine the convergence of the series*

$$\frac{3}{5}+\frac{4}{5^2}+\frac{3}{5^3}+\frac{4}{5^4}+\ldots\ldots \text{ to } \infty.$$

Sol. $$\frac{3}{5}+\frac{4}{5^2}+\frac{3}{5^3}+\frac{4}{5^4}+\ldots\ldots \text{ to } \infty$$

$$= \left(\frac{3}{5}+\frac{3}{5^3}+\ldots\ldots \text{ to } \infty\right)+\left(\frac{4}{5^2}+\frac{4}{5^4}+\ldots\ldots \text{ to } \infty\right)$$

$$= \sum_{n=1}^{\infty} a_n + \sum_{n=1}^{\infty} b_n \text{ (say)} \qquad \ldots(1)$$

Now, $\sum_{n=1}^{\infty} a_n$ is a geometric series with $r = \frac{1}{5^2} < 1$.

$\therefore$ $\sum_{n=1}^{\infty} a_n$ converges to $\dfrac{\frac{3}{5}}{1-\frac{1}{5^2}} = \dfrac{5}{8}$

$\sum_{n=1}^{\infty} b_n$ is a geometric series with $r = \dfrac{1}{5^2} < 1$

$\therefore \quad \sum_{n=1}^{\infty} b_n$ is also convergent to $\dfrac{\frac{4}{5^2}}{1-\frac{1}{5^2}} = \dfrac{1}{6}$

$\therefore$ from (1), the given series is also convergent to

$$\frac{5}{8} + \frac{1}{6} = \frac{19}{24}.$$

Example 7. *Show that the series*

$$\frac{1}{2^2+1} + \frac{\sqrt{2}}{3^2+1} + \frac{\sqrt{3}}{4^2+1} + \ldots\ldots \textit{ is convergent.}$$

Sol. Here, $a_n = \dfrac{\sqrt{n}}{(n+1)^2+1} < \dfrac{n^{1/2}}{(n+1)^2} < \dfrac{n^{1/2}}{n^2} = \dfrac{1}{n^{3/2}}$

and $\sum_{n=1}^{\infty} \dfrac{1}{n^{3/2}}$ is convergent.

$\therefore$ by comparison test the given series is convergent.

Example 8. *Show that* $\sum_{n=1}^{\infty} \dfrac{1}{n} \sin \dfrac{1}{n}$ *converges.*

Sol. Here, $$a_n = \frac{1}{n} \sin \frac{1}{n}$$

Since $$\lim_{n \to \infty} \frac{\sin \frac{1}{n}}{\frac{1}{n}} = 1,$$

Take $$b_n = \frac{1}{n^2}.$$

Then, $$\lim_{n \to \infty} \frac{a_n}{b_n} = \lim_{n \to \infty} \frac{\frac{1}{n} \sin \frac{1}{n}}{\frac{1}{n^2}} = \lim_{n \to \infty} \frac{\sin \frac{1}{n}}{\frac{1}{n}} = 1,$$ which is finite and non-zero.

Hence, the series $\sum_{n=1}^{\infty} a_n$ and $\sum_{n=1}^{\infty} b_n$ converge or diverge together.

Since $\sum_{n=1}^{\infty} b_n$ is convergent, the given series is convergent.

Example 9. *Test for convergence the series* $\sum_{n=2}^{\infty} \frac{1}{\log n}$.

Sol. Since $\log n < n \quad \forall\ n \geq 2$,

$$\frac{1}{\log n} > \frac{1}{n} \quad \forall\ n \geq 2.$$

Also, $\sum_{n=1}^{\infty} \frac{1}{n}$ is a divergent series.

$\therefore$ by comparison test, $\sum_{n=1}^{\infty} \frac{1}{\log n}$ is divergent.

Example 10. *Show that the series*

$$\sum_{n=1}^{\infty} \left[\frac{1}{n} - \log\left(\frac{n+1}{n}\right)\right] \text{ is convergent.}$$

Sol. Here, $a_n = \frac{1}{n} - \log\left(\frac{n+1}{n}\right) = \frac{1}{n} - \log\left(1 + \frac{1}{n}\right)$

$$= \frac{1}{n} - \left[\frac{1}{n} - \frac{1}{2}\cdot\frac{1}{n^2} + \frac{1}{3}\cdot\frac{1}{n^3} - \ldots\ldots\right] \quad \left(\because \log(1+x) = x - \frac{x^2}{2} + \frac{x^3}{3} - \ldots\ldots\right)$$

$$= \frac{1}{2}\cdot\frac{1}{n^2} - \frac{1}{3}\cdot\frac{1}{n^3} + \ldots\ldots$$

$\therefore$ for large n, a_n behaves like $\frac{1}{2}\cdot\frac{1}{n^2}$.

Take $\quad b_n = \frac{1}{n^2}$

$$\therefore \quad \lim_{n\to\infty} \frac{a_n}{b_n} = \lim_{n\to\infty} \left(\frac{1}{2} - \frac{1}{3n} + \ldots\ldots\right) = \frac{1}{2}, \text{ which is finite and non-zero.}$$

Hence, by the comparison test, $\sum_{n=1}^{\infty} a_n$ and $\sum_{n=1}^{\infty} b_n$ behave alike. But $\sum_{n=1}^{\infty} b_n$ is convergent.

Hence, the given series is convergent.

Example 11. *Test the convergence of the series*

$$\sum_{n=1}^{\infty} \frac{x^{n-1}}{1+x^n} = \frac{1}{1+x} + \frac{x}{1+x^2} + \frac{x^2}{1+x^3} + \ldots\ldots,\ x > 0. \qquad \text{(K.U. BCA (II) 2008)}$$

Sol. Here, $\quad a_n = \frac{x^{n-1}}{1+x^n}$

Case (*i*) $\quad 0 < x < 1.$

Then, $\quad x^n \to 0$ as $n \to \infty$.

Take $\quad b_n = x^{n-1}$

Now, $$\lim_{n\to\infty}\frac{a_n}{b_n}=\lim_{n\to\infty}\frac{1}{1+x^n}=\frac{1}{1+0}=1,$$

which is finite and non-zero.

$\therefore$ by comparison test, the series $\sum_{n=1}^{\infty} a_n$ and $\sum_{n=1}^{\infty} b_n$ converge or diverge together.

But, $\sum_{n=1}^{\infty} b_n$ being a geometric series with common ratio $x<1$, is convergent. Therefore, the given series is convergent.

Case *(ii)* $x = 1$.

Then, $a_n \to \frac{1}{1+1}=\frac{1}{2}(\neq 0)$ as $n\to\infty$. Therefore, the series is divergent.

Case *(iii)* $x > 1$

Then, $0<\frac{1}{x}<1$ and $\frac{1}{x^n}\to 0$ as $n\to\infty$

$$\therefore \quad \lim_{n\to\infty} a_n=\lim_{n\to\infty}\frac{x^n}{x(1+x^n)}=\lim_{n\to\infty}\frac{1}{x\left(1+\frac{1}{x^n}\right)}=\frac{1}{x}\neq 0.$$

Hence, the series is divergent.

Thus, the series is convergent if $x<1$ and divergent if $x\geq 1$.

Example 12. *If* $\sum_{n=1}^{\infty} a_n$ *is a convergent series of positive terms, then the series* $\sum_{n=1}^{\infty} a_n^2$ *is convergent. Is the converse true ?*

Sol. Since $\sum_{n=1}^{\infty} a_n$ is convergent, $\lim_{n\to\infty} a_n = 0$.

$\therefore$ taking ε such that $0<\varepsilon<1$, we can find a positive integer m such that

$$\begin{aligned} & |a_n - 0| < \varepsilon && \forall\, n \geq m \\ \Rightarrow\ & -\varepsilon < a_n < \varepsilon && \forall\, n \geq m \\ \Rightarrow\ & 0 < a_n < \varepsilon < 1 && \forall\, n \geq m \\ \Rightarrow\ & 0 < a_n < 1 && \forall\, n \geq m \\ \Rightarrow\ & a_n^2 < a_n && \forall\, n \geq m \end{aligned}$$

Since $\sum_{n=1}^{\infty} a_n$ is convergent, therefore, by the comparison test, the series $\sum_{n=1}^{\infty} a_n^2$ is also convergent.

Converse is false.

For example, let $a_n=\frac{1}{n}$, then $a_n^2=\frac{1}{n^2}$.

Here, $\sum_{n=1}^{\infty} a_n$ is divergent but $\sum_{n=1}^{\infty} a_n^2$ is convergent.

EXERCISE 5 (c)

Test the following series for convergence :

1. $1+\dfrac{1}{2\sqrt{2}}+\dfrac{1}{3\sqrt{3}}+\ldots\ldots+\dfrac{1}{n\sqrt{n}}+\ldots\ldots$

2. $\dfrac{1}{3}+\dfrac{1}{5}+\dfrac{1}{11}+\ldots\ldots+\dfrac{1}{3^{n-1}+2}+\ldots\ldots$

3. $\dfrac{1}{1.2.4}+\dfrac{1}{2.3.5}+\dfrac{1}{3.4.6}+\ldots\ldots$

4. $\dfrac{1}{\sqrt{2.3}}+\dfrac{1}{\sqrt{3.4}}+\dfrac{1}{\sqrt{4.5}}+\ldots\ldots$

5. $\dfrac{\sqrt{2}-1}{3^3-1}+\dfrac{\sqrt{3}-1}{4^3-1}+\dfrac{\sqrt{4}-1}{5^3-1}+\dfrac{\sqrt{5}-1}{6^3-1}+\ldots\ldots$

6. $\sum\limits_{n=1}^{\infty}\dfrac{1}{5^n+K}$, for all positive values of K.

7. $\sum\limits_{n=1}^{\infty}\dfrac{n}{\sqrt{n^2+1}}$

8. $\sum\limits_{n=1}^{\infty}\cos\left(\dfrac{1}{n}\right)$

9. $\sum\limits_{n=1}^{\infty}\left(\dfrac{1}{n}\right)^{1/n}$

10. $\sum\limits_{n=1}^{\infty}\sqrt[3]{n^3+1}-n.$

11. $\sum\limits_{n=1}^{\infty}\dfrac{n^2}{n!}.$

[**Hint.** Since the behaviour of a series does not depend upon the first few terms of a series, we write

$$\sum_{n=3}^{\infty}\frac{n^2}{n!}=\sum_{n=1}^{\infty}\frac{(n+2)^2}{(n+2)!}$$

Now $$a_n=\frac{(n+2)^2}{(n+2)!}$$

Take $$b_n=\frac{1}{n!}$$

$\therefore$ $$\frac{a_n}{b_n}=\frac{\dfrac{(n+2)^2}{(n+2)!}}{\dfrac{1}{n!}}=\frac{(n+2)^2}{(n+2)(n+1)}\to 1 \text{ as } n\to\infty.$$

Since $\sum\limits_{n=1}^{\infty} b_n$ is convergent, the given series is convergent.

A simple method is by ratio test.]

12. $\sum\limits_{n=1}^{\infty}\dfrac{n}{n!}$

13. $\dfrac{1}{2}+\dfrac{1}{3^2}+\dfrac{1}{2^3}+\dfrac{1}{3^4}+\ldots\ldots$ to ∞

14. $\dfrac{1}{6}+\dfrac{5}{10}+\dfrac{13}{18}+\ldots\ldots+\dfrac{2^{n+1}-3}{2^{n+1}+2}+\ldots\ldots$

15. Show that the series $\sum\limits_{n=1}^{\infty}(a^{\frac{1}{n}}-1)$, $a>1$ is divergent.

[**Sol.** $a_n = a^{\frac{1}{n}} - 1$

Take $b_n = \dfrac{1}{n}$.

Then, $\lim\limits_{n \to \infty} \dfrac{a_n}{b_n} = \lim\limits_{n \to \infty} \dfrac{a^{\frac{1}{n}} - 1}{\frac{1}{n}} = \log a > 0$ $\left[\because \lim\limits_{x \to 0} \dfrac{a^x - 1}{x} = \log a\right]$

Since $\sum\limits_{n=1}^{\infty} b_n$ is divergent, the series $\sum\limits_{n=1}^{\infty} a_n$ is divergent.]

16. $\dfrac{a}{6} + \dfrac{5a^2}{10} + \ldots\ldots + \dfrac{2^{n+1} - 3}{2^{n+1} + 2} a^n + \ldots\ldots, (a \geq 0)$.

17. $\sum\limits_{n=1}^{\infty} \dfrac{1}{n^{1+1/n}}$. $\left[\textbf{Hint.} \lim\limits_{n \to \infty} n^{1/n} = 1. \text{ Take } b_n = \dfrac{1}{n}\right]$

18. $\sum\limits_{n=1}^{\infty} [\log (n + 1) - \log n]$. $\left[\textbf{Hint.} \text{ Take } b_n = \dfrac{1}{n}\right]$

19. $1 + \dfrac{1}{2^2} + \dfrac{2^2}{3^3} + \dfrac{3^3}{4^4} + \ldots\ldots$ $\left[\textbf{Hint.} \text{ Take } b_n = \dfrac{1}{n}\right]$

20. $1 + \dfrac{1}{2^2} + \dfrac{1}{3^3} + \dfrac{1}{4^4} + \ldots\ldots + \dfrac{1}{n^n} + \ldots\ldots$ [**Hint.** $n^n > 2^n$, for $n \geq 3$]

21. $\sum\limits_{n=1}^{\infty} (\sqrt{(n^4 + 1)} - n^2)$.

22. $\sum\limits_{n=1}^{\infty} \dfrac{|\cos n|}{n^{3/2}}$.

23. $\sum\limits_{n=2}^{\infty} \dfrac{1}{(\log n)^{\log n}}$.

[**Hint.** We can find n so large that $\log (\log n) > 2$

$\Rightarrow \quad \log n (\log (\log n)) > 2 \log n$

$\Rightarrow \quad \log (\log n)^{\log n} > \log n^2$

$\Rightarrow \quad (\log n)^{\log n} > n^2$, etc.]

24. $\sum\limits_{n=2}^{\infty} \dfrac{1}{n^2 \log n}$. $\left[\textbf{Hint.} \dfrac{1}{n^2 \log n} < \dfrac{1}{n^2} \text{ for } n \geq 3\right]$

25. $\sum\limits_{n=1}^{\infty} \dfrac{1}{e^{n^3}}$. [**Hint.** Use $e^x > x$ for $x > 0$]

26. $\dfrac{2}{1^p} + \dfrac{3}{2^p} + \dfrac{4}{3^p} + \dfrac{5}{4^p} + \ldots\ldots$ $\left[\textbf{Hint.} \text{ Here } a_n = \dfrac{n+1}{n^p} = \dfrac{n\left(1 + \frac{1}{n}\right)}{n^p} = \dfrac{1 + \frac{1}{n}}{n^{p-1}}\right]$

27. Assuming that $\sin x < x$, for $0 < x < \frac{\pi}{2}$, show that $\sum_{n=1}^{\infty} \sin\left(\frac{1}{n^2}\right)$ converges. What can you say about $\sum_{n=1}^{\infty} \sin \frac{1}{n}$?

28. If $\sum_{n=1}^{\infty} a_n$ is a convergent series of positive terms such that $a_n \neq \pm 1$, then show that

(i) $\sum_{n=1}^{\infty} \frac{a_n}{1-a_n}$

(ii) $\sum_{n=1}^{\infty} \frac{a_n}{1+a_n}$ are both convergent series.

[**Hint.** (i) $\sum_{n=1}^{\infty} a_n$ converges $\Rightarrow$ $a_n \to 0$ as $n \to \infty$

Let $b_n = \frac{a_n}{1-a_n}$

Then, $\lim_{n\to\infty} \frac{a_n}{b_n} = \lim_{n\to\infty} (1-a_n) = 1 \neq 0.$]

29. If $\sum_{n=1}^{\infty} a_n^2$ and $\sum_{n=1}^{\infty} b_n^2$ are both convergent, prove that $\sum_{n=1}^{\infty} a_n b_n$ is also convergent.

[**Hint.** $\sum_{n=1}^{\infty} a_n^2$ and $\sum_{n=1}^{\infty} b_n^2$ are convergent

$\Rightarrow \frac{1}{2}\sum_{n=1}^{\infty} (a_n^2 + b_n^2)$ is convergent $\Rightarrow \sum_{n=1}^{\infty} \frac{a_n^2 + b_n^2}{2}$ is convergent

Since G.M. between a_n^2 and b_n^2 < A.M. between a_n^2 and b_n^2,

$\therefore \quad \sqrt{a_n^2 b_n^2} < \frac{a_n^2 + b_n^2}{2} \Rightarrow a_n b_n < \frac{a_n^2 + b_n^2}{2}$

The result follows from the comparison test.]

30. Show that $\sum_{n=1}^{\infty} \frac{2^n + 3^n}{6^n}$ converges.

31. If $\sum_{n=1}^{\infty} a_n$ is a convergent series of positive terms, does it follow that

(i) $\sum_{n=1}^{\infty} \sqrt{a_n}$ and (or) (ii) $\sum_{n=1}^{\infty} n\, a_n$ are convergent series ?

32. For what values of q does the series $\sum_{n=1}^{\infty} \frac{\log n}{n^q}$ converge ?

33. For what values of p and q does the series $\sum_{n=1}^{\infty} \frac{(\log n)^p}{n^q}$ converge ?

[**Hint.** Let $\frac{(\log n)^p}{n^q} = \frac{1}{n^\alpha}$ ...(1)

$\Rightarrow$ $(\log n)^p = n^{q-\alpha}$

Taking logarithm on both sides, we have

$$p \log(\log n) = (q - \alpha) \log n$$

or $$\alpha = \frac{q \log n - p \log(\log n)}{\log n} = q - \frac{p \log(\log n)}{\log n} \quad ...(2)$$

Now, $$\lim_{n \to \infty} \frac{p \log(\log n)}{\log n} \left(\frac{\infty}{\infty}\right) = p \,.\, \lim_{n \to \infty} \frac{\frac{1}{\log n} \cdot \frac{1}{n}}{\frac{1}{n}} = p \,.\, \lim_{n \to \infty} \frac{1}{\log n} = 0$$

$\therefore$ from (2), for large n, $\alpha = q$. Hence, from (1), the given series is convergent for $q > 1$ and p arbitrary.]

5.9. D' ALEMBERT'S RATIO TEST*

5.9.1. Theorem. *Let* $\sum_{n=1}^{\infty} a_n$ *be a series of positive terms such that*

(*a*) $\lim_{n \to \infty} \frac{a_n}{a_{n+1}} = l.$

Then (*i*) *the series is convergent if* $l > 1$,

(*ii*) *the series is divergent if* $l < 1$.

(*b*) $\lim_{n \to \infty} \frac{a_n}{a_{n+1}} = \infty.$

Then, $\sum_{n=1}^{\infty} a_n$ *is convergent.*

Proof. (*a*) **Case** (*i*) $l > 1$

Choose $\varepsilon > 0$ such that $l - \varepsilon > 1$.

(In fact, any number lying between 0 and $l - 1$ can be taken as ε.)

Since $\lim_{n \to \infty} \frac{a_n}{a_{n+1}} = l$, therefore, there exists a positive integer m such that

$$\left| \frac{a_n}{a_{n+1}} - l \right| < \varepsilon, \quad \forall\, n \geq m$$

*After the name of a French mathematician Jean Le Rond, D' Alembert (1717–1783).

$$\Rightarrow \qquad l - \varepsilon < \frac{a_n}{a_{n+1}} < l + \varepsilon, \quad \forall\, n \geq m$$

Putting $n = m, m + 1, \ldots\ldots, n - 1$, in the first part of the above inequality, we have

$$l - \varepsilon < \frac{a_m}{a_{m+1}}$$

$$l - \varepsilon < \frac{a_{m+1}}{a_{m+2}}$$

$$\vdots \qquad \vdots$$

$$l - \varepsilon < \frac{a_{n-1}}{a_n}.$$

Multiplying the above $n - m$ inequalities, we have

$$(l - \varepsilon)^{n-m} < \frac{a_m}{a_{m+1}} \cdot \frac{a_{m+1}}{a_{m+2}} \cdots \frac{a_{n-1}}{a_n}$$

or
$$(l - \varepsilon)^{n-m} < \frac{a_m}{a_n}$$

or
$$a_n < \frac{1}{(l - \varepsilon)^{n-m}} \cdot a_m$$

or
$$a_n < (l - \varepsilon)^m \cdot a_m \frac{1}{(l - \varepsilon)^n} \qquad \forall\, n \geq m$$

or
$$a_n < \text{K} \cdot \frac{1}{(l - \varepsilon)^n} \qquad (\text{where K} = (l - \varepsilon)^m \cdot a_m), \quad \forall\, n \geq m.$$

Now, $\sum_{n=1}^{\infty} \frac{1}{(l - \varepsilon)^n}$ being a geometric series with common ratio $\frac{1}{(l - \varepsilon)} < 1$, is convergent.

Therefore, by comparison test, the series $\sum_{n=1}^{\infty} a_n$ is convergent.

Case *(ii)* $l < 1$

Choose $\varepsilon > 0$ such that $l + \varepsilon < 1$.

(In fact, any number lying between 0 and $1 - l$ can be taken as ε.)

Since
$$\lim_{n \to \infty} \frac{a_n}{a_{n+1}} = l,$$

therefore, there exists a positive integer m such that

$$\left| \frac{a_n}{a_{n+1}} - l \right| < \varepsilon \quad \forall\, n \geq m$$

$$\Rightarrow \qquad l - \varepsilon < \frac{a_n}{a_{n+1}} < l + \varepsilon \quad \forall\, n \geq m.$$

Putting $n = m, m+1, \ldots\ldots, n-1$, in the second half of the above inequality, we have

$$\frac{a_m}{a_{m+1}} < l + \varepsilon$$

$$\frac{a_{m+1}}{a_{m+2}} < l + \varepsilon$$

$$\vdots \quad \vdots$$

$$\frac{a_{n-1}}{a_n} < l + \varepsilon.$$

Multiplying the above $n - m$ inequalities, we have

$$\frac{a_m}{a_{m+1}} \cdot \frac{a_{m+1}}{a_{m+2}} \cdots \frac{a_{n-1}}{a_n} < (l+\varepsilon)^{n-m}$$

or $$\frac{a_m}{a_n} < (l+\varepsilon)^{n-m} \qquad \text{or} \qquad \frac{a_n}{a_m} > \frac{1}{(l+\varepsilon)^{n-m}}$$

or $$a_n > a_m (l+\varepsilon)^m \cdot \frac{1}{(l+\varepsilon)^n}, \quad \forall\, n \geq m \qquad \text{or} \quad a_n > K \cdot \frac{1}{(l+\varepsilon)^n}$$

where $K = a_m (l+\varepsilon)^m$, $\forall\, n \geq m$.

Now, $\sum_{n=1}^{\infty} \frac{1}{(l+\varepsilon)^n}$ being a geometric series with common ratio $\frac{1}{l+\varepsilon} > 1$, is divergent.

Therefore, by comparison test the series $\sum_{n=1}^{\infty} a_n$ is divergent.

(b) $\lim_{n \to \infty} \frac{a_n}{a_{n+1}} = \infty$.

Then, there exists a positive integer m such that

$$\frac{a_n}{a_{n+1}} > 2 \quad \forall\, n \geq m.$$

Putting $n = m, m+1, \ldots\ldots, n-1$, we have

$$\frac{a_m}{a_{m+1}} > 2$$

$$\frac{a_{m+1}}{a_{m+2}} > 2$$

$$\vdots \quad \vdots$$

$$\frac{a_{n-1}}{a_n} > 2.$$

Multiplying the above $n - m$ inequalities, we have

$$\frac{a_m}{a_n} > 2^{n-m} \qquad \text{or} \qquad \frac{a_n}{a_m} < \frac{1}{2^{n-m}}$$

or $$a_n < (a_m \cdot 2^m) \cdot \frac{1}{2^n} \quad \text{or} \quad a_n < K \cdot \frac{1}{2^n}$$

(where $K = a_m \cdot 2^m$), $\forall\ n \geq m$.

Since $\sum_{n=1}^{\infty} \frac{1}{2^n}$ is convergent, being a geometric series with common ratio $\frac{1}{2} < 1$, therefore, by the comparison test, the series $\sum_{n=1}^{\infty} a_n$ is convergent.

Remark. If $\lim_{n \to \infty} \frac{a_n}{a_{n+1}} = 1$, then no information can be obtained regarding the convergence or divergence of the series $\sum_{n=1}^{\infty} a_n$.

For example, consider the p-series $\sum_{n=1}^{\infty} \frac{1}{n^p}$ which we have seen is convergent for $p > 1$ and divergent for $p \leq 1$.

Now $$\lim_{n \to \infty} \frac{a_n}{a_{n+1}} = \lim_{n \to \infty} \frac{(n+1)^p}{n^p} = \lim_{n \to \infty} \left(\frac{n+1}{n}\right)^p = \lim_{n \to \infty} \left(1 + \frac{1}{n}\right)^p$$
$$= 1 \text{ . regardless of the value of } p.$$

Thus, there are both convergent and divergent series for which the limit of the ratio of successive general terms is unity, so in this case the test fails.

Note. The ratio test is in fact a development of the comparison test of second type when we adopt a geometric series for comparison purpose.

Example 13. *Does the series* $\sum_{n=1}^{\infty} \frac{n!}{n^n}$ *converge or diverge ?*

Sol. $$a_n = \frac{n!}{n^n}$$

$$a_{n+1} = \frac{(n+1)!}{(n+1)^{n+1}}$$

$$\therefore \quad \frac{a_n}{a_{n+1}} = \frac{(n+1)^{n+1}}{(n+1)!} \cdot \frac{n!}{n^n} = \frac{(n+1)^n}{n!} \cdot \frac{n!}{n^n} = \left(\frac{n+1}{n}\right)^n = \left(1 + \frac{1}{n}\right)^n$$

$$\therefore \quad \lim_{n \to \infty} \frac{a_n}{a_{n+1}} = \lim_{n \to \infty} \left(1 + \frac{1}{n}\right)^n = e > 1.$$

Therefore, by ratio test, the series is convergent.

Example 14. *Test for convergence the series*

$$\frac{1}{2} + \frac{1.3}{2.5} + \frac{1.3.5}{2.5.8} + \ldots\ldots$$

Sol. Here

$$a_n = \frac{1.3.5 \ldots\ldots (2n-1)}{2.5.8 \ldots\ldots (3n-1)}$$

and $$a_{n+1} = \frac{1.3.5......(2n-1)(2n+1)}{2.5.8......(3n-1)(3n+2)}$$

$$\therefore \quad \frac{a_n}{a_{n+1}} = \frac{3n+2}{2n+1} \quad \therefore \quad \lim_{n\to\infty} \frac{a_n}{a_{n+1}} = \lim_{n\to\infty} \frac{3n+2}{2n+1}$$

$$= \lim_{n\to\infty} \frac{3+\frac{2}{n}}{2+\frac{1}{n}} = \frac{3}{2} > 1.$$

Therefore, by ratio test, the series is convergent.

Example 15. *Test the series*

$$1 + \frac{x^2}{2} + \frac{x^4}{4} + \frac{x^6}{6} + ...$$

for convergence for $x > 0$.

Sol. Neglecting first term, we have

$$a_n = \frac{x^{2n}}{2n} \text{ and } a_{n+1} = \frac{x^{2n+2}}{2n+2}$$

$$\lim_{n\to\infty} \frac{a_n}{a_{n+1}} = \lim_{n\to\infty} \left(\frac{x^{2n}}{2n} \cdot \frac{2n+2}{x^{2n+2}}\right)$$

$$= \lim_{n\to\infty} \left(\frac{2n+2}{2n} \cdot \frac{1}{x^2}\right) = \lim_{n\to\infty} \frac{2+\frac{2}{n}}{2} \cdot \frac{1}{x^2} = \frac{1}{x^2}.$$

Therefore, by ratio test, the given series is convergent for $\frac{1}{x^2} > 1$, *i.e.*, for $x^2 < 1$ and divergent for $\frac{1}{x^2} < 1$, *i.e.*, for $x^2 > 1$. Since $x > 0$, this implies that the series is convergent if $x < 1$ and divergent if $x > 1$. When $x = 1$, $\lim_{n\to\infty} \frac{a_n}{a_{n+1}} = 1$ and the test fails. But, in this case, the series becomes

$$1 + \frac{1}{2} + \frac{1}{4} + \frac{1}{6} + ...$$

Again neglecting first term, $a_n = \frac{1}{2n}$

Take $$b_n = \frac{1}{n}$$

Then, $\lim_{n\to\infty} \frac{a_n}{b_n} = \frac{1}{2}$, which is finite and non-zero.

Therefore, by the comparison test the series $\sum_{n=1}^{\infty} a_n$ and $\sum_{n=1}^{\infty} b_n$ converge or diverge together.

Since $\sum_{n=1}^{\infty} b_n$ is divergent, therefore, the given series is divergent.

Example 16. *Test the series*

$$1 + \frac{2^p}{2!} + \frac{3^p}{3!} + \frac{4^p}{4!} + \ldots\ldots \textit{ for convergence.}$$

Sol. Here,

$$a_n = \frac{n^p}{n!} \quad \text{and} \quad a_{n+1} = \frac{(n+1)^p}{(n+1)!}$$

$$\therefore \quad \lim_{n \to \infty} \frac{a_n}{a_{n+1}} = \lim_{n \to \infty} \left(\frac{n^p}{n!} \cdot \frac{(n+1)!}{(n+1)^p} \right)$$

$$= \lim_{n \to \infty} \left(\frac{n^p}{n!} \cdot \frac{n!}{(n+1)^{p-1}} \right) = \lim_{n \to \infty} n \cdot \left(\frac{n}{n+1} \right)^{p-1}$$

$$= \lim_{n \to \infty} n \cdot \frac{1}{\left(1 + \frac{1}{n}\right)^{p-1}} = \infty.$$

Therefore, by ratio test, the series is convergent.

EXERCISE 5 (*d*)

Test for convergence the following series :

1. $\frac{1}{5} + \frac{2!}{5^2} + \frac{3!}{5^3} + \ldots + \frac{n!}{5^n} + \ldots\ldots$ **2.** $1 + \frac{1}{2!} + \frac{1}{3!} + \frac{1}{4!} + \ldots + \frac{1}{n!} + \ldots\ldots$

3. $\sum_{n=1}^{\infty} \frac{\alpha^n}{n!}, \alpha > 0.$ **4.** $\frac{1}{2} + \frac{2!}{2^3} + \frac{3!}{2^5} + \frac{4!}{2^7} + \ldots\ldots$

5. $\sum_{a=1}^{\infty} \frac{5^n}{n^2 + 7}$ **6.** $\sum_{n=1}^{\infty} \frac{2^{n-1}}{3^n + 1}$ **7.** $\sum_{n=1}^{\infty} \frac{2^n n!}{n^n}$.

Discuss the convergence or divergence of the following series with $x > 0$.

8. $\frac{x}{1.3} + \frac{x^2}{3.5} + \frac{x^3}{5.7} + \ldots\ldots$

9. $\frac{1}{\sqrt{2}} + \frac{x}{2\sqrt{3}} + \frac{x^2}{3\sqrt{4}} + \frac{x^3}{4\sqrt{5}} + \ldots\ldots$

[**Hint.** Neglect first term.]

10. $1 + \frac{x}{1!} + \frac{x^2}{2!} + \frac{x^3}{3!} + \ldots\ldots$ *(K.U. BCA (II) 2001)*

11. $\frac{x}{1!} + \frac{x^3}{3!} + \frac{x^5}{5!} + \ldots\ldots$ **12.** $\sum_{n=1}^{\infty} \frac{\sqrt{n}}{\sqrt{n^2+1}} x^n$ **13.** $1 + 3x + 5x^2 + 7x^3 + 9x^4 + \ldots\ldots$

14. $1 + \frac{\alpha+1}{\beta+1} + \frac{(\alpha+1)(2\alpha+1)}{(\beta+1)(2\beta+1)} + \frac{(\alpha+1)(2\alpha+1)(3\alpha+1)}{(\beta+1)(2\beta+1)(3\beta+1)} + \ldots\ldots (\alpha > 0, \beta > 0)$

15. Suppose that the series $\sum_{n=1}^{\infty} a_n$ is shown to be convergent by the ratio test. Show that $\sum_{n=1}^{\infty} n\, a_n$ is also convergent. $\left[\text{Use } \lim_{n\to\infty} \frac{a_n}{a_{n+1}} > 1\right]$

16. Examine the convergent : $\frac{a}{3} + \frac{a^2}{6} + \frac{a^3}{11} + \ldots + \frac{a^n}{n^2+2} + \ldots\ldots$ $(a \geq 0)$.

*__17.__ Examine the convergent :

$$\frac{1}{3} + \frac{1.2}{3.5} + \frac{1.2.3}{3.5.7} + \ldots\ldots$$ *(K.U. BCA (II) 2001)*

5.10. CAUCHY'S ROOT TEST

5.10.1. Theorem (*a*) *Let* $\sum_{n=1}^{\infty} a_n$ *be a series of positive terms such that* $\lim_{n\to\infty} a_n^{\frac{1}{n}*} = l$.

Then,

(*i*) *the series is convergent if* $l < 1$,

(*ii*) *the series is divergent if* $l > 1$.

(*b*) *If* $\lim_{n\to\infty} a_n^{\frac{1}{n}} = \infty$, *then also the series is divergent.*

Proof. (*a*) **Case** (*i*) $\lim_{n\to\infty} a_n^{\frac{1}{n}} = l < 1$

Choose $\varepsilon > 0$ such that $l + \varepsilon < 1$.

Since $\lim_{n\to\infty} a_n^{\frac{1}{n}} = l$, therefore, there exists a positive integer m such that

$$\left| a_n^{\frac{1}{n}} - l \right| < \varepsilon \quad \forall\ n \geq m$$

$\Rightarrow$ $\quad l - \varepsilon < a_n^{\frac{1}{n}} < l + \varepsilon \quad \forall\ n \geq m$

Let $\quad r = l + \varepsilon < 1$. Then,

$$(l - \varepsilon)^n < a_n < r^n \quad \forall\ n \geq m \qquad \ldots(1)$$

Now, $\sum_{n=1}^{\infty} r^n$, being a geometric series with common ratio $r < 1$, is convergent. Therefore, from the second part of the inequality in (1) and using comparison test, the series $\sum_{n=1}^{\infty} a_n$ is convergent.

---

*Here $a_n^{\frac{1}{n}}$ stands for the positive nth root of a_n.

Case (*ii*) $\lim\limits_{n \to \infty} a_n^{\frac{1}{n}} = l > 1$

Choose $\varepsilon > 0$ such that $l - \varepsilon > 1$.

Since $\lim\limits_{n \to \infty} a_n^{\frac{1}{n}} = l$, therefore, there exists a positive integer m such that

$$\left| a_n^{\frac{1}{n}} - l \right| < \varepsilon \quad \forall \; n \geq m$$

$$\Rightarrow \qquad l - \varepsilon < a_n^{\frac{1}{n}} < l + \varepsilon \;\; \forall \; n \geq m.$$

Let $\quad r = l - \varepsilon > 1$. Then, $r^n < a_n < (l + \varepsilon)^n \;\; \forall \; n \geq m$ ...(1)

Now $\sum\limits_{n=1}^{\infty} r^n$, being a geometric series with common ratio $r > 1$, is divergent. Therefore, from first part of the inequality in (1) and using comparison test, $\sum\limits_{n=1}^{\infty} a_n$ is divergent.

(*b*) Since $\lim\limits_{n \to \infty} a_n^{\frac{1}{n}} = \infty$, therefore, there exists a positive integer m such that

$$a_n^{\frac{1}{n}} > 2 \;\; \forall \; n \geq m \;\; \Rightarrow \;\; a_n > 2^n \;\; \forall \; n \geq m$$

Now $\sum\limits_{n=1}^{\infty} 2^n$, being a geometric series with common ratio 2, is divergent. Therefore, by comparison test, the series $\sum\limits_{n=1}^{\infty} a_n$ is divergent.

Remark. If $\lim\limits_{n \to \infty} a_n^{\frac{1}{n}} = 1$, then no information can be obtained regarding the convergence or divergence of the series $\sum\limits_{n=1}^{\infty} a_n$.

For example, (*i*) let $a_n = \dfrac{1}{n}$.

Then, $\lim\limits_{n \to \infty} a_n^{\frac{1}{n}} = \lim\limits_{n \to \infty} \left(\dfrac{1}{n}\right)^{\frac{1}{n}} = 1$ and $\sum\limits_{n=1}^{\infty} a_n$ is divergent.

(*ii*) Let $a_n = \dfrac{1}{n^2}$.

Then, $\lim\limits_{n \to \infty} a_n^{\frac{1}{n}} = \lim\limits_{n \to \infty} \left(\dfrac{1}{n^n}\right)^{\frac{1}{n}} = \left(\lim\limits_{n \to \infty} \dfrac{1}{n^{1/n}}\right)^2 = 1$

and $\sum\limits_{n=1}^{\infty} a_n$ is convergent.

Thus, we find that $\lim\limits_{n \to \infty} a_n^{\frac{1}{n}} = 1$ in both the above examples and in one example, the series is convergent while in the other, the series is divergent. Hence, the test fails in this case.

Note. Cauchy's root test is more general than D' Alembert's ratio test in the sense that the former may succeed while the latter fails and the former must succeed whenever the latter succeeds.

For, by Cauchy's second theorem on limits, for a sequence $< a_n >$ of positive terms, if $\lim\limits_{n \to \infty} \frac{a_{n+1}}{a_n}$ exists, then $\lim\limits_{n \to \infty} a_n^{\frac{1}{n}}$ exists and both are equal. Thus, whenever ratio test is applicable, root test is also applicable.

Now, consider the series

$$\frac{1}{2} + \frac{1}{3} + \frac{1}{2^2} + \frac{1}{3^2} + \frac{1}{2^3} + \frac{1}{3^3} + \ldots\ldots$$

Here, $$a_{2n-1} = \frac{1}{2^n},\ a_{2n} = \frac{1}{3^n},\ a_{2n+1} = \frac{1}{2^{n+1}}$$

$$\therefore \quad \lim_{n \to \infty} \frac{a_{2n-1}}{a_{2n}} = \lim_{n \to \infty} \frac{3^n}{2^n} = \lim_{n \to \infty} \left(\frac{3}{2}\right)^n = \infty$$

and $$\lim_{n \to \infty} \frac{a_{2n}}{a_{2n+1}} = \lim_{n \to \infty} \frac{2^{n+1}}{3^n} = 2 \lim_{n \to \infty} \left(\frac{2}{3}\right)^n = 0 \qquad \left(\because \frac{2}{3} < 1\right)$$

Thus, $\lim\limits_{n \to \infty} \frac{a_n}{a_{n+1}}$ does not exist as we have different results for n even and odd.

$\therefore$ ratio test fails.

Let us try root test.

$$\lim_{n \to \infty} (a_{2n})^{\frac{1}{2n}} = \lim_{n \to \infty} \left(\frac{1}{3^n}\right)^{\frac{1}{2n}} = \frac{1}{\sqrt{3}} < 1$$

and $$\lim_{n \to \infty} (a_{2n+1})^{\frac{1}{2n+1}} = \lim_{n \to \infty} \left(\frac{1}{2^{n+1}}\right)^{\frac{1}{2n+1}} = \lim_{n \to \infty} \frac{1}{2^{\frac{n+1}{2n+1}}}$$

$$= \lim_{n \to \infty} \frac{1}{2^{\frac{1+\frac{1}{n}}{2+\frac{1}{n}}}} = \frac{1}{\sqrt{2}} < 1$$

Thus, $\lim\limits_{n \to \infty} a_n^{\frac{1}{n}} < 1$ whether n is even or odd.

$\therefore$ by root test, $\sum\limits_{n=1}^{\infty} a_n = \frac{1}{2} + \frac{1}{3} + \frac{1}{2^2} + \frac{1}{3^2} + \ldots\ldots$ is convergent.

Example 17. *Test for convergence the series*

$$\sum_{n=1}^{\infty} \frac{n^{n^2}}{(n+1)^{n^2}}.$$

Sol. Here $$a_n = \frac{n^{n^2}}{(n+1)^{n^2}} = \left(\frac{n}{n+1}\right)^{n^2}$$

$$\therefore \lim_{n\to\infty} a_n^{\frac{1}{n}} = \lim_{n\to\infty}\left(\frac{n}{n+1}\right)^{\frac{n^2}{n}} = \lim_{n\to\infty}\left(\frac{1}{1+\frac{1}{n}}\right)^n = \frac{1}{\lim_{n\to\infty}\left(1+\frac{1}{n}\right)^n} = \frac{1}{e} < 1$$

$\therefore$ by Cauchy's root test, the series is convergent.

Example 18. *Discuss the convergence of the series*

$$\sum_{n=1}^{\infty}\left[\frac{n+1}{n}\left\{\left(\frac{n+1}{n}\right)^n - 1\right\}\right]^{-n}.$$

Or

$$\left(\frac{2^2}{1^2}-\frac{2}{1}\right)^{-1} + \left(\frac{3^3}{2^3}-\frac{3}{2}\right)^{-2} + \left(\frac{4^4}{3^4}-\frac{4}{3}\right)^{-3} + \ldots\ldots$$

Sol. Here, $$a_n = \left[\frac{n+1}{n}\left\{\left(\frac{n+1}{n}\right)^n - 1\right\}\right]^{-n}$$

$$= \left(\frac{n+1}{n}\right)^{-n}\left\{\left(\frac{n+1}{n}\right)^n - 1\right\}^{-n} = \left(1+\frac{1}{n}\right)^{-n}\left\{\left(1+\frac{1}{n}\right)^n - 1\right\}^{-n}$$

$$\therefore \quad \lim_{n\to\infty} a_n^{1/n} = \lim_{n\to\infty}\left[\left(1+\frac{1}{n}\right)^{-1}\left\{\left(1+\frac{1}{n}\right)^n - 1\right\}^{-1}\right]$$

$$= \lim_{n\to\infty}\left(1+\frac{1}{n}\right)^{-1} . \lim_{n\to\infty}\left\{\left(1+\frac{1}{n}\right)^n - 1\right\}^{-1}$$

$$= (1+0)^{-1} . (e-1)^{-1} = \frac{1}{e-1} < 1 \qquad (\because \quad 2 < e < 3)$$

$\therefore$ by Cauchy's root test, the series $\sum_{n=1}^{\infty} a_n$ is convergent.

Example 19. *Test the convergence of* $\sum_{n=1}^{\infty}\left(1+\frac{1}{n}\right)^n x^n, x > 0.$

Sol. Here, $$a_n = \left(1+\frac{1}{n}\right)^n x^n.$$

$$\therefore \quad \lim_{n \to \infty} a_n^{1/n} = \lim_{n \to \infty} \left[\left(1+\frac{1}{n}\right)^n x^n\right]^{1/n} = \lim_{n \to \infty} \left(1+\frac{1}{n}\right) x = x$$

$\therefore$ by Cauchy's root test, the series is convergent if $x < 1$ and divergent if $x > 1$.

Case $x = 1$.

Now, Cauchy's root test fails. In this case, the series is $\sum_{n=1}^{\infty} \left(1+\frac{1}{n}\right)^n$.

$$\therefore \quad \lim_{n \to \infty} a_n = \lim_{n \to \infty} \left(1+\frac{1}{n}\right)^n = e \neq 0$$

$\therefore$ the series is divergent.

Hence, the series is convergent if $x < 1$ and divergent if $x \geq 1$.

EXERCISE 5 (*e*)

Using Cauchy's root test discuss the convergence of the series 1 – 7 for $x > 0$.

1. $\sum_{n=2}^{\infty} \frac{1}{(\log n)^n}$

2. $\sum_{n=1}^{\infty} \frac{[(n+1)x]^n}{n^{n+1}}$, $x > 0$

3. $\sum_{n=2}^{\infty} \frac{1}{[\log (\log n)]^n}$

4. $\sum_{n=1}^{\infty} \frac{x^n}{n^n}$

5. $\sum \left(1+\frac{1}{\sqrt{n}}\right)^{-n^{3/2}}$ $\quad\left[\textbf{Hint. } \lim_{n \to \infty} a_n^{1/n} = \lim_{n \to \infty} \left(1+\frac{1}{\sqrt{n}}\right)^{-\sqrt{n}} = \frac{1}{e} < 1\right]$

6. $\frac{1}{2} + \left(\frac{2}{3}\right) x + \left(\frac{3}{4}\right)^2 . x^2 + \left(\frac{4}{5}\right)^3 x^3 + \ldots\ldots$

7. $\left(\frac{2^2}{1^2} - \frac{2}{1}\right)^{-1} + \left(\frac{3^3}{2^3} - \frac{3}{2}\right)^{-2} + \left(\frac{4^4}{3^4} - \frac{4}{3}\right)^{-3} + \ldots\ldots$ *(K.U. BCA (II) 2001)*

8. (*a*) Using $e^x > x$ for $x > 0$, show that $\sum_{n=1}^{\infty} e^{-n^2}$ is convergent.

   (*b*) Test the series $\sum_{n=1}^{\infty} \frac{1}{e^{n^3}}$ for convergence. *(K.U. BCA (II) 2007)*

9. Assuming $\lim_{n \to \infty} \left(\frac{n^n}{n!}\right)^{1/n} = e$, show that $\sum_{n=1}^{\infty} \frac{n^n}{n!}$ diverges.

10. Show that $\sum_{n=1}^{\infty} 3^{-2n-5(-1)^n}$ is convergent.

$$\left[\textbf{Hint. } \lim_{n \to \infty} a_n^{1/n} = \lim_{n \to \infty} 3^{\frac{-2n-5(-1)^n}{n}} = \lim_{n \to \infty} 3^{-2} . 3^{\frac{-5(-1)^n}{n}} = \frac{1}{9}.\right]$$

5.11. RAABE'S TEST*

5.11.1. Theorem. *Let* $\sum_{n=1}^{\infty} a_n$ *be a series of positive terms such that*

$$\lim_{n\to\infty} n\left(\frac{a_n}{a_{n+1}} - 1\right) = l.$$

Then the series is

(i) convergent if $l > 1$ *(ii) divergent if* $l < 1$.

Proof. Case *(i)* $l > 1$

Choose a number p such that $l \geq p > 1$.

Take $b_n = \frac{1}{n^p}$ so that $\sum_{n=1}^{\infty} b_n$ is convergent.

∴ by comparison test 5.7.7, the series $\sum_{n=1}^{\infty} a_n$ is convergent

if $$\frac{a_n}{a_{n+1}} \geq \frac{b_n}{b_{n+1}} \quad \forall\ n$$

i.e., if $$\frac{a_n}{a_{n+1}} \geq \frac{(n+1)^p}{n^p}$$

i.e., if $$\frac{a_n}{a_{n+1}} \geq \left(1 + \frac{1}{n}\right)^p$$

i.e., if $$\frac{a_n}{a_{n+1}} \geq 1 + p\,.\,\frac{1}{n} + \frac{p(p-1)}{2!}\,\frac{1}{n^2} + \ldots\ldots$$

i.e., if $$n\left(\frac{a_n}{a_{n+1}} - 1\right) \geq p + \frac{p(p-1)}{2!}\,\frac{1}{n} + \ldots\ldots$$

(Taking limits on both sides as $n \to \infty$)

i.e., if $$\lim_{n\to\infty} n\left(\frac{a_n}{a_{n+1}} - 1\right) \geq \lim_{n\to\infty}\left(p + \frac{p(p-1)}{2!}\,\frac{1}{n} + \ldots\ldots\right)$$

i.e., if $$\lim_{n\to\infty} n\left(\frac{a_n}{a_{n+1}} - 1\right) \geq p$$

i.e., if $l \geq p > 1$, which is given. Hence the result.

Case *(ii)* $l < 1$

Choose a number p such that $l \leq p < 1$.

Take $b_n = \frac{1}{n^p}$ so that $\sum_{n=1}^{\infty} b_n$ is divergent.

---

*After the name of a German mathematician Raabe (1801–1859).

$\therefore$ by comparison test 5.7.7, the series $\sum_{n=1}^{\infty} a_n$ is divergent

if $$\frac{a_n}{a_{n+1}} \le \frac{b_n}{b_{n+1}} \quad \forall\ n$$

i.e., if $$\frac{a_n}{a_{n+1}} \le \frac{(n+1)^p}{n^p}$$ [Proceeding as in case (*i*)]

i.e., if $$\lim_{n \to \infty} n\left(\frac{a_n}{a_{n+1}} - 1\right) \le p$$

i.e., if $l \le p < 1$, which is given. Hence the result.

Remark. Raabe's test is inconclusive when

$$\lim_{n \to \infty} n\left(\frac{a_n}{a_{n+1}} - 1\right) = 1.$$

For example, consider the series $\sum_{n=1}^{\infty} \frac{1}{n}$.

Then, $$a_n = \frac{1}{n}$$

$$\therefore \quad \lim_{n \to \infty} n\left(\frac{a_n}{a_{n+1}} - 1\right) = \lim_{n \to \infty} n\left(\frac{n+1}{n} - 1\right) = \lim_{n \to \infty} n\left(\frac{1}{n}\right) = 1$$

and $\sum_{n=1}^{\infty} a_n$ is divergent.

Again consider the series $\sum_{n=2}^{\infty} \frac{1}{n(\log n)^2}$

(we shall see later that this series is convergent)

Here, $$a_n = \frac{1}{n(\log n)^2}$$

$$\therefore \quad \lim_{n \to \infty} n\left(\frac{a_n}{a_{n+1}} - 1\right) = \lim_{n \to \infty} n\left[\frac{(n+1)(\log(n+1))^2}{n(\log n)^2} - 1\right]$$

$$= \lim_{n \to \infty} \frac{(n+1)(\log(n+1))^2 - n(\log n)^2}{(\log n)^2}$$

$$= \lim_{n \to \infty} \frac{n(\log(n+1))^2 + (\log(n+1))^2 - n(\log n)^2}{(\log n)^2}$$

$$= \lim_{n \to \infty} \left[\left\{\frac{\log(n+1)}{\log n}\right\}^2 + \frac{n[(\log(n+1))^2 - (\log n)^2]}{(\log n)^2}\right]$$

$$= \lim_{n\to\infty}\left[\left[\frac{\log(n+1)}{\log n}\right]^2 + \frac{n\left[\log n\,(n+1)\log\dfrac{n+1}{n}\right]}{(\log n)^2}\right]$$

$$= \lim_{n\to\infty}\left[\left(\frac{\log(n+1)}{\log n}\right)^2 + \frac{n[\log n+\log(n+1)]\log\left(1+\dfrac{1}{n}\right)}{(\log n)^2}\right]$$

$$= \lim_{n\to\infty}\left[\frac{\log(n+1)}{\log n}\right]^2 + \lim_{n\to\infty}\frac{n\log n\,.\log\left(1+\dfrac{1}{n}\right)}{(\log n)^2}$$

$$+ \lim_{n\to\infty}\frac{n\log(n+1)\log\left(1+\dfrac{1}{n}\right)}{(\log n)^2}$$

$$= \left(\lim_{n\to\infty}\frac{\log(n+1)}{\log n}\right)^2 + \lim_{n\to\infty}\frac{\log\left(1+\dfrac{1}{n}\right)^n}{\log n}$$

$$+ \lim_{n\to\infty}\frac{\log(n+1)}{\log n}\,.\log\left(1+\frac{1}{n}\right)^n\cdot\frac{1}{\log n}$$

$$\left[\text{Now}\lim_{n\to\infty}\frac{\log(n+1)}{\log n}\right. \qquad \left(\frac{\infty}{\infty}\right)$$

$$\left.= \lim_{n\to\infty}\frac{\dfrac{1}{n+1}}{\dfrac{1}{n}} = \lim_{n\to\infty}\frac{n}{n+1} = 1,\ \lim_{n\to\infty}\frac{1}{\log n} = 0,\ \lim_{n\to\infty}\left(1+\frac{1}{n}\right)^n = e\right]$$

$$= 1 + 0 + 1\,.\,1\,.\,0 = 1.$$

Thus, $\lim\limits_{n\to\infty} n\left(\dfrac{a_n}{a_{n+1}} - 1\right) = 1$ in both the cases, and the series is convergent in one case and divergent in the second case.

Note. Raabe's test is more general then D' Alembert's ratio test. In fact, if a series $\sum\limits_{n=1}^{\infty} a_n$ is convergent by D' Alembert's ratio test, then

$$\lim_{n\to\infty}\frac{a_n}{a_{n+1}} > 1 \quad\Rightarrow\quad \lim_{n\to\infty} n\left(\frac{a_n}{a_{n+1}} - 1\right) = \infty$$

$\Rightarrow \sum\limits_{n=1}^{\infty} a_n$ is convergent by Raabe's test also.

If a series $\sum\limits_{n=1}^{\infty} a_n$ is divergent by D' Alembert's ratio test, then

$$\lim_{n\to\infty}\frac{a_n}{a_{n+1}} < 1 \;\Rightarrow\; \lim_{n\to\infty} n\left(\frac{a_n}{a_{n+1}} - 1\right) = -\infty$$

$\Rightarrow \sum_{n=1}^{\infty} a_n$ is divergent by Raabe's test.

Now, consider the series $\sum_{n=1}^{\infty} \frac{1}{n^2}$.

Here, $a_n = \frac{1}{n^2}$

$\therefore \lim_{n \to \infty} \frac{a_n}{a_{n+1}} = 1$

$\therefore$ ratio test fails.

But, $\lim_{n \to \infty} n\left(\frac{a_n}{a_{n+1}} - 1\right) = \lim_{n \to \infty} n\left[\left(\frac{n+1}{n}\right)^2 - 1\right] = \lim_{n \to \infty} \frac{2n+1}{n} = 2 > 1$

$\therefore$ by Raabe's test, the series $\sum_{n=1}^{\infty} a_n$ is convergent.

Hence, Raabe's test is conclusive if D' Alembert's ratio test is so and Raabe's test may be conclusive if D' Alembert's ratio test fails.

Example 20. *Test for convergence the series*

$$\sum_{n=1}^{\infty} \frac{1.3.5\ldots(2n-1)}{2.4.6\ldots(2n)} \cdot \frac{x^{2n}}{2n}, x > 0.$$

Sol. Here, $a_n = \frac{1.3.5\ldots(2n-1)}{2.4.6\ldots(2n)} \cdot \frac{x^{2n}}{2n}$

and $a_{n+1} = \frac{1.3.5\ldots(2n-1)(2n+1)}{2.4.6\ldots(2n)(2n+2)} \cdot \frac{x^{2n+2}}{2n+2}$

$\therefore \lim_{n \to \infty} \frac{a_n}{a_{n+1}} = \lim_{n \to \infty} \left(\frac{2n+2}{2n+1} \cdot \frac{2n+2}{2n} \cdot \frac{1}{x^2}\right) = \frac{1}{x^2}$

$\therefore$ by D' Alembert's ratio test, the series is convergent if $x^2 < 1$ and divergent if $x^2 > 1$. Since $x > 0$, this implies that the series is convergent if $x < 1$ and divergent if $x > 1$. If $x = 1$, then D' Alembert's ratio test fails.

Now, $\lim_{n \to \infty} n\left(\frac{a_n}{a_{n+1}} - 1\right) = \lim_{n \to \infty} n\left(\frac{2n+2}{2n+1} \cdot \frac{2n+2}{2n} - 1\right) = \lim_{n \to \infty} \frac{n(6n+4)}{2n(2n+1)} = \frac{3}{2} > 1$

by Raabe's test, the series is convergent.

Hence, the series is convergent if $x \le 1$ and divergent if $x > 1$.

Example 21. *Test the series*

$$1 + \frac{a}{1!} + \frac{a(a+1)}{2!} + \frac{a(a+1)(a+2)}{3!} + \ldots\ldots$$

for convergence. *(K.U. BCA (II) 2003)*

Sol. Neglecting first term, we have

$$a_n = \frac{a(a+1)(a+2)\ldots(a+n-1)}{n!}$$

and
$$a_{n+1} = \frac{a(a+1)(a+2)\ldots(a+n)}{(n+1)!}$$

$$\therefore \quad \frac{a_n}{a_{n+1}} = \frac{n+1}{a+n}$$

$$\therefore \quad \lim_{n\to\infty} \frac{a_n}{a_{n+1}} = \lim_{n\to\infty} \frac{n+1}{a+n} = \lim_{n\to\infty} \frac{1+\frac{1}{n}}{\frac{a}{n}+1} = 1$$

$\therefore$ the ratio test fails.

Let us try Raabe's test.

$$\lim_{n\to\infty} n\left(\frac{a_n}{a_{n+1}} - 1\right) = \lim_{n\to\infty} n\left(\frac{n+1}{a+n} - 1\right) = \lim_{n\to\infty} \frac{n(1-a)}{a+n} = \lim_{n\to\infty} \frac{1-a}{\frac{a}{n}+1} = 1-a$$

$\therefore$ by Raabe's test the series is convergent if $1 - a > 1$, *i.e.*, $a < 0$ and divergent if $1 - a < 1$, *i.e.*, $a > 0$.

If $1 - \alpha = 1$, *i.e.*, $a = 0$, the test fails.

However, in this case, the series become $1 + 0 + 0 + 0 + \ldots\ldots$ and is obviously convergent.

EXERCISE 5 (*f*)

Test for convergence the series, $(x > 0)$:

1. $\sum_{n=1}^{\infty} \frac{3.6.9\ldots 3n}{7.10.13\ldots(3n+4)} x^n$

2. $\sum_{n=1}^{\infty} \frac{1.3.5\ldots(2n-1)}{2.4.6\ldots 2n} \frac{x^{2n+1}}{2n+1}$

3. $x^2 + \frac{2^2}{3.4} x^4 + \frac{2^2.4^2}{3.4.5.6} x^6 + \frac{2^2.4^2.6^2}{3.4.5.6.7.8} x^8 + \ldots$

   $\left[\textbf{Hint.} \text{ For } n > 1,\ a_n = \frac{2^2.4^2\ldots\ldots(2n-2)^2}{3.4.5.6\ldots\ldots 2n} x^{2n}\right]$

4. $(\log 2)^p x^2 + (\log 3)^p x^3 + (\log 4)^p x^4 + \ldots\ldots$

5. (*i*) $1 + \frac{a}{b} + \frac{a(1+a)}{b(1+b)} + \ldots\ldots + \frac{a(1+a)\ldots\ldots(n+a)}{b(1+b)\ldots\ldots(n+b)} + \ldots\ldots\ (a > 0, b > 0)$

   (*ii*) $1 + \frac{a}{b} x + \frac{a(a+1)}{b(b+1)} \cdot \frac{x^2}{2!} + \frac{a(a+1)(a+2)}{b(b+1)(b+2)} \frac{x^3}{3!} + \ldots\ldots\ (a > 0, b > 0)$

[**Hint.** (*i*) Neglecting first term, we have

$$\lim_{n\to\infty} \frac{a_n}{a_{n+1}} = 1$$

$$\therefore \quad \lim_{n\to\infty} n\left(\frac{a_n}{a_{n+1}} - 1\right) = \lim_{n\to\infty} \frac{b-a}{1+\frac{1+a}{n}} = b - a$$

$\therefore$ by Raabe's test, $\sum_{n=1}^{\infty} a_n$ is convergent if $b - a > 1$ and divergent if $b - a < 1$.

Case. $b - a = 1$. Then, $a_n = \dfrac{a(1+a)\ldots\ldots[(n-1)+a]}{(1+a)(2+a)\ldots\ldots[(n-1)+a](n+a)} = \dfrac{a}{n+a}.$]

6. $\sum_{n=1}^{\infty} \dfrac{(n!)^2}{(2n)!} x^{2n}.$

5.12. LOGARITHMIC TEST*

5.12.1. Theorem. *Let $\sum_{n=1}^{\infty} a_n$ be a series of positive terms such that $\lim_{n \to \infty} n \log \dfrac{a_n}{a_{n+1}} = l$. Then, the series is convergent if $l > 1$ and divergent if $l < 1$.*

Proof. Proof is exactly on the lines on that of Raabe's test and is left as an exercise for the readers.

Remarks (*i*) The test fails if $\lim_{n \to \infty} n \log \dfrac{a_n}{a_{n+1}} = 1$.

(*ii*) The test is applied after the failure of ratio test and when in the ratio test, $\dfrac{a_n}{a_{n+1}}$ involves e.

Example 22. *Test for convergence the series*

$$1 + \frac{x}{2} + \frac{2!}{3^2} x^2 + \frac{3!}{4^3} x^3 + \ldots.$$

Sol. Neglecting first term, we have

$$a_n = \frac{n!}{(n+1)^n} . x^n$$

and
$$a_{n+1} = \frac{(n+1)!}{(n+2)^{n+1}} . x^{n+1}$$

$$\therefore \quad \lim_{n\to\infty} \frac{a_n}{a_{n+1}} = \lim_{n\to\infty} \frac{n!}{(n+1)^n} . \frac{(n+2)^{n+1}}{(n+1)!} . \frac{1}{x}$$

$$= \lim_{n\to\infty} \frac{(n+2)^{n+1}}{(n+1)^n . (n+1)} . \frac{1}{x} = \lim_{n\to\infty} \frac{\left(1+\frac{2}{n}\right)^{n+1}}{\left(1+\frac{1}{n}\right)^n . \left(1+\frac{1}{n}\right)} . \frac{1}{x}$$

$$= \lim_{n\to\infty} \frac{\left(1+\frac{2}{n}\right)^n . \left(1+\frac{2}{n}\right)}{\left(1+\frac{1}{n}\right)^n . \left(1+\frac{1}{n}\right)} . \frac{1}{x} = \frac{e^2 . 1}{e . 1} . \frac{1}{x} = \frac{e}{x}$$

$$\left[\because \lim_{n\to\infty} \left(1+\frac{a}{n}\right)^n = \lim_{n\to\infty} \left[\left(1+\frac{a}{n}\right)^{n/a}\right]^a = e^a\right]$$

*Not included in K.U.

$\therefore$ by D' Alembert's ratio test, the series is convergent if $\frac{e}{x} > 1$, *i.e.*, if $x < e$ and is divergent if $\frac{e}{x} < 1$, *i.e.*, if $x > e$.

If $x = e$, the ratio test fails.

Now, when $x = e$, $\quad \frac{a_n}{a_{n+1}} = \frac{\left(1+\frac{2}{n}\right)^{n+1}}{\left(1+\frac{1}{n}\right)^{n+1}} \cdot \frac{1}{e}$

Let us apply logarithmic test.

$$\therefore \quad \log \frac{a_n}{a_{n+1}} = (n+1)\log\left(1+\frac{2}{n}\right) - (n+1)\log\left(1+\frac{1}{n}\right) - \log e$$

$$= (n+1)\left[\log\left(1+\frac{2}{n}\right) - \log\left(1+\frac{1}{n}\right)\right] - 1$$

$$= (n+1)\left[\left(\frac{2}{n} - \frac{1}{2}\cdot\frac{4}{n^2} + \frac{1}{3}\cdot\frac{8}{n^3} - \ldots\ldots\right) - \left(\frac{1}{n} - \frac{1}{2n^2} + \frac{1}{3n^3} - \ldots\ldots\right)\right] - 1$$

$$= (n+1)\left(\frac{1}{n} - \frac{3}{2n^2} + \ldots\ldots\right) - 1 = -\frac{1}{2n} - \frac{3}{2n^2} + \ldots\ldots$$

$$\therefore \quad \lim_{n\to\infty} n \log \frac{a_n}{a_{n+1}} = \lim_{n\to\infty}\left(-\frac{1}{2} - \frac{3}{2n} + \ldots\ldots\right) = -\frac{1}{2} < 1$$

$\therefore$ by the logarithmic test, the series is divergent.

EXERCISE 5 (*g*)

Show that the series $\quad 1 + \frac{2x}{2!} + \frac{3^2x^2}{3!} + \frac{4^3 \cdot x^3}{4!} + \ldots\ldots$

is convergent if $x \le \frac{1}{e}$ and divergent if $x > \frac{1}{e}$.

5.13. DE MORGAN AND BERTRAND TEST*

5.13.1. Theorem. *Let* $\sum_{n=1}^{\infty} a_n$ *be a series of positive terms such that* $\lim_{n\to\infty}\left\{n\left(\frac{a_n}{a_{n+1}} - 1\right) - 1\right\}\log n = l.$

Then the series is

(i) convergent if $l > 1$ *and* *(ii) divergent if* $l < 1$.

*Not included in K.U.

Proof. Case (*i*) $l > 1$. Choose p such that $l > p > 1$.

Consider $\sum_{n=2}^{\infty} b_n$ where $b_n = \dfrac{1}{n\,(\log n)^p}$

We shall show later that $\sum_{n=2}^{\infty} b_n$ is convergent if $p > 1$ (See example 28)

Now, by comparison test, $\sum_{n=1}^{\infty} a_n$ is convergent if

$$\frac{a_n}{a_{n+1}} \geq \frac{b_n}{b_{n+1}}, \qquad \forall\, n \geq 2$$

i.e., if

$$\frac{a_n}{a_{n+1}} \geq \frac{(n+1)(\log(n+1))^p}{n\,(\log n)^p}, \qquad \forall\, n \geq 2$$

$$= \frac{n+1}{n} \cdot \left[\frac{\log(n+1)}{\log n}\right]^p, \qquad \forall\, n \geq 2$$

$$= \left(1+\frac{1}{n}\right) \cdot \left[\frac{\log n\left(1+\frac{1}{n}\right)}{\log n}\right]^p, \qquad \forall\, n \geq 2$$

$$= \left(1+\frac{1}{n}\right) \cdot \left[\frac{\log n + \log\left(1+\frac{1}{n}\right)}{\log n}\right]^p, \qquad \forall\, n \geq 2$$

$$= \left(1+\frac{1}{n}\right)\left[1+\frac{1}{\log n}\log\left(1+\frac{1}{n}\right)\right]^p, \qquad \forall\, n \geq 2$$

$$= \left(1+\frac{1}{n}\right)\left[1+\frac{1}{\log n}\left(\frac{1}{n}-\frac{1}{2n^2}+\ldots\ldots\right)\right]^p, \qquad \forall\, n \geq 2$$

$$\left(\because\ \log(1+x) = x - \frac{x^2}{2} + \frac{x^3}{3} - \ldots\ldots\right)$$

i.e., if

$$\frac{u_n}{a_{n+1}} \geq \left(1+\frac{1}{n}\right)\left[1+\frac{p}{\log n}\left(\frac{1}{n}-\frac{1}{2n^2}+\frac{1}{3n^3}\right)+\ldots\ldots\right], \qquad \forall\, n \geq 2$$

i.e., if

$$\frac{a_n}{a_{n+1}} > 1 + \frac{1}{n} + \frac{p}{n\log n}, \qquad \forall\, n \geq 2$$

neglecting higher powers of $1/n$

i.e., if

$$n\left(\frac{a_n}{a_{n+1}} - 1\right) > 1 + \frac{p}{\log n}, \qquad \forall\, n \geq 2$$

i.e., if

$$\lim_{n\to\infty}\left[n\left(\frac{a_n}{a_{n+1}} - 1\right) - 1\right]\log n > p > 1, \text{ which is given.}$$

Exactly on the same lines, one can show that the series $\sum_{n=1}^{\infty} a_n$ is divergent if $p < 1$.

Remark. De Morgan and Bertrand's test fails if $l = 1$.

Example 23. *Discuss the convergence of the series*

$$\frac{1^2}{2^2} + \frac{1^2 . 3^2}{2^2 . 4^2} x + \frac{1^2 . 3^2 . 5^2}{2^2 . 4^2 . 6^2} . x^2 + ..., x > 0.$$

Sol. Here,
$$a_n = \frac{1^2 . 3^2 . 5^2 ...... (2n-1)^2}{2^2 . 4^2 . 6^2 ...... (2n)^2} x^{n-1}$$

and
$$a_{n+1} = \frac{1^2 . 3^2 . 5^2 ...... (2n-1)^2 . (2n+1)^2}{2^2 . 4^2 . 6^2 ...... (2n)^2 . (2n+2)^2} x^n$$

$$\therefore \quad \frac{a_n}{a_{n+1}} = \frac{(2n+2)^2}{(2n+1)^2} . \frac{1}{x}$$

$$\therefore \quad \lim_{n \to \infty} \frac{a_n}{a_{n+1}} = \lim_{n \to \infty} \left(\frac{2n+2}{2n+1} \right)^2 . \frac{1}{x} = \frac{1}{x}$$

$\therefore$ by D' Alembert's ratio test, the series is convergent if $\frac{1}{x} > 1$, *i.e.*, $x < 1$ and divergent if $\frac{1}{x} < 1$, *i.e.*, $x > 1$.

If $x = 1$, the ratio test fails.

However, in this case,

$$\frac{a_n}{a_{n+1}} = \frac{(2n+2)^2}{(2n+1)^2} = \frac{4n^2 + 8n + 4}{4n^2 + 4n + 1}$$

$$\therefore \quad \frac{a_n}{a_{n+1}} - 1 = \frac{4n^2 + 8n + 4 - 4n^2 - 4n - 1}{4n^2 + 4n + 1}$$

$$\therefore \quad n\left(\frac{a_n}{a_{n+1}} - 1 \right) = n . \left(\frac{4n+3}{4n^2 + 4n + 1} \right)$$

$$\therefore \quad \lim_{n \to \infty} n\left(\frac{a_n}{a_{n+1}} - 1 \right) = \lim_{n \to \infty} \frac{4n^2 + 3n}{4n^2 + 4n + 1} = 1$$

$\therefore$ Raabe's test is also inconclusive.

Let us apply De Morgan and Bertrand's test.

Now,
$$n\left(\frac{a_n}{a_{n+1}} - 1 \right) - 1 = \frac{4n^2 + 3n - 4n^2 - 4n - 1}{4n^2 + 4n + 1}$$

$$\therefore \quad \left[n\left(\frac{a_n}{a_{n+1}} - 1 \right) - 1 \right] \log n = \left(\frac{-n-1}{4n^2 + 4n + 1} \right) \log n$$

$$\therefore \quad \lim_{n\to\infty}\left[n\left(\frac{a_n}{a_{n+1}}-1\right)-1\right]\log n = \lim_{n\to\infty}\left(\frac{-n^2-n}{4n^2+4n+1}\cdot\frac{\log n}{n}\right)$$

$$= -\frac{1}{4}\times 0 = 0 \qquad \left(\because \lim_{n\to\infty}\frac{\log n}{n}=0\right)$$

Hence, by De Morgan and Bertrand's test, the series is divergent.

EXERCISE 5 (h)

Using De Morgan and Bertrand's test, show that the series

$$1+\frac{a(1-a)}{1.2}+\frac{(1+a)a(1-a)(2-a)}{1^2\,.\,2^2}+\frac{(2+a)(1+a)a(1-a)(2-a)(3-a)}{1^2\,.\,2^2\,.\,3^2}+ \ldots\ldots$$ **is divergent.**

*5.14. MODIFIED FORMS FOR APPLICATIONS

Various tests of convergence, involving limits, can be modified in terms of limit **superior** and limit inferior. A few modification are given below :

1. **Comparison Tests**

Let $\sum_{n=1}^{\infty} b_n$ and $\sum_{n=1}^{\infty} c_n$ be two series of positive terms such that $\sum_{n=1}^{\infty} b_n$ **is convergent** and $\sum_{n=1}^{\infty} c_n$ is divergent. Then, a positive term series $\sum_{n=1}^{\infty} a_n$ is

convergent if $\overline{\lim_{n\to\infty}}\ \dfrac{a_n}{b_n}=\lambda\neq\infty$

and divergent if $\underline{\lim_{n\to\infty}}\ \dfrac{a_n}{c_n}=\mu\neq 0.$

2. Let $\sum_{n=1}^{\infty} a_n$ be a series of positive terms. Then the series $\sum_{n=1}^{\infty} a_n$ is

(*i*) convergent if $\underline{\lim_{n\to\infty}}\left(\dfrac{a_n}{a_{n+1}}\right)>1$

and divergent if $\overline{\lim_{n\to\infty}}\left(\dfrac{a_n}{a_{n+1}}\right)<1$ — D' Alembert's Ratio Test

(*ii*) convergent if $\overline{\lim_{n\to\infty}}\ a_n^{\frac{1}{n}}<1$

and divergent if $\overline{\lim_{n\to\infty}}\ a_n^{\frac{1}{n}}>1$ — Cauchy's Root Test

*May be omitted on first reading.

$$\left.\begin{aligned} &(iii) \text{ convergent if } \underline{\lim}_{n\to\infty}\, n\left(\frac{a_n}{a_{n+1}} - 1\right) > 1 \\ \text{and} \quad &\text{divergent if } \overline{\lim}_{n\to\infty}\, n\left(\frac{a_n}{a_{n+1}} - 1\right) < 1 \end{aligned}\right] \text{Raabe's Test}$$

$$\left.\begin{aligned} &(iv) \text{ convergent if } \underline{\lim}_{n\to\infty}\, n \log \frac{a_n}{a_{n+1}} > 1 \\ \text{and} \quad &\text{divergent if } \overline{\lim}_{n\to\infty}\, n \log \frac{a_n}{a_{n+1}} < 1 \end{aligned}\right] \text{Logarithmic Test}$$

$$\left.\begin{aligned} &(v) \text{ convergent if } \underline{\lim}_{n\to\infty} \left[n\left(\frac{a_n}{a_{n+1}} - 1\right) - 1\right] \log n > 1 \\ \text{and} \quad &\text{divergent if } \overline{\lim}_{n\to\infty} \left[n\left(\frac{a_n}{a_{n+1}} - 1\right) - 1\right] \log n < 1 \end{aligned}\right] \text{D' Morgan and Bertrand's Test}$$

The proofs of the above results are directly based on the definitions of the limit superior and limit inferior.

Proof. (1) Let $\lambda' > \lambda$ be any real number. Then, there exists a natural number k such that

$$\frac{a_n}{b_n} < \lambda', \quad \forall\, n \geq k$$

$$\Rightarrow \qquad a_n < \lambda' b_n, \quad \forall\, n \geq k \qquad \text{...}(i)$$

Since $\sum_{n=1}^{\infty} b_n$ is convergent, the sequence of its partial sums is bounded. Therefore Eq. (*i*) implies that the sequence of partial sums of $\sum_{n=1}^{\infty} a_n$ is also bounded and hence $\sum_{n=1}^{\infty} a_n$ is convergent.

Since the series are of positive terms and $\mu \neq 0$,

$\therefore \quad \mu > 0$. Take μ' such that $\mu > \mu' > 0$. Then, there exists a positive integer k such that

$$\frac{a_n}{c_n} > \mu', \quad \forall\, n \geq k$$

$$\Rightarrow \qquad a_n > \mu' c_n, \quad \forall\, n \geq k \qquad \text{...}(ii)$$

Since $\sum_{n=1}^{\infty} c_n$ is divergent and $c_n > 0$, the sequence of its partial sums is unbounded. Therefore, Eq. (*ii*) implies that the sequence of partial sums of $\sum_{n=1}^{\infty} a_n$ is unbounded. Hence $\sum_{n=1}^{\infty} a_n$ is divergent.

2. (*ii*) Let $\overline{\lim_{n \to \infty}} a_n^{\frac{1}{n}} < 1$

Choose $\rho > 0$ such that $\overline{\lim_{n \to \infty}} a_n^{\frac{1}{n}} < \rho < 1.$

Then, there exists a positive integer k such that

$$a_n^{\frac{1}{n}} < \rho, \quad \forall\, n \geq k$$

$$\Rightarrow \qquad a_n < \rho^n, \quad \forall\, n \geq k$$

Since $0 < \rho < 1$, $\sum_{n=1}^{\infty} \rho^n$, being a geometric series, is convergent and hence $\sum_{n=1}^{\infty} a_n$ is convergent.

Now, let $\overline{\lim_{n \to \infty}} a_n^{\frac{1}{n}} > 1$. Then, for infinitely many n, $a_n^{\frac{1}{n}} > 1$, *i.e.*, $a_n > 1$. This implies a_n does not tend to zero as $n \to \infty$. Hence, the series $\sum_{n=1}^{\infty} a_n$ is not convergent. Being a positive term series, $\sum_{n=1}^{\infty} a_n$ is divergent.

Proofs of others are left as an exercise to the reader.

Remark. The power of modified form is clear when we apply tests to series of the type :

$$1 + \frac{1}{2} + \frac{1}{2^2} + \frac{1}{2^4} + \frac{1}{2^5} + \frac{1}{2^7} + \frac{1}{2^8} + \frac{1}{2^{10}} + \frac{1}{2^{11}} + \ldots\ldots$$

$$1 + 2 + 2^2 + 2^4 + 2^5 + 2^7 + 2^8 + 2^{10} + 2^{11} + \ldots\ldots$$

and

$$1 + \frac{1}{2} + \frac{1}{3} + \frac{1}{2^2} + \frac{1}{3^2} + \frac{1}{2^3} + \frac{1}{3^3} + \ldots\ldots$$

$$1 + 2 + 3 + 2^2 + 3^2 + 2^3 + 3^3 + \ldots\ldots$$

We are now, going to prove a very powerful test—known as *Gauss's test*, which can be directly applied after the failure of D' Alembert's ratio test.

5.15. GAUSS'S TEST*

5.15.1. Theorem. *Let* $\sum_{n=1}^{\infty} a_n$ *be a series of positive terms such that*

$$\frac{a_n}{a_{n+1}} = 1 + \frac{\lambda}{n} + \frac{\alpha_n}{n^{1+\delta}}$$

where $\delta > 0$ *and* $< a_n >$ *is a bounded sequence. Then, the series* $\sum_{n=1}^{\infty} a_n$ *is*

(*i*) *convergent if* $\lambda > 1$ *and* (*ii*) *divergent if* $\lambda \leq 1$. (K.U. BCA (II) 2003)

Proof. Since $< \alpha_n >$ is a bounded sequence, there exists a positive number k such that

$$|\,\alpha_n\,| \leq k \quad \forall\, n \geq 1$$

---

*After the name of the German mathematician Gauss (1777—1855).

$$\therefore \quad \lim_{n\to\infty} \frac{|a_n|}{n^\delta} \le \lim_{n\to\infty} \frac{k}{n^\delta} = 0 \qquad \text{...(1)}$$

$$\therefore \quad \lim_{n\to\infty} \frac{a_n}{n^\delta} = 0$$

Now $\lim_{n\to\infty} n\left(\frac{a_n}{a_{n+1}} - 1\right) = \lim_{n\to\infty}\left(\lambda + \frac{\alpha_n}{n^\delta}\right) = \lambda + \lim_{n\to\infty} \frac{\alpha_n}{n^\delta} = \lambda.$ [By (1)]

$\therefore$ **by Raabe's test, the series is convergent if $\lambda > 1$ and divergent if $\lambda < 1$.**

If $\lambda = 1$, then

$$\lim_{n\to\infty}\left[n\left(\frac{a_n}{a_{n+1}} - 1\right) - 1\right]\log n = \lim_{n\to\infty} \frac{a_n \,.\log n}{n^\delta}$$

$$\le k \lim_{n\to\infty} \frac{\log n}{n^\delta}$$

$$= k \lim_{n\to\infty} \frac{1/n}{\delta \,.\, n^{\delta-1}} = \frac{k}{\delta} \lim_{n\to\infty} \frac{1}{n^\delta} = 0 < 1 \qquad \left(\frac{\infty}{\infty}\right)$$

$\therefore$ **by D' Morgan and Bertrand's test, the series is divergent.**

Remark. We, now discuss the case $\lambda = 1$ using comparison test.

Compare the series $\sum_{n=1}^{\infty} a_n$ with $\sum_{n=1}^{\infty} b_n$ where $b_n = \frac{1}{n \log n}$.

We have

$$\frac{\alpha_n}{a_{n+1}} - \frac{b_n}{b_{n+1}} = 1 + \frac{1}{n} + \frac{\alpha_n}{n^{1+\delta}} - \frac{(n+1)\log(n+1)}{n \log n}$$

$$= \frac{1}{n^{1+\delta}}\left[(n+1)\,n^\delta + a_n - \frac{(n+1)\,n^\delta \log(n+1)}{\log n}\right]$$

$$= \frac{1}{n^{1+\delta}}\left[a_n - (n+1)\log\left(1+\frac{1}{n}\right) . \frac{n^\delta}{\log n}\right] \qquad \text{...(1)}$$

But, $= \lim_{n\to\infty} (n+1)\log\left(1+\frac{1}{n}\right) = \lim_{n\to\infty}\left[\log\left(1+\frac{1}{n}\right)^n + \log\left(1+\frac{1}{n}\right)\right] = \log e + \log 1 = 1$

and $\lim_{n\to\infty} \frac{n^\delta}{\log n}\left(\frac{\infty}{\infty}\right) = \lim_{n\to\infty} \frac{\delta\, n^{\delta-1}}{\frac{1}{n}} = \lim_{n\to\infty} \delta \,.\, n^\delta = \infty$

$$\therefore \quad \lim_{n\to\infty} \frac{1}{n^{1+\delta}}\left[\alpha_n - (n+1)\log\left(1+\frac{1}{n}\right)\frac{n^\delta}{\log n}\right] = -\infty \qquad (\because \langle \alpha_n \rangle \text{ is bounded})$$

$\therefore$ $\exists$ **a positive integer m such that**

$$\alpha_n - (n+1)\log\left(1+\frac{1}{n}\right) . \frac{n^\delta}{\log n} < 0. \quad \forall\, n \ge m$$

$\therefore$ **from (1), for $n \ge m$,**

$$\frac{a_n}{a_{n+1}} - \frac{b_n}{b_{n+1}} < 0$$

$$\Rightarrow \qquad \frac{a_n}{a_{n+1}} < \frac{b_n}{b_{n+1}} \quad \forall\, n \geq m$$

$\Rightarrow$ by comparison test, the series $\sum_{n=1}^{\infty} a_n$ is divergent.

Cor. Gauss's test can also be stated as

If $a_n > 0\ \forall\, n$ and $\dfrac{a_n}{a_{n+1}} = 1 + \dfrac{\lambda}{n} + O\left(\dfrac{1}{n^{1+\delta}}\right)$ where $\delta > 0$ and $O\left(\dfrac{1}{n^{1+\delta}}\right)$ stands for the terms containing higher powers of $\dfrac{1}{n}$, then the series $\sum_{n=1}^{\infty} a_n$ is

(*i*) convergent if $\lambda > 1$

and (*ii*) divergent if $\lambda \leq 1$.

Note. For application of Gauss's test, expand

$$\frac{a_n}{a_{n+1}} \text{ in powers of } \frac{1}{n} \text{ as}$$

$$\frac{a_n}{a_{n+1}} = 1 + \frac{\lambda}{n} + \frac{\mu}{n^2} + \frac{v}{n^3} + \ldots\ldots = 1 + \frac{\lambda}{n} + O\left(\frac{1}{n^2}\right)$$

where $O\left(\dfrac{1}{n^2}\right)$ stands for terms of order $\dfrac{1}{n^2}$ and higher powers of $\dfrac{1}{n}$

and conclude the result. There is no need of applying first ratio and Raabe's test.

Example 24. *Test for convergence the series*

$$1 + \left(\frac{2}{3}\right)^p + \left(\frac{2\,.\,4}{3\,.\,5}\right)^p + \left(\frac{2\,.\,4\,.\,6}{3\,.\,5\,.\,7}\right)^p + \ldots\ldots$$

Sol. Neglecting first term, we have

$$a_n = \left[\frac{2\,.\,4\,.\,6 \ldots\ldots (2n)}{3\,.\,5\,.\,7 \ldots\ldots (2n+1)}\right]^p$$

and

$$a_{n+1} = \left[\frac{2\,.\,4\,.\,6 \ldots\ldots (2n)(2n+2)}{3\,.\,5\,.\,7 \ldots\ldots (2n+1)(2n+3)}\right]^p$$

$$\therefore \qquad \frac{a_n}{a_{n+1}} = \left(\frac{2n+3}{2n+2}\right)^p = \frac{\left(1+\dfrac{3}{2n}\right)^p}{\left(1+\dfrac{2}{2n}\right)^p} = \left(1+\frac{3}{2n}\right)^p \left(1+\frac{1}{n}\right)^{-p}$$

$$= \left[1 + p\,.\,\frac{3}{2n} + O\left(\frac{1}{n^2}\right)\right]\left[1 - \frac{p}{n} + O\left(\frac{1}{n^2}\right)\right]$$

$$= 1 + \left(\frac{3}{2} - 1\right)\frac{p}{n} + O\left(\frac{1}{n^2}\right) = 1 + \frac{\frac{1}{2}p}{n} + O\left(\frac{1}{n^2}\right)$$

by Gauss's test, the series is convergent if $\dfrac{p}{2} > 1$, *i.e.*, $p > 2$ and divergent if $\dfrac{p}{2} \leq 1$, *i.e.*, if $p \leq 2$.

Example 25. *Discuss the convergence of the series*

$$\frac{1^2}{2^2}+\frac{1^2.3^2}{2^2.4^2}x+\frac{1^2.3^2.5^2}{2^2.4^2.6^2}x^2+\ldots, x>0.$$

Sol.
$$a_n=\frac{1^2.3^2.5^2\ldots(2n-1)^2}{2^2.4^2.6^2\ldots(2n)^2}x^{n-1}$$

$$a_{n+1}=\frac{1^2.3^2.5^2\ldots(2n-1)^2\,(2n+1)^2}{2^2.4^2.6^2\ldots(2n)^2\,(2n+2)^2}x^n.$$

$$\therefore \quad \frac{a_n}{a_{n+1}}=\frac{(2n+2)^2}{(2n+1)^2}\frac{1}{x}.$$

$$\therefore \quad \lim_{n\to\infty}\frac{a_n}{a_{n+1}}=\lim_{n\to\infty}\left(\frac{2n+2}{2n+1}\right)^2.\frac{1}{x}=\frac{1}{x}.$$

$\therefore$ by ratio test, the given series is

convergent if $\frac{1}{x}>1$, *i.e.*, $x<1$

and divergent if $\frac{1}{x}<1$, *i.e.*, $x>1$.

If $x=1$, then

$$\frac{a_n}{a_{n+1}}=\frac{(2n+2)^2}{(2n+1)^2}=\left(1+\frac{1}{n}\right)^2\left(1+\frac{1}{2n}\right)^{-2}$$

$$=\left[1+\frac{2}{n}+O\left(\frac{1}{n^2}\right)\right]\left[1+(-2)\frac{1}{2n}+O\left(\frac{1}{n^2}\right)\right]=1+\frac{1}{n}+O\left(\frac{1}{n^2}\right)$$

Here $\lambda=1$

$\therefore$ by Gauss's test, the series is divergent.

Example 26. *If x, α, β, γ are all positive, discuss the convergence of hypergeometric series*

$$1+\frac{\alpha.\beta}{1.\gamma}.x+\frac{\alpha(\alpha+1)\beta(\beta+1)}{1.2.\gamma.(\gamma+1)}x^2+\frac{\alpha(\alpha+1)(\alpha+2)\beta(\beta+1)(\beta+2)}{1.2.3.\gamma(\gamma+1)(\gamma+2)}x^3+\ldots\ldots$$

Sol. Since x, α, β, γ are all positive, the given series is a series of positive terms.

Neglecting first term, we have

$$a_n=\frac{\alpha(\alpha+1)\ldots(\alpha+n-1).\beta(\beta+1)\ldots(\beta+n-1)}{1.2\ldots n.\gamma.(\gamma+1)\ldots(\gamma+n-1)}.x^n$$

and
$$a_{n+1}=\frac{\alpha(\alpha+1)\ldots(\alpha+n-1)(\alpha+n).\beta(\beta+1)\ldots(\beta+n-1)(\beta+n)\,x^{n+1}}{1.2\ldots n.(n+1).\gamma(\gamma+1)\ldots(\gamma+n-1)(\gamma+n)}$$

$$\therefore \quad \frac{a_n}{a_{n+1}}=\frac{(n+1)(\gamma+n)}{(\alpha+n)(\beta+n)}.\frac{1}{x}$$

$$\therefore \quad \lim_{n\to\infty}\frac{a_n}{a_{n+1}}=\lim_{n\to\infty}\frac{(n+1)(\gamma+n)}{(\alpha+n)(\beta+n)}\frac{1}{x}=\lim_{n\to\infty}\frac{\left(1+\frac{1}{n}\right)\left(\frac{\gamma}{n}+1\right)}{\left(\frac{\alpha}{n}+1\right)\left(\frac{\beta}{n}+1\right)}.\frac{1}{x}=\frac{1}{x}$$

$\therefore$ by D' Alembert's ratio test, the series is convergent if $\frac{1}{x} > 1$, *i.e.*, $x < 1$ and divergent if $\frac{1}{x} < 1$, *i.e.*, if $x > 1$.

When $x = 1$, the ratio test fails.

In this case,

$$\frac{a_n}{a_{n+1}} = \frac{(n+1)(n+\gamma)}{(n+\alpha)(n+\beta)} = \frac{\left(1+\frac{1}{n}\right)\left(\frac{\gamma}{n}+1\right)}{\left(1+\frac{\alpha}{n}\right)\left(1+\frac{\beta}{n}\right)}$$

$$= \left[\left(1+\frac{1}{n}\right)\left(1+\frac{\gamma}{n}\right)\right]\left(1+\frac{\alpha}{n}\right)^{-1}\left(1+\frac{\beta}{n}\right)^{-1}$$

$$= \left[1+(1+\gamma)\frac{1}{n}+O\left(\frac{1}{n^2}\right)\right]\left[1-\frac{\alpha}{n}+O\left(\frac{1}{n^2}\right)\right]$$

$$\times\left[1-\frac{\beta}{n}+O\left(\frac{1}{n^2}\right)\right] = 1+\frac{1+\gamma-\alpha-\beta}{n}+O\left(\frac{1}{n^2}\right)$$

$\therefore$ by Gauss's test, the series is convergent if $1+\gamma-\alpha-\beta > 1$ and divergent if $1+\gamma-\alpha-\beta \le 1$, *i.e.*, the series is convergent if $\gamma > \alpha + \beta$ and divergent if $\gamma \le \alpha + \beta$.

EXERCISE 5 (*i*)

Discuss the convergence of the following series, $x > 0$:

1. $\frac{1}{2}+\frac{1.3}{2.4}+\frac{1.3.5}{2.4.6}+\ldots\ldots$

2. $1+\left(\frac{1}{2}\right)^{\alpha}+\left(\frac{1.3}{2.4}\right)^{\alpha}+\left(\frac{1.3.5}{2.4.6}\right)^{\alpha}+\ldots.$

3. $x+\frac{1}{2}\cdot\frac{x^3}{3}+\frac{1.3}{2.4}\cdot\frac{x^5}{5}+\ldots..$

4. $1+\frac{3}{7}x+\frac{3.6}{7.10}x^2+\frac{3.6.9}{7.10.13}x^3+\ldots..$

5. $1+\frac{x^2}{2}+\frac{x^4}{4}+\frac{x^6}{6}+\ldots\ldots$

6. $\frac{1^2}{2^2}+\frac{1^2.3^2}{2^2.4^2}+\frac{1^2.3^2.5^2}{2^2.4^2.6^2}+\ldots\ldots$

7. $\sum_{n=1}^{\infty}\frac{(n!)^2}{(2n)!}x^n$

8. $\sum_{n=1}^{\infty}\frac{n!}{x(x+1)(x+2)\ldots(x+n-1)}$.

We shall now, give below a test in which a '*thin*' subsequence of $< a_n >$ determines the convergence or divergence of $\sum_{n=1}^{\infty} a_n$.

5.16. CAUCHY'S CONDENSATION TEST

5.16.1. Theorem. *If $a_n \ge a_{n+1} \ge 0$ for each n, then the series $\sum_{n=1}^{\infty} a_n$ and $\sum_{n=1}^{\infty} 2^n a_{2^n}$ converge or diverge together.*

Proof. Let $S_n = a_1 + a_2 + \ldots\ldots + a_n$

and $\sigma_n = 2a_2 + 4a_4 + \ldots\ldots. + 2^n . a_{2^n}$ denote, respectively, the nth partial sums of the two series

$$\sum_{n=1}^{\infty} a_n \quad \text{and} \quad \sum_{n=1}^{\infty} 2^n a_{2^n}.$$

Case (*i*) Suppose that $\sum_{n=1}^{\infty} 2^n a_{2^n}$ is convergent.

Then, the sequence $< \sigma_n >$ is bounded. ...(1)

Now, $S_n = a_1 + a_2 + a_3 + \ldots.. + a_n$

$= a_1 + a_2 + a_3 + a_4 + a_5 + a_6 + a_7 + \ldots.. + a_{2^k} + \ldots.. + a_{2^{k+1}-1} + a_{2^{k+1}},$

(where $n = 2^{k+1}$ so that as $n \to \infty$, $k \to \infty$)

$= a_1 + (a_2 + a_3) + (a_4 + a_5 + a_6 + a_7) + \ldots\ldots + (a_{2^k} + \ldots\ldots. + a_{2^{k+1}-1}) + a_{2^{k+1}}$

$\le 2a_1 + 2a_2 + 4a_4 + \ldots.. + 2^k a_{2^k}$

$(\because\ a_{n+1} \le a_n \quad \forall\ n$

$\therefore\ a_3 \le a_2$, etc. and $a_{2^{k+1}} \le a_1) = 2a_1 + \sigma_k$ ...(2)

Thus, $S_n \le 2a_1 + \sigma_k$

Therefore, from (1) and (2), we have

the sequence $< S_n >$ is bounded. Hence, the series $\sum_{n=1}^{\infty} a_n$ is convergent.

Case (*ii*) Suppose that $\sum_{n=1}^{\infty} 2^n a_{2^n}$ is divergent.

Then, the sequence $< \sigma_n >$ diverges to ∞ ...(3)

Now, $S_{2^n} = a_1 + a_2 + a_3 + a_4 + \ldots\ldots + a_{2^{n-1}+1} + \ldots + a_{2^n}$

$\ge a_2 + a_3 + a_4 + \ldots + a_{2^{n-1}+1} + \ldots + a_{2^n}$

$= a_2 + (a_3 + a_4) + \ldots + (a_{2^{n-1}+1} + \ldots + a_{2^n})$

$\ge a_2 + 2a_4 + \ldots + 2^{n-1} a_{2^n}$ $\quad (\because\ a_n \ge a_{n+1}\ \forall\ n\ \therefore\ a_3 \ge a_4$, etc.)

$= \frac{1}{2}(2a_2 + 4a_4 + \ldots + 2^n a_{2^n}) = \frac{1}{2}\sigma_n$

Thus, $S_{2^n} \ge \frac{1}{2}\sigma_n$ ...(4)

Therefore, from (3) and (4), we have, the sequence $< S_{2^n} >$ diverges to ∞.

So, the sequence $< S_n >$ is unbounded.

Hence, the series $\sum_{n=1}^{\infty} a_n$ is divergent.

Remark. In fact, if $a_n \geq a_{n+1} \geq 0 \ \forall \ n$ and k is any positive integer, then the series $\sum_{n=1}^{\infty} a_n$ and $\sum_{n=1}^{\infty} k^n \, a_{k^n}$ converge or diverge together.

Example 27. *Using Cauchy's condensation test, show that the series* $\sum_{n=1}^{\infty} \frac{1}{n^p}$ *is convergent if $p > 1$ and divergent if $p \leq 1$.*

Sol. Suppose that $p > 0$, because otherwise the series is obviously divergent.

Here, $$a_n = \frac{1}{n^p}$$

Since $\frac{1}{n^p} > \frac{1}{(n+1)^p} > 0 \ \forall \ n$, we have

$$a_n > a_{n+1} > 0$$

$\therefore$ by Cauchy's condensation test, the series $\sum_{n=1}^{\infty} a_n$ and $\sum_{n=1}^{\infty} 2^n a_{2^n}$ converge or diverge together.

Now, $$\sum_{n=1}^{\infty} 2^n \, a_{2^n} = \sum_{n=1}^{\infty} 2^n \cdot \frac{1}{(2^n)^p} = \sum_{n=1}^{\infty} \left(\frac{1}{2^{p-1}}\right)^n$$

is a geometric series with common ratio $\frac{1}{2^{p-1}}$ and is therefore convergent if $\frac{1}{2^{p-1}} < 1$, *i.e.*, if $p > 1$ and divergent if $p \leq 1$.

Hence, the given series is convergent, if $p > 1$ and divergent if $p \leq 1$.

Example 28. *Show that* $\sum_{n=2}^{\infty} \frac{1}{n(\log n)^p}$ *is convergent if $p > 1$ and divergent if $p \leq 1$.*

(K.U. BCA (II) 2006)

Sol. Case (*i*) $p \leq 0$. Let $p = -q$, $q \geq 0$

Then, $$\frac{1}{n(\log n)^p} = \frac{1}{n} \cdot (\log n)^q \geq \frac{1}{n} \text{ for } n \geq 3$$

Since $\sum_{n=1}^{\infty} \frac{1}{n}$ is divergent, therefore, by comparison test the given series is divergent.

Case (*ii*) $\quad p > 0$

Here, $$a_n = \frac{1}{n(\log n)^p}$$

Since $$\frac{1}{n(\log n)^p} > \frac{1}{(n+1)[\log(n+1)]^p} > 0$$

$\therefore$ $$a_n > a_{n+1} > 0 \ \ \forall \ n \geq 2$$

$\because$ by Cauchy's condensation test, the series $\sum_{n=2}^{\infty} a_n$ and $\sum_{n=2}^{\infty} 2^n a_{2^n}$ converge or diverge together.

$$\text{Now } \sum_{n=2}^{\infty} 2^n a_{2^n} = \sum_{n=2}^{\infty} 2^n . \frac{1}{2^n . (\log 2^n)^p}$$

$$= \sum_{n=2}^{\infty} \frac{1}{(n \log 2)^p} = \sum_{n=2}^{\infty} \frac{1}{n^p . (\log 2)^p} = \frac{1}{(\log 2)^p} . \sum_{n=2}^{\infty} \frac{1}{n^p}$$

which is convergent if $p > 1$ and divergent if $p \leq 1$.

Hence, the given series is convergent if $p > 1$ and divergent if $p \leq 1$.

Example 29. *Using Cauchy's condensation test, discuss the convergence of the series* $\sum_{n=3}^{\infty} \frac{1}{n \log n (\log \log n)}$.

Sol. Here, $a_n = \frac{1}{n \log n \log (\log n)}$

Since $\frac{1}{n \log n \log (\log n)} > \frac{1}{(n+1) \log (n+1) \log \log (n+1)}$ for $n \geq 3$

$\therefore$ $a_n > a_{n+1} > 0 \quad \forall\ n \geq 3.$

$\therefore$ by Cauchy's condensation test, the series $\sum_{n=2}^{\infty} a_n$ and $\sum_{n=2}^{\infty} 2^n a_{2^n}$ converge or diverge together.

$$\text{Now, } \sum_{n=3}^{\infty} 2^n a_{2^n} = \sum_{n=3}^{\infty} 2^n . \frac{1}{2^n . \log 2^n \log (\log 2^n)}$$

$$= \sum_{n=3}^{\infty} \frac{1}{\log 2^n \log (\log 2^n)} = \sum_{n=3}^{\infty} \frac{1}{(n \log 2) \log (n + \log 2)}$$

$$= \frac{1}{\log 2} \sum_{n=3}^{\infty} \frac{1}{n \log (n + \log 2)},$$

which for large n, behaves like $\sum_{n=3}^{\infty} \frac{1}{n \log n}$, which is divergent. (see above example)

Hence, the given series is divergent.

EXERCISE 5 (*j*)

Using Cauchy's condensation test, discuss the convergence of the following series :

1. $\sum_{n=2}^{\infty} \frac{1}{\log n}$
2. $\sum_{n=1}^{\infty} \left(\frac{\log n}{n}\right)^2$
3. $\sum_{n=1}^{\infty} \frac{\log n}{n}$
4. $\sum_{n=2}^{\infty} \frac{1}{(\log n)^p}$
5. $\sum_{n=3}^{\infty} \frac{1}{n \log n (\log \log n)^2}$.

5.17. CAUCHY'S INTEGRAL TEST

5.17.1. Theorem. *Let f be a continuous, non-negative and a decreasing function of x, for all $x \geq 1$.*

Let $a_n = f(n) \quad \forall \; n \geq 1.$

Then, the series $\sum_{n=1}^{\infty} a_n$ *and the sequence* $< I_n >$,

where $I_n = \int_1^n f(x)\, dx$, *both converge or diverge together.*

Proof. For a natural number n, choose a real number x such that

$$n + 1 \geq x > n$$

Since f is a decreasing function of x.

$$\therefore \qquad f(n + 1) \leq f(x) < f(n)$$

$$\Rightarrow \quad \int_n^{n+1} f(n + 1)\, dx \leq \int_n^{n+1} f(x)\, dx < \int_n^{n+1} f(n)\, dx$$

$$\Rightarrow \quad f(n + 1) \leq \int_n^{n+1} f(x)\, dx < f(n)$$

$$\Rightarrow \quad a_{n+1} \leq \int_n^{n+1} f(x)\, dx < a_n$$

Putting $n = 1, 2, 3, ..., n - 1$, we have

$$a_2 \leq \int_1^2 f(x)\, dx < a_1$$

$$a_3 \leq \int_2^3 f(x)\, dx < a_2$$

....

$$a_n \leq \int_{n-1}^n f(x)\, dx < a_{n-1}$$

Adding the above inequalities, we have

$$a_2 + a_3 + ... + a_n \leq \int_1^n f(x)\, dx < a_1 + a_2 + ... + a_{n-1} \qquad ...(1)$$

If $$S_n = a_1 + a_2 + \dots + a_n,$$
then from (1), we have

$$S_n - a_1 \le I_n < S_n - a_n \le S_n \qquad (\because \; a_n \ge 0)$$

$$\Rightarrow \quad S_n \le I_n + a_1 \qquad \dots(i)$$

and $$I_n < S_n \qquad \dots(ii)$$

Case (*i*) Let $\sum_{n=1}^{\infty} a_n$ be convergent.

Then, the sequence $< S_n >$ is bounded above.
Therefore, from (*ii*), I_n is bounded above.
Hence, the sequence $< I_n >$ is convergent.

Case (*ii*) Let $\sum_{n=1}^{\infty} a_n$ be divergent.

Then, the sequence $< S_n >$ is unbounded above.
Therefore, from (*i*) I_n is unbounded above. Hence the sequence $< I_n >$ is divergent.

Example 30. *Using integral test, test the behaviour of the series* $\sum_{n=1}^{\infty} \frac{1}{n^p}$, $p > 0$.

Sol. Consider the function $f(x) = \frac{1}{x^p}$, $x \ge 1$. Then, f is a non- negative and monotonically decreasing function.

$\therefore$ Cauchy's integral test is applicable.

Case (*i*) $p \ne 1$.

$$I_n = \int_1^n \frac{dx}{x^p} = \int_1^n x^{-p}\, dx = \left[\frac{x^{1-p}}{1-p}\right]_1^n = \frac{n^{1-p}}{1-p} - \frac{1}{1-p}$$

Let p > 1.

Then, $$\lim_{n\to\infty} I_n = \frac{1}{1-p} \lim_{n\to\infty} \frac{1}{n^{p-1}} - \frac{1}{1-p} = 0 - \frac{1}{1-p} = \frac{1}{p-1} \text{ (finite)}$$

$\Rightarrow$ the sequence $< I_n >$ converges $\Rightarrow \sum_{n=1}^{\infty} \frac{1}{n^p}$ converges

Let p < 1.

Then, $$\lim_{n\to\infty} I_n = \frac{1}{1-p} \lim_{n\to\infty} n^{1-p} - \frac{1}{1-p} = \infty - \frac{1}{1-p} = \infty.$$

$\Rightarrow$ the sequence $< I_n >$ diverges. $\Rightarrow \sum_{n=1}^{\infty} \frac{1}{n^p}$ diverges.

Case (*ii*) $p = 1$

$$I_n = \int_1^n \frac{dx}{x} = \Big[\log x\Big]_1^n = \log n - \log 1$$

$$\therefore \quad \lim_{n\to\infty} I_n = \lim_{n\to\infty} (\log n - \log 1) = \infty$$

$\Rightarrow$ the sequence $< I_n >$ diverges

$$\Rightarrow \sum_{n=1}^{\infty} \frac{1}{n^p} \text{ diverges}$$

Hence, $\sum_{n=1}^{\infty} \frac{1}{n^p}$ converges if $p > 1$ and diverges if $p \le 1$.

Example 31. *Using integral test, discuss the convergence of*

$$\sum_{n=2}^{\infty} \frac{1}{n(\log n)^p}, \; p > 0. \qquad \textit{(K.U. BCA (II) 2006)}$$

Sol. Consider $\quad f(x) = \frac{1}{x(\log x)^p}$ for $x \ge 2$.

Then, $\quad f(x) \ge 0$ and is a monotonically decreasing function of x.

$\therefore$ Cauchy's integral test is applicable.

Case (*i*) $p \neq 1$

$$I_n = \int_2^n (\log x)^{-p} \cdot \frac{1}{x}\, dx = \left[\frac{(\log x)^{1-p}}{1-p}\right]_2^n$$

$$= \frac{1}{1-p} [(\log n)^{1-p} - (\log 2)^{1-p}]$$

Let p > 1.

$$\text{Then,} \quad \lim_{n\to\infty} I_n = \frac{1}{1-p}\left[\lim_{n\to\infty} \frac{1}{(\log n)^{p-1}} - \frac{1}{(\log 2)^{p-1}}\right]$$

$$= \frac{1}{1-p}\left[0 - \frac{1}{(\log 2)^{p-1}}\right] = \frac{1}{(p-1)(\log 2)^{p-1}} \text{ (finite)}$$

$\Rightarrow$ the sequence $< I_n >$ converges

$$\Rightarrow \sum_{n=2}^{\infty} \frac{1}{n(\log n)^p} \text{ converges.}$$

Let p < 1.

$$\text{Then,} \quad \lim_{n\to\infty} I_n = \frac{1}{1-p}\left[\lim_{n\to\infty} (\log n)^{1-p} - (\log 2)^{1-p}\right] = \frac{1}{1-p} [\infty - (\log 2)^{1-p}] = \infty$$

$\Rightarrow$ the sequence $< I_n >$ diverges

$$\Rightarrow \sum_{n=2}^{\infty} \frac{1}{n(\log n)^p} \text{ diverges.}$$

Case (*ii*) $p = 1$

Then, $\quad I_n = \int_2^n \frac{dx}{x \log x} = \int_2^n \frac{\frac{1}{x}}{\log x}\, dx = \Big[\log(\log x)\Big]_2^n = \log(\log n) - \log(\log 2)$

$\therefore \quad \lim_{n\to\infty} I_n = \lim_{n\to\infty} \log(\log n) - \log(\log 2) = \infty - \log(\log 2) = \infty$

$\Rightarrow$ the sequence $< I_n >$ diverges.

$\Rightarrow \quad \sum_{n=2}^{\infty} \frac{1}{n \log n}$ diverges.

Hence, $\sum_{n=2}^{\infty} \frac{1}{n(\log n)^p}$ converges if $p > 1$ and diverges if $p \le 1$.

EXERCISE 5 (*k*)

Using the integral test, discuss the convergence of the following series :

1. $\sum_{n=1}^{\infty} \frac{1}{\sqrt{n}}$ **2.** $\sum_{n=1}^{\infty} \frac{1}{n^2+1}$ **3.** $\sum_{n=1}^{\infty} \frac{1}{(n+1)\log(n+1)}$

4. $\sum_{n=2}^{\infty} \frac{1}{n(\log n)^2}$ **5.** $\sum_{n=1}^{\infty} ne^{-n^2}$

6. $\sum_{n=3}^{\infty} \frac{1}{n \log n (\log \log n)^p}, \; p > 0.$

$\Big[$**Hint.** Here $f(x) = \frac{1}{x \log x (\log \log x)^p}, \quad \forall\, x \ge 3, p > 0.$

$f(x)$ is continuous, non-negative and decreasing.

Case (*i*) $p \ne 1$.

Then, $\quad I_n = \int_3^n f(x)\, dx = \int_3^n \frac{1}{x \log x (\log \log x)^p}\, dx$

$$= \int_3^n \frac{1}{x \log x} . (\log \log x)^{-p}\, dx = \left[\frac{\log(\log x)^{1-p}}{1-p}\right]_3^n$$

$$= \frac{1}{1-p}\left[(\log \log n)^{1-p} - (\log \log 3)^{1-p}\right]$$

Case (*ii*) $p = 1$.

Then, $\quad I_n = \int_3^n f(x)\, dx = \int_3^n \frac{1}{(x \log x)(\log \log x)}\, dx$

$$= \int_3^n \frac{\frac{1}{x \log x}}{\log \log x}\, dx = \Big[\log (\log \log x)\Big]_3^n$$

$$= \log (\log \log n) - \log (\log \log 3) \to \infty \text{ as } n \to \infty$$

Now, proceed as in the solved example 30.]

5.18. SERIES OF ARBITRARY TERMS

In the preceding section, we restricted the discussion of series to series of positive terms only. A series with all terms negative can also be handled, since the removal of the factor – 1 from each term of the series does not affect the behaviour of the series. Moreover, series in which only a finite number of terms are of one sign and the remaining ones are of a different sign, can also be handled by the preceeding theorems on series, as the behaviour of a series is not changed by the deletion of a finite number of terms of the series. With the exception of series for which the limit of the nth term is not zero, we are not in a position at present, to handle series having infinitely many terms of both signs (positive and negative). Such series are called *'series of arbitrary terms'* or *series with mixed signs.*

Some examples of such series are

$$1 - \tfrac{1}{2} + \tfrac{1}{3} - \tfrac{1}{4} \ \ldots\ldots ;$$

$$1 - 1 + 1 - 1 + 1 - 1 + \ \ldots\ldots ;$$

$$1 + \tfrac{1}{2} - \tfrac{1}{3} + \tfrac{1}{4} - \tfrac{1}{5} - \tfrac{1}{6} - \tfrac{1}{7} + \tfrac{1}{8} - \ldots\ldots ; \text{ etc.}$$

5.19. ALTERNATING SERIES

One of the simplest type among the series with mixed signs is the series with terms alternately positive and negative and is called an *alternating series.*

Thus, *a series of the type* $a_1 - a_2 + a_3 - a_4 + \ldots..$, *where* $a_n > 0$ *for each n, is an alternating series*. For example, the series.

$$1 - \tfrac{1}{2} + \tfrac{1}{3} - \tfrac{1}{4} \ \ldots\ldots \text{ is an alternating series.}$$

We give below a test for the convergence of the alternating series.

5.20. LEIBNITZ'S TEST*

5.20.1. Theorem. *The alternating series*

$$\sum_{n=1}^{\infty} (-1)^{n-1} a_n = a_1 - a_2 + a_3 - a_4 + \ldots.. \ (a_n > 0 \text{ for each } n) \text{ such that}$$

(i) $a_{n+1} \le a_n$, *for each n*

and *(ii)* $a_n \to 0$ *as* $n \to \infty$, *is convergent.*

Proof. Let S_n denote the nth partial sum of the series

$$\sum_{n=1}^{\infty} (-1)^{n-1} a_n.$$

*After the name of a German mathematician Leibnitz (1646—1716).

$\therefore \quad S_{2n} = a_1 - a_2 + a_3 - a_4 + \ldots.. + a_{2n-1} - a_{2n}$

$= a_1 - [(a_2 - a_3) + (a_4 - a_5) + \ldots + (a_{2n-2} - a_{2n-1}) + a_{2n}]$

$\leq a_1$ for each n $\quad (\because a_{n+1} \leq a_n$ and $a_n > 0 \ \forall \ n)$

$\therefore$ the sequence $< S_{2n} >$ is bounded above.

Also, $\quad S_{2n+2} - S_{2n} = a_{2n+1} - a_{2n+2} \geq 0 \ \forall \ n.$

$\Rightarrow \quad S_{2n+2} \geq S_{2n} \quad \forall \ n$

$\Rightarrow$ the sequence $< S_{2n} >$ is monotonically increasing.

Since every monotonically increasing bounded above sequence is convergent, therefore the sequence $< S_{2n} >$ is convergent.

Let $\lim\limits_{n \to \infty} S_{2n} = S.$

Since $\quad S_{2n+1} = S_{2n} + a_{2n+1}$

$\therefore \quad \lim\limits_{n \to \infty} S_{2n+1} = \lim\limits_{n \to \infty} (S_{2n} + a_{2n+1}) = \lim\limits_{n \to \infty} S_{2n} + \lim\limits_{n \to \infty} a_{2n+1} = S + 0$ [By (*ii*)]

$= S.$

Thus, the sequences, $< S_{2n} >$ and $< S_{2n+1} >$ converge to the same limit S.

We shall show that $< S_n >$ is also convergent to S, for each n.

Let $\varepsilon > 0$ be given.

Since $S_{2n} \to S$, $\exists$ a positive integer m_1 such that

$| S_{2n} - S | < \varepsilon$, whenever $2n \geq m_1$ ...(1)

Since $S_{2n+1} \to S$, $\exists$ a positive integer m_2 such that

$| S_{2n+1} - S | < \varepsilon$, whenever $2n + 1 \geq m_2$ ...(2)

Let $m = \max. \{m_1, m_2\}$.

From (1) and (2),

$| S_n - S | < \varepsilon$, whenever $n \geq m$.

Therefore, the sequence $< S_n >$ converges to S. Hence, the given series is convergent.

Example 32. *Test the convergence of the series*

$$\sum_{n=1}^{\infty} \frac{(-1)^{n-1}}{\sqrt{n}}.$$

Sol. The given series is an alternating series.

Here, $\quad a_n = \dfrac{1}{\sqrt{n}}.$

Since $\quad \dfrac{1}{\sqrt{n+1}} < \dfrac{1}{\sqrt{n}}$

$\therefore \quad a_{n+1} < a_n, \quad \forall \ n$

Also, $\quad a_n = \dfrac{1}{\sqrt{n}} \to 0$ as $n \to \infty$.

$\therefore$ by Leibnitz's test, the series is convergent.

Example 33. *Determine the behaviour of the series* $\sum_{n=1}^{\infty} \frac{(-1)^{n-1}\, n}{10n - 1}$.

Sol. The series is an alternating series.

Here $$a_n = \frac{n}{10n-1}$$

Now, $$a_{n+1} < a_n \quad \forall\ n$$

if $$\frac{n+1}{10n+9} < \frac{n}{10n-1}$$

i.e., if $$(n+1)(10n-1) < n(10n+9)$$

i.e., if $$10n^2 + 9n - 1 < 10n^2 + 9n$$

which is true.

But, the series does not converge because $a_n = \frac{n}{10n-1} \to \frac{1}{10} \neq 0$ as $n \to \infty$.

Remark. The condition $a_n \to 0$ as $n \to \infty$ is absolutely essential for the convergence of any series.

Example 34. *Show that the series* $\frac{\log 2}{2^2} - \frac{\log 3}{3^2} + \frac{\log 4}{4^2}$ ...... *converges.*

Sol. The series is an alternating series with $a_n = \frac{\log (n+1)}{(n+1)^2}$

Now, $$\lim_{n\to\infty} a_n = \lim_{n\to\infty} \frac{\log (n+1)}{(n+1)^2} \left(\frac{\infty}{\infty}\right) = \lim_{n\to\infty} \frac{\frac{1}{1+n}}{2(n+1)} = \lim_{n\to\infty} \frac{1}{2(n+1)^2} = 0$$

To show that $a_{n+1} \le a_n$, consider the function

$$f(x) = \frac{\log x}{x^2}$$

Then, $$f'(x) = \frac{x^2 \cdot \frac{1}{x} - 2x \cdot \log x}{x^4} = \frac{1 - 2\log x}{x^3}$$

$$< 0 \text{ if } 2\log x > 1$$

i.e., if $\log x > \frac{1}{2}$ *i.e.,* if $x > e^{1/2}$

$\therefore$ $f(x)$ is monotonically decreasing for $x > e^{1/2}$

In particular, $f(n+2) \le f(n+1)\ \forall\ n \ge 1$

i.e., $$a_{n+1} \le a_n \quad \forall\ n.$$

Hence, the series is convergent by Leibnitz's test.

5.21. ABSOLUTE AND CONDITIONAL CONVERGENCE

Recall that a series $\sum_{n=1}^{\infty} a_n$ is said to be absolutely convergent if the series $\sum_{n=1}^{\infty} |a_n|$ is convergent and a series $\sum_{n=1}^{\infty} a_n$ is said to be conditionally convergent (or non-absolutely convergent or semi-convergent) if it is convergent but not absolutely.

For example,

(*i*) the series $\sum_{n=1}^{\infty} \frac{(-1)^{n-1}}{n^2}$ is absolutely convergent because the series

$$\sum_{n=1}^{\infty} \left| \frac{(-1)^{n-1}}{n^2} \right| = \sum_{n=1}^{\infty} \frac{1}{n^2} \text{ is convergent.}$$

(*ii*) The series $\sum_{n=1}^{\infty} \frac{(-1)^{n-1}}{\sqrt{n}}$ is convergent, by Leibnitz's test. But,

$$\sum_{n=1}^{\infty} \left| \frac{(-1)^{n-1}}{\sqrt{n}} \right| = \sum_{n=1}^{\infty} \frac{1}{\sqrt{n}}$$

is not convergent. Hence, the given series is conditionally convergent.

Note. Since the series $\sum_{n=1}^{\infty} | a_n |$ is of positive terms, therefore, all the tests established for testing the convergence of series of positive terms, will also be the tests for determining the absolute convergence of the series $\sum_{n=1}^{\infty} a_n$.

The idea of absolute convergence is important because, it provides a mechanism for saying something about the convergence of a series with positive and negative terms, by considering the convergence of a series of positive terms. The relevant theorem is given in 5.21.1.

Significance of absolute convergence becomes more visible in the discussion of rearrangement of series in the next chapter—so much that we can operate on absolutely convergent series in the way we operate on sums of a finite number of terms, whereas in the case of conditionally convergent series, this is, in general, not possible.

5.21.1. Theorem. *Every absolutely convergent series is convergent.*

Proof. Let $\sum_{n=1}^{\infty} a_n$ be an absolutely convergent series

$$\Rightarrow \qquad \sum_{n=1}^{\infty} | a_n | \text{ is convergent.}$$

⇒ by Cauchy's general principle of convergence, that given any $\varepsilon > 0$, there exists a positive integer m such that

$$|\, | a_{m+1} | + | a_{m+2} | + \ldots\ldots + | a_n | \,| < \varepsilon, \quad \forall\ n > m$$

i.e.,

$$| a_{m+1} | + | a_{m+2} | + \ldots\ldots + | a_n | < \varepsilon, \quad \forall\ n > m \qquad \ldots(1)$$

By the triangle inequality,

$$\begin{aligned} | a_{m+1} + a_{m+2} + \ldots\ldots + a_n | & \\ \le | a_{m+1} | + | a_{m+2} | + \ldots\ldots + | a_n | & \\ < \varepsilon, \ \forall\ n > m & \qquad \text{[By (1)]} \end{aligned}$$

∴ by Cauchy's general principle of convergence, the series $\sum_{n=1}^{\infty} a_n$ is convergent.

Remark. The divergence of $\sum_{n=1}^{\infty} | a_n |$ does not imply the divergence of $\sum_{n=1}^{\infty} a_n$. For example, if $a_n = \dfrac{(-1)^{n-1}}{n}$, then $\sum_{n=1}^{\infty} | a_n |$ is divergent whereas $\sum_{n=1}^{\infty} a_n$ is convergent.

5.21.2. Theorem. *If $\sum_{n=1}^{\infty} a_n$ is an absolutely convergent series, then the series of its positive and the series of its negative terms are both convergent.*

Proof. Let $S_n = a_1 + a_2 + \ldots\ldots + a_n$

and $\sigma_n = | a_1 | + |a_2 | + \ldots\ldots + |a_n |$

Let P_n and $- Q_n$ denote the sum of positive and negative terms in S_n.

Then, $S_n = P_n - Q_n$ and $\sigma_n = P_n + Q_n$

$$\Rightarrow \quad \left.\begin{aligned} P_n &= \tfrac{1}{2}(\sigma_n + S_n) \\ \text{and} \quad Q_n &= \tfrac{1}{2}(\sigma_n - S_n) \end{aligned}\right\} \quad \ldots(1)$$

Since the series $\sum_{n=1}^{\infty} a_n$ is absolutely convergent, both S_n and σ_n tend to finite limits as $n \to \infty$.

Let $\lim\limits_{n \to \infty} S_n = l$ and $\lim\limits_{n \to \infty} \sigma_n = l'$

$\therefore$ from (1), $\lim\limits_{n \to \infty} P_n = \lim\limits_{n \to \infty} \frac{1}{2} (\sigma_n + S_n) = \frac{1}{2} (l' + l)$

and $\lim\limits_{n \to \infty} Q_n = \lim\limits_{n \to \infty} \frac{1}{2} (\sigma_n - S_n) = \frac{1}{2} (l' - l)$

Hence, the series of positive and negative terms are separately convergent.

Cor. 1. *If $\sum_{n=1}^{\infty} a_n$ is conditionally convergent, then the series of its positive terms and the series of its negative terms are both divergent.*

[For, let $S_n = a_1 + a_2 + \ldots\ldots + a_n$

and $\sigma_n = | a_1 | + | a_2 | + \ldots\ldots + | a_n |$

Now, $\sum_{n=1}^{\infty} a_n$ convergent $\Rightarrow$ $S_n \to$ a limit l (say)

and $\sum_{n=1}^{\infty} | a_n |$ divergent $\Rightarrow$ $\sigma_n \to \infty$

Since $P_n = \frac{1}{2} (\sigma_n + S_n)$ and $Q_n = \frac{1}{2} (\sigma_n - S_n)$

$\therefore$ $P_n \to \infty$ and $Q_n \to \infty$]

Cor. 2. *A series with mixed signs cannot converge, if the series of its positive terms is convergent (divergent) and the series of its negative terms is divergent (convergent).*

Example 35. *Discuss the convergence of the series*

$$1 + \frac{x}{1!} + \frac{x^2}{2!} + \ldots\ldots \text{ for all values of } x.$$

Sol. Here, $a_n = \dfrac{x^{n-1}}{(n-1)!}$ and $a_{n+1} = \dfrac{x^n}{n!}$

$$\therefore \quad \frac{|a_n|}{|a_{n+1}|} = \frac{|x|^{n-1}}{(n-1)!} \cdot \frac{n!}{|x|^n} = \frac{n}{|x|}, \text{ for } x \neq 0$$

$$\therefore \quad \lim_{n \to \infty} \frac{|a_n|}{|a_{n+1}|} = \lim_{n \to \infty} \frac{n}{|x|} = \infty \text{ for } x \neq 0.$$

$\therefore$ by the ratio test, the series $\sum_{n=1}^{\infty} |a_n|$ is convergent when $x \neq 0$. Thus, $\sum_{n=1}^{\infty} a_n$ is absolutely convergent. If $x = 0$, then the series becomes

$1 + 0 + 0 + \ldots\ldots$ and so is convergent.

Thus, the given series is absolutely convergent.

Since every absolutely convergent series is convergent therefore, the given series is convergent for every value of x.

Note. Since for a convergent series $\sum_{n=1}^{\infty} a_n$, $\lim_{n \to \infty} a_n = 0$.

$$\therefore \quad \lim_{n \to \infty} \frac{x^n}{n!} = 0 \text{ is a useful result.}$$

Example 36. *Discuss the convergence and absolute convergence of the series*

$$x - \frac{x^3}{3} + \frac{x^5}{5} - \ldots\ldots, \; x \text{ being real.}$$

Sol. Here, $a_n = (-1)^{n-1} \dfrac{x^{2n-1}}{2n-1}$ and $a_{n+1} = (-1)^n \dfrac{x^{2n+1}}{2n+1}$

$$\therefore \quad \frac{|a_n|}{|a_{n+1}|} = \frac{|x|^{2n-1}}{2n-1} \cdot \frac{2n+1}{|x|^{2n+1}}$$

$$\therefore \quad \lim_{n \to \infty} \frac{|a_n|}{|a_{n+1}|} = \lim_{n \to \infty} \frac{2n+1}{2n-1} \cdot \frac{1}{x^2} = \lim_{n \to \infty} \frac{2 + \frac{1}{n}}{2 - \frac{1}{n}} \cdot \frac{1}{x^2} = \frac{1}{x^2}.$$

$\therefore$ by ratio test, the series $\sum_{n=1}^{\infty} |a_n|$ is convergent if $\dfrac{1}{x^2} > 1$, *i.e.*, if $x^2 < 1$, *i.e.*, if $|x| < 1$ and divergent if $\dfrac{1}{x^2} < 1$, *i.e.*, $|x| > 1$.

If $|x| = 1$, then the series $\sum_{n=1}^{\infty} |a_n|$ is $1 + \dfrac{1}{3} + \dfrac{1}{5} + \ldots\ldots$ which is divergent as can be seen by comparison test.

Hence, the given series is not absolutely convergent when $|x| = 1$.

Since every absolutely convergent series is convergent, therefore, the given series is convergent if $|x| < 1$, *i.e.*, $-1 < x < 1$.

When $x = 1$, the series becomes $1 - \frac{1}{3} + \frac{1}{5} - \ldots\ldots$

which is

(*i*) an alternating series with $a_n = \frac{1}{2n-1}$

(*ii*) $a_{n+1} \le a_n \ \forall \ n$ $\left(\because \frac{1}{2n+1} < \frac{1}{2n-1} \ \forall \ n\right)$

and (*iii*) $a_n \to 0$ as $n \to \infty$.

Hence, by Leibnitz test, the series is convergent.

When $x = -1$, the series becomes $-1 + \frac{1}{3} - \frac{1}{5} + \ldots\ldots$ and is, again, convergent by Leibnitz's test.

When $x > 1$ or $x < -1$, a_n does not tend to zero as $n \to \infty$, therefore, the series $\sum_{n=1}^{\infty} a_n$ does not converge when $|x| > 1$.

Hence, (*i*) $\sum_{n=1}^{\infty} a_n$ converges if and only if $-1 \le x \le 1$.

(*ii*) $\sum_{n=1}^{\infty} a_n$ converges absolutely if and only if $|x| < 1$.

Example 37. *Show that the series* $\sum_{n=1}^{\infty} (-1)^{n-1} \sin \frac{1}{n}$ *is conditionally convergent.*

Sol. The given series is an alternating series with

$$a_n = \sin \frac{1}{n}$$

Here, $$\lim_{n \to \infty} a_n = \lim_{n \to \infty} \sin \frac{1}{n} = 0$$

and $$a_{n+1} - a_n = \sin \frac{1}{n+1} - \sin \frac{1}{n} < 0 \ \forall n \ge 1$$

i.e., $$a_{n+1} < a_n \ \forall \ n$$

$\therefore$ by Leibnitz's test, the series is convergent.

After making all the terms positive, the series become

$$\sum_{n=1}^{\infty} \sin \frac{1}{n}.$$

Here, $$a_n = \sin \frac{1}{n}$$

Take $$b_n = \frac{1}{n}$$

Then, $$\lim_{n\to\infty} \frac{a_n}{b_n} = \lim_{n\to\infty} \frac{\sin\frac{1}{n}}{\frac{1}{n}} = 1$$

which is finite and non-zero.

Since $\sum_{n=1}^{\infty} \frac{1}{n}$ is divergent, therefore, by comparison test, the given series is divergent.

Hence, the given series is convergent but not absolutely, *i.e.*, it is conditionally convergent.

Example 38. *Test for absolute convergence of*

$$\sum_{n=1}^{\infty} (-1)^{n+1} \frac{2^n}{n!}.$$

Sol. Here $$| a_n | = \frac{2^n}{n!} \quad \text{and} \quad | a_{n+1} | = \lim_{n\to\infty} \frac{2^{n+1}}{(n+1)!}$$

$$\therefore \quad \lim_{n\to\infty} \frac{|a_n|}{|a_{n+1}|} = \lim_{n\to\infty} \left(\frac{2^n}{n!} \frac{(n+1)!}{2^{n+1}}\right) = \lim_{n\to\infty} \frac{n+1}{2} = \infty.$$

Hence, by the ratio test, the series is absolutely convergent.

Example 39. *Test for absolute convergence*

$$\sum_{n=1}^{\infty} (-1)^n \frac{(2n)!}{n^{100}}.$$

Sol. Here $$| a_n | = \frac{2n!}{n^{100}} \quad \text{and} \quad | a_{n+1} | = \frac{(2n+2)!}{(n+1)^{100}}$$

$$\therefore \quad \lim_{n\to\infty} \frac{|a_n|}{|a_{n+1}|} = \lim_{n\to\infty} \frac{(2n)!}{n^{100}} \cdot \frac{(n+1)^{100}}{(2n+2)!}$$

$$= \lim_{n\to\infty} \left(\frac{n+1}{n}\right)^{100} \cdot \frac{1}{(2n+1)(2n+2)}$$

$$= \lim_{n\to\infty} \left(1+\frac{1}{n}\right)^{100} \cdot \frac{1}{(2n+1)(2n+2)} = 0.$$

Hence, by ratio test, the series $\sum_{n=1}^{\infty} | a_n |$ is divergent.

Example 40. *Find the values of x for which the series*

$$\sum_{n=1}^{\infty} \frac{(-1)^n (x+1)^n}{2^n n^2} \text{ converges.}$$

Sol. Here, $$a_n = \frac{(-1)^n (x+1)^n}{2^n n^2} \quad \text{and} \quad a_{n+1} = \frac{(-1)^{n+1} (x+1)^{n+1}}{2^{n+1} .(n+1)^2}$$

Therefore, $$\lim_{n\to\infty} \frac{|a_n|}{|a_{n+1}|} = \lim_{n\to\infty} \left(\frac{|x+1|^n}{2^n . n^2} \cdot \frac{2^{n+1} .(n+1)^2}{|x+1|^{n+1}}\right)$$

$$= \lim_{n \to \infty} \left(\frac{2}{|x+1|} \cdot \left(\frac{n+1}{n} \right)^2 \right) = \lim_{n \to \infty} \left(\frac{2}{|x+1|} \cdot \left(1 + \frac{1}{n} \right)^2 \right) = \frac{2}{|x+1|}$$

$\therefore$ by ratio test, the series is absolutely convergent (and hence convergent) if

$$\frac{2}{|x+1|} > 1,$$

i.e., $2 > | x + 1 |$, *i.e.*, $| x + 1 | < 2$

i.e., $-2 < x + 1 < 2 \Leftrightarrow -3 < x < 1$

The series is divergent if $\frac{2}{|x+1|} < 1$, *i.e.*, for $x < -3$ or $x > 1$.

When, $| x + 1 | = 2$, the ratio test fails.

But, this is the case when $x = -3$ and $x = 1$.

The corresponding series are

$$\sum_{n=1}^{\infty} \frac{(-1)^n (-2)^n}{2^n . n^2} = \sum_{n=1}^{\infty} \frac{1}{n^2} \quad \text{and} \quad \sum_{n=1}^{\infty} \frac{(-1)^n . 2^n}{2^n . n^2} = \sum_{n=1}^{\infty} \frac{(-1)^n}{n^2}$$

Both these series converge absolutely (and hence converge also).

Hence, the original series converges in the interval

$$-3 \le x \le 1.$$

5.22. WORKING PROCEDURE

For the guidance of the students we give below a working procedure for determining the convergence of a series, before we close the chapter.

(*i*) If in a series of a positive terms, *n*th term does not tend to zero, the series is divergent. If *n*th term tends to zero, then a comparison test may be applied when its *n*th term neither involves any power of x involving n nor involves factorials. If the *n*th term is the *n*th power of some expression, then Cauchy's root test may be applied. When the series involves increasing powers of x or involves factorials, one should start with the ratio test. If the $\lim_{n \to \infty} \frac{a_n}{a_{n+1}}$ turns out to be 1, then the ratio test fails and Raabe's test or Gauss's test is applied provided $\frac{a_n}{a_{n+1}}$ does not involve e, otherwise logarithmic test is applied.

(*ii*) For an arbitrary termed series, a good procedure is to try with the ratio test for absolute convergence. If the limit turns out to be 1, then try some other test. When the terms have alternating signs, then Leibnitz's test is suggested.

EXERCISE 5 (l)

Test the convergence and absolute convergence of the following series :

1. $1 - \dfrac{1}{2} + \dfrac{1}{3} - \dfrac{1}{4} + \dfrac{1}{5} - \ldots\ldots$

2. $\dfrac{1}{1.3} - \dfrac{1}{2.4} + \dfrac{1}{3.5} - \ldots\ldots$

3. $1 - \dfrac{1}{3!} + \dfrac{1}{5!} - \dfrac{1}{7!} - \ldots\ldots$

4. $\dfrac{1}{\log 2} - \dfrac{1}{\log 3} + \dfrac{1}{\log 4} - \dfrac{1}{\log 5} + \ldots\ldots$

5. $\dfrac{1}{5} - \dfrac{1}{7} + \dfrac{1}{9} - \ldots\ldots$

6. $1 - \dfrac{1}{2^3}(1+2) + \dfrac{1}{3^3}(1+2+3) - \ldots\ldots$

$$\left[\textbf{Hint. } a_n = \frac{1+2+3+\ldots\ldots+n}{n^3} = \frac{n(n+1)}{2n^3} = \frac{n+1}{2n^2}.\right]$$

7. $\left(\dfrac{1}{2}\right)^2 - \left(\dfrac{1.3}{2.4}\right)^2 + \left(\dfrac{1.3.5}{2.4.6}\right)^2 - \ldots\ldots$

8. $1 - \dfrac{1}{2^p} + \dfrac{1}{3^p} - \dfrac{1}{4^p} + \ldots\ldots,\ p > 0$

9. $\dfrac{1}{\sqrt{2}+1} - \dfrac{1}{\sqrt{3}+1} + \dfrac{1}{\sqrt{4}+1} - \ldots\ldots$

10. $\dfrac{1}{2(\log 2)^p} - \dfrac{1}{3(\log 3)^p} + \dfrac{1}{4(\log 4)^p} - \ldots\ldots$

11. $\displaystyle\sum_{n=1}^{\infty} \frac{(-1)^{n-1}}{n\sqrt{n}}$

12. $\displaystyle\sum_{n=1}^{\infty} (-1)^{n+1}\left(\sqrt{n^2-1} - n\right)$

13. $\displaystyle\sum_{n=1}^{\infty} \frac{(-1)^{n-1}\, n}{5^n}$ [**Hint.** Apply ratio test for absolute convergence.]

14. $\displaystyle\sum_{n=1}^{\infty} \frac{(-1)^{n+1}}{\operatorname{cosec}\left(\frac{\alpha}{n}\right)}$, α real.

15. $\displaystyle\sum_{n=1}^{\infty} \frac{(-1)^{n-1}\sin(n\alpha)}{n^2}$, α real.

$$\left[\textbf{Hint. } a_n = \frac{(-1)^{n-1}\sin(n\alpha)}{n^2} \Rightarrow |a_n| = \frac{|\sin(n\alpha)|}{n^2} \le \frac{1}{n^2} \qquad (\because\ |\sin n\alpha| \le 1)\right.$$

$$\left.\Rightarrow \sum_{n=1}^{\infty} |a_n| \text{ is convergent by comparison test.}\right]$$

16. $\displaystyle\sum_{n=1}^{\infty} \frac{(-1)^{n-1}\cos^2(n\alpha)}{n\sqrt{n}}$, α real. $\left[\textbf{Hint. } |a_n| = \dfrac{|\cos^2 n\alpha|}{n\sqrt{n}} \le \dfrac{1}{n\sqrt{n}} \text{ etc.}\right]$

17. $\displaystyle\sum_{n=1}^{\infty} (-1)^n \frac{n+2}{2^n+5}$ [**Hint.** Apply ratio test for absolute convergence.]

18. $\displaystyle\sum_{n=1}^{\infty} (-1)^{n-1} \frac{n}{5n-7}$

19. $\displaystyle\sum_{n=1}^{\infty} \frac{(-1)^{n-1}\, n^3}{(n+1)!}$

20. $\sum_{n=1}^{\infty} (-1)^{n-1} \left[\frac{1}{n^2} + \frac{1}{(n+1)^2}\right]$ **21.** $1 + x + \frac{x^2}{2!} + \ldots\ldots$

22. $1 - \frac{x^2}{2} + \frac{x^4}{4} - \frac{x^6}{6} + \ldots\ldots$ (Neglect first term)

23. Show that the series

$$\frac{\log 2}{2^3} - \frac{\log 3}{3^3} + \log \frac{4}{4^3} - \ldots\ldots \text{ is convergent.}$$

[**Hint.** This is an alternating series with $a_n = \frac{\log n}{n^3}$

$$\therefore \quad \lim_{n\to\infty} a_n = \lim_{n\to\infty} \frac{\log n}{n^3} \left(\frac{\infty}{\infty}\right) = \lim_{n\to\infty} \left(\frac{\frac{1}{n}}{3n^2}\right) = 0$$

Let $f(x) = \frac{\log x}{x^3} \Rightarrow f'(x) = \frac{1 - 3\log x}{x^4}$

$\therefore$ $f'(x) < 0$ if $x > e^{1/3}$

$\therefore$ $f(x)$ is a decreasing function of x $\forall\ x > e^{1/3}$.

In particular, $f(n+1) \geq f(n+2) \quad \forall\ n \in N$

$\Rightarrow$ $a_{n+1} \geq a_{n+2} \quad \forall\ n$ (Taking $a_1 = 0$)

Therefore by Leibnitz's test, the series is convergent.]

24. Show that the series $\sum_{n=1}^{\infty} \left(1 - \cos\frac{\pi}{n}\right)$ converges.

[**Hint.** $a_n = 1 - \cos\frac{\pi}{n} = 2\sin^2\frac{\pi}{2n}$.]

25. Use Leibnitz test to show that $\sum_{n=1}^{\infty} \frac{(-1)^n (n+5)}{n(n+1)}$ is convergent.

6

Limit and Continuity of Functions

The most important idea in Calculus is that of limit. The limit concept is at the foundation of almost all of mathematical analysis and an understanding of it is absolutely essential. In this chapter, we give some important results on limits. We shall deal with properties of continuous functions and uniform continuity.

6.1. PRELIMINARIES AND RESULTS ON LIMIT

(*i*) Let X by any set.

A function $f : X \to R$, *the set of real numbers, is called a* **real valued** *function.*

A function $f : A\ (\subseteq R) \to X$ *is called a function of a* **real variable.**

A function from a subset of R into R, is called a **real valued function** *of a* **real variable.**

In this chapter, we shall deal with only real valued functions of a real variable.

(*ii*) Let x be a real variable and a be a fixed finite number. Let x pass through an infinite number of values according to some rule. If the successive values approach a in such a way that $|\ x - a\ |$ becomes and remains smaller than any preassigned number $\varepsilon > 0$ (however small), then we say that '*x tends to a*' or '*x has the limit a*' and we write $x \to a$ or $\lim x = a$.

The definition implies that for each $\varepsilon > 0$, there exists a stage in the successive values of x such that all the values of x after this stage lie in the interval $(a - \varepsilon, a + \varepsilon)$. The definition does not imply that x must take the value a.

(*iii*) If x approaches to a taking all values less than a, then we say that *x tends to a from the left* and write $x \to a^-$.

(*iv*) If x approaches to a, taking all values greater than a, then we say that x tends to a from the right and write $x \to a^+$.

(*v*) If a real variable x takes successively values which ultimately become and remain greater than every positive number (however large), then *we say that x tends to plus infinity* and we write $x \to +\infty$ or just $x \to \infty$. That is, given any positive number k (however large), there exists a stage in the succession of values of x after which all values of x are larger than x.

(*vi*) If the successive values of x become and remain smaller than every negative number (however small), then *we say that x tends to minus infinity and write* $x \to \infty$.

6.1.1. Limit of a function. Let f be a function defined in some interval containing the point a, but may or may not be defined at a itself. We consider the behaviour of $f(x)$ as $x \to a$. It may happen that the values of f become closer and closer to a number l as $x \to a$, *i.e.*, the absolute value of the difference $f(x) - l$ can be made smaller than any pre-assigned positive number ε, however small, by taking x sufficiently close to a. In such a case, *we say that $f(x)$ approaches or converges or tends to the limit l as $x \to a$* and we write

$\lim_{x \to a} f(x) = l$ or $f(x) \to l$ as $x \to a$. Formally, we define :

Definition. *Let f be a function defined in a neighbourhood of a except possible at a. Then, a real number l is said to be a limit of f as x tends to a if given any $\varepsilon > 0$, however small, there exists $\delta > 0$ (depending upon ε) such that*

$$| f(x) - l | < \varepsilon, \text{ whenever } 0 < | x - a | < \delta.$$

i.e., $$l - \varepsilon < f(x) < l + \varepsilon, \text{ whenever } x \in (a - \delta, a) \cup (a, a + \delta)$$

We write $\lim_{x \to a} f(x) = l$ or $f(x) - l$ as $x \to a$.

Remarks 1. The limit of f at a, if exists, will continue to exist and be the same if we change the value of f at a only.

2. In order to show that $\lim_{x \to a} f(x) \neq l$, it is enough to produce one $\varepsilon > 0$, such that for each $\delta > 0$, there is some x for which

$$0 < | x - a | < \delta \text{ and } | f(x) - l | \geq \varepsilon.$$

6.1.2. Right hand and left hand limits. The above definition of limit implies that $f(x)$ approaches the same limit l irrespective of the manner in which x approaches a whether from right or from left. However, $f(x)$ may tend to the limit l as $x \to a$ from the right only, *i.e.*, $x \to a^+$, then we call this the right hand limit of f at $x = a$, *i.e, a real number l is said to be the right hand limit of f as $x \to a^+$ if* given any $\varepsilon > 0$, however small, there exists a $\delta > 0$ (depending upon ε) such that

$| f(x) - l | < \varepsilon$ whenever $a < x < a + \delta$ *i.e.*, $l - \varepsilon < f(x) < l + \varepsilon$ whenever $a < x < a + \delta$. We write $\lim_{x \to a^+} f(x) = l$ or $f(a^+) = l$. Likewise, a real number l is said to be the left hand limit of f as $x \to a^-$ if given any $\varepsilon > 0$, however small ; there exists a $\delta > 0$ (depending upon ε) such that

$$| f(x) - l | < \varepsilon \quad \text{whenever } a - \delta < x < a$$

i.e., $$l - \varepsilon < f(x) < l + \varepsilon \text{ ,whenever } a - \delta < x < a.$$

We write $\lim_{x \to a^-} f(x) = l$ or $f(a^-) = l$

Clearly, if $\lim_{x \to a} f(x) = l$, then both the left hand and right hand limits exist and each is equal to l. Conversely, if both the right hand and left hand limits exist and are equal, then it can be easily seen that $\lim_{x \to a} f(x)$ exists and is equal to them. Hence, a necessary and sufficient condition for limit of a function to exist is that

$$\lim_{x \to a^-} f(x) = \lim_{x \to a^+} f(x)$$

6.1.3. Theorem. *Limit of a function at a point, if exists, is unique.*

Proof. Let $\lim_{x \to a} f(x)$ exist.

Let, if possible, $f(x)$ tend to two different limits l and l' as $x \to a$.

Take $\varepsilon = \frac{1}{2} \mid l - l' \mid > 0$.

Since $f(x) \to l$ as $x \to a$, there exists $\delta_1 > 0$ such that

$$| f(x) - l | < \varepsilon \text{ whenever } 0 < | x - a | < \delta_1$$

Since $f(x) \to l'$ as $x \to a$ there exists $\delta_2 > 0$ such that

$$| f(x) - l' | < \varepsilon \text{ whenever } 0 < | x - a | < \delta_2$$

Let $\delta = \min. \{\delta_1, \delta_2\}$. Then

$$\begin{aligned} | l - l' | &= | l - f(x) + f(x) - l' | \\ &\le | f(x) - l | + | f(x) - l' | \\ &< \varepsilon + \varepsilon \text{ whenever } 0 < | x - a | < \delta \\ &= | l - l' | \end{aligned}$$

This is a contradiction.

Hence, $l = l'$.

Example 1. *Show by definition that* $\lim_{x \to 2} \frac{x^2 - 4}{x - 2} = 4$.

Sol. Here the function $f(x)$ is defined for $x \ne 2$.

Let $\varepsilon > 0$ be given

Now,
$$\begin{aligned} | f(x) - 4 | &= \left| \frac{x^2 - 4}{x - 2} - 4 \right| \\ &= | x + 2 - 4 | = | x - 2 | \end{aligned}$$

$\therefore$ we need to prove that

$$| x - 2 | < \varepsilon \quad \text{for } 0 < | x - 2 | < \delta$$

Take $\delta = \varepsilon$. Then $| f(x) - 4 | < \varepsilon$ whenever $0 < | x - 2 | < \delta$

Hence $\lim_{x \to 2} f(x) = 4$.

Example 2. *Show that the function* $f(x) = x \sin \frac{1}{x}$ *for* $x \ne 0$, *has limit 0 as* $x \to 0$.

Sol. Let $\varepsilon > 0$ be given

$$\begin{aligned} \left| x \sin \frac{1}{x} - 0 \right| &= \left| x \sin \frac{1}{x} \right| \\ &\le | x | \qquad \left(\because \left| \sin \frac{1}{x} \right| \le 1 \right) \\ &< \varepsilon \text{ if } | x | < \varepsilon \end{aligned}$$

$\therefore$ taking $\delta = \varepsilon$, we see that

$$\left| x \sin \frac{1}{x} - 0 \right| < \varepsilon, \text{ whenever } 0 < | x | < \delta$$

Hence $\lim_{x \to 0} x \sin \frac{1}{x} = 0$.

Example 3. *Discuss the behaviour as $x \to 2$, of the function $f(x)$, where*

$$f(x) = \begin{cases} 2x + 1 & \text{when } x < 2 \\ 3x + 5 & \text{when } x > 2. \end{cases}$$

Sol. Here $$f(2^-) = \lim_{x \to 2^-} (2x + 1) = 5$$

$$f(2^+) = \lim_{x \to 1^+} (3x + 5) = 11$$

Since $f(2^-) \neq f(2^+)$, $\lim_{x \to 2} f(x)$ does not exist.

6.1.4. Infinite Limits. Consider the behaviour of a function f which increases continuously as $x \to a$. If the values of f become and remain greater than any positive number, however large, for all x sufficiently close to a, we say that $f(x)$ increases beyond limit or that it tends to ∞. If as $x \to a$, $f(x)$ decreases continuously beyond any limit, we say that $f(x)$ tends to $-\infty$. Formally we define :

Definition. *A function f is said to tend to ∞ (or diverge to ∞) if given any positive number k however large, there exists a $\delta > 0$ such that $f(x) > k$ whenever $0 < | x - a | < \delta$.*

We write $$f(x) \to \infty \text{ as } x \to a \text{ or } \lim_{x \to a} f(x) = \infty.$$

A function f is said to tend to $-\infty$ (or diverges to $-\infty$) if given any positive number k, however large, there exists a $\delta > 0$ such that $f(x) < -k$ whenever $0 < | x - a | < \delta$.

We write $$f(x) \to -\infty \text{ as } x \to a \text{ or } \lim_{x \to a} f(x) = -\infty.$$

If a function f does not tend to a finite limit or to ∞ or $-\infty$, then (*i*) if it is bounded in a neighbourhood of a, it is said to *oscillate finitely*, (*ii*) if it is unbounded in a neighbourhood of a, it is said to *oscillate infinitely*.

Examples. (*i*) *Let $f(x) = \frac{1}{x}$, $x \neq 0$.*

Here $f(x)$ is not defined at $x = 0$.

Let $x \to 0^+$. As x becomes smaller and smaller, $\frac{1}{x}$ becomes larger and larger and crosses all bounds. In fact, if k is any positive number, however large, $\frac{1}{x} > k$ for $x < \frac{1}{k}$ $(= \delta)$.

Hence, whatever k is given, we can find a $\delta > 0$ such that $f(x) = \frac{1}{x} > k$ for $0 < x < \delta \left(= \frac{1}{k} \right)$.

Hence $$\lim_{x \to 0^+} f(x) = \infty$$

Similarly, $$\lim_{x \to 0^-} f(x) = -\infty$$

Since right hand limit is not equal to left hand limit, it follows that $\frac{1}{x}$ does not tend to any limit finite or infinite. As $\frac{1}{x}$ does not tend to any limit and is also not bounded in any neighbourhood of 0, therefore, $\frac{1}{x}$ oscillates infinitely as $x \to 0$.

(*ii*) Let $$f(x) = \frac{1}{x^2}, x \neq 0.$$

Here $$\lim_{x \to 0^+} f(x) = \infty = \lim_{x \to 0^-} f(x).$$

6.1.5. Limit of a function as $x \to \infty$ or $x \to -\infty$. *A function f is said to tend to the limit l as $x \to \infty$ if given $\varepsilon > 0$, however small, there exists a positive number k (depending upon ε) such that*

$$| f(x) - l | < \varepsilon \quad \forall\, x \geq k$$

We write $$\lim_{x \to +\infty} f(x) = l \quad \text{or} \quad f(x) \to l \text{ as } x \to \infty.$$

A function f is said to tend to $+\infty$ as x tends to ∞ if given any positive number k, however large, there exists a positive number k′ such that $f(x) > k \quad \forall\, x > k'$.

We write $$\lim_{x \to \infty} f(x) = \infty \quad \text{or} \quad f(x) \to \infty \text{ as } x \to \infty$$

In a similar way, we can define

$$\lim_{x \to -\infty} f(x) = l \quad \text{and} \quad \lim_{x \to -\infty} f(x) = \infty, \text{ etc.}$$

Examples. (*i*) Let $f(x) = \frac{1}{x}, x \neq 0$. Then $f(x) \to 0$ as $x \to \infty$

(*ii*) Let $f(x) = \frac{1}{x^2}, x \neq 0$.

Then
$$f(x) \to 0 \text{ as } x \to +\infty,$$
$$f(x) \to 0 \text{ as } x \to -\infty$$
and
$$f(x) \to +\infty \text{ as } x \to 0.$$

(*iii*) Let $$f(x) = -\frac{1}{x^2}, x \neq 0$$

Then $$f(x) \to -\infty \text{ as } x \to 0.$$

6.1.6. Algebraic operations of functions

Let $f : A \to R$ and $g : B \to R$ be two functions. Let $A \cap B \neq \phi$. We define on $A \cap B$ (*i.e.*, on the common portion of the domain of the functions f and g), a new function.

(*i*) $f + g$ called the *sum of the functions f and g* by

$$(f + g)(x) = f(x) + g(x), x \in A \cap B.$$

(*ii*) $f - g$ called the difference of two functions f and g by

$$(f - g)(x) = f(x) - g(x), x \in A \cap B.$$

(*iii*) *fg called the product of two functions f and g* by

$$(fg)(x) = f(x) \,.\, g(x), x \in A \cap B.$$

(*iv*) $\frac{f}{g}$ *called the quotient of the functions f and g* by

$$\left(\frac{f}{g}\right)(x) = \frac{f(x)}{g(x)}, x \in A \cap B \text{ provided } g(x) \neq 0 \text{ for } x \in A \cap B.$$

If c is a real number, then *the scalar product of f by c* is the function cf defined by

$$(cf)(x) = c \cdot f(x), \quad \forall\, x \in A.$$

If $g(x) \neq 0 \quad \forall\, x \in B$, *then the reciprocal of g is the function* $\frac{1}{g}$ defined on B by

$$\left(\frac{1}{g}\right)(x) = \frac{1}{g(x)}.$$

Remark. Product of two functions is different from the composite of two functions. For example, if $f(x) = \sin x$ and $g(x) = x^2$ then $(fog)(x) = f(g(x)) = \sin x^2$ whereas $(fg)(x) = f(x)\, g(x) = x^2 \sin x$.

We shall now study the relation between the limits of functions and the limits of their sum, product, etc.

6.1.7. Algebra of limits of functions : Theorem. *Let f and g be two functions defined in some deleted neighbourhood of a such that* $\lim_{x \to a} f(x) = l$ *and* $\lim_{x \to a} g(x) = m$.

Then,

(*i*) $\lim_{x \to a} (f+g)(x) = l + m$ (*ii*) $\lim_{x \to a} (f-g)(x) = l - m$

(*iii*) $\lim_{x \to a} (fg)(x) = l \cdot m$ (*iv*) $\lim_{x \to a} (cf)(x) = cl$

(*v*) $\lim_{x \to a} \left(\frac{1}{g}\right)(x) = \frac{1}{m}$ if $g(x) \neq 0, m \neq 0$

(*vi*) $\lim_{x \to a} \left(\frac{f}{g}\right)(x) = \frac{l}{m}$ > if $g(x) \neq 0, m \neq 0$.

Proof. (*i*) Let $\varepsilon > 0$ be given.

Now
$$\begin{aligned} |(f+g)(x) - (l+m)| &= |f(x) + g(x) - (l+m)| \\ &= |(f(x) - l) + g(x) - m| \\ &\leq |f(x) - l| + |g(x) - m| \qquad \ldots(1) \end{aligned}$$

We have to show that $|f(x) - l| + |g(x) - l| < \varepsilon$ for some δ.

$\lim_{x \to a} f(x) = l \Rightarrow \exists\, \delta_1 > 0$ such that

$$|f(x) - l| < \frac{\varepsilon}{2} \text{ whenever } 0 < |x - a| < \delta_1 \qquad \ldots(2)$$

$\lim_{x \to a} g(x) = m \Rightarrow \exists\, \delta_2 > 0$ such that

$$|g(x) - m| < \frac{\varepsilon}{2} \text{ whenever } 0 < |x - a| < \delta_2 \qquad \ldots(3)$$

Let $\delta = \min. \{\delta_1, \delta_2\} > 0$.

Using (2) and (3) in (1), we have

$$|(f+g)(x) - (l+m)| < \frac{\varepsilon}{2} + \frac{\varepsilon}{2} = \varepsilon \text{ whenever } 0 < |x - a| < \delta$$

$$\Rightarrow \qquad \lim_{x \to a} (f+g)(x) = l + m.$$

(*ii*) Proof is similar to (*i*).

(*iii*) Let $\varepsilon > 0$ be given.

Now $\quad | (fg)(x) - lm | = | f(x)\, g(x) - lm |$

$$= | f(x)\, g(x) - lg(x) + lg(x) - lm |$$

$$= | g(x)\, (f(x) - l) + l(g(x) - m) |$$

$$\leq | g(x) |\; | f(x) - l | + | l |\; | g(x) - m | \qquad \text{...(1)}$$

$$\lim_{x \to a} g(x) = m \;\Rightarrow\; \exists\, \delta_1 > 0 \text{ such that}$$

$$| g(x) - m | < 1,\; 0 < | x - a | < \delta_1$$

$$\Rightarrow \quad | g(x) | = |g(x) - m + m |$$

$$\leq | g(x) - m | + | m |$$

$$< 1 + | m |,\; 0 < | x - a | < \delta_1 \qquad \text{...(2)}$$

$$\lim_{x \to a} f(x) = l \;\Rightarrow\; \exists\, \delta_2 > 0 \text{ such that}$$

$$| f(x) - l | < \frac{\varepsilon}{2(1+|m|)},\; 0 < | x - a | < \delta_2 \qquad \text{...(3)}$$

$$\lim_{x \to a} g(x) = m \;\Rightarrow\; \exists\, \delta_3 > 0 \text{ such that}$$

$$| g(x) - m | < \frac{\varepsilon}{2(|l|+1)},\; 0 < | x - a | < \delta_3 \qquad \text{...(4)}$$

Let $\delta = \min. \{\delta_1, \delta_2, \delta_3\}$. Using (2) to (4) in (1), we have

$$| (fg)(x) - lm | < \frac{\varepsilon}{2}\frac{(1+|m|)}{1+|m|} + |l|\frac{\varepsilon}{2(|l|+1)} < \frac{\varepsilon}{2} + \frac{\varepsilon}{2} = \varepsilon,\; 0 < | x - a | < \delta$$

$$\Rightarrow \quad \lim_{x \to a} (fg)(x) = lm.$$

(*iv*) Proof is simple and is left as an exercise for the reader.

(*v*) Let $\varepsilon > 0$ be given

Now $$\left| \frac{1}{g(x)} - \frac{1}{m} \right| = \frac{|g(x) - m|}{|g(x)|\;|m|} \qquad \text{...(1)}$$

We have to show that $\dfrac{|g(x) - m|}{|g(x)|\;|m|} < \varepsilon$ for some δ.

Since $\quad \lim_{x \to a} g(x) = m,$

We can find $\delta_1 > 0$ such that

$$| g(x) - m | < \frac{|m|}{2},\; 0 < | x - a | < \delta_1$$

Now $\quad | m | = | m - g(x) + g(x) |$

$$\leq | m - g(x) | + | g(x) | < \frac{|m|}{2} + | g(x) |$$

$$\Rightarrow \quad | g(x) | > \tfrac{1}{2} | m |,\; 0 < | x - a | < \delta_1 \qquad \text{...(2)}$$

Again, $\lim_{x \to a} g(x) = m \;\Rightarrow\; \exists\, \delta_2 > 0$ such that

$$| g(x) - m | < \tfrac{1}{2} | m |^2 \varepsilon,\; 0 < | x - a | < \delta_2 \qquad \text{...(3)}$$

Let $\delta = \min. \{\delta_1, \delta_2\}$. Using (2) and (3) in (1), we have

$$\left|\frac{1}{g(x)} - \frac{1}{m}\right| < \frac{2|m|^2 \varepsilon}{2|m|.|m|} = \varepsilon, \; 0 < |x - a| < \delta$$

$$\Rightarrow \qquad \lim_{x \to a} \frac{1}{g(x)} = \frac{1}{m}.$$

(*vi*) By case (*v*) above,

$$\lim_{x \to a} g(x) = m, \; m \neq 0 \;\Rightarrow\; \lim_{x \to a} \frac{1}{g(x)} = \frac{1}{m}$$

Now $$\lim_{x \to a} f(x) = l \quad \text{and} \quad \lim_{x \to a} \frac{1}{g(x)} = \frac{1}{m}$$

$\therefore$ by case (*iii*),

$$\lim_{x \to a} \left\{f(x) . \frac{1}{g(x)}\right\} = l . \frac{1}{m} = \frac{1}{m}.$$

$$\Rightarrow \qquad \lim_{x \to a} \left(\frac{f}{g}\right)(x) = \frac{1}{m}.$$

Remark. $\lim_{x \to a} (f \pm g)(x)$, $\lim_{x \to a} (fg)(x)$ and $\lim_{x \to a} \left(\frac{f}{g}\right)(x)$ may exist even if neither $\lim_{x \to a} f(x)$ nor $\lim_{x \to a} g(x)$ exists.

For example, let f and g be defined as follows :

$$f(x) = \begin{cases} -1 & \text{if } x < a \\ 1 & \text{if } x > a \end{cases}$$

$$g(x) = \begin{cases} 1 & \text{if } x < a \\ -1 & \text{if } x > a \end{cases}$$

Then $(f + g)(x) = 0 \; \forall \; x \neq a$

and $(fg)(x) = -1 = \left(\frac{f}{g}\right)(x) \; \forall \; x \neq a$

$$\lim_{x \to a} (f + g)(x) = 0, \; \lim_{x \to a} (fg)(x) = -1 = \lim_{x \to a} \left(\frac{f}{g}\right)(x)$$

But $\lim_{x \to a^-} f(x) = -1$ and $\lim_{x \to a^+} f(x) = 1$

$\therefore$ $\lim_{x \to a} f(x)$ does not exist

Similarly, $\lim_{x \to a} g(x)$ does not exist.

Again, let f and g be defined as follows :

$$f(x) = \begin{cases} -1 & \text{if } x < a \\ 1 & \text{if } x > a \end{cases}$$

$$g(x) = \begin{cases} -1 & \text{if } x < a \\ 1 & \text{if } x > a \end{cases}$$

Then $\quad (f-g)(x) = 0 \quad \forall\, x \neq a$

$\Rightarrow \lim_{x \to a} (f-g)(x) = 0$, but $\lim_{x \to a} f(x)$ and $\lim_{x \to a} g(x)$ do not exist.

6.1.8. Theorem. *If $\lim_{x \to a} f(x) = l$, then $\lim_{x \to a} |f(x)| = |l|$.*

Proof. Let $\varepsilon > 0$ be given

$$\lim_{x \to a} f(x) = l \Rightarrow \exists\, \delta > 0 \text{ such that}$$

$$|f(x) - l| < \varepsilon,\ 0 < |x - a| < \delta$$

Now $\quad ||f(x)| - |l|| \leq |f(x) - l| < \varepsilon,\ 0 < |x - a| < \delta$

$$(\because \ |a - b| \geq ||a| - |b||)$$

$$\Rightarrow \quad \lim_{x \to a} |f(x)| = |l|.$$

Remark 1. Converse of above theorem need not to true.

For example, let $f(x) = \begin{cases} -1 & \text{if } x < a \\ 1 & \text{if } x > a \end{cases}$

Then $\quad |f(x)| = 1 \quad \forall\, x \neq a.$

$$\lim_{x \to a} |f(x)| = 1 \text{ but } \lim_{x \to a^-} f(x) = -1 \text{ and } \lim_{x \to a^+} f(x) = 1$$

$\therefore \lim_{x \to a} f(x)$ does not exist.

2. Converse of above theorem is true if $l = 0$.

6.1.9. Theorem. (*Squeeze principle*). *If $f(x) \leq h(x) \leq g(x)$ for $0 < |x - a| < \delta$ and*

$$\lim_{x \to a} f(x) = \lim_{x \to a} g(x) = l, \text{ then } \lim_{x \to a} h(x) = l.$$

Proof. Let $\varepsilon > 0$ be given. Then

$$\lim_{x \to a} f(x) = l \Rightarrow \exists\, \delta_1 > 0 \text{ such that}$$

$$|f(x) - l| < \varepsilon,\ 0 < |x - a| < \delta_1$$

i.e., $\quad l - \varepsilon < f(x) < l + \varepsilon,\ 0 < |x - a| < \delta_1.$

$$\lim_{x \to a} g(x) = l \Rightarrow \exists\, \delta_2 > 0 \text{ such that}$$

$$|g(x) - l| < \varepsilon,\ 0 < |x - a| < \delta_2$$

i.e., $\quad l - \varepsilon < g(x) < l + \varepsilon,\ 0 < |x - a| < \delta_2$

Let $\quad \delta_3 = \min\{\delta, \delta_1, \delta_2\}.$

Then $\quad l - \varepsilon < f(x) \leq h(x) \leq g(x) < l + \varepsilon,\ 0 < |x - a| < \delta_3$

i.e., $\quad l - \varepsilon < h(x) < l + \varepsilon,\ 0 < |x - a| < \delta_3$

$$\Rightarrow \quad \lim_{x \to a} h(x) = l.$$

6.1.10. We state below some results on limits which can be easily proved.

(*i*) If $f(x) \geq 0\ \forall\, x$ in a deleted neighbourhood of a, and $\lim_{x \to a} f(x)$ exists, then $\lim_{x \to a} f(x) \geq 0$.

(*ii*) If f and g are two functions defined in a deleted neighbourhood A of a and $f(x) \geq g(x)$ $\forall\, x \in$ A, then $\lim_{x \to a} f(x) \geq \lim_{x \to a} g(x)$.

(*iii*) If $\lim_{x \to a} f(x) = l > 0$, then

$$\lim_{x \to a} [f(x)]^\lambda = [\lim_{x \to a} f(x)]^\lambda = l^\lambda,$$

$$\lim_{x \to a} b^{f(x)} = b^{\left(\lim_{x \to a} f(x)\right)} = b^l$$

and
$$\lim_{x \to a} (\log f(x)) = \log(\lim_{x \to a} f(x)) = \log l.$$

6.1.11. Some Useful Limits

(*i*) $\lim_{x \to \infty} \left(1 + \frac{1}{x}\right)^x = e,$ (*ii*) $\lim_{x \to -\infty} \left(1 + \frac{1}{x}\right)^x = e$

(*iii*) $\lim_{x \to 0} (1 + x)^{1/x} = e,$ (*iv*) $\lim_{x \to 0} \frac{a^x - 1}{x} = \log a \ (a > 0)$

(*v*) $\lim_{x \to 0} \frac{\sin x}{x} = 1$ (*vi*) $\lim_{x \to 0} \frac{\tan x}{x} = 1.$

EXERCISE 6 (*a*)

1. Verify the following limits :

(*i*) $\lim_{x \to 2} \frac{x^3 + 3}{2x^2 + 5} = \frac{11}{13}$ (*ii*) $\lim_{x \to 2} \frac{x^3 - 8}{x^2 - 4} = 3$

(*iii*) $\lim_{x \to \infty} \frac{x^3 + 5x^2 + 3x + 2}{2x^3 + 7x^2 + 4x - 3} = \frac{1}{2}.$

2. Evaluate the following limits :

(*i*) $\lim_{x \to 0} \frac{\tan x - \sin x}{x^3}$ (*ii*) $\lim_{x \to 0} \frac{\sin 5x^\circ}{x}$

(*iii*) $\lim_{x \to 0} \frac{e^x - e^{-x}}{2x}.$

3. Show that as $x \to 0$, $\left(\frac{2}{3}\right)^{\frac{1}{x^2}} \to 0$ but $\left(\frac{2}{3}\right)^{\frac{1}{x}}$ does not tend to any limit.

4. Find the limit of $\frac{a_0 x^n + a_1 x^{n-1} + \ldots + a_n}{b_0 x^m + b_1 x^{m-1} + \ldots + b_m}$ when $x \to \infty$ for (*i*) $m > n$, (*ii*) $m < n$ and (*iii*) $m = n$.

5. Justify or falsify the following statements :

(*i*) If limits of two function f and g do not exist as $x \to a$, then the limits of $f + g$ and fg exist as $x \to a$.

(*ii*) If limits of $f + g$ and f exist as $x \to a$, then the limit of g exists as $x \to a$.

(*iii*) If $\lim_{x \to a} f(x)$ and $\lim_{x \to a} f(x) \cdot g(x)$ both exist, then $\lim_{x \to a} g(x)$ exists.

6. If $f(x) = g(x)$ for $0 < |x| < \delta$ for some $\delta > 0$, then prove that $\lim_{x \to 0} f(x) = \lim_{x \to 0} g(x)$. Is the converse true ?

7. Give examples of two functions f and g to show that $f(x) > g(x)$ but $\lim_{x \to a} f(x) = \lim_{x \to a} g(x)$.

8. If $f(x) \le g(x)$ and $\lim_{x \to a} g(x)$ exists, does it follow that $\lim_{x \to a} f(x)$ exist ?

9. If $f(x) = \begin{cases} 1 & \text{if } x \text{ is rational} \\ -1 & \text{if } x \text{ is irrational} \end{cases}$

then show that $\lim_{x \to a} f(x)$ does not exist for any $a \in \mathbf{R}$.

6.2. CONTINUOUS FUNCTIONS

While defining $\lim_{x \to a} f(x)$, the function f may or may not be defined at $x = a$. Even if f is defined at $x = a$, $\lim_{x \to a} f(x)$ may or may not be equal to the value of the function at $x = a$. If $\lim_{x \to a} f(x) = f(a)$, then we say that f is continuous at $x = a$.

A formal definition is :

Definition. *Let f be a function defined in a neighbourhood of a. Then f is said to be continuous at a if given any $\varepsilon > 0$, however small, there exists a $\delta > 0$ (depending upon ε) such that*

$$| f(x) - f(a) | < \varepsilon \text{ whenever } | x - a | < \delta.$$

Since $\lim_{x \to a} f(x)$ exists if and only if $\lim_{x \to a^-} f(x) = \lim_{x \to a^+} f(x)$, therefore, a function f is continuous at $x = a$ if and only if

$$\lim_{x \to a^-} f(x) = f(a) = \lim_{x \to a^+} f(x).$$

This form of definition is often useful.

If $\lim_{x \to a^-} f(x) = f(a)$, then we say that f *is continuous to the left of a (or left continuous at a).*

If $\lim_{x \to a^+} f(x) = f(a)$, then we say *that f is continuous to the right of a (or right continuous at a).*

A function f is said to be *continuous in an open interval (a, b)* if it is continuous at every point of (a, b).

A function f *is said to be continuous in a closed* interval $[a, b]$ if it is

(*i*) right continuous at a

(*ii*) continuous at every point of (a, b)

(*iii*) left continuous at b.

A function f is said to be continuous in a semi-closed interval [a, b) if it is

(i) right continuous at a

(ii) continuous at every point of (a, b).

Likewise a function f is continuous in a semi-closed interval (a, b] if it is

(i) continuous at every point of (a, b)

(ii) left continuous at b.

A function f is said to be continuous on an arbitrary set $S(\subseteq R)$ if for each $\varepsilon > 0$ and $a \in S$, there exists $\delta > 0$ such that

$$| f(x) - f(a) | < \varepsilon, \text{ whenever } x \in S \text{ and } | x - a | < \delta.$$

6.2.1. Examples on continuous functions.

1. *Every constant function $f : x \to c$, is continuous on R.*

For, let $\varepsilon > 0$ be given and $a \in R$. Then,

$$| f(x) - f(a) | = | c - c | = 0 < \varepsilon,$$

whenever $| x - a | < \delta$.

2. *The identity function $f : x \to x, x \in R$ is continuous on R.*

For, let $\varepsilon > 0$ be given and $a \in R$.

Then, for $\delta = \varepsilon$, $\quad | f(x) - f(a) | = | x - a | < \varepsilon$,

whenever $| x - a | < \delta$.

3. The function $f : x \to x^n, n \in N$, is continuous on R, because, for any $a \in R$,

$$\lim_{x \to a} x^n = a^n = f(a).$$

4. We know : $| \sin x | \le | x |$ and $| \cos x | \le 1, \forall x \in R$. Now, for any $a \in R$,

$$| \sin x - \sin a | = \left| 2 \sin \frac{x-a}{2} \cos \frac{x+a}{2} \right|$$

$$= 2 \left| \sin \frac{x-a}{2} \right| \left| \cos \frac{x+a}{2} \right|$$

$$\le 2 \left| \frac{x-a}{2} \right| = | x - a |$$

$\therefore$ given any $\varepsilon > 0$, $\exists\, \delta (= \varepsilon) > 0$ such that

$$| \sin x - \sin a | < \varepsilon, \text{ whenever } | x - a | < \delta$$

$\sin x$ is continuous at every point $a \in R$.

Similarly, $\cos x$ is continuous on R.

6.2.2. Discontinuity of a function. *A function f which is not continuous at a point 'a' is said to be discontinuous at the point 'a'.* 'a' is called *the point of discontinuity of f* or f is said to have a discontinuity at a.

A function which is discontinuous even at a single point of an interval, is said to be discontinuous in the interval.

A function f can be discontinuous at a point $x = a$ because of anyone of the following reasons :

(*i*) $f(x)$ is not defined at $x = a$

(*ii*) $\lim_{x \to a} f(x)$ does not exist

(*iii*) $\lim_{x \to a} f(x)$ and $f(a)$ both exist but are not equal.

(*i*) If $\lim_{x \to a} f(x)$ exists but is not equal to $f(a)$, then we say that f has a *removable discontinuity*, as just by changing the value of f only at a from $f(a)$ to $\lim_{x \to a} f(x)$, the new function can be made continuous at a.

(*ii*) If $\lim_{x \to a} f(x)$ does not exist, then the function cannot be made continuous, no matter how we define $f(a)$. In this case.

(*a*) If $\lim_{x \to a^-} f(x)$ and $\lim_{x \to a^+} f(x)$ both exist but are not equal, then we say that f has a *discontinuity of first kind (or that f has a jump discontinuity at a)*. It cannot be removed.

The function f is said to be left discontinuous or right discontinuous at a according as

$$\lim_{x \to a^-} f(x) \neq f(a) = \lim_{x \to a^+} f(x)$$

or

$$\lim_{x \to a^-} f(x) = f(a) \neq \lim_{x \to a^+} f(x)$$

(*b*) If either $\lim_{x \to a^-} f(x)$ or $\lim_{x \to a^+} f(x)$ does not exist, then we say that *f has a discontinuity of second kind.*

f is said to have a discontinuity of second kind from the left or right according as $\lim_{x \to a^-} f(x)$ or $\lim_{x \to a^+} f(x)$ does not exist.

(*iii*) If a function f has a discontinuity of second kind on one side of a and on the other side, it may be continuous or may have discontinuity of first kind, then f is said to have *mixed discontinuity* at a.

(*iv*) If either of the limits $\lim_{x \to a^-} f(x)$ or $\lim_{x \to a^+} f(x)$ is infinitely large, then *a is said to be a point of infinite discontinuity.*

6.2.3. Examples on discontinuities

1. The function $f(x) = \dfrac{x^2 - 4}{x - 2}$ is continuous for $x \neq 2$.

At $x = 2$:

$$\lim_{x \to 2} f(x) = \lim_{x \to 2} \frac{x^2 - 4}{x - 2} = \lim_{x \to 2} \frac{(x-2)(x+2)}{x-2}$$

$$= \lim_{x \to 2} (x + 2) = 4$$

But, the value of the function is not defined at $x = 2$. Therefore, the function has a removable discontinuity. It can be made continuous at $x = 2$, by redefining it as follows :

$$f(x) = \begin{cases} \dfrac{x^2 - 4}{x - 2}, & x \neq 2 \\ 4 \quad , & x = 2 \end{cases}$$

2. Consider the function defined by

$$f(x) = \begin{cases} \dfrac{\sin 2x}{x}, & x \neq 0 \\ 1 \quad , & x = 0 \end{cases}$$

Now, $$\lim_{x \to 0} f(x) = \lim_{x \to 0} \frac{\sin 2x}{2x} \times 2 = 2$$

so that $$\lim_{x \to 0} f(x) \neq f(0).$$

Therefore, the function is discontinuous at $x = 0$. The function has a removable discontinuity at the origin as the discontinunity can be removed by redefining the function at the origin such that $f(0) = 2$.

3. Consider the function defined by

$$f(x) = \begin{cases} x^2, & \text{if } x \leq 0 \\ 4x + 3, & \text{if } x > 0 \end{cases}$$

Now, $$\lim_{x \to 0^-} f(x) = \lim_{x \to 0^-} (x^2) = 0$$

$$\lim_{x \to 0^+} f(x) = \lim_{x \to 0^+} (4x + 3) = 3$$

$\therefore$ $\lim_{x \to 0^-} f(x)$ and $\lim_{x \to 0^+} f(x)$ both exist but are not equal.

Thus, the function has a discontinuity of first kind or that the function has a jump discontinuity.

4. Since in the example (3) above,

$$\lim_{x \to 0^-} f(x) = f(0) \neq \lim_{x \to 0^+} f(x)$$

therefore, the function is left continuous at 0 or the function is right discontinuous at 0.

5. Consider the function

$$f(x) = \begin{cases} x^2, & \text{if } x < 0 \\ 4x + 3, & \text{if } x \geq 0 \end{cases}$$

Then, $$\lim_{x \to 0^-} f(x) = 0, \ \lim_{x \to 0^+} f(x) = 3 \text{ and } f(0) = 3$$

$\therefore$ $$\lim_{x \to 0^-} f(x) \neq f(0) = \lim_{x \to 0^+} f(x)$$

$\therefore$ the function has a discontinuity of first kind and the function is right continuous or left discontinuous

6. For the function $f(x) = e^{\frac{1}{x}} \sin \dfrac{1}{x}$

$\lim_{x \to 0^-} f(x) = 0$, $\lim_{x \to 0^+} f(x)$ does not exist and the function is not defined at $x = 0$. Therefore, the function has a discontinuity of first kind from the left and a discontinuity of the second kind from the right at $x = 0$. Thus, the function has a mixed discontinuity at $x = 0$.

7. If $f(x) = \dfrac{1}{x - a}$, then $f(x)$ is continuous for each $x \neq a$. For $x = a$, $\lim_{x \to a^-} f(x) = -\infty$, $\lim_{x \to a^+} f(x) = \infty$ and $f(x)$ is not defined at $x = a$.

Thus, the function has an infinite discontinuity at a.

8. Consider the function

$$f(x) = \begin{cases} \dfrac{1}{x-a} \operatorname{cosec}(x-a), & \text{if } x \neq a \\ 0 \quad , & \text{if } x = a \end{cases}$$

The function has infinite discontinuity at $x = a$.
For, let $x = a - h$, $(h > 0)$ so that as $x \to a^-$, $h \to 0^+$.

$$\therefore \qquad \lim_{x \to a^+} f(x) = \lim_{x \to a^-} \frac{1}{x-a} \operatorname{cosec}(x-a) = \lim_{h \to 0^+} \frac{\operatorname{cosec}(a-h-a)}{a-h-a}$$

$$= \lim_{h \to 0^+} \frac{\operatorname{cosec}(-h)}{-h} = \lim_{h \to 0^+} \frac{1}{h} \times \frac{1}{\sin h} = \infty$$

Since $\lim\limits_{x \to a^-} f(x) \neq f(a)$, the function is discontinuous at a.

Further, one can see that $\lim\limits_{x \to a^+} f(x) = \infty$

$\therefore$ f has infinite discontinuity at a.

9. Consider the function

$$f(x) = \begin{cases} \dfrac{x}{|x|}, & x \neq 0 \\ 1 \quad , & x = 0 \end{cases}$$

in [0, 1] and in [– 1, 1].

Here, $f(x)$ is continuous everywhere on [0, 1]. However, the function defined on [– 1, 1] is not continuous at $x = 0$.

For, $\qquad \lim\limits_{x \to 0^-} f(x) = -1, \ \lim\limits_{x \to 0^+} f(x) = 1, f(0) = 1$

$\therefore \qquad \lim\limits_{x \to 0^-} f(x) \neq \lim\limits_{x \to 0^+} f(x)$

So, the function is left discontinuous but right continuous at 0.

10. The function

$$f(x) = \begin{cases} \dfrac{e^{\frac{1}{x}} - 1}{e^{\frac{1}{x}} + 1}, & x \neq 0 \\ 0 \quad , & x = 0 \end{cases}$$

is discontinuous at $x = 0$

For, As $x \to 0^+$, $\dfrac{1}{x} \to \infty$

$\therefore \quad e^{\frac{1}{x}} \to \infty$ and $e^{-\frac{1}{x}} \to 0$

Again, when $x \to 0^-$, $\dfrac{1}{x} \to -\infty$

$\therefore \quad e^{\frac{1}{x}} \to 0$

Hence, $$\lim_{x \to 0^-} f(x) = \lim_{x \to 0^-} \frac{e^{\frac{1}{x}} - 1}{e^{\frac{1}{x}} + 1} = \frac{0-1}{0+1} = -1$$

and $$\lim_{x \to 0^+} f(x) = \lim_{x \to 0^+} \frac{e^{\frac{1}{x}} - 1}{e^{\frac{1}{x}} + 1} = \lim_{x \to 0^+} \frac{1 - e^{-\frac{1}{x}}}{1 + e^{-\frac{1}{x}}} = \frac{1-0}{1+0} = 1$$

$\therefore$ $$\lim_{x \to 0^-} f(x) \neq \lim_{x \to 0^+} f(x)$$

$\therefore \quad \lim_{x \to 0} f(x)$ does not exist.

Thus, the function is discontinuous and the discontinuity is of first kind.

6.3. BOUNDS OF A FUNCTION

Let f be a function defined on a closed interval $[a, b]$. As x varies in this interval, $f(x)$ assumes varying values. If there exists a real number k such that $f(x) \geq k$, $\forall\, x \notin [a, b]$, we say that the function f is bounded below and k is called a lower bound of f. If there exists a real number K such that $f(x) \leq K$, $\forall\, x \in [a, b]$, we say that the function f is bounded above and K is called an upper bound of f.

If there are two numbers k and K such that $k \leq f(x) \leq K$ for every value of x in the interval $[a, b]$, then we say that f is bounded on $[a, b]$ and k and K are called **lower** and **upper** bounds of f over $[a, b]$.

If for any given $\varepsilon > 0$ (however small), there exists at least one $x \in [a, b]$ such that $f(x) < k + \varepsilon$ and at least one $x \in [a, b]$ such that $f(x) > K - \varepsilon$, then k and K are called the greatest lower bound (g.l.b.) and the least upper bound (l.u.b.) of f in the interval $[a, b]$. k and K are also called **supremum** and **infimum** of f on $[a, b]$.

For example,

(*i*) The function $\sin x$ defined on $[0, 2\pi]$ is bounded and has g.l.b. $= -1$ and l.u.b. $= 1$.

(*ii*) The function f defined on $[0, 1]$ by

$$f(x) = \begin{cases} \frac{1}{x}, & x \neq 0 \\ 0, & x = 0 \end{cases}$$

is not bounded on $[0, 1]$.

(*iii*) The function f defined by

$$f(x) = \frac{x}{x+1}, \; x \in [0, \infty)$$

is bounded and has g.l.b. $= 0$ and l.u.b. $= 1$.

6.4. THEOREMS OF CONTINUOUS FUNCTIONS

6.4.1. Theorem. *A function f is continuous at a if and only if for every sequence $< x_n >$ converging to a ($x_n \in D$, domain of the function f), the sequence $< f(x_n) >$ converges to $f(a)$.*

Proof. (*i*) Let f be continuous at a and the sequence $< x_n >$ converges to a.
Since f is continuous at a.

$\therefore$ given any $\varepsilon > 0$, $\exists\ \delta > 0$, such that

$$| f(x) - f(a) | < \varepsilon, \text{ whenever } | x - a | < \delta \qquad ...(1)$$

Since $< x_n >$ converges to a, $\exists$ a positive integer m such that

$$| x_n - a | < \delta, \quad \forall\ n \geq m \qquad ...(2)$$

From (1) and (2), we have

$$| f(x_n) - f(a) | < \varepsilon \quad \forall\ n \geq m$$

$\Rightarrow$ $< f(x_n) >$ converges to $f(a)$.

(*ii*) We are given that whenever $x_n \to a$, then $< f(x_n) >$ converges to $f(a)$. We are to show that f is continuous at a.

Let if possible, f be not continuous at a. Then, $\exists\ \varepsilon > 0$, such that for every $\delta > 0$, there exists a point x such that

$$| x - a | < \delta \text{ and } | f(x) - f(a) | \geq \varepsilon$$

Taking $\delta = \frac{1}{n}$, we can find a point x_n such that

$$| x_n - a | < \frac{1}{n} \text{ and } | f(x_n) - f(a) | \geq \varepsilon \quad \forall\ n$$

$\Rightarrow$ $< x_n >$ converges to a and $< f(x_n) >$ does not converge to $f(a)$. This contradiction proves that f is continuous at $x = a$.

Remark. *The choice of $< x_n >$ in the δ-neighbourhood of a such that $< x_n >$ converges to a is infinite. For example, $x_n = a + \frac{\delta_1}{2^n}$ where $0 < \delta_1 < \delta$, is such a choice.*

6.4.2. Theorem. *If f and g are two continuous functions at a, then*

(*i*) *$f + g$ is continuous at a*

(*ii*) *$f - g$ is continuous at a*

(*iii*) *fg is continuous at a.*

(*iv*) *$\frac{f}{g}$ is continuous at a, provided $g(a) \neq 0$*

(*v*) *cf is continuous at a, where c is constant.*

Proof. Let $< a_n >$ be a sequence converging to a.

(*i*) Since f is continuous at a, $< f(a_n) >$ converges to $f(a)$.

Since g is continuous at a, $< g(a_n) >$ converges to $g(a)$.

$$\therefore \quad \lim_{n \to \infty} (f + g)(a_n) = \lim_{n \to \infty} (f(a_n) + g(a_n))$$

$$= \lim_{n \to \infty} f(a_n) + \lim_{n \to \infty} g(a_n) = f(a) + g(a) = (f + g)(a)$$

$\Rightarrow$ the sequence $< (f + g)(a_n) >$ converges to $(f + g)(a)$.

$\Rightarrow f + g$ is continuous at a.

(*ii*) As of part (*i*).

(*iii*) Since f is continuous at a, $< f(a_n) >$ converges to $f(a)$.

Since g is continuous at a, $< g(a_n) >$ converges to $g(a)$.

$$\therefore \quad \lim_{n \to \infty} (fg)(a_n) = \lim_{n \to \infty} (f(a_n)\, g(a_n))$$

$$= \lim_{n \to \infty} f(a_n) \lim_{n \to \infty} g(a_n) = f(a)\, g(a) = (fg)(a)$$

$\Rightarrow$ the sequence $< (fg)(a_n) >$ converges to $(fg)(a)$.

$\Rightarrow$ fg is continuous at a.

(*iv*) Since f is continuous at a, $< f(a_n) >$ converges to $f(a)$.

Since g is continuous at a, $< g(a_n) >$ converges to $g(a)$.

Since $g(a) \neq 0$, there exists a positive integer m such that $g(a_n) \neq 0, \forall\, n \geq m$.

$$\therefore \qquad \lim_{n \to \infty} \left(\frac{f}{g}\right)(a_n) = \lim_{n \to \infty} \left(\frac{f(a_n)}{g(a_n)}\right) = \frac{\lim\limits_{n \to \infty} f(a_n)}{\lim\limits_{n \to \infty} g(a_n)} = \frac{f(a)}{g(a)}$$

$\Rightarrow$ the sequence $< \left(\frac{f}{g}\right)(a_n) >$ converges to $\left(\frac{f}{g}\right)(a)$.

$\Rightarrow$ $\frac{f}{g}$ is continuous at a.

(*v*) Left to the reader as an exercise.

Remark. *Since a function f is continuous at a if and only if* $\lim\limits_{x \to a} f(x) = f(a)$, *the proofs of the above theorem can also be deduced from the corresponding results on limits.*

Note. (1) Every polynomial function is continuous for every real number.

(2) Every rational function is continuous for every real number other than zeros of the denominator.

6.4.3. Theorem. *If a function f is continuous at a, then* $|\, f \,|$ *is also continuous at a.*

Proof. Let $\varepsilon > 0$ be given.

Since f is continuous at a, therefore, there exists $\delta > 0$ such that

$$|\, f(x) - f(a) \,| < \varepsilon, \text{ whenever } |\, x - a \,| < \delta$$

Now, $|\, |\, f(x) \,| - |\, f(a) \,|\, | \leq |\, f(x) - f(a) \,| < \varepsilon$, whenever $|\, x - a \,| < \delta$

$$(\because\ |\, a - b \,| \geq |\, |\, a \,| - |\, b \,|\, |)$$

$\Rightarrow$ $|\, f \,|$ is continuous at a.

Remark. Converse of the above theorem is false. For example, if $f(x) = \begin{cases} 1 \text{ if } x \text{ is rational} \\ -1 \text{ if } x \text{ is irrational} \end{cases}$ and a is any real number, then in the interval $(a - d, a)$, there lie infinitely many rationals as well as irrationals. Therefore, $f(x)$ oscillates between -1 and 1. Therefore, f cannot tend to a definite value as $x \to a^-$.

Thus $\lim\limits_{x \to a^-} f(x)$ does not exist.

Similarly, $\lim\limits_{x \to a^+} f(x)$ does not exist.

Hence f is not continuous at a.

But $\qquad |\, f(x) \,| = 1 \ \forall\, x \in \mathrm{R}$

Hence $|\, f \,|$ is continuous everywhere and in particular at a also.

Cor. *If f and g are two continuous functions at a, then the functions max.* $\{f, g\}$ *and min.* $\{f, g\}$ *are both continuous at a.*

For, $\qquad \max.\ \{f, g\} = \frac{1}{2}\,(f + g) + \frac{1}{2}\,|\, f - g \,|$

and $\qquad \min.\ \{f, g\} = \frac{1}{2}\,(f + g) - \frac{1}{2}\,|\, f - g \,|$.

6.4.4. Theorem. *Composite of two continuous functions is a continuous function, i.e., if f is a continuous function at $x = a$ and g is a continuous function at $f(a)$, then gof is a continuous function at $x = a$.*

Proof. Let $\varepsilon > 0$ be given.

Since g is continuous at $f(a)$

$\therefore$ $\exists\ \delta > 0$ such that

$$| g(f(x)) - g(f(a)) | < \varepsilon \text{ whenever } | f(x) - f(a) | < \delta \qquad ...(1)$$

Since f is continuous at $x = a$,

$\therefore$ $\exists\ \delta_1 > 0$ (depending upon δ) such that

$$| f(x) - f(a) | < \delta \text{ whenever } | x - a | < \delta_1 \qquad ...(2)$$

$\therefore$ from (1) and (2), we have, for any $\varepsilon > 0$, $\exists\ \delta_1 > 0$ such that

$$| (gof)(x) - (gof)(a) | < \varepsilon \text{ whenever } | x - a | < \delta_1$$

$\Rightarrow$ gof is continuous at $x = a$.

6.4.5. Theorem. *A function f is continuous at a if and only if for each $\varepsilon > 0$, there exists $\delta > 0$ such that $| f(x_1) - f(x_2) | < \varepsilon$, whenever $x_1, x_2 \in (a - \delta, a + \delta)$.*

Proof. Let f be continuous at a.

Then, for $\varepsilon > 0$, $\exists\ \delta > 0$ such that

$$| f(x) - f(a) | < \frac{\varepsilon}{2}, \text{ whenever } | x - a | < \delta.$$

$\therefore$ for $x_1, x_2 \in (a - \delta, a + \delta)$,

$$| f(x_1) - f(x_2) | = | f(x_1) - f(a) + f(a) - f(x_2) | \le | f(x_1) - f(a) | + | f(x_2) - f(a) |$$

$$< \frac{\varepsilon}{2} + \frac{\varepsilon}{2} = \varepsilon$$

Conversely, let for each $\varepsilon > 0$, $\exists\ \delta > 0$ such that

$$| f(x_1) - f(x_2) | < \varepsilon, \text{ whenever } x_1, x_2 \in (a - \delta, a + \delta).$$

Taking $x_1 = x$ and $x_2 = a$, we have

$$| f(x) - f(a) | < \varepsilon, \text{ whenever } x \in (a - \delta, a + \delta)$$

$\Rightarrow$ f is continuous at a.

6.4.6. Theorem. *If a function f is continuous at a, then it is bounded in some neighbourhood of a, i.e., there exists $\delta > 0$ and $k > 0$ such that $| f(x) | \le k$, for every x in $(a - \delta, a + \delta)$ at which f is defined.*

Proof. Take $\varepsilon = 1$. Since f is continuous at a, $\exists\ \delta > 0$, such that

$$| f(x) - f(a) | < 1, \text{ whenever } | x - a | < \delta, x \in D_f$$

$\Rightarrow$ $f(a) - 1 < f(x) < f(a) + 1$, whenever $| x - a | < \delta$

$\Rightarrow$ $f(x)$ is bounded.

6.4.7. Theorem. *Every function continuous on a closed interval is bounded in that interval.* (K.U. BCA (II) 2003, 2006)

Proof. Let f be a continuous function on I ($= [a, b]$). Let f be not bounded above in $[a, b]$.

Let c be the mid-point of $[a, b]$ $\left(i.e., c = \dfrac{a + b}{2} \right)$. Then f must be unbounded above in at least one of the intervals $[a, c]$ and $[c, b]$. Let us call that interval I_1 and rewrite it as $[a_1, b_1]$ (if f is unbounded in both the subintervals, then we take $[a, c]$ as $[a_1, b_1]$). Let c_1 be the mid-point of

$[a_1, b_1]$. Then f must be unbounded above in at least one of the intervals $[a_1, c_1]$ and $[c_1, b_1]$. Call that interval I_2 and rewrite it as $[a_2, b_2]$. (If f is unbounded in both the intervals, then we call $[a_1, c_1]$ as $[a_2, b_2]$). Continuing like this, we get a nested sequence of closed intervals $I \supseteq I_1 \supseteq I_2 \supseteq I_3 \supseteq \ldots \supseteq I_n \supseteq \ldots$ such that

(*i*) $\quad$ length $(I_n) = b_n - a_n = \left(\frac{1}{2^n}\right)(b - a)$ so that

$$\lim_{n \to \infty} (\text{length } I_n) = 0$$

(*ii*) f is unbounded in each of the intervals I_n.

By Nested Interval Property $\bigcap_{n=1}^{\infty} I_n$ is a singleton $= \{\xi\}$, say. Since f is continuous on I, f is continuous at ξ also. Therefore, given any $\varepsilon > 0$, $\exists\ \delta > 0$ such that

$$| f(x) - f(\xi) | < \varepsilon \text{ whenever } | x - \xi | < \delta.$$

Since $\lim_{n \to \infty}$ (length of I_n) $= 0$, $\exists$ a positive integer m such that $I_m \subseteq (\xi - \delta, \xi + \delta)$. If x is any point of I_m, then

$$| f(x) | = |f(x) - f(\xi) + f(\xi) |$$
$$\leq | f(x) - f(\xi) | + | f(\xi) | < \varepsilon + | f(\xi) |,$$

showing that f is bounded in I_m. This contradicts the fact that f is unbounded above in each subinterval I_n. Hence f must be bounded above in I. Similarly, f is bounded below in $[a, b]$. Hence f is bounded in $[a, b]$.

Remark. The above theorem need not be true if the interval is not closed.

For example :

(*i*) Let $f(x) = \frac{1}{x}$, $x \in (0, 1]$.

Then f is continuous on $(0, 1]$.

As $x \to 0^+$, $\frac{1}{x} \to \infty$. Therefore, f is not bounded on $(0, 1]$.

(*ii*) $f(x) = \log x$, $x \in (0, \infty)$

f is continuous on $(0, \infty)$, but it is not bounded.

6.4.8. Theorem. *Every function continuous on a closed interval attains its bounds (at least once on the interval).*

Proof. Let f be a continuous function defined on a closed interval I.

$\therefore$ f is bounded.

$\therefore$ l.u.b. and g.l.b. of f exist.

Let u = l.u.b. of f and l = g.l.b. of f.

We shall show that there exist real numbers ξ and η in I such that $f(\xi) = u$ and $f(\eta) = l$.

Let, if possible, $f(x) < u\ \forall\ x \in$ I.

$$\Rightarrow \quad u - f(x) > 0 \quad \forall\ x \in \text{I}.$$

Since f is continuous on I, $\frac{1}{u - f(x)}$ is also continuous on I

$$\Rightarrow \quad \frac{1}{u - f(x)} \text{ is bounded.}$$

$\therefore$ $\exists$ a positive number k such that

$$\frac{1}{u - f(x)} \leq k \quad \forall\, x \in \text{I} \;\Rightarrow\; f(x) \leq u - \frac{1}{k} \quad \forall\, x \in \text{I}$$

$\Rightarrow$ $u - \frac{1}{k}$ is an upper bound of f

This contradicts the fact that u is the l.u.b. of f. Hence there must exist some $\xi \in \text{I}$ such that

$$u - f(\xi) = 0, \text{ i.e., } f(\xi) = u.$$

Similarly, there must exist at least one point $\eta \in \text{I}$ such that $f(\eta) = l$.

Remark. The above theorem need not be true if the interval is not closed. For example :

1. The function f defined on $(0, 1]$ or ($[0, 1)$ or $(0, 1)$) by $f(x) = x$ is continuous and bounded on $(0, 1]$ or ($[0, 1)$ or $(0, 1)$) with g.l.b. of $f = 0$ and l.u.b. of $f = 1$. g.l.b. of f is not attained on $(0, 1]$, l.u.b. of f is not attained on $[0, 1)$ and neither is attained on $(0, 1)$.

2. The function $y = x$ is continuous on R but is not bounded on any infinite interval.

3. The function $f(x) = \frac{x^2}{x^2 + 1}$ is continuous and bounded on R. It attains its g.l.b. = 0 but not the l.u.b. = 1 on any interval. Here $f(x)$ is not continuous on the closed interval (R is open). So, the above theorem is not applicable.

6.4.9. Theorem. *Let f be a continuous function on $[a, b]$ and $a < c < b$. Then*

(i) $f(a) > 0$ ($f(a) < 0$) implies that there exists $\delta > 0$ such that $f(x) > 0$ (< 0) for all x in $[a, a + \delta)$.

(ii) $f(c) > 0$ ($f(c) < 0$) implies that there exists $\delta > 0$ such that $f(x) > 0$ ($f(x) < 0$) for all x in $(c - \delta, c + \delta)$.

(iii) $f(b) > 0$ ($f(b) < 0$) implies that there exists $\delta > 0$ such that $f(x) > 0$ ($f(x) < 0$) for all x in $(b - \delta, b]$.

Proof. (*i*) *Let $f(a) > 0$*. Since f is right continuous at a, corresponding to any $\varepsilon > 0, \exists\, \delta > 0$ such that

$$| f(x) - f(a) | < \varepsilon, x \in [a, a + \delta)$$

$\Rightarrow$ $$f(a) - \varepsilon < f(x) < f(a) + \varepsilon, x \in [a, a + \delta)$$

Taking $\varepsilon = f(a)$, $$0 < f(x) < 2f(a), x \in [a, a + \delta)$$

$\Rightarrow$ $$f(x) > 0. \qquad (\because\; f(a) > 0)$$

Proof for the case $f(a) < 0$ is similar.

(*ii*) *Let $f(c) > 0$.*

Take $$\varepsilon = \frac{f(c)}{2} > 0$$

Then $\exists\, \delta > 0$ such that

$$| f(x) - f(c) | < \varepsilon, \; | x - c | < \delta$$

$\Rightarrow$ $$f(c) - \varepsilon < f(x) < f(c) + \varepsilon, \; | x - c | < \delta$$

$\Rightarrow$ $$f(c) - \tfrac{1}{2} f(c) < f(x) < f(c) + \tfrac{1}{2} f(c), \; | x - c | < \delta$$

$\Rightarrow$ $$f(x) > \tfrac{1}{2} f(c) > 0, \; | x - c | < \delta$$

Proof for the case $f(c) < 0$ is similar.

(*iii*) Proof is left for the reader as an exercise.

Remark. The above theorem asserts that if a function is continuous at a point, then its sign is invariable rear the point.

6.4.10. *Let f be a continuous function on R. Then,*

(i) the set $A = [x \in R \mid f(x) > 0]$ *is an open set*

(ii) the set $B = [x \in R \mid f(x) < 0]$ *is an open set*

(iii) the set $C = [x \in R \mid f(x) = 0]$ *is a closed set.*

Proof. (*i*) Let $a \in A$.

$\Rightarrow \quad f(a) > 0.$

Since f is continuous at a, there exists $\delta > 0$ such that

$$f(x) > 0 \text{ for } x \in (a - \delta, a + \delta)$$

$\Rightarrow \quad x \in (a - \delta, a + \delta) \subseteq A$

$\Rightarrow$ A is a neighbourhood of a.

Since a is arbitrary, this implies that A is an open set.

(*ii*) Proof is similar to (*i*).

(*iii*) Since $C = R - (A \cup B)$, it follows that C is a closed set.

Cor. *Let f be a continuous function on R and c any real number. Then,*

(i) the set $A = \{x \in R \mid f(x) < c\}$ *is an open set*

(ii) the set $B = \{x \in R \mid f(x) > c\}$ *is an open set*

(iii) the set $C = \{x \in R \mid f(x) = c\}$ *is a closed set.*

Proof. (*i*) Define a function g as :

$$g(x) = f(x) - c, \quad \forall x \in R.$$

Since f is continuous on R, g is continuous on R.

$\therefore$ the set $\{x \in R \mid g(x) < 0\}$ is an open set

$\Rightarrow$ the set $\{x \in R \mid f(x) - c < 0\}$ is an open set

$\Rightarrow$ the set $A = \{x \in R \mid f(x) < c\}$ is an open set

Proofs of (*ii*) and (*iii*) are on similar lines.

6.4.11. Theorem. *A function* $f : R \longrightarrow R$ *is continuous on R if and only if for every open set A in R,* $f^{-1}(A)$ *is open in R.*

Proof. Let f be continuous on R.

Let A be an open subset of R.

If $f^{-1}(A) = \phi$, then it is open

So, let $f^{-1}(A) \neq \phi$

Let $a \in f^{-1}(A) \Rightarrow f(a) \in A$

Since A is open, A is a neighbourhood of each of its points.

$\therefore \quad \exists\, \varepsilon > 0$ such that $(f(a) - \varepsilon, f(a) + \varepsilon) \subseteq A$.

Since f is continuous at a, (for above $\varepsilon > 0$) $\exists\, \delta > 0$ such that

$$| f(x) - f(a) | < \varepsilon, \text{ for } | x - a | < \delta \text{ i.e., } x \in (a - \delta, a + \delta)$$

Hence, $x \in (a - \delta, a + \delta) \Rightarrow f(x) \in (f(a) - \varepsilon, f(a) + \varepsilon) \subseteq A$

$\Rightarrow \quad x \in f^{-1}(A)$

$\Rightarrow \quad a \in (a - \delta, a + \delta) \subseteq f^{-1}(A)$

$\Rightarrow \quad f^{-1}(A)$ is a neighbourhood of each of its points

$\Rightarrow \quad f^{-1}(A)$ is open.

Conversely, let f be not continuous at a.

Then, there exists an $\varepsilon > 0$, such that for any $\delta > 0$, there exists x_0 such that $| x_0 - a | < \delta$ but

$$| f(x_0) - f(a) | \nless \varepsilon \quad i.e., \quad f(x_0) \notin (f(a) - \varepsilon, f(a) + \varepsilon).$$

Therefore, every open interval $(a - \delta, a + \delta)$ containing a, contains an x_0 such that $f(x_0) \notin (f(a) - \varepsilon, f(a) + \varepsilon)$

$\Rightarrow f^{-1}(f(a) - \varepsilon, f(a) + \varepsilon)$ is not an open set. This contradicts the given hypothesis ($\because$ $(f(a) - \varepsilon, f(a) + \varepsilon)$ is open). Hence, f is continuous at a. Since a is arbitrary, f is continuous at every point of R.

Cor. *A function $f : R \longrightarrow R$ is continuous on R if and only if for every closed set A in R, f^{-1} (A) is closed in R.*

Proof. Let f be continuous on R.

Let A be a closed subset of R

$\Rightarrow$ A^c is an open subset of R

$\Rightarrow$ $f^{-1}(A^c)$ is open

$\Rightarrow$ $(f^{-1}(A))^c$ is open

$\Rightarrow$ $f^{-1}(A)$ is closed.

Conversely, for a closed set A, let $f^{-1}(A)$ be closed.

$\Rightarrow$ A^c and $(f^{-1}(A))^c$ are open

$\Rightarrow$ A^c and $f^{-1}(A^c)$ are open

$\Rightarrow$ f is continuous on R.

6.4.12. Theorem. *If a function f is continuous on a closed interval $[a, b]$, $f(a)$ and $f(b)$ are of opposite signs, then there exists at least one point $c \in (a, b)$ such that $f(c) = 0$.*

Proof. By considering, if necessary, the function $-f$, we can assume without loss of generality that $f(a) < 0$ and $f(b) > 0$.

Let $S = \{x : x \in [a, b] \text{ and } f(x) < 0\}$, $S \neq \phi$ $\qquad (\because a \in S)$

Also S is bounded above $\qquad (\because b = \text{u.b. } S)$

$\therefore$ by completeness property of reals, S has the l.u.b. c (say). We shall show that $f(c) = 0$. Suppose that $f(c) < 0$.

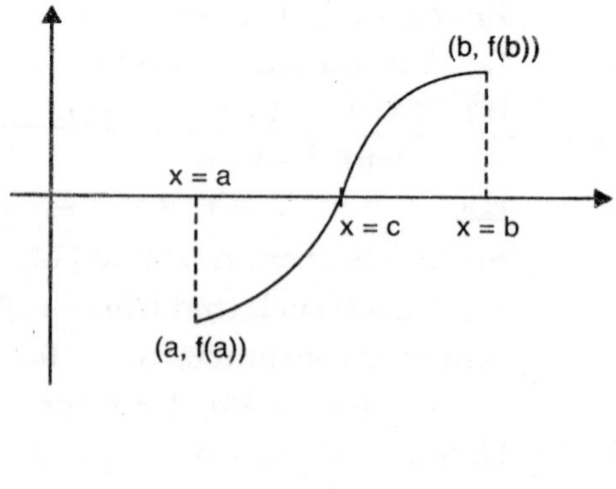

Then $c \neq b$ ($\because$ $f(b) > 0$) and by the continuity of f at c, $\exists\ \delta > 0$ such that

$$f(x) < 0, x \in [c, c + \delta).$$

$$\Rightarrow \qquad f\left(c + \frac{\delta}{2}\right) < 0.$$

$\Rightarrow$ $c + \dfrac{\delta}{2} \in S$, which contradicts the fact that

$$c = \text{l.u.b. } S.$$

Hence, $f(c) \nless 0$.

Suppose that $f(c) > 0$. Then $c \neq a$ $\qquad (\because f(a) < 0)$

By continuity of f at c, $\exists\ \delta > 0$ such that

$$f(x) > 0 \text{ for } x \in (c - \delta, c].$$

Since $c = \text{l.u.b. } S$, there exists $t \in S$ such that $c - \delta < t < c$.

Now $t \in S \Rightarrow f(t) < 0$

Since $t \in (c - \delta, c]$, $f(t) > 0$.

This gives a contradiction. Hence $f(c) = 0$ and $c \neq a$, $c \neq b$.

Cor. 1. *(Intermediate Value Theorem)*

If a function f is continuous on a closed interval [a, b] and f(a) ≠ f(b), then f assumes every value between f(a) and f(b).

Proof. W.l.o.g, we assume that $f(a) < f(b)$. Let k be any number such that

$$f(a) < k < f(b).$$

Consider a function g defined on $[a, b]$ such that

$$g(x) = f(x) - k.$$

Now g is continuous on $[a, b]$ and $g(a) = f(a) - k$,

$g(b) = f(b) - k$ are of opposite signs.

Therefore, $\exists\ c \in (a, b)$ such that $g(c) = 0$, *i.e.*, $f(c) = k$.

Remark. Converse of Intermediate Value Theorem need not be true. For example,

Let
$$f(x) = \begin{cases} x, & 0 \leq x < 1 \\ x - 1, & 1 \leq x \leq 3 \end{cases}$$

Then $f(x)$ assumes every value between $f(0) = 0$ and $f(3) = 2$ as x moves through 0 to 3 but f is not continuous on $[0, 3]$.

Cor. 2. *A function f, which is continuous on a closed interval [a, b], assumes every value between its bounds.*

Proof. Since the function f is continuous on $[a, b]$, it is bounded and attains its bounds on $[a, b]$.

$\therefore$ $\exists$ two numbers α and β in $[a, b]$ such that $f(\alpha) = M$ and $f(\beta) = m$, when M and m are the l.u.b. and the g.l.b. of f.

Since f is continuous on $[a, b]$, it is continuous on $[\beta, \alpha]$ or $[\alpha, \beta]$ (according as $\alpha > \beta$ and $\alpha < \beta$).

Hence, f assumes every value between $f(\alpha)$ and $f(\beta)$ (*i.e.*, the function assumes every value between its bounds).

We also say that the range of a continuous function on a closed interval, is a closed interval, or that the image of a closed interval under a continuous function is a closed interval.

6.4.13. Monotonic Functions. Let f be a function defined on $[a, b]$. Then,

(*i*) the function f is said to be monotonically increasing on $[a, b]$, if for $x_1, x_2 \in [a, b]$, $x_1 > x_2$ implies $f(x_1) \geq f(x_2)$.

(*ii*) the function f is said to be monotonically decreasing on $[a, b]$ if for $x_1, x_2 \in [a, b]$, $x_1 > x_2$ implies $f(x_1) \leq f(x_2)$.

(*iii*) the function f is said to be strictly monotonically increasing on $[a, b]$ if for x_1, $x_2 \in [a, b]$, $x_1 > x_2$ implies $f(x_1) > f(x_2)$.

(*iv*) the function, f is said to be strictly monotonically decreasing on $[a, b]$ if for x_1, $x_2 \in [a, b]$, $x_1 > x_2$ implies $f(x_1) < f(x_2)$.

(*v*) the function f is said to be monotonic if it is either monotonically increasing or monotonically decreasing.

6.4.14. Theorem. *(Continuity of inverse function). If a one to one function f is continuous and strictly monotonic on [a, b], then f^{-1} is also continuous on [f(a), f(b)] or [f(b), f(a)] according as f is increasing or decreasing.*

Proof. Since the function f is continuous and strictly monotonic on $[a, b]$, f is bounded and attains its bounds at a and b.

$\therefore$ $\exists$ real numbers m and M such that

$$f(a) = m, f(b) = \text{M} \quad \text{or} \quad f(a) = \text{M}, f(b) = m$$

according as f is strictly monotonically increasing or decreasing.

$\Rightarrow$ range of f is $[f(a), f(b)]$ or $[f(b), f(a)]$

according as f is strictly monotonically increasing or decreasing.

Since f is one to one, f^{-1} exists. $\therefore$ f^{-1} is a function with domain $[f(a), f(b)]$ or $[f(b), f(a)]$ according as f is strictly monotonically increasing or decreasing.

Let f be strictly monotonically increasing.

$\Rightarrow$ f^{-1} is a function with domain $[f(a), f(b)]$.

We shall show that f^{-1} is continuous on $[f(a), f(b)]$.

Let $\quad y_0 \in [f(a), f(b)]$

$\Rightarrow \exists$ some $x_0 \in [a, b]$ such that $f^{-1}(y_0) = x_0$, *i.e.*, $f(x_0) = y_0$

Let $\varepsilon > 0$ be given

Let $\quad f(x_0 - \varepsilon) = y_0 - \delta_1$ and $f(x_0 + \varepsilon) = y_0 + \delta_2$,

where δ_1, δ_2 are necessarily positive numbers.

Since f is strictly monotonically increasing,

$$x \in (x_0 - \varepsilon, x_0 + \varepsilon) \Rightarrow f(x) \in [f(x_0 - \varepsilon), f(x_0 + \varepsilon)]$$

i.e., $\quad f(x) \in [y_0 - \delta_1, y_0 + \delta_2]$

If $\delta = \min\{\delta_1, \delta_2\}$, then $(y_0 - \delta, y_0 + \delta) \subseteq (f(x_0 - \varepsilon), f(x_0 + \varepsilon))$

$\therefore \quad x \in (x_0 - \varepsilon, x_0 + \varepsilon)$ for $y = f(x) \in (y_0 - \delta, y_0 + \delta)$

i.e., $\quad | x - x_0 | < \varepsilon$ for $| y - y_0 | < \delta$

i.e., $\quad | f^{-1}(y) - f^{-1}(y_0) | < \varepsilon$ for $| y - y_0 | < \delta$

$\Rightarrow f^{-1}$ is continuous at $y_0 \in [f(a), f(b)]$

Since y_0 is arbitrary, f^{-1} is continuous at every point of $[f(a), f(b)]$ and hence f^{-1} is continuous on $[f(a), f(b)]$.

Similar is the case when f is strictly monotonically decreasing.

6.5. UNIFORM CONTINUITY

Recall that if a function f is continuous at a point x_0 of the closed interval $[a, b]$, then given any $\varepsilon > 0$, there exists *a* $\delta > 0$ such that $| f(x) - f(x_0) | < \varepsilon$ whenever $| x - x_0 | < \delta$. This δ depends not only on ε but also on the point x_0 at which the continuity is considered. That is, if $\varepsilon > 0$ remains the same for different points of the interval $[a, b]$, then the choice of δ may not be the same. If it is possible to find one $\delta > 0$ depending on ε such that $| f(x) - f(x_0) | < \varepsilon$ whenever $| x - x_0 | < \delta$ and $x, x_0 \in [a, b]$, then we say that the function in uniformly continuous. More precisely.

Definition. *A function f is said to be uniformly continuous on the interval [a, b] if given any $\varepsilon > 0$, there exists $\delta > 0$ (depending on ε only) such that $| f(x_1) - f(x_2) | < \varepsilon$, whenever $x_1, x_2 \in [a, b]$ and $| x_1 - x_2 | < \delta$.*

Remarks 1. The notion of uniform continuity of a function is a global concept. It does not make sense to say that a function is uniformly continuous at a point, while continuity is a local property.

2. One may think that the g.l.b. of the set of all values of δ corresponding to different points of $[a, b]$, would serve the purpose of δ for uniform continuity on $[a, b]$. But it is not so as the g.l.b. of a set of positive numbers may be zero.

3. A function f is not uniformly continuous on $[a, b]$ if $\exists$ some $\varepsilon > 0$ for which no δ works, *i.e.*, for any $\delta > 0$, $\exists\, x_1, x_2 \in [a, b]$ such that $| f(x_1) - f(x_2) | \geq \varepsilon$ and $| x_1 - x_2 | < \delta$.

6.5.1. Theorem. *If a function f is uniformly continuous on [a, b], then it is continuous on [a, b].*

Proof. Let f be uniformly continuous on $[a, b]$

Let $\varepsilon > 0$ be given.

Since f is uniformly continuous on $[a, b]$, there exists $\delta > 0$ such that

$$| f(x_1) - f(x_2) | < \varepsilon \quad \forall\, x_1, x_2 \in [a, b] \quad \text{and} \quad |x_1 - x_2| < \delta.$$

Let $\qquad c \in [a, b]$.

Taking $x_1 = x$ and $x_2 = c$, we have for $\varepsilon > 0$, $\exists\, \delta > 0$ such that

$$| f(x) - f(c) | < \varepsilon, \text{ for } | x - c | < \delta$$

$\Rightarrow f$ is continuous at c, any point of $[a, b]$. $\Rightarrow f$ is continuous at every point of $[a, b]$.

Note. In fact, we can consider any interval open or semi-open in the above theorem.

Remark. Converse of the above theorem need not be true. For example, consider the function f defined by

$$f(x) = \frac{1}{x}, \quad x \in (0, 1)$$

Since x is continuous on $(0, 1)$ and $x \neq 0$, therefore $\frac{1}{x}$ is continuous on $(0, 1)$.

Again, for any $\delta > 0$, $\exists\, m \in \mathbb{N}$ such that $\frac{1}{n} < \delta \;\forall\, n \geq m$.

Let $\qquad x_1 = \frac{1}{2m}$ and $x_2 = \frac{1}{m}$ so that $x_1, x_2 \in (0, 1)$,

$$| x_1 - x_2 | = \left| \frac{1}{2m} - \frac{1}{m} \right| = \frac{1}{2m} < \frac{1}{2}\,\delta < \delta.$$

and $\quad | f(x_1) - f(x_2) | = | 2m - m | = m$

which cannot be less than every $\varepsilon > 0$.

Hence f is not uniformly continuous on $(0, 1)$. This can also be seen geometrically as follows :

Consider the graph of the function $f(x) = \frac{1}{x}$ on $(0, 1)$. If we divide the interval $(0, 1)$ into subintervals of equal width, then the increment in the values of $f(x)$ are not equal, showing thereby that the function is not uniformly continuous on $(0, 1)$.

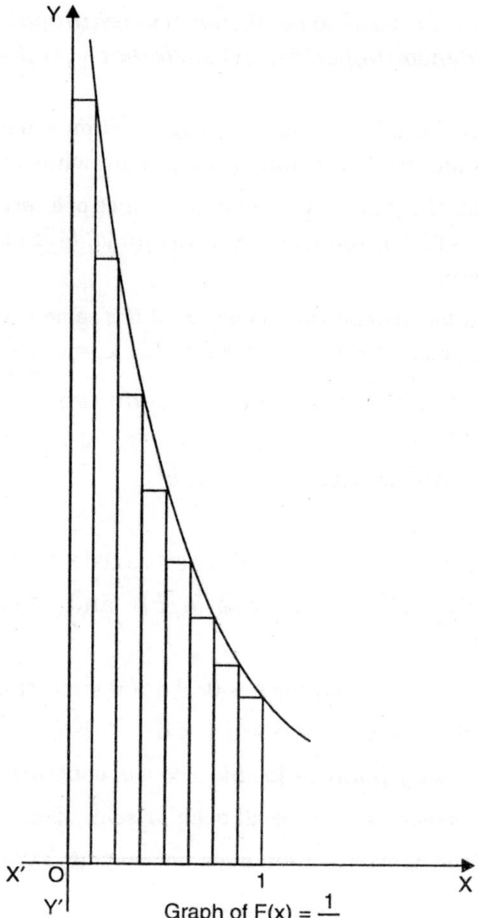

Graph of $F(x) = \frac{1}{x}$

6.5.2. Theorem. *If a function f is continuous on [a, b], then f is uniformly continuous on [a, b].*

To prove the above theorem, we first prove the following Lemma :

Lemma : If a function f : [a, b] → R is continuous on [a, b], then for given $\varepsilon > 0$, [a, b] can be divided into a finite number of sub-intervals such that $| f(x_1) - f(x_2) | < \varepsilon$ whenever x_1 and x_2 lie in the same sub-interval.

Proof of Lemma :

Let if possible, the result be false. Then, there exists none $\Sigma > 0$, such that for no subdivision of $[a, b]$, $| f(x_1) - f(x_2) | < \varepsilon$ for x_1, x_2 in the same sub-interval.

Divide $[a, b]$ into two equal sub-intervals. Then, the result is not satisfied in at least one of the two sub-intervals. Call the sub-interval $[a_1, b_1]$ in which the result is not satisfied. Again divide $[a_1, b_1]$ into two equal sub-intervals and then the result is not satisfied in at least one of the two sub-intervals, say $[a_2, b_2]$ and so on. In this way, we get a nested sequence of closed intervals with lengths :

$$b_1 - a_1 = \frac{1}{2}(b-a),\ b_2 - a_2 = \frac{1}{2^2}(b-a), \text{ ...... and so on.}$$

$$b_n - a_n = \frac{1}{2^n}(b-a) \to 0 \text{ as } n \to \infty$$

Therefore, by Nested Interval Property,

$$\bigcap_{n=1}^{\infty} [a_n, b_n] = \{x_0\} \text{ . (say) . and } x_0 \in [a, b].$$

Case (*i*) $x_0 \in (a, b)$

Since f is continuous on $[a, b]$

f is continuous at x_0 also.

$\therefore\ \exists\ \delta > 0$ such that

$$|\ f(x) - f(x_0)\ | < \frac{\varepsilon}{2},\ \forall x \in (x_0 - \delta, x_0 + \delta)$$

Also, $b_n - a_n \to 0$ as $n \to \infty$

$\therefore$ We can choose a + ve integer n such that $b_n - a_n < \delta$.

$$(a_n, b_n) \subseteq (x_0 - \delta, x_0 + \delta).$$

$$\therefore \qquad |\ f(x) - f(x_0)\ | < \frac{\varepsilon}{2},\ \forall\ x \in (a_n, b_n).$$

Let $x_1, x_2 \in (a_n, b_n)$

$$\therefore \qquad |\ f(x_1) - f(x_0)\ | < \frac{\varepsilon}{2} \text{ and } |\ f(x_2) - f(x_0)\ | < \frac{\varepsilon}{2}.$$

$$\therefore \qquad |\ f(x_1) - f(x_2)\ | = |\ f(x_1) - f(x_0) + f(x_0) - f(x_2)\ |$$

$$\leq |\ f(x_1) - f(x_0)\ | + |\ f(x_2) - f(x_0)\ | < \frac{\varepsilon}{2} + \frac{\varepsilon}{2} = \varepsilon.$$

$\therefore$ the result is satisfied in (a_n, b_n) which is a contradiction to our supposition.

Hence the Lemma is true.

Case (*ii*) $x_0 = a$ or $x_0 = b$

If $x_0 = a$, then the result follows from above by taking the interval $(a, a + \delta)$.

If $x_0 = b$, then the result follows by taking the interval $(b - \delta, b)$.

Proof of Theorem. Let $\varepsilon > 0$ be given.

Since it is continuous on $[a, b]$, therefore by the Lemma, the interval $[a, b]$ can be divided into a finite number of sub-intervals

$[a, y_1], [y_1, y_2], ..., [y_n, b]$, (say) such that whenever x_1, x_2 are in the same sub-interval,

$$|\ f(x_1) - f(x_2)\ | < \frac{\varepsilon}{2} \qquad ...(1)$$

Let $\delta = \min. \{y_1 - a, y_2 - y_1, ..., b - y_n\} > 0$

Now, if $x_1, x_2 \in [a, b]$ with $|\ x_1 - x_2\ | < \delta$ ...(2)

then either x_1, x_2 belong to the same sub-interval or they belong to two consecutive sub-intervals.

Case (*i*) $\mathbf{x_1, x_2}$ belong to the same sub-interval

$$| f(x_1) - f(x_2) | < \frac{\varepsilon}{2} < \varepsilon \qquad \text{[By (1)]}$$

Case (*ii*) $\mathbf{x_1, x_2}$ lie in consecutive intervals

Let y_i be the point of division of two sub-intervals in which x_1 and x_2 lie.

$\therefore$ x_1, y_i lie in the same sub-intervals and y_i, x_2 lie in the same sub-interval.

$$\therefore \qquad | f(x_1) - f(y_i) | < \frac{\varepsilon}{2} \qquad ...(3)$$

and

$$| f(y_i) - f(x_2) | < \frac{\varepsilon}{2} \qquad ...(4)$$

Now,

$$\begin{aligned} | f(x_1) - f(x_2) | &= | [f(x_1) - f(y_i)] + [f(y_i) - f(x_2)] | \\ &\le | f(x_1) - f(y_i) | + | f(y_i) - f(x_2) | \\ &< \frac{\varepsilon}{2} + \frac{\varepsilon}{2} = \varepsilon. \end{aligned}$$

Thus, $x_1, x_2 \in [a, b]$ with $| x_1 - x_2 | < \delta$,

$$| f(x_1) - f(x_2) | < \varepsilon.$$

Hence f is uniformly continuous on $[a, b]$.

6.5.3. Theorem. *Let K be a compact set. If a function $f : K \to R$ is continuous then f is uniformly continuous.*

***Proof.** Let $\varepsilon > 0$ be given.

Since f is continuous at a point $\xi \in K$, there exists a $\delta(\xi) > 0$ ($\delta(\xi)$ depending upon ε and ξ) such that $x \in K$ and $| x - \xi | < \delta(\xi)$

$$\Rightarrow \qquad | f(x) - f(\xi) | < \frac{\varepsilon}{2}.$$

Now, the open sets $\left(\xi - \frac{\delta(\xi)}{2}, \xi + \frac{\delta(\xi)}{2}\right)$, $\xi \in K$, form an open covering of K.

Since K is compact, by Heine-Borel Theorem, there exist finitely many points $\xi_1, \xi_2, ...,$ ξ_n such that the sets $\left(\xi_i - \frac{\delta(\xi_i)}{2}, \xi_i + \frac{\delta(\xi_i)}{2}\right)$, $1 \le i \le n$, form an open covering of K.

Let

$$\delta = \min_{1 \le i \le n} \delta(\xi_i) > 0$$

Let $x, y \in K$ such that $| x - y | < \frac{\delta}{2}$ ...(1)

Let

$$x \in \left(\xi_i - \frac{\delta(\xi_i)}{2}, \xi_i + \frac{\delta(\xi_i)}{2}\right) \qquad ...(2)$$

Now,

$$\begin{aligned} | \xi_i - y | &= | \xi_i - x + x - y | \\ &\le | \xi_i - x | + | x - y | \end{aligned}$$

*Not included in K.U. syllabus.

$$< \frac{\delta(\xi_i)}{2} + \frac{\delta}{2} \qquad \text{[By (1) and (2)]}$$

$$< \delta(\xi_i)$$

$$\Rightarrow \quad | f(\xi_i) - f(y) | < \frac{\varepsilon}{2} \qquad ...(3)$$

Since $\quad | x - \xi_i | < \dfrac{\delta(\xi_i)}{2},$ [By (2)]

$$\therefore \quad | f(x) - f(\xi_i) | < \frac{\varepsilon}{2} \qquad ...(4)$$

Now, $\quad | f(x) - f(y) | = | f(x) - f(\xi_i) + f(\xi_i) - f(y) |$

$$\leq | f(x) - f(\xi_i) | + | f(\xi_i) - f(y) |$$

$$< \frac{\varepsilon}{2} + \frac{\varepsilon}{2} = \varepsilon \qquad \text{[By (3) and (4)]}$$

Hence, f is uniformly continuous on K.

Cor. *A function which is continuous on a closed interval [a, b], in uniformly continuous on that interval.*

6.5.4. Some Solved Examples

Example 1. *Show that the function*

$$f(x) = \begin{cases} \sin \frac{1}{x}, & x \neq 0 \\ 0, & x = 0 \end{cases}$$

is discontinuous at x = 0 and the discontinuity is of second kind. (BCA (II) 2007)

Sol. We shall show that $\lim_{x \to 0} f(x)$ does not exist.

Let, if possible, $\lim_{x \to 0} f(x)$ exist and be l.

Take $\varepsilon = \frac{1}{2}$. Then $\exists\, \delta > 0$, such that

$$\left| \sin \frac{1}{x} - l \right| < \frac{1}{2}, \text{ whenever } 0 < | x | < \delta$$

Let $0 < | x_1 | < \delta$ and $0 < | x_2 | < \delta$.

Then,

$$\left| \sin \frac{1}{x_1} - \sin \frac{1}{x_2} \right| = \left| \left(\sin \frac{1}{x_1} - l \right) - \left(\sin \frac{1}{x_2} - l \right) \right|$$

$$\leq \left| \sin \frac{1}{x_1} - l \right| + \left| \sin \frac{1}{x_2} - l \right|$$

$$< \frac{1}{2} + \frac{1}{2} = 1.$$

Taking $x_1 = \frac{1}{2n\pi + \pi/2}$, $x_2 = \frac{1}{2n\pi - \pi/2}$

with n so large that $|x_1|$ and $|x_2|$ are both $< \delta$. (This is possible by Archimedean property of reals), we have

$$\left| \sin\left(2n\pi + \frac{\pi}{2}\right) - \sin\left(2n\pi - \frac{\pi}{2}\right) \right| < 1$$

i.e., $$2 < 1 \qquad \left[\because \sin\left(2n\pi + \frac{\pi}{2}\right) = \sin\frac{\pi}{2} = 1 \text{ and } \sin\left(2n\pi - \frac{\pi}{2}\right) = -1\right]$$

This contradiction implies that $\lim_{x \to 0} \sin\frac{1}{x}$ does not exist.

Example 2. *Show that the function*

$$f(x) = \begin{cases} x \sin \frac{1}{x}, & x \neq 0 \\ 0, & x = 0 \end{cases}$$

is continuous for every real x.

Sol. Let $x \in \mathbb{R}$

Case (*i*) $x = 0$.

Let $\varepsilon > 0$ be given. Then, for $\delta = \varepsilon$, we have

Now, $$|f(x) - f(0)| = \left| x \sin\frac{1}{x} - 0 \right| = \left| x \sin\frac{1}{x} \right|$$

$$= |x| \left| \sin\frac{1}{x} \right| \leq |x| < \varepsilon,$$

if $$|x| < \varepsilon \qquad \left[\because \left|\sin\frac{1}{x}\right| < 1\right]$$

Taking $\delta = \varepsilon$, $|f(x) - f(0)| < \varepsilon$, whenever $|x| < \delta$

Thus, f is continuous at $x = 0$.

Case (*ii*) $x \neq 0$.

Then, $1/x$ is continuous and $\sin 1/x$ is continuous. Since product of two continuous functions is continuous, it follows that $x \sin 1/x$ is continuous at x.

Example 3. *Dirichlet's Function. Show that the function*

$$f(x) = \begin{cases} 1 & \text{if } x \text{ is rational} \\ -1 & \text{if } x \text{ is irrational} \end{cases}$$

is discontinuous for every real x.

Sol. Case (*i*) *x is a rational number.*

Since in any interval there lie infinity many rationals as well as irrationals, for each $n \in \mathbb{N}$, there exists an irrational number x_n such that

$$x - \frac{1}{n} < x_n < x + \frac{1}{n} \quad \Rightarrow \quad |x_n - x| < \frac{1}{n}$$

$\Rightarrow$ the sequence $< x_n >$ is convergent to x.

But, $f(x_n) = -1 \quad \forall\, n$, and $f(x) = 1$

$\therefore \quad < f(x_n) >$ is convergent to $-1 \neq f$

$\therefore \quad f$ is discontinuous at x, any rational number.

Case *(ii) x is any irrational number.*

Since in any interval, there lie infinitely many rationals as well as irrationals, for each $n \in \mathrm{N}$, there exists a rational number x_n (say) such that

$$x - \frac{1}{n} < x_n < x + \frac{1}{n} \qquad \Rightarrow \quad |\, x_n - x \,| < \frac{1}{n}$$

$\Rightarrow$ the sequence $< x_n >$ is convergent to x.

But, $f(x_n) = 1 \quad \forall\, n$, and $f(x) = -1$

$\therefore \quad < f(x_n) >$ is convergent to $1 \neq f(x)$

$\therefore \quad f$ is discontinuous at x, any irrational number.

Example 4. *Show that the function f defined by*

$$f(x) = \begin{cases} x, & \text{if } x \text{ is rational} \\ -x, & \text{if } x \text{ is irrational} \end{cases}$$

continuous only at x = 0.

Sol. Case *(i) x is a non-zero rational number.*

For each $n \in \mathrm{N}$, there exists an irrational number x_n such that

$$x - \frac{1}{n} < x_n < x + \frac{1}{n} \qquad \Rightarrow \quad |\, x_n - x \,| < \frac{1}{n}$$

$\Rightarrow$ the sequence $< x_n >$ is convergent to x.

But, $f(x_n) = -x_n \quad \forall\, n \in \mathrm{N}$ and $f(x) = x$

$\therefore \quad < f(x_n) >$ is convergent to $-x \neq f(x)$

$\therefore \quad f$ is discontinuous at x, any rational number.

Case *(ii) x is an irrational number.*

Proof is similar to case *(i)*.

Case *(iii)* $x = 0$

Let $\varepsilon > 0$ be given.

Now, for a rational number x,

$$|\, f(x) - f(0) \,| = |\, x \,| < \varepsilon \quad \text{for} \quad |\, x \,| < \varepsilon$$

and for an irrational number x,

$$|\, f(x) - f(0) \,| = |\, -x \,| = |\, x \,| < \varepsilon \quad \text{for} \quad |\, x \,| < \varepsilon$$

$\therefore \quad$ for any $\varepsilon > 0$, $\exists\, \delta\, (= \varepsilon) > 0$ such that

$$|\, f(x) - f(0) \,| < \varepsilon, \text{ whenever } |\, x - 0 \,| < \delta$$

$\Rightarrow \quad f$ is continuous at $x = 0$.

Example 5. *Show that the function f defined by*

$$f(x) = 2x^2 - 3x + 5$$

is uniformly continuous on [– 2, 2].

Sol. Let $\varepsilon > 0$ be given

Let x_1, x_2 be any two points of $[-2, 2]$

$$\begin{aligned} \text{Then,} \quad |\, f(x_1) - f(x_2) \,| &= |\, 2(x_1^2 - x_2^2) - 3(x_1 - x_2) \,| = |\, (x_1 - x_2)(2(x_1 + x_2) - 3) \,| \\ &= |\, x_1 - x_2 \,|\, |\, 2(x_1 + x_2) - 3 \,| \\ &\leq |\, x_1 - x_2 \,|\, |\, 2(\,|\, x_1 \,| + |\, x_2 \,|\,) + 3 \,| \end{aligned}$$

$$\leq | x_1 - x_2 | [2(2+2)+3] \qquad (\because |x| \leq 2)$$

$$= 11 . | x_1 - x_2 | < \varepsilon \text{ for } | x_1 - x_2 | < \frac{\varepsilon}{11}$$

$\therefore$ for any $\varepsilon > 0$, $\exists\, \delta \left(= \frac{\varepsilon}{11}\right)$ such that

$$| f(x_1) - f(x_2) | < \varepsilon \text{ for } | x_1 - x_2 | < \delta.$$

$\Rightarrow$ f is uniformly continuous on $[-2, 2]$.

Example 6. *Prove that the function f defined by*

$$f(x) = \sin \frac{1}{x}, \; x \in R^+$$

is continuous but not uniformly continuous on R^+.

Sol. For $x \in R^+$, $\frac{1}{x}$ is continuous and $\sin x$ is continuous for each x.

$\therefore$ $\sin \frac{1}{x}$ is continuous for $x \in R^+$.

Now for any $\delta > 0$, $\exists\, m \in N$ such that $\dfrac{1}{2m\left(m\pi + \frac{\pi}{2}\right)} < \delta$

Take $\qquad x_1 = \dfrac{1}{m\pi + \frac{\pi}{2}}, \; x_2 = \dfrac{1}{m\pi}$

Then $x_1, x_2 \in R^+$

$$| x_1 - x_2 | = \left| \frac{1}{m\pi + \frac{\pi}{2}} - \frac{1}{m\pi} \right| = \frac{1}{2m\left(m\pi + \frac{\pi}{2}\right)} < \delta$$

but $$| f(x_1) - f(x_2) | = \left| \sin\left(m\pi + \frac{\pi}{2}\right) - \sin m\pi \right| = | \cos m\pi | = 1 \qquad (\because \sin m\pi = 0)$$

which is not less than each $\varepsilon > 0$.

Hence f is not uniformly continuous.

Example 7. *Examine for continuity the function f defined by*

$$f(x) = \lim_{n \to \infty} \frac{e^x - x^n \sin x}{1 + x^n}, \; 0 \leq x \leq \frac{\pi}{2}$$

at x = 1. Explain why the function f does not vanish anywhere on $\left[0, \frac{\pi}{2}\right]$ *although f(0).* $f\left(\frac{\pi}{2}\right) < 0$.

Sol. Since $\lim_{n \to \infty} x^n = \begin{cases} 0 & \text{if } 0 \le x < 1 \\ 1 & \text{if } x = 1 \\ \infty & \text{if } 1 < x \le \frac{\pi}{2} \end{cases}$

$$\therefore \qquad f(x) = \lim_{n \to \infty} \frac{e^x - x^n \sin x}{1 + x^n} = \begin{cases} e^x & \text{if } 0 \le x < 1 \\ \frac{e - \sin 1}{2} & \text{if } x = 1 \\ -\sin x & \text{if } 1 < x \le \frac{\pi}{2} \end{cases}$$

$$\therefore \qquad \lim_{x \to 1^-} f(x) = \lim_{x \to 1^-} e^x = e$$

$$\lim_{x \to 1^+} f(x) = \lim_{x \to 1^+} (-\sin x) = -\sin 1$$

$$\therefore \qquad \lim_{x \to 1^-} f(x) \ne \lim_{x \to 1^+} f(x)$$

$\Rightarrow \quad \lim_{x \to 1} f(x)$ does not exist

$\Rightarrow \quad f(x)$ is discontinuous at $x = 1$

Now $\qquad f(0) = e^0 = 1.$

Also $\qquad f\left(\frac{\pi}{2}\right) = -\sin \frac{\pi}{2} = -1.$

$$\therefore \qquad f(0)\, f\left(\frac{\pi}{2}\right) = -1 < 0.$$

Since 1, the point of discontinuity of f belongs to $\left[0, \frac{\pi}{2}\right]$, f is not continuous on $\left[0, \frac{\pi}{2}\right]$.

Intermediate value theorem is not applicable.

Example 8. *Prove that the function f defined by $f(x) = x^2$, $x \in R$ is uniformly continuous on every finite closed interval but is not uniformly continuous on R.* (K.U. BCA (II) 2003)

Sol. Let $[a, b]$ be a finite closed interval.

Let $\varepsilon > 0$ be given.

Let $c = \max. \{|a|, |b|\} > 0$.

Let $x_1, x_2 \in [a, b]$ be any two numbers.

$$\therefore \qquad |x_1| \le c, \ |x_2| \le c \qquad \text{...(1)}$$

Now, $|f(x_1) - f(x_2)| = |x_1^2 - x_2^2| = |x_1 + x_2|\,|x_1 - x_2|$

$\le (|x_1| + |x_2|)\,|x_1 - x_2| \le 2c\,|x_1 - x_2|$ [By (1)]

$\therefore \ |f(x_1) - f(x_2)| < \varepsilon$ whenever $|x_1 - x_2| < \frac{\varepsilon}{2c}$.

i.e., $\qquad |f(x_1) - f(x_2)| < \varepsilon$ whenever $|x_1 - x_2| < \delta$

where $\delta = \frac{\varepsilon}{2c} > 0$.

Thus $| f(x_1) - f(x_2) | < \varepsilon$ whenever $| x_1 - x_2 | < \delta$, $\forall\ x_1, x_2 \in [a, b]$

$\therefore$ f is uniformly continuous on $[a, b]$.

To show that f is not uniformly continuous on R

To show this, we have to show that for any given $\varepsilon > 0$ and any $\delta > 0$, we can find $x_1, x_2 \in$ R such that

$$| f(x_1) - f(x_2) | \geq \varepsilon \text{ but } | x_1 - x_2 | < \delta.$$

Define a sequence $< a_n >$ such that

$$a_n = \sqrt{n + 2\varepsilon} - \sqrt{n} = \frac{2\varepsilon}{\sqrt{n+2\varepsilon} + \sqrt{n}}$$

$\therefore$ as $n \to \infty$, $a_n \to 0$.

$\therefore$ for given $\delta > 0$, there exists a positive integer m such that

$$| a_n - 0 | < \delta \text{ whenever } n \geq m.$$

$\Rightarrow$ $| \sqrt{n+2\varepsilon} - \sqrt{n} \ | < \delta$ whenever $n \geq m$.

In particular, taking $n = m$,

$$| \sqrt{m+2\varepsilon} - \sqrt{m} \ | < \delta \qquad \text{...(2)}$$

Taking $x_1 = \sqrt{m + 2\varepsilon}$, and $x_2 = \sqrt{m}$ so that from (2)

$$| x_1 - x_2 | < \delta \text{ but } | f(x_1) - f(x_2) | = | m + 2\varepsilon - m | = 2\varepsilon > \varepsilon.$$

Thus, we have found $x_1, x_2 \in$ R such that $| x_1 - x_2 | < \delta$ but

$$| f(x_1) - f(x_2) | > \varepsilon.$$

Hence, f is not uniformly continuous.

EXERCISE 6 (*b*)

1. Give one example of each of the following :
   (*i*) A function on [0, 1] which is continuous everywhere except the end points.
   (*ii*) a function on [1, 2] which is continuous everywhere except at 3/2.
   (*iii*) a function continuous on (0, 1) but is not bounded.
   (*iv*) a function which is bounded on (0, 1) but is not continuous.
   (*v*) a function f continuous no where but $| f |$ continuous everywhere.
2. Examine the continuity of the following functions at the indicated point :

(*i*) $f(x) = \dfrac{x^3 - 8}{x - 2}$, at $x = 2$

(*ii*) $f(x) = \begin{cases} \dfrac{x^2 - 4}{x - 2}, & \text{if } x \neq 2 \\ 5, & \text{if } x = 2 \end{cases}$ at $x = 2$

(*iii*) $f(x) = \begin{cases} \dfrac{x^2 - 4}{x - 2}, & \text{if } x \neq 2 \\ 4, & \text{if } x = 2 \end{cases}$ at $x = 2$

(*iv*) $f(x) = \begin{cases} \dfrac{1}{x - 2}, & \text{if } x \neq 2 \\ 0, & \text{if } x = 2 \end{cases}$ at $x = 2$

(*v*) $f(x) = \begin{cases} e^{-\frac{1}{(x-2)^2}}, & \text{if } x \neq 2 \\ 0, & \text{if } x = 2 \end{cases}$ at $x = 2$

(*vi*) $f(x) = \begin{cases} x \cos \dfrac{1}{x}, & \text{if } x \neq 0 \\ 1, & \text{if } x = 0 \end{cases}$ at $x \in$ R (*vii*) $f(x) = \begin{cases} \dfrac{x - |x|}{x}, & x \neq 0 \\ 2, & x = 0 \end{cases}$ at $x \in R$ (*viii*) $f(x) = \dfrac{|x|}{x}$ at $x = 0$

(*ix*) $f(x) = \begin{cases} \dfrac{e^{\frac{1}{x}} - e^{-\frac{1}{x}}}{e^{\frac{1}{x}} + e^{-\frac{1}{x}}}, & \text{if } x \neq 0 \\ 0, & \text{if } x = 0 \end{cases}$ at $x = 0$ (*x*) $f(x) = \begin{cases} \dfrac{1 - \cos x}{x^2}, & \text{if } x \neq 0 \\ 1, & \text{if } x = 0 \end{cases}$ at $x = 0$

(*xi*) $f(x) = [x]$ on R^+ at $x = 1, 2, 3, \ldots\ldots$ (*xii*) $f(x) = \begin{cases} e^{\frac{1}{x}}; & x \neq 0 \\ 0; & x = 0 \end{cases}$ (*K.U. BCA (II) 2008*)

Also, discuss the type of discontinuity in each case.

3. Discuss the continuity at 1, 2, 3 of the following function :

$$f(x) = \begin{cases} 5x + 4, & x \le 1 \\ x^2 + 7x + 1, & 1 < x \le 2 \\ x + 3, & 2 < x < 3 \\ 5x + 2, & x \ge 3 \end{cases}$$

4. Obtain the points of discontinuity of the function defined on [0, 1] as follows :

$$f(x) = \begin{cases} 0 & \text{if } x = 0 \\ \frac{1}{2} - x & \text{if } 0 < x < \frac{1}{2} \\ \frac{1}{2} & \text{if } x = \frac{1}{2} \\ \frac{2}{3} - x & \text{if } \frac{1}{2} < x < 1 \\ 1 & \text{if } x = 1 \end{cases}$$

Also, examine the kinds of discontinuities.

5. (*a*) Show that the function f defined by

$$f(x) = \begin{cases} 1, & \text{if } x \text{ is rational} \\ 0, & \text{if } x \text{ is irrational} \end{cases}$$

is discontinuous at every point and discontinuity is of second kind.

(*b*) Show that $f(x) = \begin{cases} 4, & \text{if } x \text{ is rational} \\ -3, & \text{if } x \text{ is irrational} \end{cases}$ is discontinuous for each real x.

6. Show that the function f defined by

$$f(x) = \begin{cases} x, & \text{if } x \text{ is rational} \\ -x, & \text{if } x \text{ is irrational} \end{cases}$$

has a discontinuity of second kind at every real number except 0.

7. (*a*) Show that the function $[x]$, where $[x]$ denotes the largest integer $\le x$ is discontinuous at each integral value and the discontinuities are of first kind.

(*b*) Show that the function $x - [x]$ is discontinuous at every integral value of x and all discontinuities are of first kind from left.

8. (*a*) If a function f satisfies the inequality $f(x) \le x$, $\forall$ x, then show that f is continuous at $x = 0$.

(*b*) Let f be a continuous function at $x = 0$ and $f(0) = 0$. If g is another function such that $|g(x)| \le |f(x)|$ $\forall$ x, then g is continuous at $x = 0$.

9. Show that the function $f(x) = x^2$ is uniformly continuous on $[-1, 1]$.

10. Give an example of each of the following :

(*i*) a continuous bounded function on R may attain the l.u.b. but may not attain the g.l.b.

(*ii*) a continuous bounded function on R which attains the g.l.b. but not the l.u.b.

(*iii*) a function continuous on an open interval but may fail to be uniformly continuous.

11. If f and g are two continuous functions on $[a, b]$ such that $f(a) < g(a)$ and $f(b) > g(b)$, then show that there exists a real number $c \in (a, b)$ such that $f(c) = g(c)$.

[**Hint.** Apply Intermediate Value Theorem to the continuous function $f - g$ on $[a, b]$.]

12. If $f : (0, \infty) \to \mathrm{R}$ is a function defined by $f(x) = \dfrac{1}{x}$, prove that f is uniformly continuous on $[a, \infty)$ where $a > 0$. Show that f is continuous but not uniformly on $(0, \infty)$.

13. Show that the function :

$$\phi(y) = \begin{cases} y, & \text{if } y \text{ is irrational} \\ -y, & \text{if } y \text{ is rational} \end{cases}$$

is continuous only at $y = 0$.

[**Hint.** Proceed on lines of Example 5.]

14. If a function f is continuous and strictly monotonically decreasing in $[a, b]$, then f^{-1} is also continuous in $[f(b), f(a)]$.

[See Th. 6.4.14]

15. If a function f is continuous on R, then prove that the set $\mathrm{B} = \{x \mid x \in \mathrm{R} \text{ and } f(x) < 0\}$ is an open set.

16. (*a*) Prove that the function f, defined by $f(x) = x^2 - 3x + 5$ is uniformly continuous on $[-3, 4]$.

(*b*) Show that the function of defined by $f(x) = 2x^2 + 3x - 4$ is uniformly continuous on $[0, 2]$.

(BCA (II) 2006)

17. Show that the function $f(x) = \begin{cases} \dfrac{e^{\frac{1}{x}} - 1}{e^{\frac{1}{x}} + 1}, & x \neq 0 \\ 0, & x = 0 \end{cases}$

is discontinuous at $x = 0$. Write the type of discontinuity.

18. Let $f(x) = \begin{cases} 1 & \text{if } x \le 3 \\ ax + b & \text{if } 3 < x < 5 \\ 7 & \text{if } x \ge 5 \end{cases}$

Determine the constants a and b so that f may be continuous.

19. Prove that the function $f(x) = x^2$ is not uniformly continuous on $(-\infty, \infty)$.

[See second part of example 9]

20. If $f(x)$ and $g(x)$ are continuous functions on $[a, b]$ then $f(x) + g(x)$ is also continuous on $[a, b]$.

(K.U. BCA (II) 2001)

7

The Derivative and Mean Value Theorems

In universe, we observe several phenomena where changes are taking place. The motion of a planet around the sun, the speed of a car and the temperature at a fixed point of a place, are some such examples. In the motion of a planet around the sun, one would like to know (*i*) the speed at which it is moving at any instant (*ii*) instantaneous direction, (*iii*) its position relative to the sun after sometime, etc. The attempts of scientists to answer such questions is responsible for the origin and development of the derivative. In this chapter, we shall introduce the concept of derivative of a real valued function of a real variable and will prove Rolle's, Lagrange ; Mean Value and Taylor's Theorems.

7.1. DERIVABILITY

7.1.1. Derivability at an interior point. *A function f defined on [a, b] is said to be derivable or differentiable at* $c \in (a, b)$ *if* $\lim\limits_{h \to 0} \dfrac{f(c+h) - f(c)}{h}$ *exists.*

In case the limit exists, the same is called *the derivative of f at c or differential co-efficient of f at c* and is denoted by $f'(c)$ or $\dfrac{d}{dx} f(x) \Big|_{x=c}$. The process of getting $f'(c)$ is called the *differentiation*.

Left hand derivative

$\lim\limits_{h \to 0^-} \dfrac{f(c+h) - f(c)}{h}$, if exists is called the *left hand derivative* of f at c and is denoted by $f'(c^-)$.

Right hand derivative

$\lim\limits_{h \to 0^+} \dfrac{f(c+h) - f(c)}{h}$, if exists is called the *right hand derivative* of f at c and is denoted by $f'(c^+)$.

Clearly, $f'(c)$ exists if $f'(c^-)$ and $f(c^+)$ both exist and are equal.

7.1.2. Derivatives at the end points. *A function f defined on [a, b] is said to be derivable at a if* $\lim\limits_{h \to 0^+} \dfrac{(a+h) - f(a)}{h}$ *exists.*

The above limit is called the derivative of f at a and is denoted by $f'(a)$ (by which we mean $f'(a^+)$.

A function f defined on [a, b], is said to be derivable at b if $\lim_{h \to 0^-} \dfrac{f(b+h)-f(b)}{h}$ *exists.*

The above limit is called the derivative of f at b and is denoted by $f'(b)$ (by which we mean $f'(b^-)$).

7.1.3. Derivability in an interval. *A function f defined on an interval I is said to be derivable on I, if it is derivable at every point of the interval I.*

7.1.4. Derived function. If a function f defined on an interval I is derivable on the interval I, then there exists another function f' with domain I such that the value of f' for any $x \in$ I is $f'(x)$. The function f' is called the *derived function* of f or the derivative of f.

7.1.5. Theorem. *If a function is derivable at a point, then it is continuous at the point.*

Proof. Let f be derivable at a point a.

Then, $\lim_{h \to 0} \dfrac{f(a+h)-f(a)}{h}$ exists and is $f'(a)$

$$\text{Now} \quad \lim_{h \to 0} [f(a+h)-f(a)] = \lim_{h \to 0} \left[\frac{f(a+h)-f(a)}{h}.h\right]$$

$$= \lim_{h \to 0} \frac{f(a+h)-f(a)}{h} . \lim_{h \to 0} h = f'(a).0 = 0.$$

$$\Rightarrow \quad \lim_{h \to 0} f(a+h) = f(a)$$

$\Rightarrow$ f is continuous at a.

Remark. While continuity of a function is a necessary condition for the derivability of the function, it is not a sufficient condition as is clear from the following examples :

(*i*) Consider the function $f(x) = |x|, x \in$ R

$$\text{By definition} \quad f(x) = \begin{cases} x & \text{if } x \ge 0 \\ -x & \text{if } x < 0 \end{cases}$$

Here $f(0) = 0$.

$$\lim_{x \to 0^-} f(x) = \lim_{x \to 0^-} (-x) = 0$$

$$\lim_{x \to 0^+} f(x) = \lim_{x \to 0^+} x = 0 \quad \Rightarrow \quad f'(0^-) = f(0) = f(0^+)$$

$\Rightarrow$ the function is continuous at $x = 0$

$$\text{Now} \quad f'(0^-) = \lim_{h \to 0^-} \frac{f(h)-f(0)}{h} = \lim_{h \to 0^-} \frac{-h-0}{h} = -1$$

$$f'(0^+) = \lim_{h \to 0^+} \frac{f(h)-f(0)}{h} = \lim_{h \to 0^+} \frac{h-0}{h} = 1$$

$$\Rightarrow \quad f'(0^-) \ne f'(0^+)$$

$\Rightarrow$ $f'(0)$ does not exist.

(*ii*) Consider the function $f(x) = \begin{cases} x \sin \dfrac{1}{x} & \text{if } x \ne 0 \\ 0 & \text{if } x = 0 \end{cases}$

We have seen in previous chapter that $f(x)$ is continuous at $x = 0$.

$$\text{Now} \quad f'(0) = \lim_{h \to 0} \frac{f(0+h)-f(0)}{h} = \lim_{h \to 0} \frac{h \sin \frac{1}{h} - 0}{h}$$

$= \lim_{h \to 0} \sin \frac{1}{h}$, which does not exist as $\sin \frac{1}{h}$ oscillates between -1 and 1 infinitely many times when $h \to 0$. Hence $f'(0)$ does not exist.

Example 1. *Prove that the function f defined by*

$$f(x) = |x + 1| + |x| \quad \forall x \in R$$

is continuous but not derivable at x = – 1, 0. (K.U. BCA (II) 2006)

Sol. (*i*) *Continuity of f at x = – 1*

$$f(-1) = 0 + 1 = 1$$

$$\lim_{x \to -1^-} f(x) = \lim_{x \to -1^-} [\,|x+1| + |x|\,]$$

$$= \lim_{x \to -1^-} [-(x+1) + (-x)] = 1 \qquad \left[\because |x| = \begin{cases} x & \text{if } x > 0 \\ -x & \text{if } x \le 0 \end{cases}\right]$$

$$\lim_{x \to -1^+} f(x) = \lim_{x \to -1^+} [\,|x+1| + |x|\,]$$

$$= \lim_{x \to -1^+} |(x+1) + (-x)| = 1$$

Thus $\lim_{x \to -1^-} f(x) = f(-1) = \lim_{x \to -1^+} f(x)$

$\Rightarrow$ f is continuous at $x = -1$

Derivability of f at x = – 1

We have
$$f'(-1^-) = \lim_{h \to 0^-} \frac{f(-1+h) - f(-1)}{h}$$

$$= \lim_{h \to 0^-} \frac{|-1+h+1| + |-1+h| - 1}{h} \qquad (\because |h| = -h \text{ as } h < 0)$$

$$= \lim_{h \to 0^-} \frac{-h+1-h-1}{h} = -2$$

$$f'(-1^+) = \lim_{h \to 0^+} \frac{f(-1+h) - f(-1)}{h}$$

$$= \lim_{h \to 0^+} \frac{|-1+h+1| + |-1+h| - 1}{h} = \lim_{h \to 0^+} \frac{h+1-h-1}{h} = 0$$

$\therefore$ $f'(-1^-) \ne f'(-1^+)$

$\Rightarrow$ $f'(-1)$ does not exist

$\Rightarrow$ f is not derivable at $x = -1$.

(*ii*) The case $x = 0$ can be dealt with on the same lines.

Example 2. *Examine for continuity and derivability of the function f defined by*

$$f(x) = \begin{cases} x\,\dfrac{e^{1/x} - e^{-1/x}}{e^{1/x} + e^{-1/x}} & \text{if } x \ne 0 \\ 0 & \text{if } x = 0. \end{cases}$$

Sol. *Continuity of f at x = 0*

$$f(0) = 0$$

$$\lim_{x \to 0^-} f(x) = \lim_{x \to 0^-} x\,\frac{e^{1/x} - e^{-1/x}}{e^{1/x} + e^{-1/x}}$$

$$= \lim_{x \to 0^-} x \,.\, \lim_{x \to 0^-} \frac{e^{2/x} - 1}{e^{2/x} + 1}$$

$$= 0 \,.\, \frac{0-1}{0+1} \qquad (\because \lim_{x \to 0^-} e^{2/x} = e^{-\infty} = 0)$$

$$= 0$$

$$= \lim_{x \to 0^+} f(x) = \lim_{x \to 0^+} x \frac{e^{1/x} - e^{-1/x}}{e^{1/x} + e^{-1/x}} = \lim_{x \to 0^+} \,.\, \frac{1 - e^{-2/x}}{1 + e^{-2/x}}$$

$$= 0 \,.\, \frac{1-0}{1+0} \qquad (\because \lim_{x \to 0^+} e^{-2/x} = 0)$$

$\therefore$ $$\lim_{x \to 0^-} f(x) = 0 = \lim_{x \to 0^+} f(x)$$

$\Rightarrow$ f is continuous at $x = 0$.

Derivability of f at $x = 0$

$$f'(0^-) = \lim_{h \to 0^-} \frac{f(0+h) - f(0)}{h}$$

$$= \lim_{h \to 0^-} \frac{h\left(\dfrac{e^{1/h} - e^{-1/h}}{e^{1/h} + e^{-1/h}}\right) - 0}{h} = \lim_{h \to 0^-} \frac{e^{2/h} - 1}{e^{2/h} - 1} = \frac{0-1}{0+1} = -1$$

$$f'(0^+) = \lim_{h \to 0^+} \frac{f(0+h) - f(0)}{h}$$

$$= \lim_{h \to 0^+} \frac{h\left(\dfrac{e^{1/h} - e^{-1/h}}{e^{1/h} + e^{-1/h}}\right) - 0}{h} = \lim_{h \to 0^+} \frac{1 - e^{-2/h}}{1 + e^{-2/h}} = \frac{1-0}{1+0} = 1.$$

$\therefore$ $$f'(0^-) \neq f'(0^+)$$

$\Rightarrow$ $f'(0)$ does not exist

$\Rightarrow$ f is not derivable at $x = 0$.

Example 3. *Prove that the function f defined by*

$$f(x) = \begin{cases} x^2 \sin \dfrac{1}{x} & \text{if } x \neq 0 \\ 0 & \text{if } x = 0 \end{cases}$$

is continuous and derivable at $x = 0$ but its derivative is not continuous at $x = 0$.

Sol. *Continuity of f at $x = 0$*

$$| f(x) - f(0) | = \left| x^2 \sin \frac{1}{x} - 0 \right| = | x^2 | \left| \sin \frac{1}{x} \right|$$

$$\leq x^2 \cdot 1 < \varepsilon \text{ for } |x| < \sqrt{\varepsilon} \qquad \left(\because \left|\sin\frac{1}{x}\right| \leq 1 \text{ for } x \neq 0\right)$$

$\Rightarrow$ for any $\varepsilon > 0, \exists\ \delta\ (= \sqrt{\varepsilon}) > 0$ such that

$\Rightarrow$ $|f(x) - f(0)| < \varepsilon$ whenever $|x - 0| < \delta$.

$\Rightarrow$ f is continuous at $x = 0$.

Derivability of f at x = 0.

For $x = 0$,
$$f'(0) = \lim_{h \to 0} \frac{f(0+h) - f(0)}{h}$$

$$= \lim_{h \to 0} \frac{h^2 \sin\frac{1}{h} - 0}{h} = \lim_{h \to 0} h \sin\frac{1}{h} = \lim_{h \to 0} h \cdot \lim_{h \to 0} \sin\frac{1}{h}$$

$= 0 \times a$ finite quantity lying between -1 and 1

$= 0$

For $x \neq 0$,
$$f'(x) = 2x \sin\left(\frac{1}{x}\right) + x^2 \cos\frac{1}{x} \cdot \left(-\frac{1}{x^2}\right) = 2x \sin\frac{1}{x} - \cos\frac{1}{x}$$

$\Rightarrow$ $f'(x)$ exists $\forall\ x \in$ R and is given by

$$f'(x) = \begin{cases} 2x \sin\frac{1}{x} - \cos\frac{1}{x} & \text{if } x \neq 0 \\ 0 & \text{if } x = 0 \end{cases}$$

Continuity of f′ at x = 0.

$$\lim_{x \to 0} f'(x) = \lim_{x \to 0}\left(2x \sin\frac{1}{x} - \cos\frac{1}{x}\right)$$

$$= 2 \lim_{x \to 0} x \cdot \lim_{x \to 0} \sin\frac{1}{x} - \lim_{x \to 0} \cos\frac{1}{x} = 0 - \lim_{x \to 0} \cos\frac{1}{x},$$

which does not exist as $\cos\frac{1}{x}$ oscillates between 1 and 1 as $x \to 0$.

Hence f' is not continuous at $x = 0$.

EXERCISE 7 (*a*)

1. Prove that the following functions are continuous but not derivable :

(*i*) $f(x) = \begin{cases} 2 + x & \text{if } x \geq 0 \\ 2 - x & \text{if } x < 0 \end{cases}$ at $x = 0$

(*ii*) $f(x) = \begin{cases} x^2 - 1 & \text{if } x \geq 1 \\ 1 - x & \text{if } x < 1 \end{cases}$ at $x = 1$

(*iii*) $f(x) = \begin{cases} x \sin\frac{1}{x} & \text{if } x \neq 0 \\ 0 & \text{if } x = 0 \end{cases}$ at $x = 0$

(*iv*) $f(x) = \begin{cases} x \tan^{-1}\frac{1}{x} & \text{if } x \neq 0 \\ 0 & \text{if } x = 0 \end{cases}$

2. If $f'(x) < 0$, $\forall\ x \in [a, b]$, then show that f is strictly decreasing on $[a, b]$.

3. Give an example of a function which is continuous at a point but is not differentiable at that point.

7.2. ROLLE'S THEOREM*

If a function f defined on [a, b] is such that it is

(i) continuous on [a, b] *(ii) derivable on (a, b)*

(iii) f(a) = f(b)

then there exists at least one $c \in (a, b)$ such that $f'(c) = 0$. (K.U. BCA (II) 2006, 2007)

Proof. If f is a constant function, then $f'(x) = 0 \ \forall \ x \in (a, b)$ and we have nothing to prove. So, let f be not a constant function.

Since f is continuous on $[a, b]$, f is bounded and attains its bounds in $[a, b]$.

Let M = l.u.b. of f and m = g.l.b. of f on $[a, b]$.

Then, $\exists\, c, d \in [a, b]$, such that $f(c) = \text{M}$ and $f(d) = m$. Clearly $\text{M} \neq m$ (as otherwise $f(x) = \text{M} \ \forall \ x$ and hence would be a constant function). Therefore, either of M on m is different from $f(a)$ and $f(b)$. ($\because \ f(a) = f(b)$)

Suppose that $\text{M} = f(c)$ is different from $f(a)$ and $f(b)$. Then $c \neq a$ and $c \neq b$.

$\Rightarrow \qquad c \in (a, b)$.

Since f is derivable on (a, b), f is derivable at c also.

$$\therefore \qquad \lim_{h \to 0^-} \frac{f(c+h) - f(c)}{h} \quad \text{and} \quad \lim_{h \to 0^+} \frac{f(c+h) - f(c)}{h}$$

both exist and are equal each being equal to $f'(c)$

Since $\qquad f(c) = \text{M}$ (= l.u.b. of f)

$$\therefore \qquad \lim_{h \to 0^-} \frac{f(c+h) - f(c)}{h} \geq 0 \quad \text{and} \quad \lim_{h \to 0^+} \frac{f(c+h) - f(c)}{h} \leq 0$$

$$\Rightarrow \qquad f'(c) \geq 0 \quad \text{and} \quad f'(c) \leq 0$$

$$\Rightarrow \qquad f'(c) = 0.$$

Remarks 1. Converse of Rolle's theorem is not true, *i.e.*, $f'(x)$ may vanish at a point $c \in (a, b)$ without $f(x)$ satisfying the three conditions of Rolle's theorem.

2. There may exist more than one real number $c \in (a, b)$ such that $f'(c) = 0$ but Rolle's theorem ensures the existence of *at least* one such c.

3. The hypothesis of Rolle's theorem cannot be weakened. For exmaple, if $f(x) = 1 - |x|$, $-1 \leq x \leq 1$, then $f(-1) = f(1) = 0$ and f is continuous on $[-1, 1]$. Also, $f'(x)$ exists for all $x \in (-1, 1)$, except at $x = 0$. Thus, f satisfies all the conditions of Rolle's theorem except that f is not derivable at 0. For this f, there is no c in $(-1, 1)$ for which $f'(c) = 0$.

4. Geometrically, Rolle's theorem means that if the curve $y = f(x)$ is continuous from $x = a$ to $x = b$, has a definite tangent at each point of (a, b) and the ordinates at the extremities are equal, then there exists at least one point between a anb b at which the tangent is parallel to x-axis.

5. Algebraically, Rolle's theorem means that if $f(x)$ is a polynomial function in x and $x = a$ and $x = b$ are two roots of the equation $f(x) = 0$, then there is, at least one root of the equation $f'(x) = 0$ which lies between a and b.

Example 1. *Verify Rolle's theorem to the function*

$$f(x) = e^{-x} \sin x, \ [0, \pi].$$

Sol. Here $\qquad f(x) = e^{-x} \sin x, \ \forall \ x \in [0, \pi]$

e^{-x} and $\sin x$ are both continuous as well as derivable $\forall \ x \in \text{R}$.

---

*After the name of a British mathematician Michael Rolle (1652—1719).

$\Rightarrow$ $e^{-x} \sin x$ is continuous as well as derivable $\forall\ x \in \mathrm{R}$.

$\Rightarrow$ $e^{-x} \sin x$ is continuous as well as derivable on $[0, \pi]$ also.

Also $\qquad f(0) = e^0 \sin 0 = 0, f(\pi) = e^{-\pi} \sin \pi = 0$

$\therefore \qquad f(0) = f(\pi)$

Thus f satisfies all the 3 conditions of Rolle's theorem on $[0, \pi]$. Hence Rolle's theorem is applicable on $[0, \pi]$.

$\therefore\ \exists\ c \in (0, \pi)$ such that $\qquad f'(c) = 0.$

But $\qquad f'(x) = -e^{-x} \sin x + e^{-x} \cos x$

$\therefore\ f'(c) = 0 \Rightarrow -e^{-c} \sin c + e^{-c} \cos c = 0 \qquad \Rightarrow e^{-c}(-\sin c + \cos c) = 0$

$\Rightarrow \qquad -\sin c + \cos c = 0 \qquad (\because\ e^{-c} \neq 0)$

$$\Rightarrow \qquad \tan c = 1 \qquad \Rightarrow \qquad c = \frac{\pi}{4} \in (0, \pi).$$

Hence Rolle's theorem is verified to f on $[0, \pi]$.

Example 2. *Examine the applicability of Rolle's theorem to the function*

$$f(x) = 2 + (x - 1)^{2/3} \text{ on } [0, 2].$$

(K.U. BCA (II) 2003)

Sol. $f'(x) = \dfrac{2}{3(x-1)^{1/3}}$, which is defined for every x except at $x = 1$. So, the function is derivable on $(0, 2)$ except perhaps at $x = 1$.

Also, $\qquad f(0) = f(0^+)$ and $f(2^-) = f(2)$.

Since derivability implies continuity, therefore the function is continuous on $[0, 2]$ except possibly at $x = 1$.

Now $\qquad | f(x) - f(1) | = | (x-1)^{2/3} | < \varepsilon$ for $| x - 1 | < \varepsilon^{3/2}$

i.e., for any $\varepsilon > 0$, $\exists\ \delta\ (= \varepsilon^{3/2})$ such that

$$| f(x) - f(1) | < \varepsilon, \text{ whenever } | x - 1 | < \delta.$$

$\Rightarrow$ f is continuous at $x = 1$ also.

Let us now check the derivability of f at $x = 1$.

$$f'(1^-) = \lim_{h \to 0^-} \frac{f(1+h) - f(1)}{h} = \lim_{h \to 0^-} \frac{2 + (1+h-1)^{2/3} - 2}{h}$$

$$= \lim_{h \to 0^-} \frac{h^{2/3}}{h} = \lim_{h \to 0^-} \frac{1}{h^{1/3}} = -\infty$$

$$f'(1^+) = \lim_{h \to 0^+} \frac{f(1+h) - f(1)}{h} = \lim_{h \to 0^+} \frac{1}{h^{1/3}} = \infty$$

$\Rightarrow \qquad f'(1^-) \neq f'(1^+) \qquad \Rightarrow f'(1)$ does not exist

$\Rightarrow$ f is not derivable on $(0, 2)$

$\Rightarrow$ Rolle's theorem is not applicable on $[0, 2]$.

Example 3. *Show that for no value of k, the equation $x^3 - 3x + k = 0$ has two different roots in (0, 1).*

Sol. Let $\qquad f(x) = x^3 - 3x + k$

Let, if possible, the equation $f(x) = 0$ has two different roots α, β in $(0, 1)$, $f(x)$ being a polynomial, f is continuous and derivable $\forall\ x \in \mathrm{R}$.

$\Rightarrow$ f is continuous on $[\alpha, \beta]$ and f is derivable on (α, β).

Also $$f(\alpha) = f(\beta) = 0$$

$\therefore$ by Rolle's theorem, $\exists\, c \in (\alpha, \beta)$ such that

$$f'(c) = 0 \quad \Rightarrow \quad 3c^2 - 3 = 0 \qquad (\because \; f'(x) = 3x^2 - 3)$$

$\Rightarrow$ $c = \pm 1$ which contradicts that $c \in (\alpha, \beta) \subseteq (0, 1)$.

Hence the result.

Example 4. *Prove that if $a_0, a_1, \ldots\ldots, a_n, \in R$ are such that $\dfrac{a_0}{n+1} + \dfrac{a_1}{n} + \ldots + \dfrac{a_{n-1}}{2} + a_n = 0$, then there exists at least one $x \in (0, 1)$ such that $a_0 x^n + a_1 x^{n-1} + \ldots + a_n = 0$.*

Sol. Consider the function

$$f(x) = a_0 \frac{x^{n+1}}{n+1} + a_1 \frac{x^n}{n} + \ldots + a_{n-1} \frac{x^2}{2} + a_n x$$

Then, $f(x)$ being a polynomial, f is continuous and derivable $\forall\, x$.

$\Rightarrow$ f is continuous on $[0, 1]$ and derivable on $(0, 1)$. Also

$$f(0) = 0, f(1) = \frac{a_0}{n+1} + \frac{a_1}{n} + \ldots + \frac{a_{n-1}}{2} + a_n = 0 \text{ (given)}$$

$\therefore$ by Rolle's theorem, $\exists$ at least one $c \in (0, 1)$ such that $f'(c) = 0$.

Now $$f'(x) = a_0 x^n + a_1 x^{n-1} + \ldots + a_{n-1} x + a_n$$

$\therefore$ $f'(c) = 0$ for at least one $c \in (0, 1)$

$\Rightarrow$ $a_0 x^n + a_1 x^{n-1} + \ldots + a_n$ vanishes at least once in $(0, 1)$.

Example 5. *Using Rolle's theorem, show that between any two roots of $e^x \cos x = 1$, there exists at least one root of $e^x \sin x - 1 = 0$.*

Sol. Let α and β be any two different roots of $e^x \cos x = 1$.

$$\therefore \quad e^{\alpha} \cos \alpha - 1 = 0 \quad \text{and} \quad e^{\beta} \cos \beta - 1 = 0$$

$$\Rightarrow \quad \cos \alpha = e^{-\alpha} \quad \text{and} \quad \cos \beta = e^{-\beta} \qquad \ldots(1)$$

Consider the function f defined by

$$f(x) = e^{-x} - \cos x \text{ on } [\alpha, \beta].$$

$f(x)$ is clearly continuous on $[\alpha, \beta]$ and differentiable on (α, β).

Also, $$f(\alpha) = e^{-\alpha} - \cos \alpha = 0$$

$$f(\beta) = e^{-\beta} - \cos \beta = 0 \qquad \text{[By (1)]}$$

$\therefore$ f satisfies all the conditions of Rolle's theorem on $[\alpha, \beta]$.

$\therefore$ There exists c, $\alpha < c < \beta$ such that $f'(c) = 0$

$$\Rightarrow \quad -e^{-c} + \sin c = 0 \quad \Rightarrow \quad e^c \sin c - 1 = 0 \text{ and } c \in (\alpha, \beta)$$

$\Rightarrow$ the equation $e^x \sin x - 1 = 0$ has at least one root c between any two roots of $e^x \cos x - 1 = 0$.

EXERCISE 7 (*b*)

1. Verify Rolle's theorem for the following functions :

(*i*) $f(x) = (x - a)^m (x - b)^n$ on $[a, b]$ where $m, n \in$ N *(K.U. BCA (II) 2008)*

(*ii*) $f(x) = x^3 - 6x^2 + 11x - 6$ on $[1, 3]$. *(K.U. BCA (II) 2001)*

2. Discuss the applicability of Rolle's theorem to the function

$$f(x) = | x | \text{ on } [-1, 1].$$

3. Verify Rolle's theorem for the function $f(x) = \begin{cases} x^2 + 1, & 0 \le x \le 1 \\ 3 - x, & 1 \le x \le 2 \end{cases}$.

4. If $f(x)$ is a polynomial in x and $f'(x) \ne 0$ anywhere between a and b, show that $f(x)$ can have at the most one root in $[a, b]$. *(K.U. BCA (II) 2007)*

[**Sol.** If possible, let α, β be two different roots of $f(x) = 0$ in $[a, b]$. $f(x)$ being a polynomial in x, is continuous on $[\alpha, \beta]$ and differentiable on (α, β). Moreover $f(\alpha) = f(\beta) = 0$.

$\therefore$ by Rolle's theorem, $\exists$ at least one $c \in (\alpha, \beta)$ such that $f'(c) = 0$, which is a contradiction.]

7.3. LAGRANGE'S MEAN VALUE THEOREM* (OR FIRST MEAN VALUE THEOREM OF DIFFERENTIAL CALCULUS)

If a function f defined on [a, b] is

(i) continuous on [a, b] and *(ii) derivable on (a, b)*

then there exists at least one real number c, a < c < b such that

$$\frac{f(b) - f(a)}{b - a} = f'(c).$$ (K.U. BCA (II) 2001)

Proof. Consider a function $\phi(x) = f(x) + Ax$

where A is a constant to be determined such that $\phi(a) = \phi(b)$. This gives us

$$A = -\frac{f(b) - f(a)}{b - a} \quad \ldots(1)$$

Now, (*i*) Since f is continuous on $[a, b]$ and Ax is continuous on $[a, b]$, therefore, ϕ is continuous on $[a, b]$.

(*ii*) Since f is derivable on (a, b) and Ax is derivable on (a, b), therefore ϕ is derivable on (a, b).

(*iii*) $\phi(a) = \phi(b)$ (By def. of ϕ).

Thus, ϕ satisfies all the conditions of Rolle's theorem on $[a, b]$.

Therefore, there exists a real number c, $a < c < b$,

such that $\phi'(c) = 0$

But, $\phi'(x) = f'(x) + A$

$\therefore$ $0 = \phi'(c) = f'(c) + A \Rightarrow f'(c) = -A \quad \ldots(2)$

$\therefore$ from (1) and (2) $\dfrac{f(b) - f(a)}{b - a} = f'(c)$ or $f(b) - f(a) = (b - a) f'(c)$

Another form of statement

If in the statement of the theorem, b is replaced by a + h, then the number c between a and b may be written as a + θh, 0 < θ < 1.

$$f(a + h) - f(a) = hf'(a + \theta h)$$

or

$$f(a + h) = f(a) + hf'(a + \theta h), 0 < \theta < 1$$

Cor. 1. *If f is continuous on [a, b] and is differentiable on (a, b), then*

(*i*) $f'(x) = 0 \ \forall \ x \in (a, b) \Rightarrow f$ is constant on $[a, b]$. *(K.U. BCA (II) 2003)*

---

*After the name of French Mathematician Joseph Louis lagrange (1736–1813).

(*ii*) $f'(x) > 0 \ \forall \ x \in (a, b) \Rightarrow f$ is strictly increasing on $[a, b]$.

(*iii*) $f'(x) < 0 \ \forall \ x \in (a, b) \Rightarrow f$ is strictly decreasing on $[a, b]$.

Proof (*i*) Let x_1, x_2 (where $x_1 < x_2$) be any two distinct points of $[a, b]$. Since f is continuous on $[a, b]$, f is continuous on $[x_1, x_2]$. Since f is differentiable on (a, b), f is differentiable on (x_1, x_2).

Therefore, by Lagrange's mean value theorem, there exists at least one c, $x_1 < c < x_2$ such that

$$\frac{f(x_2) - f(x_1)}{x_2 - x_1} = f'(c).$$

But $\quad f'(c) = 0 \qquad [\because \ c \in (x_1, x_2) \subset (a, b) \text{ and } f'(x) = 0 \ \forall \ x \in (a, b)]$

$\therefore \quad f(x_2) - f(x_1) = 0$

$\Rightarrow \quad f(x_2) = f(x_1), \ \forall \ x_1, x_2 \in [a, b]$

Thus, the function has the same value and is therefore constant on $[a, b]$.

(*ii*) Let x_1, x_2 (where $x_1 < x_2$) be any two different points of $[a, b]$. Then, proceeding as in case (*i*) above, we have

$$\frac{f(x_2) - f(x_1)}{x_2 - x_1} = f'(c), \text{ for some } c, x_1 < c < x_2.$$

But $\quad f'(c) > 0 \qquad [\because \ f'(x) > 0, \ \forall \ x \in (a, b)]$

$\therefore \quad f(x_2) - f(x_1) > 0 \quad \Rightarrow \quad f(x_2) > f(x_1)$

Thus, $\quad x_2 > x_1 \quad \Rightarrow \quad f(x_2) > f(x_1), \ \forall \ x_1, x_2 \in [a, b]$

$\therefore f$ is strictly increasing on $[a, b]$.

Remark. For a strictly increasing function f, the derivative $f'(x)$ need not be strictly positive. For example, consider $f(x) = x^3$, $x \in (-1, 1)$. This function f is strictly increasing. But $f'(x) = 3x^2$ which is zero at $x = 0 \in (-1, 1)$.

(*iii*) As in Case (*ii*).

Cor. 2. *If two functions f and g are*

(*i*) *continuous on* $[a, b]$ (*ii*) *derivable on* (a, b)

(*iii*) $f'(x) = g'(x) \ \forall \ x \in (a, b)$,

then $f - g$ *is a constant function.*

Proof. Let $\quad F(x) = f(x) - g(x), \ \forall \ x \in [a, b]$

Since f and g are continuous on $[a, b]$

$\therefore$ F is continuous on $[a, b]$

Similarly F is differentiable on (a, b)

Also, $\quad F'(x) = f'(x) - g'(x) = 0, \ \forall \ x \in (a, b).$

$\Rightarrow$ F is a constant function on $[a, b]$ *i.e.*, $f - g$ is a constant function on $[a, b]$.

7.3.1. Geometrical Interpretation. Geometrically, Lagrange's Mean Value Theorem says that between two points A and B on the graph of f, there exists at least one point C where the tangent is parallel to the chord AB.

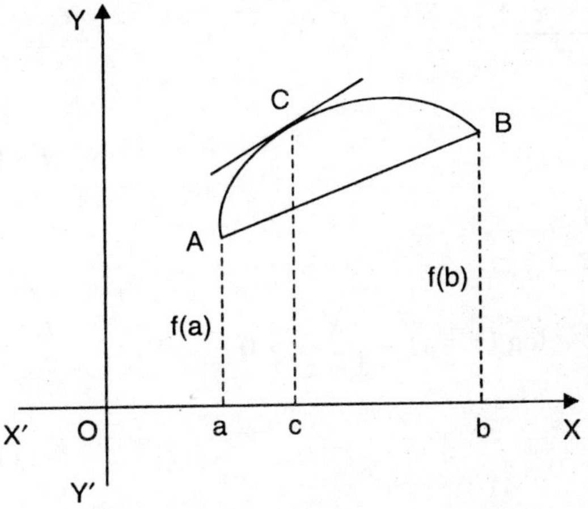

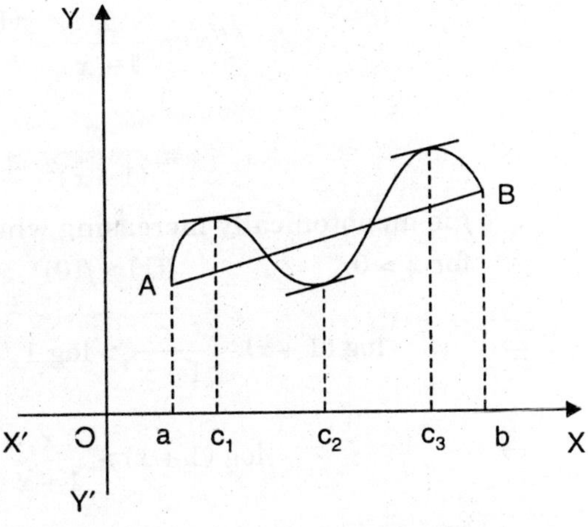

Remark. The hypothesis of Lagrange's Mean Value Theorem cannot be weakened, as is clear from the following examples :

(*i*) Consider the function f defined on [1, 2] as follows :

$$f(x) = \begin{cases} 2, & x = 1 \\ x^2, & 1 < x < 2 \\ 1, & x = 2 \end{cases}$$

Then, f is continuous as well as derivable on (1, 2) but is not continuous at 1 and 2. Thus, first condition is violated ?

Also, $\dfrac{f(2) - f(1)}{2 - 1} = -1$

$$f'(x) = 2x, \; 1 < x < 2$$

$\Rightarrow$ $f'(x) > 0$ for $1 < x < 2$

Thus, $\dfrac{f(2) - f(1)}{2 - 1} \neq f'(x)$ for any $x \in (1, 2)$.

(*ii*) Consider the function $f(x) = |\, x \,|$ defined on [– 1, 2].

Then, f is continuous on [– 1, 2] and derivable at all points of (– 1, 2) except at $x = 0$. (Thus, second condition is violated).

Now, $f'(x) = \begin{cases} -1, & \text{if } x \in (-1, 0) \\ 1, & \text{if } x \in (0, 2) \end{cases}$

Also $\dfrac{f(2) - f(-1)}{2 - (-1)} = \dfrac{1}{3}$

Thus, $\dfrac{f(2) - f(-1)}{2 - (-1)} \neq f'(x)$ for any $x \in (-1, 2)$.

Example 1. *Show that if $x > 0$, $\log (1 + x) > \dfrac{x}{1 + x}$. Also show that $\dfrac{\log (1 + x)}{x}$ monotonically decreases as x increases from 0 to ∞.*

Sol. Let $f(x) = \log (1 + x) - \dfrac{x}{1 + x}$.

$\therefore \qquad f'(x) = \dfrac{1}{1+x} - \dfrac{(1+x).1 - x.1}{(1+x)^2}$

$\qquad = \dfrac{x}{(1+x)^2} > 0 \qquad (\because x > 0)$

$\therefore$ f is monotonically increasing when $x > 0$.

$\therefore$ for $x > 0$, $\qquad f(x) > f(0)$

$\Rightarrow \qquad \log(1+x) - \dfrac{x}{1+x} > \log 1 \qquad \Rightarrow \log(1+x) - \dfrac{x}{1+x} > 0$

$\Rightarrow \qquad \log(1+x) > \dfrac{x}{1+x} \qquad ...(1)$

Let $\qquad F(x) = \dfrac{\log(1+x)}{x}$

$\therefore \qquad F'(x) = \dfrac{x \cdot \dfrac{1}{1+x} - \log(1+x)}{x^2}$

$\therefore \qquad F'(x) < 0$ for $x > 0 \qquad$ [By (1)]

$\therefore$ the function F is monotonically decreasing for $x > 0$.

Example 2. *If f is a twice differentiable function on $[a, b]$ such that $f(a) = f(b) = 0$ and $f(c) > 0$ for $a < c < b$, prove that there is at least one ξ between a and b for which $f''(\xi) < 0$.*

Sol. Since f is twice differentiable on $[a, b]$, therefore f and f' are both continuous and differentiable on $[a, b]$ and hence, in particular, continuous and differentiable on $[a, c]$ and $[c, b]$ for $a < c < b$.

Applying Lagrange's mean value theorem to f on $[a, c]$ and $[c, b]$, we have

$$\frac{f(c) - f(a)}{c - a} = f'(c_1),\ a < c_1 < c \qquad ...(1)$$

and

$$\frac{f(b) - f(c)}{b - c} = f'(c_2),\ c < c_2 < b \qquad ...(2)$$

But $\qquad f(a) = f(b) = 0 \qquad$ (given)

$\therefore$ From (1) and (2), $\qquad f'(c_1) = \dfrac{f(c)}{c - a}$ and $f'(c_2) = -\dfrac{f(c)}{b - c} \qquad ...(3)$

Applying Lagrange's mean value theorem to f' on $[c_1, c_2]$, we have

$$\frac{f'(c_2) - f'(c_1)}{c_2 - c_1} = f''(\xi), \qquad \text{where } c_1 < \xi < c_2.$$

But, from (3), $\qquad f'(c_1) > 0$ and $f'(c_2) < 0 \qquad (\because f(c) > 0$ for $a < c < b)$ (given)

$\therefore \qquad f''(\xi) < 0$, for at least one ξ between a and b.

EXERCISE 7 (c)

1. Examine the validity of the hypothesis and the conclusion of Lagrange's Mean Value Theorem for the following functions :

   (i) $f(x) = |x|$ on $x \in [-1, 1]$ (ii) $f(x) = x(x-1)(x-2), x \in [0, \frac{1}{2}]$

   (iii) $f(x) = x^{1/3}, x \in [-1, 1]$ (iv) $f(x) = 2x^2 - 7x + 10, y \in [2, 5]$.

2. If a function f is such that its derivative f', is continuous on $[a, b]$ and derivable on (a, b), then show that there exists a number c between a and b that

$$f(b) = f(a) + (b-a)f'(a) + \tfrac{1}{2}(b-a)^2 f''(c).$$

   [**Hint.** Apply Rolle's Theorem to the function

$$\phi(x) = f(b) - f(x) - (b-x)f'(x) - (b-x)^2 \text{A and get A} = \tfrac{1}{2} f''(x)].$$

3. If a function f is twice derivable on $[a, a+h]$, then show that

$$f(a+h) = f(a) + hf'(a) + \frac{h^2}{2} f''(a+\theta h), \; 0 < \theta < 1.$$

4. Use Lagrange's mean value theorem to prove that

$$1 + x < e^x < 1 + e^x, \quad \forall\, x > 0.$$

7.4. HIGHER ORDER DERIVATIVES

We know that the existence of the derivative f' of a function f at a point c implies the existence and continuity of the function in a neighbourhood of c. The derivative of the function f' at c, if exists, is called the second derivative of f at c and is denoted by $f''(c)$. The existence of $f''(c)$ implies the existence and continuity of f' in a neighbourhood of c. Similar is the case for higher order derivatives. The nth derivative of f at c is denoted by $f^n(c)$.

7.4.1. Taylor's Theorem*. *If a function f defined on [a, a +h] is such that*

(i) the (n – 1)th derivative f^{n-1} is continuous on [a, a + h] and

(ii) the nth derivative f^n exists on (a, a + h), then there exists a real number θ, 0 < θ < 1, such that

$$f(a+h) = f(a) + hf'(a) + \frac{h^2}{2!} f''(a) + \frac{h^3}{3!} f'''(a) + \ldots$$
$$+ \frac{h^{n-1}}{(n-1)!} f^{n-1}(a) + \frac{h^n (1-\theta)^{n-p}}{p((n-1)!)} f^n(a+\theta h) \quad \ldots(1)$$

where p is a given positive integer.

Proof. Since f^{n-1} exists, all the derivatives $f', f'', \ldots, f^{n-1}$ exist and continuous on $[a, a+h]$ consider *a* function ϕ defined on $[a, a+h]$ such that

$$\phi(x) = f(x) + (a+h-x)f'(x) + \frac{(a+h-x)^2}{2!} f''(x) + \ldots$$
$$+ \frac{(a+h-x)^{n-1}}{(n-1)!} f^{n-1}(x) + \text{A}(a+h-x)^p \quad \ldots(2)$$

where A is a constant to be determined such that

$$\phi(a+h) = \phi(a)$$

*After the name of an English Mathematician Brook Taylor (1685—1731).

But, $\phi(a + h) = f(a + h)$

and $$\phi(a) = f(a) + h f'(a) + \frac{h^2}{2!} f''(a) + ... + \frac{h^{n-1}}{(n-1)!} f^{n-1}(a) + Ah_p$$

$$\therefore \quad f(a + h) = f(a) + hf'(a) + \frac{h^2}{2!} f''(a) + ... + \frac{h^{n-1}}{(n-1)!} f^{n-1}(a) + Ah_p \quad ...(3)$$

(*i*) Now, $f, f', f'', ..., f^{n-1}$ being all continuous on $[a, a + h]$, the function ϕ is continuous on $[a, a + h]$;

(*ii*) the functions $f, f', ..., f^{n-1}$ and $(a + h - x)^r$ for all r, being all derivable on $(a, a + h)$, the function ϕ is derivable on $(a, a + h)$; and

(*iii*) $\phi(a + h) = \phi(a)$.

Thus, the function ϕ satisfies all the conditions of Rolle's theorem and hence $\exists$ at a real number θ, $0 < \theta < 1$, such that

$$\phi'(a + \theta h) = 0.$$

But, $$\phi'(x) = f'(x) + [-f'(x) + (a + h - x) f''(x)]$$
$$+ \frac{1}{2!} [-2(a + h - x) f''(x) + (a + h - x)^2 f'''(x)] + ...$$
$$+ \frac{1}{(n-1)!} [-(n-1)(a + h - x)^{n-2} f^{n-1}(x)$$
$$+ (a + h - x)^{n-1} f^n(x)] - Ap(a + h - x)^{p-1}$$
$$= \frac{(a + h - x)^{n-1}}{(n-1)!} f^n(x) - Ap(a + h - x)^{p-1}$$

(Other terms cancel in pairs)

$$\therefore \quad 0 = \phi(a + \theta h) = \frac{h^{n-1}(1-\theta)^{n-1}}{(n-1)!} f^n(a + \theta h) - Aph^{p-1}(1-\theta)^{p-1}$$

$$\Rightarrow \quad A = \frac{h^{n-p}(1-\theta)^{n-p}}{p(n-1)!} f^n(a + \theta h), \; h \neq 0, \; \theta \neq 1$$

Substituting A from (3) in (2), we get the required result.

7.4.2. Forms of Remainder after n terms :

(*i*) the term $R_n = \dfrac{h^n (1-\theta)^{n-p}}{p((n-1)!)} f^n(a + \theta h)$,

which occurs after n terms, is called the Taylor's remainder after n terms. The theorem with this form of remainder is called Taylor's theorem with *Schlomilch* and *Roche form of remainder*.

(*ii*) For $p = 1$, we get $\quad R_n = \dfrac{h^n (1-\theta)^{n-1}}{(n-1)!} f^n(a + \theta h)$

and is called *Cauchy's form of remainder*.

(*iii*) For $p = n$, we get $\quad R_n = \dfrac{h^n}{n!} f^n(a + \theta h)$

and is called *Lagrange's form of remainder*.

7.4.3. Second form of Taylor's Theorem. Replacing $a + h$ by x (or h by $x - a$) in (1), we get

$$f(x) = f(a) + (x-a)f'(a) + \frac{(x-a)^2}{2!}f''(a) + \ldots + \frac{(x-a)^{n-1}}{(n-1)!}f^{n-1}(a) + \frac{(x-a)^n}{p((n-1)!)}(1-\theta)^{n-p}f^n(a+\theta(x-a))$$

where $0 < \theta < 1$ ...(4)

The remainder after n terms can be written as

$$R_n = \frac{(x-a)^n (1-\theta)^{n-p}}{p((n-1)!)} f^n(c), a < c < x.$$

7.4.4. Maclaurin's Theorem. Putting $a = 0$ in (4), we have for $x \in (0, h)$,

$$f(x) = f(0) + xf'(0) + \frac{x^2}{2!}f''(0) + \ldots + \frac{x^{n-1}}{(n-1)!}f^{n-1}(0) + \frac{x^n(1-\theta)^{n-p}}{p((n-1)!)}f^n(\theta x)$$

and is called *Maclaurin's theorem with Schlomilch and Roche form of remainder.*

For $p = 1$, $R_n = \frac{x^n(1-\theta)^{n-1}}{(n-1)!} f^n(\theta x)$ and is *Cauchy's form of remainder.*

For $p = n$, $R_n = \frac{x^n}{n_1} f^n(\theta x)$ and is Lagrange's form of remainder.

EXERCISE 7 (*d*)

1. Prove Taylor's theorem with Lagrange's form of remainder by considering the function

$$\phi(x) = f(x) + (a+h-x)f'(x) + \frac{(a+h-x)^2}{2!}f''(x) + \ldots + \frac{(a+h-x)^{n-1}}{(n-1)!}f^{n-1}(x) + A(a+h-x)^n.$$

2. Prove Taylor's theorem with Cauchy's form of remainder by considering the function

$$\phi(x) = f(x) + (a+h-x)f'(x) + \frac{(a+h-x)^2}{2!}f''(x) + \ldots + \frac{(a+h-x)^{n-1}}{(n-1)!}f^{n-1}(x) + A(a+h-x).$$

3. Applying Lagrange's Mean Value Theorem to the function

$$\phi(x) = f(x) + (a+h-x)f'(x) + \frac{(a+h-x)^2}{2!}f''(x) + \ldots + \frac{(a+h-x)^{n-1}}{(n-1)!}f^{n-1}(x) + Ah_p$$

prove Taylor's theorem with Cauchy form of remainder.

Remark. *For n = 1, the theorem reduces to the Mean Value Theorem. For this reason, Taylor's theorem is also called the Generalized Mean Value Theorem.*

4. Using Taylor's theorem, prove that

$$x - \frac{x^3}{3!} \le \sin x \le x - \frac{x^3}{3!} + \frac{x^5}{5!}; x > 0.$$ *(K.U. BCA (II) 2008)*

5. Expand $\sin x$ as an infinite series. *(K.U. BCA (II) 2008)*

8

Arbitrary Series and Infinite Products

In chapter 5, we have considered the convergence and absolute convergence of a series of arbitrary terms—especially Leibnitz's test for the convergence of alternating series. In this chapter, we shall deal with tests for convergence of series of arbitrary signs, derive some conclusions regarding grouping and rearrangement of terms of a series, Cauchy product of two series and convergence of infinite products.

8.1. ABEL AND DIRICHLET'S TESTS

8.1.1. Abel's Lemma*. *Let the sequence $< S_n >$ of partial sums of the series $\sum_{n=1}^{\infty} a_n$ satisfy $m \le S_n \le M$, $(n \in N)$ and let $< b_n >$ be a non-increasing sequence of non-negative real numbers, then for $n \in N$,*

$$mb_1 \le \sum_{k=1}^{n} a_k b_k \le Mb_1.$$

Proof. Since $\quad S_1 = a_1,\ S_2 = a_1 + a_2, \ldots\ldots, S_n = a_1 + a_2 + \ldots.. + a_n$

$\therefore \quad a_1 = S_1,\ a_2 = S_2 - S_1, \ldots\ldots, a_n = S_n - S_{n-1}$

$$\therefore \quad S = \sum_{k=1}^{n} a_k b_k = a_1 b_1 + a_2 b_2 + \ldots\ldots + a_n b_n$$
$$= S_1 b_1 + (S_2 - S_1) b_2 + \ldots\ldots + (S_n - S_{n-1}) b_n$$
$$= S_1(b_1 - b_2) + S_2(b_2 - b_3) + \ldots\ldots + S_{n-1}(b_{n-1} - b_n) + S_n b_n$$

Now,
$$mb_1 = m(b_1 - b_2) + m(b_2 - b_3) + \ldots\ldots + m(b_{n-1} - b_n) + mb_n$$
$$\le S \qquad (\because \ m \le S_n \le M)$$
$$\le M(b_1 - b_2) + M(b_2 - b_3) + \ldots\ldots + M(b_{n-1} - b_n) + Mb_n = Mb_1$$

$$\Rightarrow \quad mb_1 \le S \le Mb_1.$$

8.1.2. Theorem (Abel's Test). *If $\sum_{n=1}^{\infty} a_n$ is convergent and the sequence $< b_n >$ is monotonic and bounded, then $\sum_{n=1}^{\infty} a_n b_n$ is convergent.* *(K.U. BCA (II) 2007, 08)*

Proof. Since the sequence $< b_n >$ is monotonic and bounded, it is convergent. Let $\lim_{n \to \infty} b_n = b$.

---

*After the name of a distinguished Norwegian mathematician N.H. Abel (1802—1829).

Let $$u_n = \begin{cases} b - b_n, & \text{if } < b_n > \text{ is increasing} \\ b_n - b, & \text{if } < b_n > \text{ is decreasing} \end{cases}$$

Then, $u_n \geq u_{n+1} \geq 0, n = 1, 2, \ldots..$

$$\therefore \quad a_n b_n = \begin{cases} a_n b - a_n u_n & \text{if } < b_n > \text{ is increasing} \\ a_n b + a_n u_n & \text{if } < b_n > \text{ is decreasing} \end{cases} \quad \ldots(1)$$

$$\left(\because \begin{array}{l} u_n = b - b_n \Rightarrow b_n = b - u_n \\ u_n = b_n - b \Rightarrow b_n = b + u_n \end{array} \right)$$

Now, $\sum\limits_{n=1}^{\infty} a_n$ convergent $\Rightarrow \sum\limits_{n=1}^{\infty} b\, a_n$ convergent.

$\therefore$ from (1), it is enough to show that $\sum\limits_{n=1}^{\infty} a_n u_n$ is convergent.

Let $\varepsilon > 0$ be given.

Since $\sum\limits_{n=1}^{\infty} a_n$ is convergent, there exists a natural number k such that

$$| a_{m+1} + a_{m+2} + \ldots\ldots + a_n | < \varepsilon \text{ for, } n > m \geq k.$$

Applying Abel's Lemma, we have

$$\begin{aligned} & | a_{m+1} u_{m+1} + a_{m+2} u_{m+2} + \ldots\ldots + a_n u_n | \\ & \leq \varepsilon u_{m+1} \\ & < \varepsilon u_1, n > m \geq k \end{aligned} \qquad (\because < u_n > \text{ is decreasing})$$

This implies that the series $\sum\limits_{n=1}^{\infty} a_n u_n$ is convergent.

Remark 1. *The sequence $< b_n >$ is monotonic and bounded means the sequence $< b_n >$ is either convergent or oscillates between finite limits. However, the proof when the sequence $< b_n >$ oscillates between finite limits is not given.*

2. *The above theorem states that a convergent series $\sum\limits_{n=1}^{\infty} a_n$ (which may not be absolutely convergent) remains convergent if its terms are each multiplied by a factor b_n, provided that the sequence $< b_n >$ is monotonic and bounded.*

Example 1. *Test the convergence of the series*

$$(i)\ 0 - \frac{1}{2} + \frac{1}{2^2} - \frac{1}{3} + \frac{2}{3^2} - \frac{1}{4} + \frac{3}{4^2} - \ldots\ldots \qquad (ii)\ 1 - \frac{1}{3.2^2} + \frac{1}{5.3^2} - \frac{1}{7.4^2} + \ldots\ldots$$

Sol. (*i*) The series can be considered to have arisen as a result of multiplying the terms of the series

$$1 - 1 + \frac{1}{2} - \frac{1}{2} + \frac{1}{3} - \frac{1}{3} + \frac{1}{4} - \frac{1}{4} + \ldots\ldots \quad \ldots(1)$$

with the terms of the sequence

$$0, \frac{1}{2}, \frac{1}{2}, \frac{2}{3}, \frac{2}{3}, \frac{3}{4}, \frac{3}{4}, \ldots\ldots \quad \ldots(2)$$

Now, the series in (1) is convergent and the sequence in (2) is monotonic and bounded. Therefore, by Abel's test, the given series is convergent.

(*ii*) Take $\sum_{n=1}^{\infty} a_n = \frac{1}{1^2} - \frac{1}{2^2} + \frac{1}{3^2} - \frac{1}{4^2} + \ldots\ldots$ and $< b_n > = < \frac{1}{2n-1} >$.

Then, by Leibnitz test, the series $\sum_{n=1}^{\infty} a_n$ is convergent and $< b_n >$ is a monotonically decreasing sequence of positive terms.

Therefore, by Abel's test, the given series $\sum_{n=1}^{\infty} a_n b_n$ is convergent.

Example 2. *Show that the series* $\sum_{n=1}^{\infty} \frac{(-1)^{n-1}}{n}\left(1+\frac{1}{n}\right)^{-n}$ *is convergent.*

Sol. Let $a_n = \frac{(-1)^{n-1}}{n}$, $b_n = \left(1+\frac{1}{n}\right)^{-n}$

By Leibnitz's test, the series $\sum_{n=1}^{\infty} a_n$ is convergent.

Also, the sequence $< b_n >$ is positive and monotonically decreasing (show it).

$\therefore$ by Abel's test, the series

$$\sum_{n=1}^{\infty} a_n b_n = \sum_{n=1}^{\infty} \frac{(-1)^{n-1}}{n}\left(1+\frac{1}{n}\right)^{-n} \text{ is convergent.}$$

8.1.3. Theorem (Dirichlet's Test)*. *If* $\sum_{n=1}^{\infty} a_n$ *has bounded partial sums and* $< b_n >$ *is a monotonic sequence converging to zero, then* $\sum_{n=1}^{\infty} a_n b_n$ *is convergent.*

Proof. Replacing b_n by $-b_n$, if necessary, we assume that $< b_n >$ decreases to 0.

Let $S_n = a_1 + a_2 + \ldots\ldots + a_n$, $n \in \mathbb{N}$.

Since $\sum_{n=1}^{\infty} a_n$ has bounded partial sums, there exists $K > 0$ such that $|S_n| \le K$, $n \in \mathbb{N}$.

Therefore, for $n > m$,

$$|a_{m+1} + \ldots\ldots + a_n| = |S_n - S_m| \le |S_n| + |S_m| \le 2K$$

Let $\varepsilon > 0$ be given.

Since $< b_n >$ converges to 0, there exists a positive integer m_0 such that

$$b_n < \frac{\varepsilon}{2K}, \forall n \ge m_0$$

Then, for $n > m \ge m_0$, we have by Abel's lemma,

$$|a_{m+1}b_{m+1} + \ldots\ldots + a_n b_n| \le 2Kb_{m+1} < \varepsilon.$$

---

*After the name of a distinguished German Mathematician P.G.L. Dirichlet (1805—1859).

This implies that the series $\sum_{n=1}^{\infty} a_n b_n$ is convergent.

Cor. Leibnitz's test as a particular case of Dirichlet's test

The sequence $< (-1)^{n-1} >$ has bounded partial sums. If $< a_n >$ is a positive, monotonically decreasing to 0 sequence, then by Dirichlet's test, the series $\sum_{n=1}^{\infty} (-1)^{n-1} a_n$ is convergent, *i.e.*, the alternating series $a_1 - a_2 + a_3 - a_4 + ...$ is convergent.

Example 3. *Test the convergence of the series* $\sum_{n=1}^{\infty} \frac{\cos nx}{n^p}$, $(p > 0)$. (K.U. BCA (II) 2006)

Sol. Let $a_n = \cos nx$, $b_n = \frac{1}{n^p}$, $(p > 0)$.

Here $< b_n >$ is positive and decreasing sequence with

$$\lim_{n \to \infty} b_n = \lim_{n \to \infty} \frac{1}{n^p} = 0.$$

Now,

$$S_n = \cos x + \cos 2x + \cos 3x + ... + \cos nx$$

$$= \frac{\cos \frac{x + nx}{2} \sin \left(\frac{nx}{2}\right)}{\sin \frac{x}{2}}$$

$$\left[\because \cos \alpha + \cos (\alpha + \beta) + \cos (\alpha + 2\beta) + ... + \cos (\alpha + \overline{n-1}\,\beta) = \frac{\cos \frac{\alpha + \alpha + \overline{n-1}\beta}{2} \sin \left(n\frac{\beta}{2}\right)}{\sin \frac{\beta}{2}}\right]$$

$$\therefore \quad | S_n | \le \frac{\left|\cos (n+1)\frac{x}{2}\right| . \left|\sin \left(\frac{nx}{2}\right)\right|}{\left|\sin \frac{x}{2}\right|} \le \frac{1.1}{\left|\sin \frac{x}{2}\right|}$$

$\Rightarrow$ $< S_n >$ is bounded for all x for which $\sin \frac{x}{2} \neq 0$ *i.e.*, for which $x \neq 2\pi r$, $r \in Z$.

$\Rightarrow$ by Dirichlet's test, $\sum_{n=1}^{\infty} a_n b_n = \sum_{n=1}^{\infty} \frac{\cos nx}{n^p}$, $(p > 0)$ is convergent except possibly for $x = 2\pi r$, $r \in Z$.

For $x = 2\pi r$, the series $\sum_{n=1}^{\infty} \frac{\cos nx}{n^p}$, $(p > 0)$ becomes $\sum_{n=1}^{\infty} \frac{1}{n^p}$, which is convergent if $p > 1$ and divergent if $p \le 1$.

Remark. *If $p \le 0$, then the nth term of the series $\sum_{n=1}^{\infty} \frac{\cos nx}{n^p}$ does not tend to 0, the series diverges.*

Example 4. *Discuss the convergence of the series $\sum_{n=1}^{\infty} \frac{1}{n}\left(1+\frac{1}{2}+\ldots\ldots+\frac{1}{n}\right) \sin(n\theta+\alpha)$.*

Sol. Let
$$a_n = \sin(n\theta + \alpha)$$
$$b_n = \frac{1}{n}\left(1+\frac{1}{2}+\ldots\ldots+\frac{1}{n}\right)$$

Now,
$$b_n - b_{n+1} = \frac{1}{n}\left(1+\frac{1}{2}+\ldots\ldots+\frac{1}{n}\right) - \frac{1}{n+1}\left(1+\frac{1}{2}+\ldots\ldots+\frac{1}{n+1}\right)$$
$$= \left(1+\frac{1}{2}+\ldots\ldots+\frac{1}{n}\right)\left(\frac{1}{n}-\frac{1}{n+1}\right) - \left(\frac{1}{n+1}\right)^2$$
$$\ge \frac{1}{n}-\frac{1}{n+1}-\left(\frac{1}{n+1}\right)^2 = \frac{1}{n}-\frac{1}{n+1}\left(1+\frac{1}{n+1}\right) = \frac{1}{n}-\frac{n+2}{(n+1)^2}$$
$$= \frac{(n+1)^2 - n(n+2)}{n(n+1)^2} = \frac{1}{n(n+1)^2} > 0$$

$\Rightarrow$ $< b_n >$ is a decreasing sequence.

Since $\lim_{n\to\infty} \frac{1}{n} = 0$.

$\therefore$ by Cauchy's first theorem on limits,
$$\lim_{n\to\infty} \frac{1+\frac{1}{2}+\ldots\ldots+\frac{1}{n}}{n} = 0$$

$\Rightarrow$
$$\lim_{n\to\infty} b_n = 0.$$

Also,
$$| S_n | = | \sin(\theta+\alpha) + \ldots\ldots + \sin(n\theta+\alpha) |$$
$$= \left| \frac{\sin\left(\alpha + \frac{n+1}{2}\theta\right)\sin\frac{n\theta}{2}}{\sin\frac{\theta}{2}} \right| = \frac{\left|\sin\left(\alpha+\frac{n+1}{2}\theta\right)\right|\left|\sin\frac{n\theta}{2}\right|}{\left|\sin\frac{\theta}{2}\right|} \le \frac{1}{\left|\sin\frac{\theta}{2}\right|}$$

$\Rightarrow$ $< S_n >$ is bounded for each θ except when $\sin\frac{\theta}{2} = 0$, *i.e.*, when $\theta = 2\pi r$, $r \in Z$.

Therefore, by Dirichlet's test, the series
$$\sum_{n=1}^{\infty} a_n b_n = \sum_{n=1}^{\infty} \frac{1}{n}\left(1+\frac{1}{2}+\ldots\ldots+\frac{1}{n}\right)\sin(n\theta+\alpha)$$
is convergent except when $\theta = 2\pi r$, $r \in Z$. If $\theta = 2\pi r$, $r \in Z$, then the series reduces to
$$\sum_{n=1}^{\infty} \frac{1}{n}\left(1+\frac{1}{2}+\ldots\ldots+\frac{1}{n}\right)\sin\alpha,$$
which is convergent if α is a multiple of π.

If α is not a multiple of π, then the series diverges because

$$\frac{1}{n}\left(1+\frac{1}{2}+\ldots\ldots+\frac{1}{n}\right) \geq \frac{1}{n},\ n \in \mathrm{N} \text{ and } |\sin\alpha| \leq 1.$$

Example 5. *If* $a_n = \dfrac{1}{\sqrt{n}} + \dfrac{(-1)^n}{n}$, *prove that the series* $a_1 - a_2 + a_3 - a_4 + a_5 - \ldots\ldots$ *diverges to* $-\infty$. *What condition is violated for the failure of the series to converge ?*

Sol. Consider

$$\begin{aligned} S_{2n} &= a_1 - a_2 + a_3 - a_4 + \ldots\ldots + a_{2n-1} - a_{2n} \\ &= \left(1-\frac{1}{1}\right) - \left(\frac{1}{\sqrt{2}}+\frac{1}{2}\right) + \left(\frac{1}{\sqrt{3}}-\frac{1}{3}\right) - \left(\frac{1}{\sqrt{4}}+\frac{1}{4}\right) + \ldots\ldots \\ &\quad + \left(\frac{1}{\sqrt{2n-1}} - \frac{1}{2n-1}\right) - \left(\frac{1}{\sqrt{2n}} + \frac{1}{2n}\right) \\ &= \left(1 - \frac{1}{\sqrt{2}} + \frac{1}{\sqrt{3}} - \frac{1}{\sqrt{4}} + \ldots\ldots - \frac{1}{\sqrt{2n}}\right) - \left(1 + \frac{1}{2} + \frac{1}{3} + \ldots\ldots + \frac{1}{2n}\right) \end{aligned}$$

The series $1 - \dfrac{1}{\sqrt{2}} + \dfrac{1}{\sqrt{3}} - \dfrac{1}{\sqrt{4}} + \ldots\ldots$ is convergent by Leibnitz's theorem and the series $\displaystyle\sum_{n=1}^{\infty} \frac{1}{n}$ is divergent to ∞.

$\therefore$ $\qquad S_{2n} \to -\infty \quad \text{as} \quad n \to \infty$

Also, $\qquad S_{2n+1} = S_{2n} + a_{2n+1}$

$\Rightarrow \qquad \lim\limits_{n\to\infty} S_{2n+1} = \lim\limits_{n\to\infty} S_{2n} + \lim\limits_{n\to\infty} a_{2n+1} = \lim\limits_{n\to\infty} S_{2n} \qquad (\because \lim\limits_{n\to\infty} a_{2n+1} = 0)$

$\Rightarrow \qquad \lim\limits_{n\to\infty} S_n = -\infty$

Hence, the series is divergent to $-\infty$.

In this example, $a_n \to 0$ as $n \to \infty$, but the convergence is not monotonic.

[For, $a_1 = 0,\ a_2 = \dfrac{1}{\sqrt{2}} + \dfrac{1}{2},\ a_3 = \dfrac{1}{\sqrt{3}} - \dfrac{1}{3}$

$\therefore \quad a_1 \leq a_2$ but $a_2 \geq a_3$]

Hence, Leibnitz theorem is not applicable.

EXERCISE 8 (*a*)

1. Show that the series $\displaystyle\sum_{n=2}^{\infty} \frac{(n^3+1)^{1/3} - n}{\log n}$ is convergent.

[**Hint.** The series $\displaystyle\sum_{n=2}^{\infty} (n^3+1)^{1/3} - n$ is convergent and $\left< \dfrac{1}{\log n} \right>$ is a positive monotonically decreasing sequence.]

2. Show that the convergence of $\sum_{n=1}^{\infty} a_n$ implies the convergence of each of the following series :

(*i*) $\sum_{n=1}^{\infty} \frac{1}{n} a_n$ (*ii*) $\sum_{n=1}^{\infty} \frac{1}{n^p} a_n , p \geq 0$ (*iii*) $\sum_{n=1}^{\infty} n^{\frac{1}{n}} a_n$

(*iv*) $\sum_{n=2}^{\infty} \frac{1}{\log n} a_n$ (*v*) $\sum_{n=1}^{\infty} \frac{n+1}{n} a_n$ (*vi*) $\sum_{n=1}^{\infty} \left(1+\frac{1}{n}\right)^n a_n$.

[**Hint.** (*iii*) For $n \geq 2$,

$$\left(1+\frac{1}{n}\right)^n = 1 + n \cdot \frac{1}{n} + \frac{n(n-1)}{2!} \frac{1}{n^2} + \ldots\ldots + \left(\frac{1}{n}\right)^n$$

$$= 1 + 1 + \frac{1}{2!}\left(1-\frac{1}{n}\right) + \frac{1}{3!}\left(1-\frac{1}{n}\right)\left(1-\frac{2}{n}\right)$$

$$+ \ldots\ldots + \frac{1}{n!}\left(1-\frac{1}{n}\right)\left(1-\frac{2}{n}\right)\ldots\ldots\left(1-\frac{n-1}{n}\right)$$

$$\leq 1 + 1 + \frac{1}{2!} + \frac{1}{3!} + \ldots\ldots + \frac{1}{n!} < e < 3.$$

$\therefore$ for $n \geq 3$, $n > \left(1+\frac{1}{n}\right)^n = \frac{(n+1)^n}{n^n}$

$\Rightarrow$ $n^{n+1} > (n+1)^n$ $\Rightarrow$ $n > (n+1)^{\frac{n}{n+1}}$

$\Rightarrow$ $n^{\frac{1}{n}} > (n+1)^{\frac{1}{n+1}}$ $\Rightarrow$ $< n^{\frac{1}{n}} >$ is monotonic.

Also, $\lim_{n \to \infty} n^{\frac{1}{n}} = 1$

$\therefore$ $< n^{\frac{1}{n}} >$ is bounded.

Hence, by Abel's test, $\sum_{n=1}^{\infty} n^{\frac{1}{n}} a_n$ is convergent.]

3. Show that the series

$$\sum_{n=1}^{\infty} (-1)^{n-1} \cdot \frac{1}{n-\alpha}, \ \alpha \text{ not an integer} \geq 1$$

and $$\sum_{n=1}^{\infty} (-1)^{n-1} \cdot \frac{1}{\sqrt{n+\alpha}}, \ \alpha \text{ not an integer} \leq -1 \text{ are both convergent.}$$

[**Hint.** $< (-1)^{n-1} >$ has bounded partial sums and $< \frac{1}{n-\alpha} >$ decreases to zero, etc.]

4. (*a*) Show that the series :

(*i*) $\sum_{n=1}^{\infty} \frac{\sin nx}{n^p}$, $p > 0$ converges for all real x.

(*ii*) $\sum_{n=1}^{\infty} \frac{(-1)^n \sin nx}{n^p}$, $p > 0$ converges for all real x.

(*iii*) $\sum_{n=1}^{\infty} \frac{(-1)^n \cos nx}{n^p}$, $p > 0$ converges provided x is not an odd multiple of π.

(*b*) Use Dirichlet's test to show that the series $\sum_{n=1}^{\infty} \frac{\sin (nx)}{n}$ is convergent for $x \in \mathbf{R}$.

5. If $< a_n >$ is a monotonic sequence converging to 0, then show that the following series are convergent :

(*i*) $\sum_{n=1}^{\infty} a_n \sin n\theta$ for all values of θ.

(*ii*) $\sum_{n=1}^{\infty} a_n \cos n\theta$ for all real values of $\theta \neq 2r\pi$, $r \in \mathbf{Z}$.

[Apply Dirichlet's test]

6. Show that the following series are convergent :

(*i*) $\sum_{n=1}^{\infty} \frac{(n\,!)^2}{(2n)\,!} \cos nx$ (*ii*) $\sum_{n=1}^{\infty} \frac{(n\,!)^2}{(2n)\,!} \sin n\theta$.

[**Hint.** $< \cos nx >$ has bounded partial sums.

To show : $\frac{(n\,!)^2}{(2n)\,!}$ is monotonically decreasing to 0.

$$\frac{a_n}{a_{n+1}} = \frac{(2n+2)(2n+1)}{(n+1)(n+1)} > 1 \quad \Rightarrow \quad a_n > a_{n+1} \; \forall n$$

$\Rightarrow$ $< a_n >$ is monotonically decreasing.

Also, $$a_n = \frac{(n\,!)(n\,!)}{(2n)\,!} = \frac{n\,!}{2n\,.\,(2n-1)\ldots n} \le \frac{1}{n} \quad \Rightarrow \quad a_n \to 0 \text{ as } n \to \infty.]$$

7. If $\sum a_n$ is convergent and if b_n's are positive such that $b_0 \ge b_1 \ge b_2 \ge \ldots\ldots$ prove that $\sum a_n b_n$ is convergent.

8.2. INSERTION AND REMOVAL OF PARENTHESES

It is a remarkable fact, as the following illustration shows, that parentheses may not be inserted at will in an infinite series, because such a grouping of terms can alter the convergence behaviour of the series.

Consider the series $\sum_{n=1}^{\infty} (-1)^{n-1} = 1 - 1 + 1 - 1 + \ldots\ldots$

If parentheses are inserted to obtain

$$(1-1)+(1-1)+ \ldots\ldots + (1-1)+ \ldots\ldots,$$

then the sum of each pair of terms is 0 and the series converges to 0. But, if we write

$$1-(1-1)-(1-1)- \ldots\ldots,$$

then again the sum of pairs of terms in the parentheses is 0 and the series converges to 1.

Actually, the partial sums of the series have alternately the values 0 and 1 and so, on the basis of definition of convergence, the series does not converge.

The following theorem gives the condition under which the parentheses may be inserted in a series.

8.2.1. Theorem. *If a series is convergent, then parentheses may be inserted in the series (keeping the order of terms unchanged) without altering the convergence or the value to which the series converges.*

Proof. Let $\sum_{n=1}^{\infty} a_n$ be a convergent series.

Let $\sum_{n=1}^{\infty} b_n$ be a series obtained from $\sum_{n=1}^{\infty} a_n$ by grouping of its terms. Let $< S_n >$ be the sequence of partial sums of $\sum_{n=1}^{\infty} a_n$ and $< T_n >$ be the sequence of partial sums of $\sum_{n=1}^{\infty} b_n$. Then, $< T_n >$ is a subsequence of $< S_n >$. Hence, if $< S_n >$ converges to S, then $< T_n >$ is also convergent to S.

If a series with parentheses consisting of both positive and negative terms is convergent, then the parentheses cannot be removed without affecting the convergence of the series. However, under certain conditions, the parentheses may be removed. Parentheses can be removed in the case of positive terms. For a series with mixed signs, following rule may be observed :

If $\sum_{n=1}^{\infty} a_n$ converges to S and if $\sum_{n=1}^{\infty} b_n$ obtained from it by the removal of parenthesis also converges, then $\sum_{n=1}^{\infty} b_n$ also converges to S. Moreover, if the series $\sum_{n=1}^{\infty} b_n$ with parentheses converges to S and if the sum of the absolute values of the terms in b_n tends to zero as n tends to infinity, then the series $\sum_{n=1}^{\infty} a_n$ obtained on removing the parentheses, will also converge to S.

EXERCISE 8 (*b*)

Show that the series :

$$\left(1-\frac{1}{2}\right)+\left(1-\frac{3}{4}\right)+\left(1-\frac{7}{8}\right)+\ldots\ldots$$

is convergent, but when the parentheses are removed, it oscillates.

8.3. REARRANGEMENT OF SERIES

We have seen that insertion of parentheses in an infinite series may cause drastic changes in the behaviour of the series. In a similar way, rearranging the order of the terms of a series may cause a radical change in the behaviour of the series.

8.3.1. Definition. *If two different series* $\sum_{n=1}^{\infty} a_n$ *and* $\sum_{n=1}^{\infty} b_n$ *are so related that every* a_n *is some* b_n *and every* b_n *is some* a_m, *then one series is said to be a rearrangement of the other.*

Example 6. *Show that the series :*

$$1+\frac{1}{3}-\frac{1}{2}+\frac{1}{5}+\frac{1}{7}-\frac{1}{4}+\frac{1}{9}+\frac{1}{11}-\frac{1}{6}+\ldots\ldots,$$

where $a_{3n-2}=\dfrac{1}{4n-3},\ a_{3n-1}=\dfrac{1}{4n-1},\ a_{3n}=-\dfrac{1}{2n}$ *is a rearrangement of the series*

$$1-\frac{1}{2}+\frac{1}{3}-\frac{1}{4}+\frac{1}{5}-\ldots\ldots+(-1)^{n-1}\frac{1}{n}+\ldots\ldots$$

and if $\log 2=1-\dfrac{1}{2}+\dfrac{1}{3}-\dfrac{1}{4}+\dfrac{1}{5}-\dfrac{1}{6}+\ldots\ldots,$

then find the value of the rearranged series.

Sol. To show that the first series is a rearrangement of the second, we need only observe that every term of one series occurs somewhere in the other series.

To find the value of the rearranged series, we have

$$\log 2=1-\frac{1}{2}+\frac{1}{3}-\frac{1}{4}+\frac{1}{5}-\frac{1}{6}+\ldots\ldots \qquad \ldots(1)$$

$$\Rightarrow \qquad \frac{1}{2}\log 2=\frac{1}{2}-\frac{1}{4}+\frac{1}{6}-\frac{1}{8}+\frac{1}{10}-\frac{1}{12}+\ldots\ldots$$

which can be rewritten as :

$$\frac{1}{2}\log 2=0+\frac{1}{2}+0-\frac{1}{4}+0+\frac{1}{6}+0-\frac{1}{8}+\ldots\ldots \qquad \ldots(2)$$

Adding term by term, the series (1) and (2), we have

$$\frac{3}{2}\log 2=1+0+\frac{1}{3}-\frac{1}{2}+\frac{1}{5}+0+\ldots\ldots=1+\frac{1}{3}-\frac{1}{2}+\frac{1}{5}+\ldots\ldots$$

The following theorem gives the condition under which we may rearrange the terms of the series without altering its sum.

8.3.2. Theorem (Dirichlet's Theorem). *A series obtained from an absolutely convergent series by a rearrangement of terms converges absolutely and has the same sum as that of the original series.*

Proof. Let $\sum_{n=1}^{\infty} a_n$ be an absolutely convergent series converging to a. Let $\sum_{n=1}^{\infty} b_n$ be a rearrangement of $\sum_{n=1}^{\infty} a_n$.

Case (*i*) $\sum_{n=1}^{\infty} \mathbf{a_n}$ is a series of positive terms.

Let S_n and σ_n denote respectively the partial sums of the series $\sum_{n=1}^{\infty} a_n$ and $\sum_{n=1}^{\infty} b_n$.

Since σ_n contains first n terms of $\sum_{n=1}^{\infty} b_n$, we can find an integer m such that first m terms of $\sum_{n=1}^{\infty} a_n$ contain all first n terms (and possibly some more) of the former.

Since the series are of positive terms,

$$\therefore \qquad \sigma_n \le S_m \le S, \text{ where } \lim_{n \to \infty} S_n = S.$$

$\Rightarrow$ the sequence $< \sigma_n >$ of partial sums of the positive term series $\sum_{n=1}^{\infty} b_n$ is bounded above by S.

$\Rightarrow$ the sequence $< \sigma_n >$ and hence the series $\sum_{n=1}^{\infty} b_n$ is convergent to σ (say) where $\sigma \le S$.

Considering $\sum_{n=1}^{\infty} a_n$ as a rearrangement of $\sum_{n=1}^{\infty} b_n$, we can similarly show that $S \le \sigma$.

Hence, $\sigma = S$ and the two series converge to the same sum.

Case (*ii*) $\sum_{n=1}^{\infty} \mathbf{a_n}$ is series of arbitrary terms.

Let
$$u_n = \begin{cases} a_n, & \text{if } a_n > 0 \\ 0, & \text{if } a_n \le 0 \end{cases}$$

$$v_n = \begin{cases} -a_n, & \text{if } a_n < 0 \\ 0, & \text{if } a_n \ge 0 \end{cases}$$

$$u_n' = \begin{cases} b_n, & \text{if } b_n > 0 \\ 0, & \text{if } b_n \le 0 \end{cases}$$

$$v_n' = \begin{cases} -b_n, & \text{if } b_n < 0 \\ 0, & \text{if } b_n \ge 0 \end{cases}$$

Then, u_n, v_n, u_n', v_n' are non-negative and

$$\left.\begin{aligned} a_n &= u_n - v_n, \quad & |a_n| &= u_n + v_n \\ b_n &= u_n' - v_n', \quad & |b_n| &= u_n' + v_n' \end{aligned}\right\} \qquad \ldots(1)$$

$$\Rightarrow \quad \left.\begin{aligned} u_n &= \tfrac{1}{2}(|a_n| + a_n) \\ v_n &= \tfrac{1}{2}(|a_n| - a_n) \end{aligned}\right\} \qquad ...(2)$$

and
$$\left.\begin{aligned} u_n' &= \tfrac{1}{2}(|b_n| + b_n) \\ v_n' &= \tfrac{1}{2}(|b_n| - b_n) \end{aligned}\right\} \qquad ...(3)$$

Now, $\sum_{n=1}^{\infty} a_n$ is absolutely convergent, therefore, from (2), $\sum_{n=1}^{\infty} u_n$ and $\sum_{n=1}^{\infty} v_n$ are convergent series of non-negative terms and $\sum_{n=1}^{\infty} u_n'$ and $\sum_{n=1}^{\infty} v_\iota'$ are respectively their rearrangements. Therefore, by case (*i*), $\sum_{n=1}^{\infty} u_n'$ and $\sum_{n=1}^{\infty} v_n'$ are convergent to the sums of the series $\sum_{n=1}^{\infty} u_n$ and $\sum_{n=1}^{\infty} v_r$ respectively.

Let α, β, α', β' denote the sums of the series $\sum_{n=1}^{\infty} u_n, \sum_{n=1}^{\infty} v_n, \sum_{n=1}^{\infty} u_n', \sum_{n=1}^{\infty} v_n'$ respectively, then $\alpha = \alpha'$ and $\beta = \beta'$.

$\therefore$ from (1), it follows that $\sum_{n=1}^{\infty} b_n$ and $\sum_{n=1}^{\infty} |b_n|$ are convergent and

$$\sum_{n=1}^{\infty} b_n = \alpha' - \beta' = \alpha - \beta = \sum_{n=1}^{\infty} a_n$$

and
$$\sum_{n=1}^{\infty} |b_n| = \alpha' + \beta' = \alpha + \beta = \sum_{n=1}^{\infty} |a_n|.$$

Hence the rearranged series $\sum_{n=1}^{\infty} b_n$ converges absolutely to the sum of $\sum_{n=1}^{\infty} a_n$.

Remarks 1. A positive term series, if convergent, has sum independent of the order of its terms, and if divergent, it remains divergent however its terms may be rearranged.

2. Parentheses may be inserted or removed, or terms be picked up and placed at random without changing the behaviour of a positive term series or an absolutely convergent series. In fact, they behave like finite sums.

3. Since an absolutely convergent series remains convergent with unaltered sum, under every rearrangement of terms, it is also called an unconditionally convergent.

8.3.3. Theorem (Riemann's Rearrangement Theorem)*. *By a suitable rearrangement of the terms, a conditionally convergent series* $\sum_{n=1}^{\infty} a_n$ *can be made*

(*i*) *to converge to any number* α, *or* (*ii*) *to diverge to* ∞ *or* $-\infty$ *or*
(*iii*) *to oscillate finitely or infinitely.*

---

*After the name of a distinguished German mathematician G.F.B. Riemann (1826–1866).

Let $u_1, u_2, u_3, \ldots\ldots$, be the positive terms and $-v_1, -v_2, -v_3, \ldots\ldots$, be the negative terms, of $\sum_{n=1}^{\infty} a_n$. Then, the series $\sum_{n=1}^{\infty} u_n$ and $\sum_{n=1}^{\infty} v_n$ are both divergent to ∞. [$\because$ If a series is conditionally convergent, then the series of its positive terms and the series of its negative terms are both divergent.]

Let S_n denote the sum of first n terms of $\sum_{n=1}^{\infty} u_n$

and S_n' denote the sum of first n terms of $\sum_{n=1}^{\infty} v_n$

(*i*) We first show that a rearrangement of $\sum_{n=1}^{\infty} a_n$ can be found, which converges to α. It is enough to consider the case $\alpha \geq 0$, because if $\alpha < 0$, then $-\alpha > 0$ and we can argue the case in a similar way.

Choose the first n_1 positive terms of $\sum_{n=1}^{\infty} u_n$ until the partial sum becomes $\geq \alpha$, n_1 being the smallest such number.

Then,
$$S_{n_1-1} < \alpha \leq S_{n_1}$$

Add to these n_1 positive terms, first m_1 negative terms (m_1 being the least number) until the sum is less than α. Such a choice is possible because $-\sum_{n=1}^{\infty} v_n = -\infty$.

$\therefore$
$$S_{n_1} - S'_{m_1} < \alpha \leq S_{n_1} - S'_{m_1-1}$$

Add the next n_2 positive terms of $\sum_{n=1}^{\infty} u_n$, (n_2 being the least number) so that the sum becomes $\geq \alpha$.

$\therefore$ We get
$$S_{n_1+n_2-1} - S'_{m_1} < \alpha \leq S_{n_1+n_2} - S'_{m_1}$$

Now, add m_2 negative terms (m_2 being least) so that the sum is $< \alpha$.

$\therefore$
$$S_{n_1+n_2} - S'_{m_1+m_2} < \alpha \leq S_{n_1+n_2} - S'_{m_1+m_2-1}$$

and so on. The process may be continued indefinitely. The process is always possible because of the divergence of the two series $\sum_{n=1}^{\infty} u_n$ and $\sum_{n=1}^{\infty} v_n$.

Let $\sum_{n=1}^{\infty} b_n$ be the new series so constructed and σ_n be the sum of its first n terms.

Then, $|\ \sigma_n - \alpha\ | \leq |\ a_l\ |$, where a_l is the last term added.

Since $\sum_{n=1}^{\infty} a_n$ is convergent, $| a_l | \to 0$ as $n \to \infty$.

$$\Rightarrow \qquad \sigma_n \to \alpha \quad \text{as} \quad n \to \infty$$

$\Rightarrow$ the rearranged series $\sum_{n=1}^{\infty} b_n$ is convergent to α.

(ii) Now, we show that a rearrangement of $\sum_{n=1}^{\infty} a_n$ can be found, which diverges to ∞.

Choose a positive integer n_1 so large that

$$u_1 + u_2 + \ldots\ldots + u_{n_1} > 1 + v_1$$

i.e.,

$$u_1 + u_2 + \ldots\ldots + u_{n_1} - v_1 > 1$$

Next, choose a positive integer $n_2 > n_1$ so large that

$$(u_1 + u_2 + \ldots\ldots + u_{n_1}) + (u_{n_1+1} + u_{n_1+2} + \ldots\ldots + u_{n_2}) > 2 + v_1 + v_2$$

i.e.,

$$(u_1 + u_2 + \ldots\ldots + u_{n_1} - v_1) + (u_{n_1+1} + u_{n_1+2} + \ldots\ldots + u_{n_2} - v_2) > 2$$

Proceeding in this way, we have

$$(u_1 + u_2 + \ldots\ldots + u_{n_1} - v_1) + (u_{n_1+1} + u_{n_1+2} + \ldots\ldots + u_{n_2} - v_2) + \ldots\ldots + (u_{n_{k-1}+1} + u_{n_{k-1}+2} + \ldots\ldots + u_{n_k} - v_k) > k,$$

where k is a positive integer, however large.

Let $\sum_{n=1}^{\infty} b_n$ be the new series so constructed. Then, the sequence of its partial sums σ_{n_1+1}, σ_{n_2+1}, ... is unbounded above and hence the series $\sum_{n=1}^{\infty} b_n$ is divergent to ∞.

Proceeding as in case *(ii)* above, we can show that

$$(-v_1 - v_2 - \ldots\ldots - v_{n_1} + u_1) + (-v_{n_1+1} - v_{n_1+2} - \ldots\ldots - v_{n_2} + u_2) + \ldots\ldots + (-v_{n_{k-1}+1} - v_{n_{k-1}+2} - \ldots\ldots - v_{n_k} + u_k) < -k$$

where k is a positive integer however large.

This shows that the rearranged series is divergent to $-\infty$.

(iii) Next, we show that a rearrangement of $\sum_{n=1}^{\infty} a_n$ can be found, which oscillates between two numbers α and β. To show this, take just sufficient number of positive terms so that the sum is greater than α and then take just sufficient number of negative terms so that the sum becomes $< \beta$ and so on. The process is repeated infinitely. The new series so formed, in this case will oscillate between $\beta - 1$ and $\alpha + 1$ finitely or infinitely.

Example 7. *Criticise the paradox :*

$$\log 2 = 1 - \frac{1}{2} + \frac{1}{3} - \frac{1}{4} + \frac{1}{5} - \ldots\ldots$$

$$= \left(1 + \frac{1}{2} + \frac{1}{3} + \frac{1}{4} + \frac{1}{5} + \ldots\ldots\right) - 2\left(\frac{1}{2} + \frac{1}{4} + \frac{1}{6} + \frac{1}{8} + \frac{1}{10} + \ldots\ldots\right)$$

$$= \left(1+\frac{1}{2}+\frac{1}{3}+\frac{1}{4}+\frac{1}{5}+......\right)-\frac{2}{2}\left(1+\frac{1}{2}+\frac{1}{3}+\frac{1}{4}+\frac{1}{5}+......\right)$$

$$= \left(1+\frac{1}{2}+\frac{1}{3}+\frac{1}{4}+\frac{1}{5}+......\right)-\left(1+\frac{1}{2}+\frac{1}{3}+\frac{1}{4}+\frac{1}{5}+......\right)$$

$$= 0.$$

$\Rightarrow$ *log 2 = 0, which is not true.*

Sol. The mistake lies in the fact that the given series is only convergent and not absolutely convergent. A conditionally convergent series can be made to converge to any real number by a suitable rearrangement of the terms.

8.4. CAUCHY PRODUCT OF SERIES

If we multiply two power series $\sum_{n=1}^{\infty} a_n x^{n-1}$ and $\sum_{n=1}^{\infty} b_n x^{n-1}$, we obtain the series $\sum_{n=1}^{\infty} c_n x^{n-1}$. where $c_n = a_1 b_n + a_2 b_{n-1} + ... + a_n b_1 \quad \forall n \in$ N. Motivated from this, we define product of two series as follows.

8.4.1. Definition. *Let $\sum_{n=1}^{\infty} a_n$ and $\sum_{n=1}^{\infty} b_n$ be two series.*

Then, the series $\sum_{n=1}^{\infty} c_n$, where $c_n = a_1 b_n + a_2 b_{n-1} + ... + a_n b_1 = \sum_{n=1} a_k b_{n-k+1}$, for each $n \in N$,

is called the Cauchy product of the two series $\sum_{n=1}^{\infty} a_n$ and $\sum_{n=1}^{\infty} b_n$. The terms of the two series are multiplied and arranged in such a way that all the terms which have the same sum of suffixes are bracketed together.

If the indices in the two series vary over non-negative integers, i.e., the series are $\sum_{n=0}^{\infty} a_n$ and $\sum_{n=0}^{\infty} b_n$, then their Cauchy product is defined as $\sum_{n=0}^{\infty} c_n$, where $c_n = \sum_{k=0}^{n} a_k b_{n-k} = a_0 b_n + a_1 b_{n-1} + ... + a_n b_0$, $\forall n \in N$.

Note that $c_n = \sum_{k=0}^{n} a_k b_{n-k} = \sum_{k=0}^{n} a_{n-k} b_k$.

It may happen that the series $\sum_{n=1}^{\infty} a_n$ and $\sum_{n=1}^{\infty} b_n$ both converge but the Cauchy product $\sum_{n=1}^{\infty} c_n$ of these two series may not converge, as is clear from the following example.

Example 8. *Show that the Cauchy product of the convergent series* $\sum_{n=1}^{\infty} \frac{(-1)^{n-1}}{\sqrt{n}}$ *with itself is not convergent.*

Sol. Let $a_n = b_n = \frac{(-1)^{n-1}}{\sqrt{n}}$, $n \in$ N. By Leibnitz's test the series $\sum_{n=1}^{\infty} a_n$ and $\sum_{n=1}^{\infty} b_n$ are both convergent. The Cauchy product of the two series is $\sum_{n=1}^{\infty} c_n$, where

$$c_n = a_1 b_n + a_2 b_{n-1} + \ldots\ldots + a_n b_1$$

$$= \frac{(-1)^0}{\sqrt{1}} \frac{(-1)^{n-1}}{\sqrt{n}} + \frac{(-1)^1}{\sqrt{2}} \frac{(-1)^{n-2}}{\sqrt{n-1}} + \ldots\ldots + \frac{(-1)^{n-1}}{\sqrt{n}} \frac{(-1)^0}{\sqrt{1}}$$

$$= (-1)^{n-1} \left[\frac{1}{\sqrt{1n}} + \frac{1}{\sqrt{2(n-1)}} + \ldots\ldots + \frac{1}{\sqrt{n\,.\,1}} \right]$$

$$\geq (-1)^{n-1} \left[\frac{1}{\sqrt{n\,.\,n}} + \frac{1}{\sqrt{n\,.\,n}} + \ldots\ldots + \frac{1}{\sqrt{n\,.\,n}} \right]$$

(Replacing each natural number under the radical sign by the largest of them, namely n)

$$= (-1)^{n-1} \frac{n}{\sqrt{n\,.\,n}} = (-1)^{n-1}$$

$\therefore$ c_n does not tend to 0 as $n \to \infty$.

$\Rightarrow$ $\sum_{n=1}^{\infty} c_n$ is not convergent.

Following example shows that the product of two divergent series may not be divergent.

Example 9. *Show that the Cauchy product of two divergent series*

$$\sum_{n=1}^{\infty} a_n = 2 + 2 + 2^2 + 2^3 + \ldots\ldots + 2^{n+1} + \ldots\ldots$$

and

$$\sum_{n=1}^{\infty} b_n = -1 + 1 + 1 + 1 + \ldots\ldots$$

is convergent.

Sol. The Cauchy product of the two given series is $\sum_{n=1}^{\infty} c_n$,

where
$$c_n = a_1 b_n + a_2 b_{n-1} + \ldots\ldots + a_n b_1$$
$$= 2.1 + 2.1 + 2^2.1 + 2^3.1 + \ldots\ldots + 2^{n-2}.1 + 2^{n-1}.(-1)$$
$$= 2 + (2 + 2^2 + 2^3 + \ldots\ldots + 2^{n-2}) - 2^{n-1}$$
$$= 2 + \frac{2(2^{n-2} - 1)}{2 - 1} - 2^{n-1}$$
$$= 2 + 2^{n-1} - 2 - 2^{n-1}$$
$$= 0 \quad \forall n \geq 2$$

and
$$c_1 = a_1 b_1 = -2$$

$\Rightarrow \sum_{n=1}^{\infty} c_n$ is convergent.

8.4.2. Theorem. *Let* $\sum_{n=1}^{\infty} a_n$ *be a non-negative term series convergent to A and* $\sum_{n=1}^{\infty} b_n$ *another non-negative term series convergent to B. Then, the Cauchy product series* $\sum_{n=1}^{\infty} c_n$, *converges to AB.*

Proof. Let $A_n = a_1 + a_2 + \ldots\ldots + a_n$

and $B_n = b_1 + b_2 + \ldots\ldots + b_n$

Then, $\lim_{n \to \infty} A_n = A$ and $\lim_{n \to \infty} B_n = B$

Let $C_n = c_1 + c_2 + \ldots\ldots + c_n$

$$\therefore \quad \begin{aligned} C_n = {} & a_1b_1 \\ & + a_1b_2 + a_2b_1 \\ & + a_1b_3 + a_2b_2 + a_3b_1 \\ & + \ldots\ldots \\ & + a_1b_n + a_2b_{n-1} + \ldots\ldots + a_nb_1 \end{aligned}$$

$$\begin{aligned} C_{2n} = {} & a_1b_1 \\ & + a_1b_2 + a_2b_1 \\ & + a_1b_3 + a_2b_2 + a_3b_1 \\ & + \ldots\ldots \\ & + a_1b_n + a_2b_{n-1} + \ldots\ldots + a_nb_1 \\ & + \ldots\ldots \\ & + a_1b_{2n} + a_2b_{2n-1} + \ldots\ldots + a_{2n}b_1 \end{aligned}$$

Also,
$$\begin{aligned} A_nB_n &= (a_1 + a_2 + \ldots\ldots + a_n)(b_1 + b_2 + \ldots\ldots + b_n) \\ &= a_1b_1 + a_2b_1 + \ldots\ldots + a_nb_1 \\ & \quad + a_1b_2 + a_2b_2 + \ldots\ldots + a_nb_2 \\ & \quad + a_1b_3 + a_2b_3 + \ldots\ldots + a_nb_3 \\ & \quad + \ldots\ldots + \ldots\ldots + \ldots\ldots + \ldots\ldots \\ & \quad + a_1b_n + a_2b_n + \ldots\ldots + a_nb_n \end{aligned}$$

$$\therefore \quad C_n \le A_nB_n \le C_{2n} \qquad \ldots(1)$$

The sequences $< A_n >$ and $< B_n >$ being convergent, are bounded.

$\Rightarrow \quad < C_n >$ is bounded.

$\therefore \quad < C_n >$, being monotonically increasing, is convergent to C (say).

$\Rightarrow \quad \sum_{n=1}^{\infty} c_n$ is convergent to C.

Further, from (1)

$$\lim_{n \to \infty} C_n \le \lim_{n \to \infty} (A_nB_n) \le \lim_{n \to \infty} C_{2n}$$

$$\Rightarrow \quad C \le AB \le C \quad \Rightarrow \quad C = AB.$$

8.4.3. Theorem *(Cauchy's Theorem). Let* $\sum_{n=1}^{\infty} a_n$ *and* $\sum_{n=1}^{\infty} b_n$ *to two absolutely convergent series such that* $\sum_{n=1}^{\infty} a_n = A$ *and* $\sum_{n=1}^{\infty} b_n = B$*. Then the Cauchy product series* $\sum_{n=1}^{\infty} c_n$ *is also absolutely convergent and* $\sum_{n=1}^{\infty} c_n = AB$*.* (K.U. BCA (II) 2003)

Proof. Let $A_n = \sum_{n=1}^{n} a_n$ and $B_n = \sum_{n=1}^{n} b_n$,

$\therefore$ $\lim_{n \to \infty} A_n = A$ and $\lim_{n \to \infty} B_n = B$.

Let $A_n' = \sum_{n=1}^{n} |a_n|$ and $B_n' = \sum_{n=1}^{n} |b_n|$.

Let $\lim_{n \to \infty} A_n' = A'$ and $\lim_{n \to \infty} B_n' = B'$.

By definition of Cauchy product,

$$c_n = a_1 b_n + a_2 b_{n-1} + \ldots\ldots + a_n b_1.$$

Let $d_n = |a_1||b_n| + |a_2||b_{n-1}| + \ldots\ldots + |a_n||b_1|$

Then, $\sum_{n=1}^{\infty} d_n$ is the Cauchy product of the series $\sum_{n=1}^{\infty} |a_n|$ and $\sum_{n=1}^{\infty} |b_n|$.

Now,
$$\begin{aligned} |c_n| &= |a_1 b_n + a_2 b_{n-1} + \ldots\ldots + a_n b_1| \\ &\le |a_1||b_n| + |a_2||b_{n-1}| + \ldots\ldots + |a_n||b_1| \\ &= d_n, \forall n \end{aligned} \qquad \ldots(1)$$

By the above theorem, $\sum_{n=1}^{\infty} d_n$ is convergent to A′B′ ...(2)

$\therefore$ from (1), by comparison test, $\sum_{n=1}^{\infty} |c_n|$ is convergent, *i.e.*, $\sum_{n=1}^{\infty} c_n$ is absolutely convergent.

It remains to show that $\sum_{n=1}^{\infty} c_n = AB$.

Now,
$$\begin{aligned} A_n B_n - C_n &= (a_1 + a_2 + \ldots\ldots + a_n)(b_1 + b_2 + \ldots\ldots + b_n) - (c_1 + c_2 + \ldots\ldots + c_n) \\ &= a_1 b_1 + a_2 b_1 + \ldots\ldots + a_n b_1 \\ &\quad + a_1 b_2 + a_2 b_2 + \ldots\ldots + a_n b_2 \\ &\quad + \ldots\ldots \\ &\quad + a_1 b_n + a_2 b_n + \ldots\ldots + a_n b_n - (a_1 b_1 \\ &\quad + a_1 b_2 + a_2 b_1 \\ &\quad + \ldots\ldots \\ &\quad + a_1 b_n + a_2 b_{n-1} + \ldots\ldots + a_n b_1) \end{aligned}$$

$$= a_2b_n + a_3(b_{n-1} + b_n) + a_4(b_{n-2} + b_{n-1} + b_n) + \ldots\ldots + a_n(b_2 + b_3 + \ldots\ldots + b_n)$$

$$\Rightarrow \quad |A_nB_n - C_n| \le |a_2|\,|b_n| + |a_3|\,(|b_{n-1}| + |b_n|) + \ldots\ldots + |a_n|\,(|b_2| + |b_3| + \ldots\ldots + |b_n|)$$

$$= A_n'B_n' - \sum_{n=1}^{n} d_n \qquad \ldots(3)$$

Also, $\lim_{n \to \infty} A_n' = A'$, $\lim_{n \to \infty} B_n' = B'$

and $\lim_{n \to \infty} \left(\sum_{n=1}^{n} d_n \right) = A'B'$ [By (2)]

$$\therefore \quad A_n'B_n' - \sum_{n=1}^{n} d_n \to 0 \quad \text{as} \quad n \to \infty$$

$\therefore$ from (3), $|A_nB_n - C_n| \to 0$ as $n \to \infty$.

Also, $|A_nB_n - AB| \to 0$ as $n \to \infty$

$$\therefore \quad |C_n - AB| = |C_n - A_nB_n + A_nB_n - AB| \le |C_n - A_nB_n| + A_nB_n - AB|$$

$$\Rightarrow \quad |C_n - AB| \to 0 \quad \text{as} \quad n \to \infty$$

$$\Rightarrow \quad \lim_{n \to \infty} C_n = AB \quad \Rightarrow \quad \sum_{n=1}^{\infty} c_n = AB.$$

8.4.4. Theorem (*Mertin's Theorem*). *Let* $\sum_{n=1}^{\infty} a_n$ *and* $\sum_{n=1}^{\infty} b_n$ *be two convergent series and* $\sum_{n=1}^{\infty} a_n$ *convergent absolutely. Then, the Cauchy product series* $\sum_{n=1}^{\infty} c_n$ *converges to AB, where A* $= \sum_{n=1}^{\infty} a_n$ *and* $B = \sum_{n=1}^{\infty} b_n$.

Proof. Let $A_n = a_1 + a_2 + \ldots\ldots + a_n$,

$B_n = b_1 + b_2 + \ldots\ldots + b_n$

Then, $\lim_{n \to \infty} A_n = A$ and $\lim_{n \to \infty} B_n = B$.

Let $C_n = c_1 + c_2 + \ldots\ldots + c_n$

We are to show that $\lim_{n \to \infty} C_n = AB$.

Since $\sum_{n=1}^{\infty} a_n$ is absolutely convergent, for a small $\varepsilon > 0$, there exists a positive integer m_0 such that

$$|a_{m_0+1}| + |a_{m_0+2}| + \ldots\ldots + |a_n| < \varepsilon \quad \forall\, n > m_0 \qquad \ldots(1)$$

Since $< B_n >$ is convergent, it is bounded.

$\therefore$ there exists a positive number k such that $| B_n | < k, \forall n$ ...(2)

Since every convergent sequence is a Cauchy sequence,

$\therefore$ $< B_n >$ is a Cauchy sequence.

$\therefore$ there exists a positive integer m_1 such that

$$| B_n - B_m | < \varepsilon \quad \forall n, m > m_1 \qquad ...(3)$$

$$\begin{aligned}
\text{Now, } A_nB_n - C_n &= (a_1 + a_2 + \ldots\ldots + a_n)(b_1 + b_2 + \ldots\ldots + b_n) - (c_1 + c_2 + \ldots\ldots + c_n) \\
&= (a_1 + a_2 + \ldots\ldots + a_n)(b_1 + b_2 + \ldots\ldots + b_n) \\
&\quad - (a_1b_1 + a_1b_2 + a_2b_1 + a_1b_3 + a_2b_2 + a_3b_1 + \ldots\ldots \\
&\quad + a_1b_n + a_2b_{n-1} + \ldots\ldots + a_nb_1) \\
&= a_2b_n + a_3(b_{n-1} + b_n) + a_4(b_{n-2} + b_{n-1} + b_n) + \ldots\ldots + a_n(b_2 + b_3 + \ldots\ldots + b_n) \\
&= a_2(B_n - B_{n-1}) + a_3(B_n - B_{n-2}) + \ldots\ldots + a_n(B_n - B_1) \qquad (\because b_1 = B_1)
\end{aligned}$$

Choose a positive integer $n \geq m_0 + m_1$.

Then,

$$\begin{aligned}
A_nB_n - C_n &= [a_2(B_n - B_{n-1}) + \ldots\ldots + a_{m_0}(B_n - B_{n-m_0+1})] \\
&\quad + [a_{m_0+1}(B_n - B_{n-m_0}) + \ldots\ldots + a_n(B_n - B_1)]
\end{aligned}$$

$$\begin{aligned}
\Rightarrow \quad | A_nB_n - C_n | &\leq [| a_2 | \, | B_n - B_{n-1} | + \ldots\ldots + |a_{m_0}| \, | B_n - B_{n-m_0+1} |] \\
&\quad + [|a_{m_0+1}| \, | B_n - B_{n-m_0} | + \ldots\ldots + | a_n | \, | B_n - B_1 |] \\
&\leq [| a_2 | \varepsilon + \ldots\ldots + |a_{m_0}| \varepsilon] + [|a_{m+1}| \,.\, 2k + \ldots\ldots + | a_n | \,.\, 2k] \\
&= \varepsilon(| a_2 | + \ldots\ldots + |a_{m_0}|) + 2k(|a_{m_0+1}| + \ldots\ldots + | a_n |)
\end{aligned}$$

[By (1) to (3) and $|B_n - B_{n-m_0}| \leq | B_n | + |B_{n-m_0}| \leq 2k$]

$$\Rightarrow \quad |A_nB_n - C_n| < \varepsilon(| a_2 | + \ldots\ldots + | a_{m0} |) + 2k\varepsilon$$

Since ε is arbitrary, this implies that

$$A_nB_n - C_n \to 0 \quad \text{as} \quad n \to \infty$$

$$\begin{aligned}
\text{Thus,} \quad | C_n - AB | &= | C_n - A_nB_n + A_nB_n - AB | \\
&\leq | C_n - A_nB_n | + | A_nB_n - AB |
\end{aligned}$$

As $\quad A_n \to A, B_n \to B, C_n - A_nB_n \to 0$

$\therefore \quad | C_n - AB | \to 0$ as $n \to \infty$

$$\Rightarrow \quad \lim_{n \to \infty} C_n = AB \quad \Rightarrow \quad \sum_{n=1}^{\infty} c_n = AB.$$

8.4.5. Theorem (*Abel's Theorem*). *Let* $\sum_{n=1}^{\infty} a_n$ *and* $\sum_{n=1}^{\infty} b_n$ *be two convergent series. Let* $\sum_{n=1}^{\infty} c_n$ *be the Cauchy product series. If the series* $\sum_{n=1}^{\infty} a_n$ *converges to A,* $\sum_{n=1}^{\infty} b_n$ *converges to B and* $\sum_{n=1}^{\infty} c_n$ *converges to C, then C = AB.*

Proof. Before proving the theorem, we first prove the following Lemma.

Lemma *(Cesaro's theorem). If a sequence $< a_n >$ converges to a and another sequence $< b_n >$ converges to b, then the sequence $< \frac{a_1 b_n + a_2 b_{n-1} + a_n b_1}{n} >$ converges to ab.*

Proof of Lemma. Let $a_n = a + \alpha_n \quad \forall n \in N$.

Then, $\lim_{n \to \infty} \alpha_n = 0 \quad \Rightarrow \quad \lim_{n \to \infty} |\alpha_n| = 0.$

Since $< b_n >$ converges, $< b_n >$ is bounded.

$\therefore$ there-exists K > 0 such that

$$| b_n | < K, \quad \forall n \in N \qquad \text{...(1)}$$

Now,
$$\frac{a_1 b_n + a_2 b_{n-1} + \ldots\ldots + a_n b_1}{n}$$
$$= \frac{(a + \alpha_1) b_n + (a + \alpha_2) b_{n-1} + \ldots\ldots + (a + \alpha_n) b_1}{n}$$
$$= \frac{a(b_n + b_{n-1} + \ldots\ldots + b_1)}{n} + \frac{\alpha_1 b_n + \alpha_2 b_{n-1} + \ldots\ldots + \alpha_n b_1}{n} \qquad \text{...(2)}$$

Since $b_n \to b$ as $n \to \infty$,

$\therefore$ by Cauchy's first theorem on limits,

$$\frac{b_n + b_{n-1} + \ldots\ldots + b_1}{n} \to b \quad \text{as} \quad n \to \infty \qquad \text{...(3)}$$

Also,
$$\frac{|\alpha_1 b_n + \alpha_2 b_{n-1} + \ldots\ldots + \alpha_n b_1|}{n}$$
$$\leq \frac{|\alpha_1||b_n| + |\alpha_2||b_{n-1}| + \ldots\ldots + |\alpha_n||b_1|}{n}$$
$$\leq K \frac{(|\alpha_1| + |\alpha_2| + \ldots\ldots + |\alpha_n|)}{n} \qquad \text{[By (1)]}$$

But $\quad | \alpha_n | \to 0 \quad \Rightarrow \quad \frac{|\alpha_1| + |\alpha_2| + \ldots\ldots + |\alpha_n|}{n} \to 0, \quad \text{as} \quad n \to \infty$

$$\therefore \quad \frac{|\alpha_1 b_n + \alpha_2 b_{n-1} + \ldots\ldots + \alpha_n b_1|}{n} \to 0, \quad \text{as} \quad n \to \infty$$

$$\Rightarrow \quad \frac{\alpha_1 b_n + \alpha_2 b_{n-1} + \ldots\ldots + \alpha_n b_1}{n} \to 0 \quad \text{as} \quad n \to \infty \qquad \text{...(4)}$$

Using (3) and (4) in (2), we have

$$\frac{a_1 b_n + a_2 b_{n-1} + \ldots\ldots + a_n b_1}{n} \to ab \quad \text{as} \quad n \to \infty$$

Proof of theorem. Let $A_n = a_1 + a_2 + \ldots\ldots + a_n$,

$B_n = b_1 + b_2 + \ldots\ldots + b_n$ and $C_n = c_1 + c_2 + \ldots\ldots + c_n$.

Now,
$$C_n = c_1 + c_2 + \ldots\ldots + c_n$$
$$= a_1 b_1$$
$$+ a_1 b_2 + a_2 b_1$$
$$+ a_1 b_3 + a_2 b_2 + a_3 b_1$$
$$+ \ldots\ldots$$
$$+ a_1 b_n + a_2 b_{n-1} + \ldots\ldots + a_n b_1$$

$$= a_1(b_1 + b_2 + \ldots\ldots + b_n) + a_2(b_1 + b_2 + \ldots\ldots + b_{n-1})$$
$$+ a_3(b_1 + b_2 + \ldots\ldots + b_{n-2})$$
$$+ \ldots\ldots + a_n b_1$$

$\Rightarrow$ $$C_n = a_1 B_n + a_2 B_{n-1} + \ldots\ldots + a_n B_1 \qquad (\because B_1 = b_1)$$

Changing n to $n - 1, n - 2, \ldots\ldots, 1$ successively, we have

$$C_{n-1} = a_1 B_{n-1} + a_2 B_{n-2} + \ldots\ldots + a_{n-1} B_1$$

....

$$C_1 = a_1 B_1$$

Adding, we have

$$C_1 + C_2 + \ldots\ldots + C_n = A_1 B_n + A_2 B_{n-1} + \ldots\ldots + A_n B_1 \qquad (\because A_1 = a_1)$$

$\Rightarrow$ $$\frac{C_1 + C_2 + \ldots\ldots + C_n}{n} = \frac{A_1 B_n + A_2 B_{n-1} + \ldots\ldots + A_n B_1}{n} \qquad \ldots(1)$$

But, $A_n \to A$, $B_n \to B$ and $C_n \to C$ as $n \to \infty$

$\therefore$ $$\frac{C_1 + C_2 + \ldots\ldots + C_n}{n} \to C$$

and $$\frac{A_1 B_n + A_2 B_{n-1} + \ldots\ldots + A_n B_1}{n} \to AB \qquad \text{(By Lemma)}$$

$\therefore$ from (1), it follows that $C = AB$.

Example 10. *Show that the Cauchy product of two divergent series*

$$\sum_{n=0}^{\infty} a_n = 1 - \frac{3}{2} - \left(\frac{3}{2}\right)^2 - \left(\frac{3}{2}\right)^3 \ldots\ldots$$

and $$\sum_{n=0}^{\infty} b_n = 1 + \left(2 + \frac{1}{2^2}\right) + \frac{3}{2}\left(2^2 + \frac{1}{2^3}\right) + \left(\frac{3}{2}\right)^2 \left(2^3 + \frac{1}{2^4}\right) + \ldots\ldots$$

is convergent.

Sol. Cauchy product of given series is $\sum_{n=0}^{\infty} c_n$, where

$$c_0 = a_0 b_0 = 1$$

and for $n \geq 1$,

$$c_n = a_0 b_n + a_1 b_{n-1} + \ldots\ldots + a_n b_0$$

$$= 1 \,.\, \left(\frac{3}{2}\right)^{n-1} \left(2^n + \frac{1}{2^{n+1}}\right) - \left(\frac{3}{2}\right)\left(\frac{3}{2}\right)^{n-2} \left(2^{n-1} + \frac{1}{2^n}\right) - \ldots\ldots - \left(\frac{3}{2}\right)^n \,.\, 1$$

$$= \left(\frac{3}{2}\right)^{n-1} \left[\left(2^n + \frac{1}{2^{n+1}}\right) - \left(2^{n-1} + \frac{1}{2^n}\right) - \left(2^{n-2} + \frac{1}{2^{n-1}}\right)\right.$$
$$\left. - \ldots\ldots - \left(2 + \frac{1}{2^2}\right)\right] - \left(\frac{3}{2}\right)^n$$

$$= \left(\frac{3}{2}\right)^{n-1}\left[\left(2^n + \frac{1}{2^{n+1}}\right) - (2^{n-1} + 2^{n-2} + \ldots\ldots + 2)\right.$$
$$\left. - \left(\frac{1}{2^n} + \frac{1}{2^{n-1}} + \ldots\ldots + \frac{1}{2^2}\right)\right] - \left(\frac{3}{2}\right)^n$$

$$= \left(\frac{3}{2}\right)^{n-1}\left[2^n + \frac{1}{2^{n+1}} - 2\,\frac{(2^{n-1}-1)}{2-1} - \frac{\frac{1}{2^2}\left(1-\frac{1}{2^{n-1}}\right)}{1-\frac{1}{2}}\right] - \left(\frac{3}{2}\right)^n$$

$$= \left(\frac{3}{2}\right)^{n-1}\left[2^n + \frac{1}{2^{n+1}} - 2^n + 2 - \frac{1}{2} + \frac{1}{2^n}\right] - \left(\frac{3}{2}\right)^n$$

$$= \left(\frac{3}{2}\right)^{n-1}\left[\frac{3}{2} + \frac{1}{2^{n+1}} + \frac{1}{2^n}\right] - \left(\frac{3}{2}\right)^n = \left(\frac{3}{2}\right)^{n-1}\left[\frac{3}{2} + \frac{3}{2^{n+1}}\right] - \left(\frac{3}{2}\right)^n = \left(\frac{3}{4}\right)^n$$

$$\Rightarrow \qquad \sum_{n=0}^{\infty} c_n = \sum_{n=0}^{\infty}\left(\frac{3}{4}\right)^n$$

being a geometric series with common ratio $\frac{3}{4}(< 1)$, is convergent.

Example 11. *Discuss the convergence of the Cauchy product of the series* $\sum_{n=1}^{\infty} \frac{(-1)^{n-1}}{n^p}$, *$(p > 0)$ with itself.*

Sol. Here $\qquad a_n = b_n = \frac{(-1)^{n-1}}{n^p}$

Cauchy product is the series $\sum_{n=1}^{\infty} c_n$,

where $c_n = a_1 b_n + a_2 b_{n-1} + \ldots\ldots + a_n b_1$

$$= (-1)^{n-1}\left[\frac{1}{1^p \cdot n^p} + \frac{1}{2^p(n-1)^p} + \ldots\ldots + \frac{1}{r^p(n-r+1)^p} + \ldots\ldots + \frac{1}{n^p \cdot 1^p}\right]$$

$$= (-1)^{n-1} d_n, \text{ where } d_n = \sum_{r=1}^{n} \frac{1}{r^p(n-r+1)^p} = \sum_{r=1}^{n} \frac{1}{[r(n-r+1)]^p}$$

But $\qquad r(n+1-r) = -(r^2 - (n+1)r)$

$$= \left(\frac{n+1}{2}\right)^2 - \left(r - \frac{n+1}{2}\right)^2 \le \left(\frac{n+1}{2}\right)^2$$

$$\therefore \qquad r^p(n+1-r)^p \le \left(\frac{n+1}{2}\right)^{2p} \Rightarrow \quad d_n \ge \sum_{r=1}^{n}\left(\frac{2}{n+1}\right)^{2p} = n \cdot \left(\frac{2}{n+1}\right)^{2p}$$

$$= 2^{2p} \cdot n^{1-2p} \cdot \left(\frac{n}{n+1}\right)^{2p}$$

As for $n \in \mathbf{N}$, $2^{2p}\left(\dfrac{n}{n+1}\right)^{2p} \geq k$ (some positive number).

$\therefore\quad d_n \geq kn^{1-2p}\ \forall\ n \in \mathbf{N}.$

Case (*i*) $p \leq \frac{1}{2}$.

Then, $d_n \geq k$ and the series $\sum_{n=1}^{\infty} c_n$ does not converge.

Case (*ii*) $p > \frac{1}{2}$.

Then, $2p - 1 > 0$

$$\therefore \qquad d_n \geq \frac{k}{n^{2p-1}} \geq 0 \quad \text{and} \quad n^{2p-1} \to \infty \text{ as } n \to \infty$$

$$\therefore \qquad d_n \to 0 \quad \text{as} \quad n \to \infty.$$

Clearly, $< d_n >$ is monotonically decreasing.

$\therefore$ by Leibnitz's test, the series $\sum_{n=1}^{\infty} c_n$ converges.

Example 12. *For all $x \in \mathbf{R}$, show that*

$$\left(1 + \frac{x}{1^2} + \frac{x^2}{(2\,!)^2} + \frac{x^3}{(3\,!)^2} + \ldots\ldots\right) \times \left(1 - \frac{x}{1^2} + \frac{x^2}{(2\,!)^2} - \frac{x^3}{(3\,!)^2} + \ldots\ldots\right)$$

$$= 1 - \frac{x^2}{1^2\,.\,2\,!} + \frac{x^4}{(2\,!)^2\,.\,4\,!} - \frac{x^6}{(3\,!)^2\,.\,6\,!} + \ldots\ldots$$

Sol. For $x \in \mathbf{R}$, the series

$$\sum_{n=0}^{\infty} \frac{x^n}{(n\,!)^2} = 1 + \frac{x}{1^2} + \frac{x^2}{(2\,!)^2} + \frac{x^3}{(3\,!)^2} + \ldots\ldots$$

and

$$\sum_{n=0}^{\infty} (-1)^n \frac{x^n}{(n\,!)^2} = 1 - \frac{x}{1^2} + \frac{x^2}{(2\,!)^2} - \frac{x^3}{(3\,!)^2} + \ldots\ldots$$

are both convergent absolutely. (Prove !)

$\therefore$ by Cauchy's theorem, the product of these two series $\sum_{n=0}^{\infty} c_n$ is also absolutely convergent.

Here, $c_0 = 1$ and for $n \geq 1$,

$$c_n = 1\,.\,(-1)^n \frac{x^n}{(n\,!)^2} + \frac{x}{1^2}(-1)^{n-1} \frac{x^{n-1}}{[(n-1)\,!]^2} + \frac{x^2}{(2\,!)^2}(-1)^{n-2} \frac{x^{n-2}}{[(n-2)\,!]^2} + \ldots\ldots + \frac{x^n}{(n\,!)^2}$$

$$= \frac{(-1)^n x^n}{(n\,!)^2}\left[1 - \frac{(n\,!)^2}{1^2\,.\,[(n-1)\,!]^2} + \frac{(n\,!)^2}{(2\,!)^2[(n-2)\,!]^2} - \ldots\ldots + (-1)^n \frac{(n\,!)^2}{(n\,!)^2}\right]$$

$$= \frac{(-1)^n x^n}{(n!)^2} [1 - ({}^nC_1)^2 + ({}^nC_2)^2 - ({}^nC_3)^2 + \ldots\ldots + (-1)^n({}^nC_n)^2]$$

$$\left[\because \frac{(n!)^2}{1^2 \cdot ((n-1)!)^2} = \left(\frac{n!}{1!(n-1)!}\right)^2 = ({}^nC_1)^2, \text{etc.}\right]$$

$$= \begin{cases} 0, & \text{if } n \text{ is odd} \\ \dfrac{(-1)^n \cdot x^n}{(n!)^2} \cdot (-1)^{n/2}\, {}^nC_{n/2}, & \text{if } n \text{ is even} \end{cases}$$

$[\because {}^nC_0^{\,2} - {}^nC_1^{\,2} + {}^nC_2^{\,2} - \ldots\ldots + (-1)^n\, {}^nC_n^{\,2}$

$= 0$ or $(-1)^{\frac{n}{2}}\, {}^nC_{n/2}$ according as n is odd or even$]$

$$= \begin{cases} 0, & \text{if } n \text{ is odd} \\ \dfrac{(-1)^{\frac{3n}{2}}}{(n!)^2}\, {}^nC_{n/2} \cdot x^n, & \text{if } n \text{ is even} \end{cases}$$

Thus, for $x \in \mathbb{R}$, the product of the series

$$\sum_{n=0}^{\infty} \frac{x^2}{(n!)^2} \quad \text{and} \quad \sum_{n=0}^{\infty} \frac{(-1)^n x^n}{(n!)^2}$$

is
$$\sum_{n=0}^{\infty} c_n = 1 - \frac{x^2}{2!} + \frac{x^4}{(4!)(2!)^2} - \frac{x^6}{(6!)(3!)^2} + \ldots\ldots$$

Example 13. *If for $|x| < 1$, the series*

$$a_0 + a_1x + a_2x^2 + \ldots\ldots + a_nx^n + \ldots\ldots$$

is absolutely convergent to A(x), then show that

$$(1-x)^{-1} A(x) = \sum_{n=0}^{\infty} S_n x^n, \text{ where } S_n = a_0 + a_1 + \ldots\ldots + a_n.$$

Hence, show that $\sum_{n=0}^{\infty} (n+1)x^n = (1-x)^{-2}$.

Sol. The series $\sum_{n=0}^{\infty} x^n$ converges absolutely for $|x| < 1$ and has sum

$$= \frac{1}{1-x} = (1-x)^{-1}$$

$\therefore$ by Cauchy's theorem, the Cauchy product,

$$\sum_{n=0}^{\infty} c_n \text{ of } \sum_{n=0}^{\infty} x^n \text{ and } \sum_{n=0}^{\infty} a_n x^n \text{ converges absolutely for } |x| < 1$$

and
$$\sum_{n=0}^{\infty} c_n = (1-x)^{-1} A(x), \ (|x| < 1)$$

Now, $c_0 = a_0$ and for $n \geq 1$,

$$c_n = 1 \,.\, a_n x^n + x(a_{n-1}x^{n-1}) + x^2(a_{n-2}x^{n-2}) + \ldots\ldots + a_0 x^n$$
$$= x^n (a_0 + a_1 + \ldots\ldots + a_n) = S_n \,.\, x^n.$$

Thus, for $|x| < 1$, $(1-x)^{-1} A(x) = \sum_{n=0}^{\infty} S_n x^n$ ...(1)

where $S_n = a_0 + a_1 + \ldots\ldots + a_n$.

If $a_n = 1, \forall\, n \geq 0$, then from (1),

$$\sum_{n=0}^{\infty} (n+1)x^n = (1-x)^{-1}(1-x)^{-1} = (1-x)^{-2}.$$

Example 14. *Show that*

$$\sum_{n=0}^{\infty} (-1)^n \left[\frac{1}{(n+1)\,.\,1} + \frac{1}{n\,.\,2} + \ldots\ldots + \frac{1}{1\,.\,(n+1)}\right] = (\log 2)^2.$$

(BCA (II) 2007)

Sol. We know

$$\log 2 = 1 - \frac{1}{2} + \frac{1}{3} - \frac{1}{4} + \frac{1}{5} - \ldots\ldots$$

$\therefore$ $(\log 2)^2$ is the product of the series

$$1 - \frac{1}{2} + \frac{1}{3} - \frac{1}{4} + \frac{1}{5} - \ldots\ldots$$

with itself.

Let this product be $\sum_{n=0}^{\infty} c_n$, then $c_0 = 1$ and for $n \geq 1$,

$$c_n = (-1)^n \left[\frac{1}{(n+1)\,.\,1} + \frac{1}{n\,.\,2} + \ldots\ldots + \frac{1}{1\,.\,(n+1)}\right]$$
$$= \frac{(-1)^n}{n+2}\left[\frac{n+2}{(n+1)\,.\,1} + \frac{n+2}{n\,.\,2} + \ldots\ldots + \frac{n+2}{1\,.\,(n+1)}\right]$$
$$= \frac{(-1)^n}{n+2}\left[\left(\frac{1}{1} + \frac{1}{n+1}\right) + \left(\frac{1}{2} + \frac{1}{n}\right) + \ldots\ldots + \left(\frac{1}{n+1} + \frac{1}{1}\right)\right]$$

The result will follow from Abel's theorem, if we can show that $\sum_{n=0}^{\infty} c_n$ converges.

Now, $$|c_n| = \frac{2}{n+2}\left(\frac{1}{1} + \frac{1}{2} + \ldots\ldots + \frac{1}{n+1}\right)$$
$$= 2\,.\left(\frac{n+1}{n+2}\right).\frac{\left(\frac{1}{1} + \frac{1}{2} + \ldots\ldots + \frac{1}{n+1}\right)}{n+1} \text{ which} \to 0 \text{ as } n \to \infty$$

(by Cauchy's first theorem on limits)

$$\therefore \quad |c_{n+1}| - |c_n| = \left(\frac{2}{n+3} - \frac{2}{n+2}\right)\left(1 + \frac{1}{2} + \ldots\ldots + \frac{1}{n+1}\right) + \frac{2}{(n+2)(n+3)}$$

$$= \frac{-2}{(n+2)(n+3)}\left(1 + \frac{1}{2} + \ldots\ldots + \frac{1}{n+1}\right) + \frac{2}{(n+2)(n+3)}$$

$$= \frac{-2}{(n+2)(n+3)}\left(\frac{1}{2} + \ldots\ldots + \frac{1}{n+1}\right) < 0$$

$\Rightarrow \quad |c_{n+1}| < |c_n|$ **and so** $<|c_n|>$ **is monotonically decreasing.**

$\therefore$ **by Leibnitz's test, the alternating series** $\sum_{n=0}^{\infty} c_n$ **converges.**

EXERCISE 8 (*c*)

1. (*a*) Show that the Cauchy product of the convergent series $\sum_{n=0}^{\infty} \frac{(-1)^n}{\sqrt{n+1}}$ with itself is not convergent.

(*b*) Prove that Cauchy product of the convergent series $\sum_{n=1}^{\infty} \frac{(-1)^{n-1}}{n}$ with itself is not convergent.

(*K.U. BCA (II) 2001*)

2. Using Cauchy's theorem, show that

$$\frac{1}{2}\left(x - \frac{1}{2}x^2 + \frac{1}{3}x^3 \ldots\ldots\right)^2 = \sum_{n=1}^{\infty} (-1)^{n+1}\left(1 + \frac{1}{2} + \frac{1}{3} + \ldots\ldots + \frac{1}{n}\right)\frac{x^{n+1}}{n+1}$$

when (*i*) $|x| < 1$ and (*ii*) $x = 1$.

[**Hint.** (*i*) If $|x| < 1$, then one can see applying ratio test, that the series $x - \frac{1}{2}x^2 + \frac{1}{3}x^3 - \ldots\ldots$ is absolutely convergent.

Hence, by Cauchy's theorem, the product of the series with itself is convergent. If $\sum_{n=1}^{\infty} c_n$ represents the Cauchy product series, then

$$c_n = x \,.\, (-1)^{n-1}\frac{1}{n}x^n + \left(-\frac{1}{2}x^2\right)(-1)^{n-2}\frac{1}{n-1}x^{n-1} + \ \ldots\ldots + (-1)^{n-1}\frac{1}{n}x^n \,.\, x$$

$$= (-1)^{n-1}x^{n+1}\left[\frac{1}{1.n} + \frac{1}{2.(n-1)} + \ldots\ldots + \frac{1}{n.1}\right]$$

$$= (-1)^{n-1}\frac{x^{n+1}}{n+1}\left[\frac{n+1}{1.n} + \frac{n+1}{2.(n-1)} + \ldots\ldots + \frac{n+1}{n.1}\right]$$

$$= (-1)^{n-1}\frac{x^{n+1}}{n+1}\left[\left(1 + \frac{1}{n}\right) + \left(\frac{1}{2} + \frac{1}{n-1}\right) + \ldots\ldots + \left(\frac{1}{n} + 1\right)\right]$$

$$= (-1)^{n+1}\frac{x^{n+1}}{n+1} \,.\, 2\left(1 + \frac{1}{2} + \frac{1}{3} + \ldots\ldots + \frac{1}{n}\right).\ \Big]$$

3. Show that

$$\left(1-\frac{1}{2}+\frac{1}{3}-\ldots\ldots\right)^2=\sum_{n=1}^{\infty}(-1)^{n-1}\left[\frac{1}{1.n}+\frac{1}{2.(n-1)}+\ldots\ldots+\frac{1}{n.1}\right]$$

4. If $|x|<1$, show that $\frac{1}{1-x}\log\frac{1}{1-x}=\sum_{n=1}^{\infty}\left(1+\frac{1}{2}+\frac{1}{3}+\ldots+\frac{1}{n}\right)x^n$.

5. Assuming $\tan^{-1}x=x-\frac{x^3}{3}+\frac{x^5}{5}-\ldots\ldots$, prove that

$$(\tan^{-1}x)^2=2\sum_{n=0}^{\infty}(-1)^n\frac{x^{2n+2}}{2n+2}\left(1+\frac{1}{3}+\ldots+\frac{1}{2n+1}\right).$$

Also, show that the series is absolutely convergent if $|x|<1$ and convergent if $x=1$.

Or

For $-1<x\le 1$, show that

$$\frac{1}{2}\left(x-\frac{1}{3}x^3+\frac{1}{5}x^5+\ldots\right)^2=\sum_{n=0}^{\infty}(-1)^n\frac{x^{2n+2}}{2n+2}\times\left(1+\frac{1}{3}+\ldots+\frac{1}{2n+1}\right)$$

6. Prove that the Cauchy product of the two series $3+\sum_{n=1}^{\infty}3^n$ and $-2+\sum_{n=1}^{\infty}2^n$ is absolutely convergent, although both the series are divergent.

7. Consider the validity of $\left(\sum_{n=1}^{\infty}a_n\right)\left(\sum_{n=1}^{\infty}b_n\right)=\sum_{n=1}^{\infty}c_n$

where $c_n=a_1b_n+a_2b_{n-1}+\ldots\ldots+a_nb_1$ in the three cases :

(*i*) $a_n=b_n=\frac{1}{n^2}$ (*ii*) $a_n=b_n=\frac{(-1)^n}{n}$ (*iii*) $a_n=b_n=\frac{(-1)^n}{\sqrt{n}}$.

8.5. CONVERGENCE OF INFINITE PRODUCTS

So far, we have been concerned with the convergence of infinite series. Now we, give a few notions of convergence of infinite products similar to those of infinite series.

If $<a_n>$ is a sequence, we denote by P_n the product of first n terms $a_1, a_2, \ldots\ldots, a_n$ of this sequence and write $P_n=a_1a_2\ldots\ldots a_n=\prod_{r=1}^{n}a_r$.

The product $a_1a_2\ldots$ written as $\prod_{n=1}^{\infty}a_n$ or just $\prod a_n$ is called the infinite product.

The sequence $<P_n>$ is called the sequence of partial products of $<a_n>$, the number P_n is called the nth partial product and a_n is called the nth factor of the product.

8.5.1. Definitions. *Given an infinite product* $\prod_{n=1}^{\infty} a_n$, $P_n = \prod_{r=1}^{n} a_r$

(*i*) *If no factor* a_n *is zero, we say that the product* $\prod_{n=1}^{\infty} a_n$ *converges if the sequence* $< P_n >$ *converges to a finite non-zero number P(i.e.,* $\lim_{n \to \infty} P_n = P$*). P is called the value of the product and we write* $\prod_{n=1}^{\infty} a_n = P$

If $P_n \to \infty$, *we say that the product diverges to* ∞.

If $P_n \to 0$, *we say that the product diverges to zero*.*

(*ii*) *If infinitely many factors* a_n *are zero, we say that the product diverges to zero.*

(*iii*) *If each of a finite number of factors is zero, we say that the product converges if it converges with the zero factors removed.*

(*iv*) *If a finite number of factors are negative, there exists an integer m such that n > m implies* $a_n > 0$. *We say that* $\prod_{n=1}^{\infty} a_n$ *converges if* $\prod_{n=m+1}^{\infty} a_n$ *converges and we write*

$$\prod_{n=1}^{\infty} a_n = a_1 a_2 \ldots\ldots a_m \prod_{n=m+1}^{\infty} a_n.$$

(*v*) *If* $< P_n >$ *oscillates, we say that product* $\prod_{n=1}^{\infty} a_n$ *oscillates.*

Remarks 1. *The convergence of an infinite product is not affected by inserting or removing a finite number of factors, zero or not.*

2. *Convergence of infinite products can easily be seen by converting the product into a series by taking suitably the logarithm of the product.*

Example 15. *Show that the infinite product*

$$\left(1-\frac{1}{2^2}\right)\left(1-\frac{1}{3^2}\right)\left(1-\frac{1}{4^2}\right) \ldots\ldots \text{ is convergent to } \tfrac{1}{2}.$$

**The phrase 'diverges to zero' applied to an infinite product though appears strange, yet it is quite natural when we observe that the behaviour of the product* $\prod_{n=1}^{\infty} a_n$, *when* $a_n > 1$, *is determined by the behaviour of the series.*

$\sum_{n=1}^{\infty} \log a_n$ *because* $\log P_n = \log \left(\sum_{r=1}^{n} a_r\right) = \sum_{r=1}^{n} \log a_r$.

Thus, to say that the product diverges to zero is the same thing as saying that the series $\sum_{n=1}^{\infty} \log a_n$ *diverges to* $-\infty$.

Sol. Here, $$a_n = 1 - \frac{1}{(n+1)^2} = \frac{(n+1)^2 - 1}{(n+1)^2} = \frac{n(n+2)}{(n+1)^2}$$

$$\therefore \quad P_n = \frac{1.3}{2^2} \cdot \frac{2.4}{3^2} \cdot \frac{3.5}{4^2} \cdot \frac{4.6}{5^2} \ldots\ldots \frac{(n)(n+2)}{(n+1)^2}$$

$$= \frac{1}{2}\left(\frac{n+2}{n+1}\right) = \frac{1}{2}\left(1 + \frac{1}{n+1}\right) \to \tfrac{1}{2} \text{ as } n \to \infty$$

Hence, the given infinite product is convergent to $\frac{1}{2}$.

Let us now slove the above example by converting the product into a series by taking suitably the logarithm of the product.

Example 16. *Show that the infinite product*

$$\left(1 - \frac{1}{2^2}\right)\left(1 - \frac{1}{3^2}\right)\left(1 - \frac{1}{4^2}\right) \ldots\ldots \textit{ is convergent.}$$ (K.U. BCA (II) 2008)

Sol. $$P = \left(1 - \frac{1}{2^2}\right)\left(1 - \frac{1}{3^2}\right)\left(1 - \frac{1}{4^2}\right) \ldots\ldots \text{ to } \infty$$

$$\therefore \quad \log P = \log\left(1 - \frac{1}{2^2}\right) + \log\left(1 - \frac{1}{3^2}\right) + \log\left(1 - \frac{1}{4^2}\right) + \ldots\ldots \text{ to } \infty$$

$$= \sum_{n=2}^{\infty} \log\left(1 - \frac{1}{n^2}\right) \qquad \ldots(1)$$

Now, $$a_n = \log\left(1 - \frac{1}{n^2}\right)$$

$$= -\left[\frac{1}{n^2} + \frac{1}{2n^4} + \frac{1}{3n^6} + \ldots\ldots\right] \qquad \left(\because \; -\log(1-x) = x + \frac{1}{2}x^2 + \frac{1}{3}x^3 + \ldots\ldots\right)$$

Comparing with the convergent series $\sum_{n=2}^{\infty} \frac{1}{n^2}$, we find that $\sum_{n=1}^{\infty} a_n$ is convergent.

$\therefore$ from (1), log P is convergent and hence P is convergent.

Example 17. *Show that the infinite products :*

(*i*) $\prod_{n=1}^{\infty}\left(1 + \frac{1}{n}\right)$ *and* (*ii*) $\prod_{n=2}^{\infty}\left(1 - \frac{1}{n}\right)$ *are both divergent.*

Sol. (*i*) $$P_n = \left(1 + \frac{1}{1}\right)\left(1 + \frac{1}{2}\right) \ldots\ldots \left(1 + \frac{1}{n}\right)$$

$$= \left(\frac{2}{1}\right)\left(\frac{3}{2}\right) \ldots\ldots \left(\frac{n+1}{n}\right) = n + 1 \to \infty \text{ as } n \to \infty$$

$$\therefore \quad \prod_{n=1}^{\infty}\left(1 + \frac{1}{n}\right) \text{ is divergent.}$$

(*ii*)
$$P_n = \left(1-\frac{1}{2}\right)\left(1-\frac{1}{3}\right)\left(1-\frac{1}{4}\right)\ldots\ldots\left(1-\frac{1}{n}\right)$$
$$= \frac{1}{2}\cdot\frac{2}{3}\cdot\frac{3}{4}\ \ldots\ldots\ \frac{n-1}{n} = \frac{1}{n} \to 0 \text{ as } n \to \infty$$

$\therefore\ \prod_{n=2}^{\infty}\left(1-\frac{1}{n}\right)$ is divergent to 0.

EXERCISE 8 (*d*)

1. Show that the infinite product
$$\frac{1}{2}\cdot\frac{3}{2}\cdot\frac{3}{4}\cdot\frac{5}{4}\cdot\frac{5}{6}\cdot\frac{7}{6}\ldots\ldots\frac{2n-1}{2n}\cdot\frac{2n+1}{2n}\ldots\ldots$$
tends to a finite limit as $n \to \infty$.

2. Show that the infinite product
$$\left(1+\frac{1}{1^3}\right)\left(1+\frac{1}{2^3}\right)\left(1+\frac{1}{3^3}\right)\ldots\ldots\left(1+\frac{1}{n^3}\right)\ldots\ldots$$
is convergent. (*K.U. BCA (II) 2007*)

It is usually convenient to write the factors of the infinite product as $1 + a_n$ instead of a_n. Thus, we shall write infinite product as
$$\prod_{n=1}^{\infty}(1+a_n) \text{ and } P_n = \prod_{r=1}^{n}(1+a_r).$$
Since an infinite product, having infinitely many negative factors cannot be convergent (will be clear from the following theorem), we do not consider such products. So, we shall assume throughout our discussion than $a_n > -1,\ \forall\ n$.

8.5.2. Theorem. (A necessary condition for convergence).

If the product $\prod_{n=1}^{\infty}(1+a_n)$ is convergent, then $a_n \to 0$. (K.U. BCA (II) 2001)

Proof. Let $\prod_{n=1}^{\infty}(1+a_n)$ be convergent.

Then, P_n and P_{n-1} converge to the same non-zero limit P (say) as $n \to \infty$.

Now,
$$\lim_{n\to\infty}\frac{P_n}{P_{n-1}} = \frac{\lim_{n\to\infty}P_n}{\lim_{n\to\infty}P_{n-1}} = \frac{P}{P} = 1$$
$$\Rightarrow \quad \lim_{n\to\infty}(1+a_n) = 1$$
$$\Rightarrow \quad a_n \to 0 \text{ as } n \to \infty$$

8.5.3. Cauchy's general principle of convergence. Since the convergence of the product $\prod_{n=1}^{\infty}(1+a_n)$ is determined by the convergence of the sequence $< P_n >$, we find that

'*A necessary and sufficient condition for the convergence of a product* $\prod_{n=1}^{\infty}(1+a_n)$ *is that for every* $\varepsilon > 0$, *there exists a +ve integer m such that*

$$| P_{n+p} - P_n | < \varepsilon, \quad \forall\, n \geq m, p \geq 1$$

Since $| P_{n+p} - P_n | < \varepsilon$

$$\Rightarrow \quad \left| \frac{P_{n+p}}{P_n} - 1 \right| < \frac{\varepsilon}{|P_n|} = \varepsilon_1 \quad \text{(say)}$$

$$\Rightarrow \quad | (1 + a_{n+1})(1 + a_{n+2}) \ldots\ldots (1 + a_{n+p}) - 1 | < \varepsilon_1 \quad \forall\, n \geq m, p \geq 1.$$

Thus, a product $\prod_{n=1}^{\infty}(1 + a_n)$ *is convergent if and only if for every* $\varepsilon_1 > 0$, $\exists$ *a +ve integer m such that*

$$| (1 + a_{n+1})(1 + a_{n+2}) \ldots\ldots (1 + a_{n+p}) - 1 | < \varepsilon_1, \quad \forall\, n \geq m, p \geq 1.$$

8.5.4. Theorem. *If* $a_n \geq 0$, *the series* $\sum_{n=1}^{\infty} a_n$ *and the product* $\prod_{n=1}^{\infty}(1 + a_n)$ *converge and diverge together.*

Proof. When $x \geq 0$, we have $1 + x \leq e^x$. $\left[\because\ e^x = 1 + \frac{x}{1!} + \frac{x^2}{2!} + \frac{x^3}{3!} + \ldots\right]$

Substituting $x = a_1, a_2, \ldots\ldots, a_n$ in succession and multiplying, we have

$$a_1 + a_2 + \ldots\ldots + a_n < (1 + a_1)(1 + a_2) \ldots\ldots (1 + a_n)$$
$$\leq e^{a_1 + a_2 + \ldots\ldots + a_n}$$

$$\Rightarrow \quad S_n < P_n \leq e^{S_n} \quad \ldots(1)$$

where $S_n = a_1 + a_2 + \ldots\ldots + a_n$ and $P_n = (1 + a_1)(1 + a_2) \ldots\ldots (1 + a_n)$

Since $< S_n >$ and $< P_n >$ are monotonically increasing sequences, they have a limit finite or infinite. If $\sum_{n \to 1}^{\infty} a_n$ converges to a finite limit a (say), *i.e.*, $S_n \to a$, then P_n is bounded above by e^a (By (1)) and therefore the sequence $< P_n >$ and hence the product $\prod_{n=1}^{\infty}(1 + a_n)$ converges. If $\sum_{n=1}^{\infty} a_n$ diverges to ∞, *i.e.*, $S_n \to \infty$, then $P_n \to \infty$ and hence $\prod_{n=1}^{\infty}(1 + a_n)$ diverges to ∞.

Similarly, the series $\sum_{n=1}^{\infty} a_n$ converges or diverges with $\prod_{n=1}^{\infty}(1 + a_n)$.

Remark. *Since $a_n \geq 0$, the series $\sum_{n=1}^{\infty} a_n$ and $\sum_{n=1}^{\infty} (1 + a_n)$ cannot tend to a negative limit. Since, each factor $1 + a_n$ is positive, the product $\prod_{n=1}^{\infty} (1 + a_n)$ cannot diverge to zero ; it can therefore either converge or diverge to ∞.*

For example,

(*i*) The product $\prod_{n=1}^{\infty} \left(1+\frac{1}{n^{\alpha}}\right)$, $(\alpha > 1)$ is convergent because the series $\sum_{n=1}^{\infty} \frac{1}{n^{\alpha}}$, $(\alpha > 1)$ is convergent.

(*ii*) The product $\prod_{n=1}^{\infty} \left(1+\frac{1}{n^{\alpha}}\right)$, $(0 < \alpha \leq 1)$ is divergent because the series $\sum_{n=1}^{\infty} \frac{1}{n^{\alpha}}$, $(0 < \alpha \leq 1)$ is divergent.

8.5.5. Theorem. *If $-1 < a_n \leq 0$, then the series $\prod_{n=1}^{\infty} a_n$ and the product $\prod_{n=1}^{\infty} (1 + a_n)$ converge or diverge together.*

Proof. Let $b_n = -a_n$ so that $0 \leq b_n < 1$.

The product $\prod_{n=1}^{\infty} (1 + a_n)$ becomes $\prod_{n=1}^{\infty} (1 - b_n)$, $0 \leq b_n < 1$.

Let $P_n = \prod_{r=1}^{n} (1 - b_r)$.

For $0 \leq x < 1$, $1 - x \leq e^{-x}$ $\left[\because\ e^{-x} = 1 - \frac{x}{1!} + \frac{x^2}{2!} - \frac{x^3}{3!} + \ldots\ldots\right]$

$\therefore$ substituting $x = b_1, b_2, \ldots\ldots, b_n$ successively and multiplying, we have

$$0 < (1 - b_1)(1 - b_2) \ldots\ldots (1 - b_n) \leq e^{-(b_1 + b_2 + \ldots\ldots + b_n)}$$

$$\Rightarrow \quad 0 < P_n \leq e^{-(b_1 + b_2 + \ldots\ldots + b_n)} \quad \ldots(1)$$

Case (*i*) $\sum_{n=1}^{\infty} a_n$ diverges to $-\infty$.

Then $\sum_{n=1}^{\infty} b_n$ diverges to ∞ and so from (1), $P_n \to 0$. Hence, the product $\prod_{n=1}^{\infty} (1 - b_n)$ diverges to zero.

or $\prod_{n=1}^{\infty} (1 + a_n)$ diverges to zero.

Case (*ii*) $\sum_{n=1}^{\infty} a_n$ **is convergent.**

Then, for each $\varepsilon > 0$, we can find a positive integer m such that

$$| a_{m+1} + a_{m+2} + \ldots\ldots + a_n | < \varepsilon, \quad \forall\, n > m$$

$$\Rightarrow \quad | - b_{m+1} - b_{m+2} - \ldots\ldots - b_n | < \varepsilon, \quad \forall\, n > m$$

$$\Rightarrow \quad | b_{m+1} + b_{m+2} + \ldots\ldots + b_n | < \varepsilon, \quad \forall\, n > m$$

Also,
$$(1 - b_{m+1})(1 - b_{m+2}) \geq 1 - b_{m+1} - b_{m+2}$$

$$(1 - b_{m+1})(1 - b_{m+2})(1 - b_{m+3}) \geq (1 - b_{m+1} - b_{m+2})(1 - b_{m+3}) \geq 1 - b_{m+1} - b_{m+2} - b_{m+3}$$

....

$\therefore$ for $n > m$,

$$(1 - b_{m+1})(1 - b_{m+2}) \ldots\ldots (1 - b_n) \geq 1 - b_{m+1} - b_{m+2} - \ldots - b_n > 1 - \varepsilon$$

$$\Rightarrow \quad \frac{P_n}{P_m} > 1 - \varepsilon, \quad \forall\, n > m$$

$$\Rightarrow \quad P_n > (1 - \varepsilon) P_m, \quad \forall\, n > m.$$

Since each factor of P_n is < 1, $< P_n >$ is monotonically decreasing and is bounded below by $(1 - \varepsilon) P_m$. Therefore, $< P_n >$ is convergent.

Hence, the product $\prod_{n=1}^{\infty} (1 - b_n)$ or $\prod_{n=1}^{\infty} (1 + a_n)$ is convergent.

Remark. The above theorem may be stated as :

If $0 \leq b_n < 1$, *then* $\prod_{n=1}^{\infty} (1 - b_n)$ *converges to a finite non-zero limit if* $\sum_{n=1}^{\infty} b_n$ *converges and diverges to zero if* $\sum_{n=1}^{\infty} b_n$ *diverges.*

For instance :

(*i*) $\prod_{n=2}^{\infty} \left(1 - \frac{1}{n^2}\right)$ converges because $\sum_{n=1}^{\infty} \frac{1}{n^2}$ converges

(*ii*) $\prod_{n=2}^{\infty} \left(1 - \frac{1}{n}\right)$ diverges to zero because $\sum_{n=1}^{\infty} \frac{1}{n}$ diverges.

8.5.6. Theorem. *If* $| a_n | < 1$ *and the series* $\sum_{n=1}^{\infty} a_n^2$ *is convergent, then the product* $\prod_{n=1}^{\infty} (1 + a_n)$ *and the series* $\sum_{n=1}^{\infty} a_n$ *converge or diverge together.*

Proof. Since $\sum_{n=1}^{\infty} a_n^2$ is convergent, $\Rightarrow a_n^2 \to 0 \Rightarrow a_n \to 0$ as $n \to \infty$

$\therefore$ taking $\varepsilon = \frac{1}{2}$, there exists a positive integer m such that

$$|a_n| < \tfrac{1}{2}, \quad \forall \quad n \geq m.$$

$\therefore$ for $n \geq m$,

$$|a_n - \log(1 + a_n)| = \left|\frac{a_n^2}{2} - \frac{a_n^3}{3} + \ldots\ldots\right| \qquad \left(\because \log(1+x) = x - \frac{x^2}{2} + \frac{x^3}{3} + \ldots\ldots\right)$$

$$\leq \left|\frac{a_n^2}{2}\right| + \left|\frac{a_n^3}{3}\right| + \ldots\ldots$$

$$\leq \tfrac{1}{2} a_n^2 \{1 + |a_n| + \ldots\ldots\}$$

$$= \frac{a_n^2}{2(1 - |a_n|} < a_n^2.$$

Since the series $\sum_{n=1}^{\infty} a_n^2$ is convergent, therefore, by the comparison test, the series $\sum_{n=1}^{\infty} |a_n - \log(1 + a_n)|$ is also convergent.

$\Rightarrow \sum_{n=1}^{\infty} (a_n - \log(1 + a_n))$ is convergent.

$\Rightarrow \sum_{n=1}^{\infty} a_n$ and $\sum_{n=1}^{\infty} \log(1 + a_n)$ converge or diverge together

$\Rightarrow \sum_{n=1}^{\infty} a_n$ and $\prod_{n=1}^{\infty} (1 + a_n)$ converge or diverge together.

Remark. *The above theorem gives an applied test for the convergence of an infinite product in which a_n is of either sign.*

Since $\sum_{n=1}^{\infty} a_n$ and $\sum_{n=1}^{\infty} \log(1 + a_n)$ converge or diverge together, therefore when $\sum_{n=1}^{\infty} a_n^2$ is convergent, we have

(i) if $\sum_{n=1}^{\infty} a_n$ converges, then $\sum_{n=1}^{\infty} \log(1 + a_n)$ and therefore $\prod_{n=1}^{\infty} (1 + a_n)$ converges.

(ii) if $\sum_{n=1}^{\infty} a_n$ diverges to ∞, then $\sum_{n=1}^{\infty} \log(1 + a_n)$ and therefore $\prod_{n=1}^{\infty} (1 + a_n)$ diverges to ∞.

(iii) if $\sum_{n=1}^{\infty} a_n$ *diverges to* $-\infty$, *then* $\sum_{n=1}^{\infty} \log(1+a_n)$ *diverges to* $-\infty$ *and so* $\prod_{n=1}^{\infty} (1+a_n)$ *diverges to zero.*

For instance, the product $\prod_{n=2}^{\infty}\left(1+\frac{(-1)^n}{n}\right)$ *is convergent because if* $a_n = \frac{(-1)^n}{n}$, *then* $\sum_{n=1}^{\infty} a_n$ *is convergent by Leibnitz's test and* $\sum_{n=1}^{\infty} a_n^2 = \sum_{n=1}^{\infty} \frac{1}{n^2}$ *is also convergent.*

8.5.7. Theorem. *If* $\sum_{n=1}^{\infty} a_n^2$ *is divergent and* $\sum_{n=1}^{\infty} a_n$ *is either convergent or oscillates between finite limits, then the product* $\prod_{n=1}^{\infty} (1+a_n)$ *is divergent to zero.*

Proof. Let $P_n = \prod_{r=1}^{n} (1+a_r)$. Then, $\log P_n = \sum_{r=1}^{n} \log(1+a_r)$.

Now, by Taylor's theorem with remainder after two terms,

$$\log(1+a_n) = \log 1 + a_n - \frac{a_n^2}{2(1+\theta a_n)^2}, \text{ where } 0<\theta<1.$$

[For, by Taylor's theorem,

$$f(a+h) = f(a) + \frac{h}{1!}f'(a) + \frac{h^2}{2!}f''(a) + \ldots\ldots + \frac{h^{n-1}}{(n-1)!}f^{n-1}(a)$$

$$+ \frac{h^n}{n!}f^{(n)}(a+\theta h), \text{ where } 0<\theta<1.$$

Comparing $f(a+h)$ with $\log(1+a_n)$ and substituting $a = 1$, $h = a_n$ and $n = 2$, we have

$$\log(1+a_n) = \log 1 + \frac{a_n}{1!} + \frac{a_n^2}{2!}\cdot\frac{-1}{(1+\theta a_n)^2}$$

$$= a_n - \frac{a_n^2}{2(1+\theta a_n)^2}, \; 0<\theta<1.$$

$$\Rightarrow \qquad a_n - \log(1+a_n) = \frac{a_n^2}{2(1+\theta a_n)^2} > 0.$$

Now, if $\sum_{n=1}^{\infty} a_n$ is convergent or oscillates between finite limits, then we can find a positive number k and a positive integer m such that

$$|a_n| < k, \quad \forall \quad n \geq m.$$

$$\therefore \qquad 1+\theta a_n < 1 + |a_n| < 1+k \qquad (\because \quad 0<\theta<1)$$

$$\Rightarrow \qquad \frac{1}{(1+\theta a_n)^2} > \frac{1}{(1+k)^2}$$

$$\therefore \qquad a_n - \log(1+a_n) = \frac{a_n^2}{2(1+\theta a_n)^2} > \frac{1}{2(1+k)^2}\, a_n^2$$

Since $\sum_{n=1}^{\infty} a_n^2$ is divergent $\sum_{n=1}^{\infty} [a_n - \log(1+a_n)]$ is divergent to ∞.

Since $\sum_{n=1}^{\infty} a_n$ is either convergent or oscillates between finite limits, $\sum_{n=1}^{\infty} \log(1+a_n)$ is divergent to $-\infty$.

$\therefore$ $< \log P_n >$ is divergent to $-\infty$.

$\Rightarrow$ $< P_n >$ is divergent to 0.

Hence, $\prod_{n=1}^{\infty} (1+a_n)$ is divergent to 0.

Example 18. *Discuss the convergence of the product*

$$\prod_{n=1}^{\infty}\left(1+\frac{x}{n^p}\right), \text{ where } x \text{ is a real number.}$$

Sol. Here, $a_n = \dfrac{x}{n^p}$.

$$\therefore \qquad \sum_{n=1}^{\infty} a_n = x \sum_{n=1}^{\infty} \frac{1}{n^p}$$

$\Rightarrow$ $\sum_{n=1}^{\infty} a_n$ is convergent if $p>1$ and divergent if $p \le 1$

$\Rightarrow$ $\prod_{n=1}^{\infty} (1+a_n)$ is convergent if $p>1$ and divergent if $p \le 1$.

Example 19. *Discuss the convergence of the product :*

$$\prod_{n=1}^{\infty}\left(1+n \sin\frac{\theta}{n^2}\right).$$

Sol. Here,

$$a_n = n \sin \frac{\theta}{n^2}$$

$$= n\left[\frac{\theta}{n^2} - \frac{1}{3!}\frac{\theta^3}{n^6} + \ldots\right] = \frac{\theta}{n} - \frac{1}{3!}\frac{\theta^3}{n^5} + \ldots..$$

If $\theta \neq 0$, then $a_n \simeq \frac{1}{n}$.

$\therefore$ by comparison test $\Sigma\, a_n$ is divergent.

$\Rightarrow$ the given product is divergent.

If $\theta = 0$, then the product is obviously convergent.

Example 20. *Discuss the convergence of*

$$\frac{2}{1}\cdot\frac{2}{3}\cdot\frac{4}{3}\cdot\frac{4}{5}\cdot\frac{6}{5}\cdots\frac{2n-2}{2n-3}\cdot\frac{2n-2}{2n-1}\cdot\frac{2n}{2n-1}\cdots$$

Sol. $\frac{2}{1}\cdot\frac{2}{3}\cdot\frac{4}{3}\cdot\frac{4}{5}\cdot\frac{6}{5}\cdots\frac{2n-2}{2n-3}\cdot\frac{2n-2}{2n-1}\cdot\frac{2n}{2n-1}\cdots$

$$= (1+1)\left(1-\frac{1}{3}\right)\left(1+\frac{1}{3}\right)\left(1-\frac{1}{5}\right)\left(1+\frac{1}{5}\right)\cdots\cdots$$

$$\cdots\left(1+\frac{1}{2n-3}\right)\left(1-\frac{1}{2n-1}\right)\left(1+\frac{1}{2n-1}\right)\cdots\cdots$$

Comparing with $\prod_{n=1}^{\infty}(1+a_n)$, we have

$$\sum_{n=1}^{\infty} a_n = 1-\frac{1}{3}+\frac{1}{3}-\frac{1}{5}+\frac{1}{5} \;\cdots\cdots \text{ and is convergent.}$$

Here, $-1 < a_n < 1, \quad \forall\, n$

Also, $\sum_{n=1}^{\infty} a_n^{\,2} = 1+\frac{1}{3^2}+\frac{1}{3^2}+\frac{1}{5^2}+\frac{1}{5^2}+\cdots\cdots$, which is also convergent. Hence,

$\prod_{n=1}^{\infty}(1+a_n)$ is convergent.

8.5.8. Absolute convergence. *The product* $\prod_{n=1}^{\infty}(1+a_n)$ *is said to be absolutely convergent if the product* $\prod_{n=1}^{\infty}(1+|\,a_n\,|)$ *is convergent.*

Note that the product $\prod_{n=1}^{\infty}(1+a_n)$ is absolutely convergent if and only if $\sum_{n=1}^{\infty} a_n$ is absolutely convergent.

For, $\sum_{n=1}^{\infty} a_n$ is absolutely convergent.

$$\Leftrightarrow \sum_{n=1}^{\infty} |a_n| \text{ is convergent} \Leftrightarrow \prod_{n=1}^{\infty} (1 + |a_n|) \text{ is convergent.}$$

$$\Leftrightarrow \prod_{n=1}^{\infty} (1 + a_n) \text{ is absolutely convergent.}$$

The following theorem is useful in checking the absolute convergence of infinite product $\prod_{n=1}^{\infty} (1 + a_n)$.

8.5.9. Theorem. *The series* $\sum_{n=1}^{\infty} |\log(1 + a_n)|$ *is convergent if and only if* $\sum_{n=1}^{\infty} |a_n|$ *is convergent.* (K.U. BCA (II) 2003)

Proof. Let $\sum_{n=1}^{\infty} |\log(1 + a_n)|$ be convergent.

$\Rightarrow$ $|\log(1 + a_n)| \to 0$ and hence $\log(1 + a_n) \to 0$ as $n \to \infty$.

$\Rightarrow$ $1 + a_n \to 1$ as $n \to \infty$ and hence $a_n \to 0$ as $n \to \infty$.

$\therefore$ there exists a positive integer m such that

$$|a_n| < \tfrac{1}{2}, \quad \forall\, n \geq m. \qquad \ldots(1)$$

Now,
$$\log(1 + a_n) = a_n - \frac{a_n^2}{2} + \frac{a_n^3}{3} - \ldots$$

$$\Rightarrow \quad \frac{\log(1+a_n)}{a_n} - 1 = -\frac{1}{2} a_n + \frac{1}{3} a_n^2 - \ldots..$$

$$\Rightarrow \quad \left|\frac{\log(1+a_n)}{a_n} - 1\right| = \left|-\frac{1}{2} a_n + \frac{1}{3} a_n^2 - \ldots..\right|$$

$$\leq \frac{1}{2} |a_n| + \frac{1}{3} |a_n|^2 + \ldots\ldots$$

$$< \frac{1}{4} + \frac{1}{3} \cdot \frac{1}{4} + \ldots., \forall\, n \geq m \qquad \text{[By (1)]}$$

$$< \frac{1}{4} + \frac{1}{8} + \ldots\ldots \qquad \left(\because \frac{1}{3} < \frac{1}{2}, \frac{1}{4} < \frac{1}{2} \text{ etc.}\right)$$

$$= \frac{\frac{1}{4}}{1 - \frac{1}{2}} = \frac{1}{2}$$

$$\therefore \quad 1 - \frac{1}{2} < \frac{\log(1+a_n)}{a_n} < 1 + \frac{1}{2}$$

$$\Rightarrow \quad \frac{1}{2} < \frac{\log(1+a_n)}{a_n} < \frac{3}{2}$$

$$\Rightarrow \qquad \frac{1}{2} < \frac{|\log(1+a_n)|}{|a_n|} < \frac{3}{2}$$

$$\Rightarrow \qquad \frac{1}{2}\,|\,a_n\,| < |\,\log(1+a_n)\,| < \frac{3}{2}\,|\,a_n\,|, \quad \forall\, n \ge m$$

$\therefore$ by comparison test, $\sum_{n=1}^{\infty} |\,\log(1+a_n)\,|$ is convergent if and only if $\sum_{n=1}^{\infty} a_n$ is convergent.

8.5.10. Theorem. *Every absolutely convergent infinite product is convergent.*

Proof. Let $\prod_{n=1}^{\infty} (1+a_n)$ be absolutely convergent.

$\therefore$ by definition $\prod_{n=1}^{\infty} (1+|\,a_n\,|)$ is convergent.

Let $\varepsilon > 0$ be given

Then, there exists a positive integer m such that

$|\,(1+|\,a_{n+1}\,|)(1+|\,a_{n+2}\,|)\,.....\,(1+|\,a_{n+p}\,|)-1\,| < \varepsilon,\ n \ge m,\ \forall\, p \ge 1$

But, $|\,(1+a_{n+1})(1+a_{n+2})\,.....\,(1+a_{n+p})-1\,|$

$\le |\,(1+|\,a_{n+1}\,|)(1+|\,a_{n+2}\,|)\,.....\,(1+|\,a_{n+p}\,|)-1\,|$

$\therefore \quad |\,(1+a_{n+1})(1+a_{n+2})\,.....\,(1+a_{n+p})-1\,| < \varepsilon,\ \forall\, n \ge m,\ p \ge 1$

$\Rightarrow$ by Cauchy's general principle of convergence that $\prod_{n=1}^{\infty} (1+a_n)$ is convergent.

Note. $\prod_{n=1}^{\infty} (1+a_n)$ converges absolutely if and only if $\sum_{n=1}^{\infty} a_n$ converges absolutely. However, the following example shows that $\prod_{n=1}^{\infty} (1+a_n)$ may converge and $\sum_{n=1}^{\infty} a_n$ diverge.

For example, if $a_n = \dfrac{(-1)^n}{\sqrt{n}}$, $n \in \mathbb{N}$, then $\prod_{n=1}^{\infty} (1+|\,a_n\,|)$ diverges but $\sum_{n=1}^{\infty} a_n$ converges (by Leibnitz's test).

Lastly, we close the section with the following result :

'The factors of an absolutely convergent infinite product may be rearranged in any order without affecting its convergence or its value.'

Example 21. *Discuss the convergence of the product*

$$\prod_{n=1}^{\infty} \left(1+\left(\frac{nx}{n+1}\right)^n\right),\ |\,x\,| < 1.$$

Sol. Here, $a_n = \left(\frac{nx}{1+n}\right)^n$, $|x| < 1$

$$\therefore \quad \lim_{n \to \infty} |a_n^{1/n}| = \lim_{n \to \infty} \frac{|x|}{1+\frac{1}{n}} = |x| < 1$$

$\therefore$ by Cauchy's root test, $\sum_{n=1}^{\infty} |a_n|$ is convergent.

$\Rightarrow$ $\sum_{n=1}^{\infty} a_n$ is absolutely convergent.

$\Rightarrow$ the given product is absolutely convergent if $|x| < 1$.

Example 22. *Prove that* $\prod_{n=1}^{\infty} \left(1+\frac{x}{n}\right) e^{-x/n}$ *is absolutely convergent for any real x.*

Sol. $\prod_{n=1}^{\infty} \left(1+\frac{x}{n}\right) e^{-x/n}$

$$= \prod_{n=1}^{\infty} \left(1+\frac{x}{n}\right)\left(1-\frac{x}{n}+\frac{x^2}{2n^2}-\ldots..\right) \qquad \left(\because e^{-x} = 1-\frac{x}{1!}+\frac{x^2}{2!}-\ldots\right)$$

$$\sim \prod_{n=1}^{\infty} \left(1-\frac{x^2}{2n^2}\right), \text{ for a fixed } x \text{ and sufficiently large } n.$$

Since the series $\sum_{n=1}^{\infty} \frac{x^2}{n^2}\left(= x^2 \sum_{n=1}^{\infty} \frac{1}{n^2}\right)$ is absolutely convergent, therefore $\prod_{n=1}^{\infty} \left(1-\frac{x^2}{n^2}\right)$ and hence $\prod_{n=1}^{\infty} \left(1+\frac{x}{n}\right) e^{-x/n}$ is absolutely convergent.

Example 23. *Discuss the convergence of the product* $\prod_{n=1}^{\infty} \left(\frac{x+x^{2n}}{1+x^{2n}}\right)$.

Sol. Here, $$1 + a_n = \frac{x+x^{2n}}{1+x^{2n}}$$

$$\Rightarrow \quad a_n = \frac{x+x^{2n}}{1+x^{2n}} - 1 = \frac{x-1}{1+x^{2n}}$$

Now, $$\sum_{n=1}^{\infty} |a_n| = \sum_{n=1}^{\infty} \left|\frac{x-1}{1+x^{2n}}\right|$$

$$= |x-1| \sum_{n=1}^{\infty} \left| \frac{1}{1+x^{2n}} \right|$$

$$\leq |x-1| \sum_{n=1}^{\infty} \frac{1}{|x^{2n}|}$$

If $|x| > 1$, $\sum_{n=1}^{\infty} \frac{1}{x^{2n}}$ is convergent,

$\Rightarrow \sum_{n=1}^{\infty} a_n$ is absolutely convergent.

Hence, $\prod_{n=1}^{\infty} (1+a_n)$ is absolutely convergent.

If $x = 1$, every factor is unity and hence the product is convergent.
If $x = -1$, every factor is zero and $a_n \to -1 \neq 0$

$\therefore \sum_{n=1}^{\infty} a_n$ is divergent. $\Rightarrow \prod_{n=1}^{\infty} (1+a_n)$ is divergent.

If $|x| < 1$, $1 + a_n = \dfrac{x(1+x^{2n-1})}{1+x^{2n}} \to x$ and $a_n \nrightarrow 0$ as $n \to \infty$.

Hence, the product is divergent.

Example 24. *Test the absolute convergence of*

$$\prod_{n=1}^{\infty} \left(1 + \frac{x^n}{x^{2n}+1}\right).$$ (K.U. BCA (II) 2003)

Sol. Here, $1 + a_n = 1 + \dfrac{x^n}{x^{2n}+1}$

$$\Rightarrow \qquad a_n = \frac{x^n}{x^{2n}+1}$$

$$\therefore \qquad \frac{|a_n|}{|a_{n+1}|} = \left| \frac{x^n}{x^{2n}+1} \cdot \frac{x^{2n+2}+1}{x^{n+1}} \right| = \left| \frac{x^{2n+2}+1}{x^{2n}+1} \right| \frac{1}{|x|}$$

If $|x| < 1$, then $\lim_{n \to \infty} \dfrac{|a_n|}{|a_{n+1}|} = \dfrac{1}{|x|} > 1$ $(\because x^n \to 0 \text{ as } n \to \infty)$

$\therefore$ by ratio test, the series $\sum_{n=1}^{\infty} |a_n|$ is convergent.

Hence, the given infinite product is convergent absolutely.

If $|x| > 1$, then $\lim\limits_{n\to\infty} \dfrac{|a_n|}{|a_{n+1}|}$

$$= \lim_{n\to\infty} \left| \frac{x + \dfrac{1}{x^{2n+1}}}{1 + \dfrac{1}{x^{2n+1}}} \right| = |x| > 1 \qquad \left(\because \frac{1}{x^n} \to 0 \text{ as } n \to \infty\right)$$

$\therefore$ by ratio test, $\sum\limits_{n=1}^{\infty} |a_n|$ is convergent.

Hence, the given infinite product is convergent absolutely.

If $x = 1$, then $a_n = \dfrac{1}{2}$, $\forall\, n$.

$\therefore$ $\sum\limits_{n=1}^{\infty} a_n$ is divergent and therefore, the product is divergent.

If $x = -1$, $a_n = \dfrac{(-1)^n}{2}$ which oscillates finitely and $\sum\limits_{n=1}^{\infty} a_n^2$ is divergent. Hence $\prod\limits_{n=1}^{\infty} (1 + a_n)$ is divergent to 0.

EXERCISE 8 (e)

1. Discuss the convergence of the following infinite products :

(i) $\left(1+\dfrac{1}{1^2}\right)\left(1+\dfrac{1}{2^2}\right)\left(1+\dfrac{1}{3^2}\right)\cdots\cdots\left(1+\dfrac{1}{n^2}\right)\cdots\cdots$

(ii) $\prod\limits_{n=1}^{\infty}\left(1+\sin^2\dfrac{\theta}{n}\right)$ $\left[\textbf{Hint } (ii) \sin^2\dfrac{\theta}{n} = \left(\dfrac{\theta}{n} - \dfrac{\theta^3}{3!\,n^3} + \cdots\cdots\right)^2 = \dfrac{\theta^2}{n^2} - \dfrac{\theta^4}{3n^4} + \cdots\cdots\right]$

(iii) $\prod\limits_{n=1}^{\infty}\left(1+\dfrac{1}{\sqrt{n}}\right)$

(iv) $\dfrac{3}{4}\cdot\dfrac{6}{7}\cdot\dfrac{9}{10}\cdots\cdots\dfrac{3n}{3n+1}\cdots\cdots$

[**Hint.** The infinite product can be written as

$$\left(1-\frac{1}{4}\right)\left(1-\frac{1}{7}\right)\left(1-\frac{1}{10}\right)\cdots\cdots\left(1-\frac{1}{3n+1}\right)\cdots\cdots\Bigg]$$

2. Show that the following products are divergent to 0

(i) $\prod\limits_{n=1}^{\infty}\left(1-\dfrac{1}{n}\right)$

(ii) $\prod\limits_{n=1}^{\infty}\left(1-\dfrac{(-1)^{n-1}}{\sqrt{n}}\right)$

(iii) $\prod\limits_{n=1}^{\infty}[1+(-1)^n]$.

[**Hint.** (*i*) Here $a_n = -\dfrac{1}{n}$. So $\sum_{n=1}^{\infty} a_n$ is divergent to $-\infty$ and $\sum_{n=1}^{\infty} a_n^2 = \sum_{n=1}^{\infty} \dfrac{1}{n^2}$ is convergent.

(*ii*) Here $\sum_{n=1}^{\infty} a_n$ is convergent by Leibnitz's test but $\sum_{n=1}^{\infty} a_n^2$ is divergent.

(*iii*) Here a_n oscillates between -1 and 0, and $\sum_{n=1}^{\infty} a_n$ is divergent.]

3. Show that $(1+x)\left(1+\dfrac{x}{2}\right)\left(1+\dfrac{x}{3}\right)$ ...... diverges to ∞ or to 0 according as $x > 0$ or $x < 0$.

[**Hint.** Here $\sum_{n=1}^{\infty} a_n = \sum_{n=1}^{\infty} \dfrac{x}{n}$ which is divergent to ∞ if $x > 0$ and divergent to $-\infty$ if $x < 0$.]

4. Discuss the convergence of $\left(1-\dfrac{1}{2}\right)\left(1+\dfrac{1}{3}\right)\left(1-\dfrac{1}{4}\right)$ ......

5. Show that the infinite product $\prod_{n=2}^{\infty}\left(1+\dfrac{(-1)^n}{n^{\alpha}}\right)$ is convergent if $\alpha > \dfrac{1}{2}$.

[**Hint.** Here, $a_n = \dfrac{(-1)^n}{n^{\alpha}}$

$\therefore \quad \sum_{n=1}^{\infty} a_n$ is convergent if $\alpha > 0$ by Leibnitz's test.

Also $\sum_{n=1}^{\infty} a_n^2 = \sum_{n=1}^{\infty} \dfrac{1}{n^{2\alpha}}$ is convergent if $2\alpha > 1$, *i.e.*, $\alpha > \dfrac{1}{2}$.]

6. Show that $\prod_{n=1}^{\infty}\left(1-\dfrac{1}{n^{2/3}}\right)e^{\frac{1}{n^{2/3}}}$ is absolutely convergent.

[**Hint.** $1 + a_n = \left(1-\dfrac{1}{n^{2/3}}\right)e^{\frac{1}{n^{2/3}}} = \left(1-\dfrac{1}{n^{2/3}}\right)\left(1+\dfrac{1}{n^{2/3}}+\dfrac{1}{n^{4/3}\,.\,2!}+......\right) \simeq 1-\dfrac{1}{2n^{4/3}}$.]

7. Show that each of the following product diverges to 0

(*i*) $\left(1-\dfrac{1}{2}\right)\left(1+\dfrac{1}{2}\right)\left(1-\dfrac{1}{2}\right)\left(1+\dfrac{1}{2}\right)$ ......

(*ii*) $\left(1+\dfrac{1}{\sqrt{2}}\right)\left(1-\dfrac{1}{\sqrt{3}}\right)\left(1+\dfrac{1}{\sqrt{4}}\right)\left(1-\dfrac{1}{\sqrt{5}}\right)$ ......

[**Hint.** (*i*) The series $\sum_{n=1}^{\infty} a_n = \sum_{n=1}^{\infty} \dfrac{(-1)^n}{2}$ oscillates finitely and

$\sum_{n=1}^{\infty} a_n^2$ is divergent.]

8. **Prove that** $\prod_{n=1}^{\infty}\left[\left(1-\frac{x}{\alpha+n}\right)e^{x/n}\right]$ **is convergent absolutely for any value of** x**, provided** α **is not a negative integer and that** $\prod_{n=1}^{\infty}\left[1-\left(\frac{nx}{n+1}\right)^n\right]$ **is absolutely convergent if** $|x|<1$.

[**Hint.** Here, $\left(1-\frac{x}{\alpha+n}\right)e^{x/n}=\left(1-\frac{x}{\alpha+n}\right)\left[1+\frac{x}{n}+\frac{x^2}{2!n^2}+...\right]$

$$=1+\frac{\alpha}{n(n+\alpha)}x+\frac{n-\alpha}{2n^2(n+\alpha)}x^2+.....$$

$$\therefore \qquad a_n=\frac{\alpha}{n(n+\alpha)}x+\frac{(n-\alpha)x^2}{2n^2(n+\alpha)}+.....$$

Taking $b_n=\frac{1}{n^2}$, we find that $\lim_{n\to\infty}\left|\frac{a_n}{b_n}\right|=|cx|$, a finite quantity.

$\therefore \quad \sum_{n=1}^{\infty}a_n$ is absolutely convergent.

$\Rightarrow \quad \prod_{n=1}^{\infty}(1+a_n)$ is absolutely convergent, etc.]

9. **Show the absolute convergence of the following products :**

(*i*) $\prod_{n=1}^{\infty}\cos\frac{\theta}{n}$ (*ii*) $\prod_{n=1}^{\infty}\left[\left(1-\frac{1}{n}\right)^x\left(1+\frac{x}{n}\right)\right]$

(*iii*) $\prod_{n=1}^{\infty}\left[\frac{\sin\frac{x}{n}}{\frac{x}{n}}\right]$ (*iv*) $\prod_{n=1}^{\infty}\frac{n^{x+1}}{(n-1)^x(n+x)}$

[**Hint.** (*i*) Here, $1+a_n=\cos\frac{\theta}{n}=1-\frac{\theta^2}{2!n^2}+\frac{\theta^4}{4!n^4}-...$

$$\Rightarrow \qquad |a_n|=\frac{1}{n^2}\left|\frac{-\theta^2}{2!}+\frac{\theta^4}{4!n^2}-...\right|$$

$\Rightarrow \quad \lim_{n\to\infty}\frac{|a_n|}{\frac{1}{n^2}}=\frac{\theta^2}{2}$, a finite quantity.

But, $\sum_{n=1}^{\infty}\frac{1}{n^2}$ is convergent.

$\Rightarrow \quad \sum_{n=1}^{\infty}|a_n|$ is convergent.

(*ii*) $$1 + a_n = \left(1 - \frac{1}{n}\right)^x \left(1 + \frac{x}{n}\right)$$

$$= \left(1 - \frac{x}{n} + \frac{x(x-1)}{2!}\frac{1}{n^2} - \dots\right)\left(1 + \frac{x}{n}\right) = 1 - \frac{x^2 + x}{2n^2} + \dots$$

Comparing a_n with $\frac{1}{n^2}$, we find $\lim_{n \to \infty} \frac{a_n}{\frac{1}{n^2}} = \frac{1}{2} \mid x^2 + x \mid$, a finite quantity and $\sum_{n=1}^{\infty} \frac{1}{n^2}$ is convergent.

(*iii*) $$1 + a_n = \frac{\sin \frac{x}{n}}{\frac{x}{n}} = 1 - \frac{x^2}{3!\,n^2} + \frac{x^4}{5!\,n^2} + \dots$$

Compare a_n with $\frac{1}{n^2}$.

(*iv*) $$1 + a_n = \frac{n^{x+1}}{(n-1)^x (n+x)} = \frac{1}{\left(1 - \frac{1}{n}\right)^x \left(1 + \frac{x}{n}\right)} = \left(1 - \frac{1}{n}\right)^{-x} \left(1 + \frac{x}{n}\right)^{-1}$$

$$= \left(1 + \frac{x}{n} + \frac{x(x+1)}{2!}\frac{1}{n^2} + \dots\right)\left(1 - \frac{x}{n} + \frac{x^2}{n^2} - \dots\right) = 1 + \frac{x^2 + x}{2!}\frac{1}{n^2}.$$

Now, proceed as in part (*i*).]

9

Riemann Integration

In elementary calculus, the process of integration is generally introduced in two ways—one as 'inverse of differentiation' and second 'as limit of a sum' when the number of terms in the sum tends to infinity each term tending to zero. 'To integrate' literally means 'to give the sum of'. So, the name Integral Calculus had its origin in the process of summation. It was only afterwards that it was realized that integration can also be viewed as inverse of differentiation. The reference to integration from summation point of view was always associated with the geometrical concepts. It was George Friedrich Bernhard Riemann (1826–1966), a distinguished German mathematician, who gave the first satisfactory rigorous arithmetic treatment, free from geometrical concepts. Many refinements and generalizations of the subject have appeared since then and the most noteworthy of these is the theory of integration by Lebesque (1875–1941), a well known French mathematician. However, in this chapter, we shall confine ourselves to the discussion of Riemann integration only.

Here, we shall be dealing with closed finite intervals $[a, b]$ and, unless otherwise stated, all functions denoted by $f, g, ...,$ etc., will be assumed to be real-valued functions defined and bounded on $[a, b]$.

9.1. RIEMANN INTEGRAL

9.1.1. Definitions. *Let $[a, b]$ be a finite closed interval. By a partition (sub-division) P of $[a, b]$, we mean a finite ordered set $\{x_0, x_1, ..., x_n\}$ of the points of $[a, b]$ such that*

$$a = x_0 < x_1 < x_2 < ...... < x_n = b,$$

The partition P consists of $n + 1$ points and does a partitioning of $[a, b]$ into n sub-intervals :

$$I_1 = [x_0, x_1], I_2 = [x_1, x_2], ..., I_n = [x_{n-1}, x_n] \text{ such that}$$

$$\bigcup_{i=1}^{n} I_i = \bigcup_{i=1}^{n} [x_{i-1}, x_i] = [a, b].$$

Clearly, there can be any number of partitions of $[a, b]$. We shall use the same symbol δ_i to denote the ith sub-interval $[x_{i-1}, x_i]$ as well as its length $x_i - x_{i-1}$.

Let f be a real-valued and bounded function on $[a, b]$. Then f is also bounded on each of sub-intervals corresponding to each partition P.

Let $\qquad m_i = \text{g.l.b. } f \text{ in } [x_{i-1}, x_i], (i = 1, 2, ..., n)$

$\qquad M_i = \text{l.u.b. } f \text{ in } [x_{i-1}, x_i], (i = 1, 2, ..., n)$

$\qquad m = \text{g.l.b. } f \text{ in } [a, b]$

and $\qquad M = \text{l.u.b. } f \text{ in } [a, b].$

Clearly, $m \le m_i \le f(x) \le M_i \le M, (i = 1, 2, ..., n)$

Form the sums :

$$m_1\delta_1 + m_2\delta_2 + \ldots\ldots + m_n\delta_n = \sum_{i=1}^{n} m_i\delta_i$$

and

$$M_1\delta_1 + M_2\delta_2 + \ldots\ldots + M_n\delta_n = \sum_{i=1}^{n} M_i\delta_i$$

These are respectively called the lower and upper (Darboux) sums of f corresponding to the partition P and are denoted by L(f, P) and U(f, P) respectively.*

Clearly, these sums depend upon the function and the partition and do exist for any bounded function.

Now, $m \le m_i \le M_i \le M$

$\Rightarrow$ $m\delta_i \le m_i\delta_i \le M_i\delta_i \le M\delta_i$

Putting $i = 1, 2, \ldots, n$ and adding, we have

$$m(b-a) \le L(f, P) \le U(f, P) \le M(b-a),\ b \ge a \qquad \ldots(1)$$

Now, corresponding to each partition P of $[a, b]$, we have lower and upper sums of f. Let $\mathcal{P}$ denote the set of all partitions of $[a, b]$. Considering all partitions of $[a, b]$, we get a set $\{L(f, P)\}_{P\in\mathcal{P}}$ of lower sums and a set $\{U(f, P)\}_{P\in\mathcal{P}}$ of upper sums of f. The inequality (1) shows that these two sets are bounded and hence each has the g.l.b. and l.u.b. The *l.u.b. of the set* $\{L(f, P)\}_{P\in\mathcal{P}}$ *of lower sums is called the lower integral and the g.l.b. of the set* $\{U(f, P)\}_{P\in\mathcal{P}}$ *of upper sums is called the upper integral of f on* $[a, b]$*. These are respectively denoted by*

$$\underline{\int_a^b} f\,dx \quad \text{and} \quad \overline{\int_a^b} f\,dx$$

A function f is said to be Riemann integrable or just integrable on [a, b] if its lower and upper integrals are equal. The common value is called the Riemann integral or just integral of f on [a, b] and is denoted by

$$\int_a^b f\,dx \quad \text{or} \quad \int_a^b f(x)\,dx.$$

The numbers a and b are respectively called the lower and upper limits of integration.

Note 1. The statement that $\int_a^b f\,dx$ exists is equivalent to saying that the function f is bounded and integrable on $[a, b]$.

2. If f is integrable, then

$$m(b-a) \le L(f, P) \le \int_a^b f\,dx \le U(f, P) \le M(b-a),\ b \ge a. \quad \textit{(K.U. BCA (II) 2008)}$$

[For, from (1), we have

$$m(b-a) \le L(f, P) \le u(f, P) \le M(b-a),\ b \ge a.$$

But $\int_a^b f\,dx$ = the l.u.b. of the set of all possible lower sums.

$$\therefore \qquad L(f, P) \le \int_a^b f\,dx \Rightarrow L(f, P) \le k\int_a^b f\,dx \qquad (\because\ f \text{ is integrable})$$

---

* After the name of a distinguished French mathematican J.G. Darboux (1842—1917).

Similarly, $\int_a^b f\,dx \le U(f, P)$.

$$\therefore \quad m(b-a) \le L(f, P) \le \int_a^b f\,dx \le U(f, P) \le M(b-a)].$$

3. The upper and lower integrals are defined for every bounded function but they may not necessarily be equal for every bounded function. Thus, a bounded function may not be integrable. (See example 1, below).

4. The concept of integrability of a function on an interval, as introduced here, is subject to two limitations : (*i*) the function is bounded, (*ii*) the interval is finite.

5. If f is bounded and integrable on $[b, a]$, $a \ge b$, then we define

$$\int_a^b f\,dx = -\int_b^a f\,dx, \text{ if } a > b \quad \text{and} \quad \int_a^b f\,dx = 0 \text{ if } a = b.$$

Example 1. *Show that the function* :*

$$f(x) = \begin{cases} 0, & \textit{if } x \textit{ is rational} \\ 1, & \textit{if } x \textit{ is irrational} \end{cases}$$

is not integrable on any interval.

Sol. Let $[a, b]$ be any interval.

Let $P = \{x_0, x_1, \ldots\ldots, x_n\}$ be a partition of $[a, b]$.

Since each non-empty interval contains rationals as well as irrationals, we have

$$m_i = \text{g.l.b. } f \text{ in } [x_{i-1}, x_i] = 0$$

and $$M_i = \text{l.u.b. } f \text{ in } [x_{i-1}, x_i] = 1, \ (i = 1, 2, \ldots\ldots, n)$$

$$\therefore \quad L(f, P) = \sum_{i=1}^{n} m_i \delta_i = 0$$

and $$U(f, P) = \sum_{i=1}^{n} M_i \delta_i$$

$$= 1\delta_1 + 1\delta_2 + \ldots\ldots + 1\delta_n = b - a$$

This holds for all partitions P of $[a, b]$.

Thus, $$\underline{\int_a^b} f\,dx = 0 \quad \text{and} \quad \overline{\int_a^b} f\,dx = b - a$$

$$\Rightarrow \quad \underline{\int_a^b} f\,dx \ne \overline{\int_a^b} f\,dx$$

$\Rightarrow$ f is not integrable.

Remark. The above is an example of a bounded function which is not integrable.

Example 2. *Show that every constant function is integrable on a closed and bounded interval.*

Sol. Let $f(x) = k, \ \forall\ x \in [a, b]$.

Let $P = \{x_0, x_1, \ldots\ldots, x_n\}$ be any partition of $[a, b]$.

Then, $$m_i = \text{g.l.b. } f \text{ in } [x_{i-1}, x_i] = k$$

and $$M_i = \text{l.u.b. } f \text{ in } [x_{i-1}, x_i] = k, \ i = 1, 2, \ldots\ldots, n$$

---

*When a function is defined in this manner, it is referred to as the characteristic function of the set of all rationals.

$$\Rightarrow \qquad L(f, P) = \sum_{i=1}^{n} m_i \delta_i = k(\delta_1 + \delta_2 + ... + \delta_n) = k(b-a)$$

and $$U(f, P) = \sum_{i=1}^{n} M_i \delta_i = k(\delta_1 + \delta_2 + ... + \delta_n) = k(b-a)$$

This holds for all partitions P of $[a, b]$.

Thus, $$\underline{\int_a^b} f\,dx = k(b-a) = \overline{\int_a^b} f\,dx.$$

$\Rightarrow$ f is integrable and $\int_a^b f\,dx = k(b-a)$.

Example 3. *Show that the function f defined by $f(x) = x^2$, $x \in [0, a]$, $a > 0$, is integrable on $[0, a]$ and $\int_0^a x^2\,dx = \frac{a^3}{3}$.* (K.U. BCA (II) 2001)

Sol. Consider a partition

$$P = \left\{0, \frac{a}{n}, \frac{2a}{n}, ..., \frac{na}{n}\right\}$$

dividing the interval $[0, a]$ into n equal parts. Clearly,

$$\left[(i-1)\left(\frac{a}{n}\right)\right]^2 \text{ and } \left[i\left(\frac{a}{n}\right)\right]^2$$

are the lower and upper bounds of the function f in the interval δ_i and length of each sub-interval is $\frac{a}{n}$.

$$\therefore \qquad L(f, P) = \sum_{i=1}^{n} m_i \delta_i = \sum_{i=1}^{n} \left((i-1)\frac{a}{n}\right)^2 \left(\frac{a}{n}\right)$$

$$= \frac{a^3}{n^3}\{0^2 + 1^2 + 2^2 + ... + (n-1)^2\}$$

$$= \frac{a^3}{n^3}\,\frac{(n-1)\,n(2n-1)}{6} = \frac{a^3}{6}\left(1-\frac{1}{n}\right)\left(2-\frac{1}{n}\right)$$

and $$U(f, P) = \sum_{i=1}^{n} M_i\,\delta_i = \sum_{i=1}^{n} \left(i\frac{a}{n}\right)^2 \frac{a}{n} = \frac{a^3}{n^3}\{1^2 + 2^2 + ... + n^2\}$$

$$= \frac{a^3}{n^3}\,\frac{n(n+1)(2n+1)}{6} = \frac{a^3}{6}\left(1+\frac{1}{n}\right)\left(2+\frac{1}{n}\right)$$

Considering all partitions of $[0, a]$,

$$\text{l.u.b. } \{L(f, P)\}_{P \in \mathcal{P}} = \frac{a^3}{3} = \text{g.l.b. } \{U(f, P)\}_{P \in \mathcal{P}}$$

$$\Rightarrow \qquad \underline{\int_0^a} f\,dx = \frac{a^3}{3} = \overline{\int_0^a} f\,dx$$

$\Rightarrow$ f is integrable on $[0, a]$ and

$$\int_0^a x^2\,dx = \frac{a^3}{3}.$$

Example 4. *Show that the function f defined on* $\left[0, \frac{\pi}{4}\right]$ *by*

$$f(x) = \begin{cases} \cos x & \text{if } x \text{ is rational} \\ \sin x & \text{if } x \text{ is irrational} \end{cases}$$

is not integrable on $\left[0, \frac{\pi}{4}\right]$.

Sol. Consider a partition

$$P = \left[0, \frac{\pi}{4n}, \frac{2\pi}{4n}, \frac{3\pi}{4n}, \ldots\ldots, \frac{(i-1)\pi}{4n}, \frac{i\pi}{4n}, \ldots\ldots, \frac{n\pi}{4n} = \frac{\pi}{4}\right]$$

of $\left[0, \frac{\pi}{4}\right]$. Then, for any sub-interval

$\left[\frac{(i-1)\pi}{4n}, \frac{i\pi}{4n}\right]$, $i = 1, 2, \ldots\ldots, n$, we have

$$M_i = \cos\frac{(i-1)\pi}{4n} \text{ and } m_i = \sin\frac{(i-1)\pi}{4n}.$$

$$\left(\because \cos x \geq \sin x \text{ over } \left[0, \frac{\pi}{4}\right]\right).$$

Also, $\delta_i = \frac{\pi}{4n}, \forall_i.$

$$\therefore \quad L(f, P) = \sum_{i=1}^{n} m_i\delta_i = \sum_{i=1}^{n} \frac{\pi}{4n}\sin\frac{(i-1)\pi}{4n}$$

$$= \frac{\pi}{4n}\left[\sin 0 + \sin\frac{\pi}{4n} + \sin\frac{2\pi}{4n} + \ldots + \sin\frac{(n-1)\pi}{4n}\right]$$

$$= \frac{\pi}{4n} \cdot \frac{\sin\left(\dfrac{0 + \frac{(n-1)\pi}{4n}}{2}\right)\sin\left(\frac{n}{2}\cdot\frac{\pi}{4n}\right)}{\sin\frac{\pi}{8n}}$$

$$= \frac{\pi}{4n} \cdot \frac{\sin\frac{(n-1)\pi}{8n}\sin\frac{\pi}{8}}{\sin\frac{\pi}{8n}} = \frac{\frac{\pi}{8n}}{\sin\frac{\pi}{8n}} \cdot 2\sin\left(\frac{\pi}{8} - \frac{\pi}{8n}\right)\sin\frac{\pi}{8}.$$

$$\therefore \quad \text{l.u.b. } L(f, P)_{P\in\mathcal{P}} = \lim_{n\to\infty}\left(\frac{\frac{\pi}{8n}}{\sin\frac{\pi}{8n}} \cdot 2\sin\left(\frac{\pi}{8} - \frac{\pi}{8n}\right)\sin\frac{\pi}{8}\right)$$

$$= 2\sin^2\frac{\pi}{8} = 1 - \cos\frac{\pi}{4} = 1 - \frac{1}{\sqrt{2}}.$$

$$\therefore \quad \underline{\int_0}^{\pi/4} f\,dx = 1 - \frac{1}{\sqrt{2}}.$$

$$U(f, P) = \sum_{i=1}^{n} M_i \delta_i = \sum_{i=1}^{n} \frac{\pi}{4n} \cdot \cos \frac{(n-1)\pi}{4n}$$

$$= \frac{\pi}{4n}\left[\cos 0 + \cos\frac{\pi}{4n} + \cos\frac{2\pi}{4n} + \ldots.. + \cos\frac{(n-1)\pi}{4n}\right]$$

$$= \frac{\frac{\pi}{8n}}{\sin\frac{\pi}{8n}} \cdot 2\cos\left(\frac{\pi}{8} - \frac{\pi}{8n}\right)\sin\frac{\pi}{8} \qquad \text{(As above)}$$

$$\therefore \qquad \text{g.l.b. } U(f, P)_{P \in \mathcal{P}} = \lim_{n\to\infty}\left(\frac{\frac{\pi}{8n}}{\sin\frac{\pi}{8n}} 2\cos\left(\frac{\pi}{8} - \frac{\pi}{8n}\right)\sin\frac{\pi}{8}\right)$$

$$= 2 \cdot \cos\frac{\pi}{8} \cdot \sin\frac{\pi}{8} = \sin\frac{\pi}{4} = \frac{1}{\sqrt{2}}.$$

$$\therefore \qquad \overline{\int_0^{\pi/4}} f\,dx = \frac{1}{\sqrt{2}}$$

Since $\underline{\int_0^{\pi/4}} f\,dx \neq \overline{\int_0^{\pi/4}} f\,dx$, f is not integrable on $[0, \pi]$.

EXERCISE 9 (*a*)

1. (*a*) Compute $L(f, P)$ and $U(f, P)$ for the function f defined by $f(x) = x$, $x \in [0, 3]$ and the partition $P = \{0, 1, 2, 3\}$. *(BCA (II) 2007)*

(*b*) Let $f(x) = x$ for $x \in [0, 1]$ and let $P = \left\{0, \frac{1}{3}, \frac{2}{3}, 1\right\}$ be the partition of $[0, 1]$. Compute $U(f, P)$ and $L(f, P)$.

(*c*) Compute $L(f, P)$ and $U(f, P)$ for the function $f(x) = x$, $x \in [0, 1]$ and $P = \left\{0, \frac{1}{5}, \frac{2}{5}, \frac{3}{5}, \frac{4}{5}, 1\right\}$ is a partition of $[0, 1]$.

2. Show that the function f defined by $f(x) = x$, $x \in [0, 1]$, is integrable and $\int_0^1 f(x)\,dx = \frac{1}{2}$.

3. (*a*) Show that the function f defined on $[0, 1]$ by

$$f(x) = \begin{cases} 1, & \text{if } x \text{ is rational} \\ -1, & \text{if } x \text{ is irrational} \end{cases}$$

is bounded but not Riemann integrable.

(*b*) If
$$f(x) = \begin{cases} 1 \text{ when } x \text{ is rational} \\ -1 \text{ when } x \text{ is irrational} \end{cases}$$

then show that $|f|$ is integrable but f is not integrable.

4. If a function f is defined on $[0, a]$, $a > 0$ by $f(x) = x^3$, then show that f is Riemann integrable on $[0, a]$ and $\int_0^a f(x)\,dx = \frac{a^4}{4}$.

5. Give an example to show that every bounded function is not integrable.

6. Show that $f(x) = 2x + 1$ is integrable on [1, 2] and $\int_1^2 (2x + 1)\, dx = 4$.

7. A function f is bounded on $[a, b]$. Show that

(*i*) when k is a positive constant,

$$\overline{\int_a^b} k\, f\, dx = k \overline{\int_a^b} f\, dx \;;\; \underline{\int_a^b} k\, f\, dx = k \underline{\int_a^b} f\, dx \text{ and}$$

(*ii*) when k is a negative constant,

$$\overline{\int_a^b} k\, f\, dx = k \underline{\int_a^b} f\, dx \;;\; \underline{\int_a^b} k\, f\, dx = k \overline{\int_a^b} f\, dx$$

Also, deduce that if f is bounded and integrable on $[a, b]$, then so is kf, where k is a constant and

$$\int_a^b k\, f\, dx = k \int_a^b f\, dx.$$

[**Hint.** If m_i and M_i are g.l.b. and l.u.b. of f in $[x_{i-1}, x_i]$, then km_i and kM_i are g.l.b. and l.u.b. of f in $[x_{i-1}, x_i]$, if k is positive. But, if k is negative, then km_i and kM_i are l.u.b. and g.l.b. of kf in $[x_{i-1}, x_i]$.]

9.1.2. Norm of a partition. *For a partition P, the length of the largest sub-interval is called the norm or mesh of the partition P and is denoted by* $\| P \|$.

Thus, $\| P \| = max. \{x_i - x_{i-1} \mid 1 \le i \le n\}$.

9.1.3. Refinement. *A partition P′ is called a refinement of a partition P of [a, b] if* $P' \supseteq P$, *i.e.*, every point of P is a point of P′ and may have some other points of $[a, b]$ in it.

If P_1 *and* P_2 *are two partitions of [a, b], then* $P^* = P_1 \cup P_2$ *is called their common refinement.*

In the next theorem, we prove that any refinement of P does not lower the lower sum and does not raise the upper sum.

9.1.4. Theorem. *If f is a bounded function and P* is a refinement of a partition P, then*

(*i*) $L(f, P^*) \ge L(f, P)$ (K.U. BCA (II) 2001)

and (*ii*) $U(f, P^*) \le U(f, P)$.

Proof. Let P* contain just one point ξ (say) more than $P = \{a = x_0, x_1, x_2, \ldots\ldots, x_{i-1}, x_i, \ldots\ldots, x_n = b\}$ and $x_{i-1} < \xi < x_i$.

Let m_i', m_i'' and m_i be the g.l.b. of f in the intervals $[x_{i-1}, \xi]$, $[\xi, x_i]$ and $[x_{i-1}, x_i]$ respectively. Then $m_i \le m_i'$ and $m_i \le m_i''$.

$$\begin{aligned} \therefore \quad L(f, P^*) - L(f, P) &= m_i' (\xi - x_{i-1}) + m_i'' (x_i - \xi) - m_i (x_i - x_{i-1}) \\ &= m_i' (\xi - x_{i-1}) + m_i'' (x_i - \xi) - m_i (x_i - \xi + \xi - x_{i-1}) \\ &= (m_i' - m_i)(\xi - x_{i-1}) + (m_i'' - m_i)(x_i - \xi) \ge 0 \end{aligned}$$

$$\therefore \quad L(f, P^*) \ge L(f, P).$$

If P* contains p points more than P, then repeating the above argument p times, we get the result.

Similarly, $U(f, P^*) \le U(f, P)$.

Cor. *If P* is a refinement of P containing p points more than P and* $| f(x) | \le k$, *for* $x \in [a, b]$, *then*

(*i*) $L(f, P) \le L(f, P^*) \le L(f, P) + 2pk\delta$

and (*ii*) $U(f, P) \ge U(f, P^*) \ge U(f, P) - 2pk\delta$

when $\| P \| = \delta$.

For, Proceeding as in the above theorem, when P* contains one point ξ more than P, we get

$$L(f, P^*) - L(f, P) = (m_i' - m_i)(\xi - x_{i-1}) + (m_i'' - m_i)(x_i - \xi) \quad ...(1)$$

But, $\quad | f(x) | \le k, x \in [a, b]$

$\Rightarrow \quad -k \le m_i \le m_i' \le k$

$\Rightarrow \quad 0 \le m_i' - m_i \le 2k$

Similarly, $\quad 0 \le m_i'' - m_i \le 2k$

$\therefore$ from (1),

$$L(f, P^*) - L(f, P) \le 2k\,(\xi - x_{i-1}) + 2k\,(x_i - \xi) = 2k\,(x_i - x_{i-1}) \le 2k\delta.$$

If P* contains p additional points, then introducing the points one by one and proceeding as above p times, we get

$$L(f, P^*) \le L(f, P) + 2pk\delta.$$

Also, $\quad L(f, P) \le L(f, P^*)$

$\therefore \quad L(f, P) \le L(f, P^*) \le L(f, P) + 2pk\delta.$

Similarly, we can prove the second result.

9.1.5. Theorem. *For any two partitions P_1 and P_2,*

$$L(f, P_1) \le U(f, P_2)$$

i.e., no lower sum can exceed any upper sum.

Proof. Let P* be a common refinement of P_1 and P_2 so that $P^* = P_1 \cup P_2$.

Then, $\quad L(f, P_1) \le L(f, P^*) \le U(f, P^*) \le U(f, P_2)$

$\Rightarrow \quad L(f, P_1) \le U(f, P_2).$

Cor. *For a bounded function f,*

$$\underline{\int_a^b} f\,dx \le \overline{\int_a^b} f\,dx$$

For any two partitions P_1, P_2,

$$L(f, P_1) \le U(f, P_2)$$

Keeping P_2 fixed and taking l.u.b. overall partitions P_1, we have

$$\underline{\int_a^b} f\,dx \le U(f, P_2).$$

Taking g.l.b. overall partitions P_2, we get

$$\underline{\int_a^b} f\,dx \le \overline{\int_a^b} f\,dx.$$

9.1.6. Darboux's Theorem. *If f is a bounded function on [a, b], then for each $\varepsilon > 0$, there corresponds a $\delta > 0$ such that*

(i) $U(f, P) < \overline{\int_a^b} f\,dx + \varepsilon$ (K.U. BCA (II) 2001, 2008)

(ii) $L(f, P) > \underline{\int_a^b} f\,dx - \varepsilon$

for each partition P of [a, b] with $\| P \| < \delta$.

Proof. (*i*) Since f is bounded on $[a, b]$, there exists $k > 0$ such that

$$| f(x) | \le k, \quad \forall\, x \in [a, b].$$

Since $\int_a^{\bar{b}} f\,dx$ is the g.l.b. of the set of upper sums, therefore to each $\varepsilon > 0$, there exists a partition $P_1 = \{a = x_0, x_1, ... x_p = b\}$ such that

$$U(f, P_1) < \int_a^{\bar{b}} f\,dx + \frac{\varepsilon}{2} \qquad ...(1)$$

The partition P_1 has $p + 1$ points, including a and b.

Let δ be a positive number such that

$$2k(p-1)\,\delta \leq \frac{\varepsilon}{2} \qquad ...(2)$$

Let P be a partition with $\| P \| < \delta$.

Let $P^* = P \cup P_1$ so that P^* is finer than P and contains at the most $p - 1$ additional points. Since P^* is finer than P and contains at the most $p - 1$ additional points, we have

$$U(f, P) - 2k(p-1)\,\delta \leq U(f, P^*) \leq U(f, P_1)$$

$$< \int_a^{\bar{b}} f\,dx + \frac{\varepsilon}{2} \qquad \text{[By (1)]}$$

$$\Rightarrow \quad U(f, P) < 2k(p-1)\,\delta + \int_a^{\bar{b}} f\,dx + \frac{\varepsilon}{2}$$

$$\Rightarrow \quad U(f, P) < \frac{\varepsilon}{2} + \int_a^{\bar{b}} f\,dx + \frac{\varepsilon}{2} \qquad \text{[By (2)]}$$

$$\Rightarrow \quad U(f, P) < \int_a^{\bar{b}} f\,dx + \varepsilon, \text{ for a partition P with } \| P \| < \delta.$$

Similarly, one can prove (*ii*).

Cor. *If f is bounded on $[a, b]$, then*

$$\lim_{\| P \| \to 0} U(f, P) = \int_a^{\bar{b}} f\,dx \text{ and } \lim_{\| P \| \to 0} L(f, P) = \int_{\underline{a}}^{b} f\,dx.$$

Remark. *$U(f, P) < \int_a^{\bar{b}} f\,dx + \varepsilon$ is clear even by the definition of g.l.b. but then it implies that there exists at least one partition P with this property. However, Darboux's theorem asserts the existence of infinitely many partitions with this property provided the norm of each partition is $< \delta$ (depending upon the choice of ε).*

9.1.7. Oscillatory sum. Corresponding to a partition

$P = \{a = x_0, x_1, ..., x_n = b\}$ of $[a, b]$, we have

$$U(f, P) - L(f, P) = \sum_{i=1}^{n} M_i\,\delta_i - \sum_{i=1}^{n} m_i\,\delta_i$$

$$= \sum_{i=1}^{n} (M_i - m_i)\delta_i = \sum_{i=1}^{n} O_i\,\delta_i$$

where M_i = l.u.b. f in $[x_{i-1}, x_i]$, m_i = g.l.b. f in $[x_{i-1}, x_i]$

and $O_i = M_i - m_i$

is the oscillation of the function f in $[x_{i-1}, x_i]$.

$\sum_{i=1}^{n} O_i \, \delta_i$ *is called the oscillatory sum of the function f in [a, b] and is denoted by* $\omega(P)$.

Since $O_i \geq 0$, $(i = 1, 2, \ldots\ldots, n)$, each oscillatory sum consists of a finite number of non-negative terms.

The following two theorems give the necessary and sufficient conditions for a bounded function to be integrable in two different forms.

9.1.8. Theorem *(First form). A bounded function f is integrable on [a, b] if and only if for each* $\varepsilon > 0$, *there exists a partition P of [a, b] such that* $U(f, P) - L(f, P) < \varepsilon$.

Proof. (*i*) Let f be integrable on $[a, b]$ so that

$$\underline{\int_a^b} f\,dx = \overline{\int_a^b} f\,dx = \int_a^b f\,dx.$$

Let $\varepsilon > 0$ be given.

Since the upper and the lower integrals are the g.l.b. and the l.u.b. respectively of the upper and the lower sums, therefore, there exist partitions P_1 and P_2 of $[a, b]$ such that

$$U(f, P_1) < \overline{\int_a^b} f\,dx + \frac{\varepsilon}{2} = \int_a^b f\,dx + \frac{\varepsilon}{2}$$

and

$$L(f, P_2) > \underline{\int_a^b} f\,dx - \frac{\varepsilon}{2} = \int_a^b f\,dx - \frac{\varepsilon}{2}$$

Let $$P = P_1 \cup P_2.$$

Then, $$U(f, P) \leq U(f, P_1) < \int_a^b f\,dx + \frac{\varepsilon}{2}$$

$$< L(f, P_2) + \frac{\varepsilon}{2} + \frac{\varepsilon}{2} \leq L(f, P) + \frac{\varepsilon}{2} + \frac{\varepsilon}{2}$$

$\Rightarrow$ $U(f, P) - L(f, P) < \varepsilon$, for a partition P.

(*ii*) Conversely, let $\varepsilon > 0$ be given. Let P be a partition of $[a, b]$ such that

$$U(f, P) - L(f, P) < \varepsilon \qquad \ldots(1)$$

Since $$L(f, P) \leq \underline{\int_a^b} f\,dx \leq \overline{\int_a^b} f\,dx \leq U(f, P)$$

$$\therefore \quad \overline{\int_a^b} f\,dx - \underline{\int_a^b} f\,dx \leq U(f, P) - L(f, P) < \varepsilon \qquad \text{[By (1)]}$$

Since $\varepsilon > 0$ is arbitrary, it implies that

$$\overline{\int_a^b} f\,dx - \underline{\int_a^b} f\,dx = 0$$

$$\Rightarrow \quad \overline{\int_a^b} f\,dx = \underline{\int_a^b} f\,dx$$

$\Rightarrow$ f is integrable.

9.1.9. Theorem *(Second form). A bounded function f is integrable on [a, b] if and only if for each* $\varepsilon > 0$, *there corresponds a* $\delta > 0$ *such that for every partition P of [a, b] with* $\| P \| < \delta$, $U(f, P) - L(f, P) < \varepsilon$. (K.U. BCA (II) 2001)

Proof. (*i*) Let f be integrable

$$\therefore \quad \underline{\int_a^b} f\,dx = \overline{\int_a^b} f\,dx = \int_a^b f\,dx$$

Let $\varepsilon > 0$ be given. By Darboux's theorem, there exists $\delta > 0$ such that for each partition P of $[a, b]$, with $\| P \| < \delta$,

$$U(f, P) < \overline{\int_a^b} f\,dx + \frac{\varepsilon}{2} = \int_a^b f\,dx + \frac{\varepsilon}{2} \quad \text{...(1)}$$

and
$$L(f, P) > \underline{\int_a^b} f\,dx - \frac{\varepsilon}{2} = \int_a^b f\,dx - \frac{\varepsilon}{2}$$

or
$$-L(f, P) < -\int_a^b f\,dx + \frac{\varepsilon}{2} \quad \text{...(2)}$$

(1) + (2) gives,

$$U(f, P) - L(f, P) < \varepsilon$$

for each partition P with $\| P \| < \delta$.

(*ii*) Conversely, let $\varepsilon > 0$ be given. Then, for each partition P with $\| P \| < \delta$ (some positive number depending upon ε),

$$U(f, P) - L(f, P) < \varepsilon \quad \text{...(1)}$$

Also, for any partition P,

$$L(f, P) \le \underline{\int_a^b} f\,dx \le \overline{\int_a^b} f\,dx \le U(f, P)$$

$$\Rightarrow \quad \overline{\int_a^b} f\,dx - \underline{\int_a^b} f\,dx \le U(f, P) - L(f, P) < \varepsilon \quad \text{[By (1)]}$$

Since $\varepsilon > 0$ is arbitrary, this implies

$$\overline{\int_a^b} f\,dx - \underline{\int_a^b} f\,dx = 0$$

$$\Rightarrow \quad \overline{\int_a^b} f\,dx = \underline{\int_a^b} f\,dx$$

$\Rightarrow$ f is integrable.

Remark. We may restate the above theorem as follows :

A bounded function f is integrable on [a, b] if and only if

$$\lim_{\| P \| \to 0} (U(f, P) - L(f, P)) = 0.$$

Cor. 1. *A bounded function f is integrable on [a, b] if and only if there exists a number I such that for each $\varepsilon > 0$, there exists a partition P of [a, b] such that*

$$| U(f, P) - I | < \varepsilon \quad \text{and} \quad | I - L(f, P) | < \varepsilon.$$

Proof. (*i*) Let f be integrable on $[a, b]$.

Let $\varepsilon > 0$ be given. Then, $\exists$ a partition P such that

$$| U(f, P) - L(f, P) | < \varepsilon$$

If I is a number lying between

L(f, P) and U(f, P),

then
$$| U(f, P) - I | < | U(f, P) - L(f, P) | < \varepsilon$$
and
$$| I - L(f, P) | < | U(f, P) - L(f, P) | < \varepsilon.$$

(*ii*) Conversely, let $\varepsilon > 0$ be given. Then there exists a partition P such that

$$| \, U(f, P) - I \, | < \frac{\varepsilon}{2} \quad \text{and} \quad | \, I - L(f, P) \, | < \frac{\varepsilon}{2}$$

$$\begin{aligned} \text{Now, } | \, U(f, P) - L(f, P) \, | &= | \, U(f, P) - I + I - L(f, P) \, | \\ &\le | \, U(f, P) - I \, | + | \, I - L(f, P) \, | \\ &< \frac{\varepsilon}{2} + \frac{\varepsilon}{2} = \varepsilon \end{aligned}$$

$\Rightarrow$ f is integrable on $[a, b]$.

Cor. 2. *A bounded function f is integrable on [a, b] if and only if there is a number I such that for each $\varepsilon > 0$, there exists a $\delta > 0$ such that for all partitions P with $\| P \| < \delta$,*

$$| \, U(f, P) - I \, | < \varepsilon \quad \text{and} \quad | \, I - L(f, P) \, | < \varepsilon.$$

Proof is simple and left to the reader as an exercise.

9.2. SOME INTEGRABLE FUNCTIONS

9.2.1. Theorem. *If a function f is continuous on [a, b], then it is integrable on [a, b].*

Proof. Since f is continuous on $[a, b]$, f is bounded on $[a, b]$ and attains its g.l.b. and l.u.b. on $[a, b]$ and on every closed sub-interval of it. Furthermore, f is uniformly continuous on $[a, b]$.

$\therefore$ for each $\varepsilon > 0$, $\exists$ a $\delta > 0$ such that

$$| \, f(x') - f(x'') \, | < \frac{\varepsilon}{b-a}, \text{ for } | \, x' - x'' \, | < \delta \qquad ...(1)$$

Choose a natural number $n > \dfrac{b-a}{\delta}$ and divide $[a, b]$ into n equal parts. Then, length of each sub-interval is less than δ. Let this partition be denoted by P.

Let $\qquad m_i = \text{g.l.b. } f \text{ in } [x_{i-1}, x_i]$

and $\qquad M_i = \text{l.u.b. } f \text{ in } [x_{i-1}, x_i], \qquad (i = 1, 2, \ldots\ldots, n)$

If $\qquad f(x_i') = m_i$ and $f(x_i'') = M_i$, then $| \, x_i' - x_i'' \, | < \delta$ and therefore,

$$M_i - m_i < \frac{\varepsilon}{b-a}, \ (i = 1, 2, 3, \ldots\ldots, n)$$

$$\begin{aligned} \text{Now, } \quad U(f, P) - L(f, P) &= \sum_{i=1}^{n} (M_i - m_i) \, \delta_i = \sum_{i=1}^{n} [f(x_i'') - f(x_i')] \, \delta_i \\ &< \sum_{i=1}^{n} \left(\frac{\varepsilon}{b-a} \right) \delta_i \qquad \text{[By (1)]} \\ &= \frac{\varepsilon}{b-a} \sum_{i=1}^{n} \delta_i = \frac{\varepsilon}{b-a} (b-a) = \varepsilon \end{aligned}$$

$\Rightarrow \qquad U(f, P) - L(f, P) < \varepsilon$

$\Rightarrow$ f is integrable on $[a, b]$.

Remark. *Continuity of a function is a sufficient condition for its integrability but is not a necessary condition. For example, the function*

$$f(x) = \begin{cases} 2, & x \in [0, 1] \\ 3, & x \in [1, 2] \end{cases}$$

defined on [0, 2] is discontinuous at $x = 1$. However, it is integrable on [0, 2]. For, to each $\varepsilon > 0$, we can find a partition P such that it contains a sub-interval containing 1 and whose length is less than ε. Then,

$$U(f, P) - L(f, P) < \varepsilon.$$

In fact, a function can be discontinuous at infinitely many points as long as the length of the set of points of discontinuities is zero.

9.2.2. Theorem. *A bounded function f, having a finite number of points of discontinuity on [a, b], is integrable on [a, b].*

Proof. Let $a_1, a_2, a_3, \ldots, a_p$ be the finite number of points of discontinuity of f in $[a, b]$.

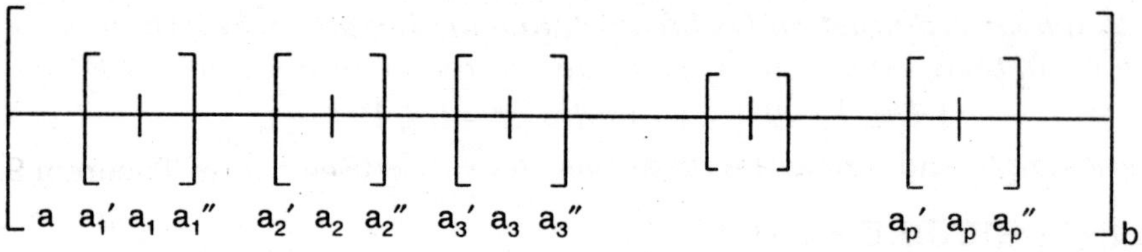

Let $\varepsilon > 0$ be given.

Enclose the points $a_1, a_2, \ldots, a_p$ into p non-overlapping intervals $[a_1', a_1'']$, $[a_2', a_2'']$, ..., $[a_p', a_p'']$ such that the total length of these intervals is $< \dfrac{\varepsilon}{2(M - m)}$

where $\begin{cases} M = \text{l.u.b. of } f \text{ in } [a, b] \\ m = \text{g.l.b. of } f \text{ in } [a, b] \end{cases}$ ...(1)

If M_i = l.u.b. of f and m_i = g.l.b. of f in $[a_i', a_i'']$, $i = 1, 2, \ldots, p$, then $M_i \le M$ and $m_i \ge m$.

$$\therefore \quad M_i - m_i \le M - m, \quad \text{for} \quad i = 1, 2, \ldots, p.$$

$\therefore$ Contribution to the oscillatory sum U $(f, P) - L(f, P)$ from these intervals,

$$= \sum_{i=1}^{p} (M_i - m_i)(a_i'' - a_i')$$

$$\le \sum_{i=1}^{p} (M - m)(a_i'' - a_i')$$

$$= (M - m) \sum_{i=1}^{p} (a_i'' - a_i')$$

$$< (M - m) \frac{\varepsilon}{2(M - m)} = \frac{\varepsilon}{2} \qquad \text{[By (1)]} \quad \ldots(2)$$

Now, since f is continuous on each of the $p + 1$ sub-intervals : $[a, a_1']$, $[a_1'', a_2']$, ..., $[a_p'', b]$, therefore, f is integrable on each of these intervals. Hence, each of these sub-intervals can be further sub-divided in such a way that the contribution to the oscillatory sum $U(f, P) - L(f, P)$ arising from each of these sub-intervals is separately $< \dfrac{\varepsilon}{2(p + 1)}$.

$\therefore$ Total contribution to the oscillatory sum $U(f, P) - L(f, P)$ from these $p + 1$ sub-intervals is

$$< \frac{\varepsilon}{2(p + 1)} \cdot (p + 1) = \frac{\varepsilon}{2} \qquad \ldots(3)$$

Now, the intervals, $[a_1', a_1'']$, $[a_2', a_2'']$, ..., $[a_p', a_p'']$ and the partitions of the intervals $[a, a_1']$, $[a_1'', a_2']$, ..., $[a_p'', b]$ together form the partition P of $[a, b]$ such that

$$U(f, P) - L(f, P) < \frac{\varepsilon}{2} + \frac{\varepsilon}{2} = \varepsilon \qquad \text{[From (2) and (3)]}$$

Hence, the function f is integrable on $[a, b]$.

9.2.3. Theorem. *A bounded function f is integrable on $[a, b]$ if the set of its points of discontinuity has a finite number of limit points.*

Proof. Let $a_1, a_2, \ldots\ldots, a_p$ be the limit points of the set of points of discontinuity of f in $[a, b]$. Enclose these points into p non-overlapping intervals :

$$[a_1', a_1''], [a_2', a_2''], \ldots, [a_p', a_p'']$$

such that the total length of these intervals is $< \dfrac{\varepsilon}{2(M-m)}$ (See Fig. of Theorem 9.2.2) ..(1)

where M = l.u.b. of f in $[a, b]$

m = g.l.b. of f in $[a, b]$

If M_i = l.u.b. of f and m_i = g.l.b. of f in $[a_i', a_i'']$, $i = 1, 2, \ldots\ldots, p$, then $M_i \le M$ and $m_i \ge m$.

$\therefore \qquad M_i - m_i \le M - m, \; i = 1, 2, \ldots\ldots, p.$

$\therefore$ Contribution to the oscillatory sum $U(f, P) - L(f, P)$ from these intervals

$$= \sum_{i=1}^{p} (M_i - m_i)(a_i'' - a_i')$$

$$\le \sum_{i=1}^{p} (M - m)(a_i'' - a_i')$$

$$= (M - m) \sum_{i=1}^{p} (a_i'' - a_i')$$

$$< (M - m) \frac{\varepsilon}{2(M - m)} = \frac{\varepsilon}{2} \qquad \text{[By (1)]} \quad \ldots(2)$$

Since only a finite number of points of discontinuity of f can lie in each of $p + 1$ intervals $[a, a_1']$, $[a_1'', a_2']$, ..., $[a_p'', b]$.

$\therefore$ f is integrable in each of these intervals. Hence, each of these sub-intervals can be further sub-divided in such a way that the contribution to the oscillating sum $U(f, P) - L(f, P)$ arising from each of these sub-intervals is separately $< \dfrac{\varepsilon}{2(p+1)}$.

$\therefore$ The total contribution to the oscillatory sum $U(f, P) - L(f, P)$ from these $p + 1$ sub-intervals is

$$< \frac{\varepsilon}{2(p+1)} (p + 1) = \frac{\varepsilon}{2} \qquad \ldots(3)$$

Now, the intervals $[a_1', a_1'']$, $[a_2', a_2'']$, ..., $[a_p', a_p'']$ and the partitions of the intervals $[a, a_1']$, $[a_1'', a_2']$, ..., $[a_p'', b]$ together form the partition P of $[a, b]$ such that

$$U(f, P) - L(f, P) < \frac{\varepsilon}{2} + \frac{\varepsilon}{2} = \varepsilon \qquad \text{[From (2) and (3)]}$$

Hence, the function f is integrable on $[a, b]$.

9.2.4. Theorem. *If a function f is monotonic on [a, b], then it is integrable on [a, b].*

(BCA (II) 2006)

Proof. We shall prove the theorem when f is monotonically increasing. The proof is similar when f is monotonically decreasing. Clearly, f is bounded on $[a, b]$.

If $f(a) = f(b)$, then f is a constant function and hence is integrable.

So, let $f(a) < f(b)$.

Let $\varepsilon > 0$ be given.

Construct a partition $P = \{a = x_0, x_1, x_2, ..., x_n = b\}$ of $[a, b]$ such that length of each sub-interval is less than

$$\frac{\varepsilon}{f(b) - f(a)}.$$

Since f is monotonically increasing,

$$m_i = f(x_{i-1}),\ M_i = f(x_i),\ (i = 1, 2, ..., n).$$

Now, $$U(f, P) - L(f, P) = \sum_{i=1}^{n} (M_i - m_i)\,\delta_i$$

$$= \sum_{i=1}^{n} [f(x_i) - f(x_{i-1})]\,\delta_i$$

But, $$\delta_i < \frac{\varepsilon}{f(b) - f(a)}$$

$\therefore$ $$U(f, P) - L(f, P) < \frac{\varepsilon}{f(b) - f(a)} \sum_{i=1}^{n} [f(x_i) - f(x_{i-1})]$$

$$= \frac{\varepsilon}{f(b) - f(a)} [f(b) - f(a)] = \varepsilon$$

Hence, f is integrable on $[a, b]$.

Similar is the case, when $f(a) > f(b)$.

Example 5. *Show that the function f defined by*

$$f(x) = \begin{cases} 0, & \text{if } x \text{ is an integer} \\ 1, & \text{otherwise} \end{cases}$$

is integrable on [0, m], m being a positive integer.

Sol. The given function is integrable because its points of discontinuity are finite in number, being situated at $x = 1, 2, ..., m$.

Example 6. *Show that the function f defined by*

$$f(x) = \frac{1}{2^n},\ \text{when}\ \frac{1}{2^{n+1}} < x \le \frac{1}{2^n},\ (n = 0, 1, 2, ...),$$

$$f(0) = 0$$

is integrable on [0, 1], although it has an infinite number of points of discontinuity.

Sol. The function f is continuous on, [0, 1] except, at the points

$$0, \frac{1}{2}, \frac{1}{2^2}, ...$$

The set of its points of discontinuity

$$\left\{0, \frac{1}{2}, \frac{1}{2^2}, \ldots\ldots\right\}$$

has 0 as its only limit point. Hence, f is integrable.

Example 7. *Show that the function f defined on [0, 1] by*

$$f(x) = \begin{cases} 1 & \text{if } x \neq \frac{1}{2} \\ 0 & \text{if } x = \frac{1}{2} \end{cases}$$

is integrable on [0, 1] and find $\int_0^1 f\,dx$.

Sol. The given function f is constant on [0, 1] except at $x = \frac{1}{2}$. Therefore, f is continuous on [0, 1] except at $x = \frac{1}{2}$. So, the set of points of discontinuity of f on [0, 1] is finite. Hence, f is integrable on [0, 1]. Now, to find $\int_0^1 f$, it is enough to find $\underline{\int_0^1} f$ or $\overline{\int_0^1} f$, we calculate $\overline{\int_0^1} f$.

For any partition $P = \{x_0, x_1, x_2, \ldots\ldots, x_n\}$ of [0, 1],

We have $M_i = \text{l.u.b. of } f \text{ in } [x_{i-1}, x_i] = 1, \forall\, i$

$$\therefore \quad U(f, P) = \sum_{i=1}^{n} 1 \,.\, \delta_i = \text{length of interval } [0, 1] = 1.$$

This holds to every partition P of [0, 1], therefore,

$$\overline{\int_0^1} f\,dx = 1.$$

Hence,

$$\int_0^1 f\,dx = 1.$$

9.3. THE INTEGRAL AS A LIMIT OF SUMS

We arrived at the integral of a function *via* the upper and the lower sums. The number M_i, m_i occurring in these sums may not be the value of the function f (In fact, M_i and m_i are values of f, if f is continuous). We shall now approach the integral $\int_a^b f\,dx$ *via* the limit of a sequence of sums in which M_i and m_i are replaced by values of f.

9.3.1. Riemann Sum. *Let f be a real valued function defined on [a, b]. Let* $P : a = x_0, x_1, x_2, \ldots\ldots, x_n = b$ *be a partition of [a, b]. Let* ξ_i *be any arbitrary point of* $\delta_i = [x_{i-1}, x_i]$, $(i = 1, 2, 3, \ldots\ldots, n)$. *Then the sum* $\sum_{i=1}^{n} f(\xi_i)\,\delta_i$ *is called a Riemann sum of f on [a, b] relative to P.*

Since ξ_i is arbitrary, therefore, corresponding to each partition P of $[a, b]$, there exist infinitely many Riemann sums.

We shall now give below another definition of a function to be integrable.

9.3.2. Definition (Second definition of integrability). *A function f is said to be integrable on $[a, b]$ if there exists a number I such that for each $\varepsilon > 0$, there exists a $\delta > 0$ such that for every partition $P : \{a = x_0, x_1, \ldots\ldots, x_n = b\}$* of $[a, b]$ with $\| P \| < \delta$ *and* $\forall\ \xi_i \in [x_{i-1}, x_i]$,

$$\left| \sum_{i=1}^{n} f(\xi_i)\, \delta_i - I \right| < \varepsilon.$$

The number I is nothing but $\int_a^b f\, dx$.

i.e., A function f is integrable over $[a, b]$ if $\lim\limits_{\| P \| \to 0} \sum\limits_{i=1}^{n} f(\xi_i)\, \delta_i$ *exists and is independent of the choice of δ_i and of the point ξ_i of δ_i. This limit I, if exists, is* $\int_a^b f\, dx$.

We now establish the equivalence of the two definitions.

(*i*) Let a bounded function f be integrable on $[a, b]$ according to the former definition, so that

$$\int_{\underline{a}}^{b} f\, dx = \int_a^{\bar{b}} f\, dx = \int_a^b f\, dx.$$

Let $\varepsilon > 0$ be given.

Then, by Darboux's theorem, there exists a $\delta > 0$ such that for every partition P with $\| P \| < \delta$,

$$U(f, P) < \int_a^{\bar{b}} f\, dx + \varepsilon = \int_a^b f\, dx + \varepsilon \qquad \ldots(1)$$

and

$$L(f, P) > \int_{\underline{a}}^{b} f\, dx - \varepsilon = \int_a^b f\, dx - \varepsilon \qquad \ldots(2)$$

Let $\xi_i \in \delta_i = [x_{i-1}, x_i]$, $(i = 1, 2, \ldots\ldots, n)$ be arbitrary. Then,

$$L(f, P) \le \sum_{i=1}^{n} f(\xi_i)\, \delta_i \le U(f, P) \qquad \ldots(3)$$

$\therefore$ from (1) to (3), we have for each $\varepsilon > 0$, there exists a $\delta > 0$ such that for each partition P, with $\| P \| < \delta$,

$$\int_a^b f\, dx - \varepsilon < \sum_{i=1}^{n} f(\xi_i)\, \delta_i < \int_a^b f\, dx + \varepsilon$$

$$\Rightarrow \qquad \left| \sum_{i=1}^{n} f(\xi_i)\, \delta_i - I \right| < \varepsilon,$$

where $I = \int_a^b f\, dx$

$\Rightarrow$ f is integrable according to the second definition.

(*ii*) Conversely, let f be integrable on $[a, b]$ according to the second definition.

We shall show that f is bounded and its upper and lower integrals on $[a, b]$ are equal.

Let, if possible, f be not bounded on $[a, b]$.

By second definition, for $\varepsilon = 1$, there exits $\delta > 0$ and a number I such that for each partition P of $[a, b]$ with $\| P \| < \delta$,

$$\left| \sum_{i=1}^{n} f(\xi_i)\,\delta_i - I \right| < 1, \forall\, \xi_i \in \delta_i$$

$$\Rightarrow \qquad |\, I \,| - 1 < \sum_{i=1}^{n} f(\xi_i)\,\delta_i < |\, I \,| + 1, \forall\, \xi_i \in \delta_i \qquad ...(1)$$

As f is not bounded on $[a, b]$, it must not be bounded on at least one sub-interval, δ_m (say).

Taking ξ_i = mid-point of ith sub-interval, when $i \neq m$, each term ξ_i except ξ_m is fixed and accordingly each term of $\sum_{i=1}^{n} f(\xi_i)\,\delta_i$ except $f(\xi_m)\,\delta_m$ is also fixed. Since f is not bounded on δ_m, we can choose a point ξ_m in δ_m such that $\sum_{i=1}^{n} f(\xi_i)\,\delta_i > |\, I \,| + 1$, which contradicts (1).

Hence, f is bounded on $[a, b]$.

Let $\varepsilon > 0$ be given. Then, there exists a $\delta > 0$ and a number I such that for every partition P with $\| P \| < \delta$,

$$\left| \sum_{i=1}^{n} f(\xi_i)\,\delta_i - I \right| < \frac{\varepsilon}{2}, \quad \forall\, \xi_i \in \delta_i$$

$$\Rightarrow \qquad I - \frac{\varepsilon}{2} < \sum_{i=1}^{n} f(\xi_i)\,\delta_i < I + \frac{\varepsilon}{2}, \quad \forall\, \xi_i \in \delta_i \qquad ...(2)$$

If m_i and M_i are the g.l.b. and the l.u.b. of f in $\delta_i = [x_{i-1}, x_i]$, then there exist points α_i, $\beta_i \in \delta_i$ such that

$$f(\alpha_i) < m_i + \frac{\varepsilon}{2(b-a)} \quad \text{and} \quad f(\beta_i) > M_i - \frac{\varepsilon}{2(b-a)}$$

$$\Rightarrow \qquad \sum_{i=1}^{n} f(\alpha_i)\,\delta_i < L(f, P) + \frac{\varepsilon}{2} \text{ and} \sum_{i=1}^{n} f(\beta_i)\,\delta_i > U(f, P) - \frac{\varepsilon}{2} \qquad ...(3)$$

$\therefore$ from (2) and (3), taking $\xi_i = \alpha_i$ and β_i, we have

$$I - \varepsilon < L(f, P) \le U(f, P) < I + \varepsilon \qquad ...(4)$$

for every partition P with $\| P \| < \delta$.

Since $$L(f, P) \le \int_{\underline{a}}^{b} f\,dx \le \int_{a}^{\bar{b}} f\,dx \le U(f, P) \qquad ...(5)$$

$\therefore$ from (4) and (5),

$$I - \varepsilon < \int_{\underline{a}}^{b} f\,dx \le \int_{a}^{\bar{b}} f\,dx \le I + \varepsilon$$

$$\Rightarrow \qquad \left| \int_{a}^{\bar{b}} f\,dx - \int_{\underline{a}}^{b} f\,dx \right| < 2\varepsilon$$

Since $\varepsilon > 0$ is arbitrary, this implies

$$\overline{\int_a^b} f\,dx = \underline{\int_a^b} f\,dx$$

$\Rightarrow$ f is integrable according to the former definition.

Example 8. *If f is integrable on* $[a, b]$, *then*

(i) $\lim\limits_{n\to\infty} \sum\limits_{r=1}^{n} hf(a+rh) = \int_a^b f\,dx$, *where* $h = \dfrac{b-a}{n}$

(ii) $\lim\limits_{n\to\infty} \sum\limits_{r=1}^{n} f(ah^r)(ah^r - ah^{r-1}) = \int_a^b f\,dx$, *where* $h = \left(\dfrac{b}{a}\right)^{\frac{1}{n}}$.

Sol. (*i*) Consider a partition P of $[a, b]$ into n equal parts so that

$$\| P \| = h = \frac{b-a}{n}$$

$\therefore$ $\| P \| \to 0$ as $n \to \infty$.

Thus,
$$\int_a^b f\,dx = \lim_{n\to\infty} \sum_{r=1}^{n} f(a+rh)\,\delta_r$$
$$= \lim_{n\to\infty} \sum_{r=1}^{n} hf(a+rh).$$

(*ii*) Consider a partition P :

$a, ah, ah^2, \ldots, ah^n = b$ of $[a, b]$. Here, points of partition of $[a, b]$ are in geometrical progression. Then,

$$\| P \| \to 0 \text{ as } n \to \infty \text{ and } h \to 1.$$

Thus,
$$\int_a^b f\,dx = \lim_{n\to\infty} \sum_{r=1}^{n} f(ah^r)\,\delta_r$$
$$= \lim_{n\to\infty} \sum_{r=1}^{n} f(ah^r)(ah^r - ah^{r-1}).$$

Notes 1. *The partition P :*

$a, a+h, a+2h, \ldots\ldots, a+(n-1)h, a+nh = b$ of $[a, b]$ is useful in the evaluation of the definite integrals $\int_a^b e^x\,dx$, $\int_a^b \cos x\,dx$, $\int_a^b \sin x\,dx$, $\int_a^b \sinh x\,dx$, $\int_a^b \cosh x\,dx$ and $\int_a^b x^m\,dx$, (where m is a small positive integer).

2. *The partition P :*

$a, ah, ah^2, \ldots\ldots, ah^n = b$ of $[a, b]$ is useful for evaluating a definite integral $\int_a^b x^m\,dx$, (where m is a positive or negative integer or a fraction).

Example 9. *Show that* $\int_1^2 f\,dx = 6$, *where* $f(x) = 2x + 3$.

Sol. Consider a partition $P = \{1 = x_0, x_1, \ldots\ldots, x_n = 2\}$, of $[1, 2]$, dividing it into n equal sub-intervals, each of length

$$\frac{2-1}{n} = \frac{1}{n} \text{ so that } \| P \| \to 0 \text{ as } n \to \infty$$

Also, $$x_i = 1 + \frac{i}{n}, \ (i = 1, 2, ..., n)$$

and $$\delta_i = \frac{1}{n}, \ (i = 1, 2, ..., n)$$

$$\therefore \quad \sum_{i=1}^{n} \delta_i = n \cdot \frac{1}{n} = 1$$

Now, $$\lim_{n \to \infty} \sum_{i=1}^{n} f(x_i)\, \delta_i \qquad [\text{Taking } \xi_i = x_i]$$

$$= \lim_{n \to \infty} \sum_{i=1}^{n} (2x_i + 3)\, \delta_i = \lim_{n \to \infty} \sum_{i=1}^{n} \left[2\left(1+\frac{i}{n}\right)+3\right] \frac{1}{n}$$

$$= \lim_{n \to \infty} \sum_{i=1}^{n} \left(\frac{5}{n} + \frac{2i}{n^2}\right) = \lim_{n \to \infty} \left(\frac{5}{n} \cdot n + \frac{2}{n^2} \sum_{i=1}^{n} i\right)$$

$$= \lim_{n \to \infty} \left[5 + \frac{2}{n^2} \frac{n(n+1)}{2}\right] = 5 + 1 = 6$$

Since, $\lim_{n \to \infty} \sum_{i=1}^{n} f(x_i)\, \delta_i$ exists, the function is integrable and $\int_1^2 f\, dx = \lim_{n \to \infty} \sum_{i=1}^{n} f(x_i)\, \delta_i = 6.$

Example 10. *Evaluate* $\int_{-1}^{1} f\, dx$, *where* $f(x) = | x |$.

Sol. Since $$f(x) = \begin{cases} -x, & x \le 0 \\ x, & x > 0 \end{cases}$$

$\therefore$ f is bounded and continuous on $[-1, 1]$.

Hence, f is integrable on $[-1, 1]$.

Consider a partition

$$P = \{-1 = x_0, x_1, ..., x_n = 0, x_{n+1}, x_{n+2}, ..., x_{2n} = 1\}$$

dividing $[-1, 1]$ into $2n$ equal sub-intervals each of length $\frac{1}{n}$ so that $\| P \| \to 0$ as $n \to \infty$.

Also, $$x_i = \begin{cases} -1 + \frac{i}{n}, & (i = 1, 2, ..., n) \\ \frac{i-n}{n}, & (i = n+1, ..., 2n) \end{cases}$$

Taking $\xi_i = x_i$, we have

$$\lim_{n \to \infty} \sum_{i=1}^{2n} f(x_i)\, \delta_i = \lim_{n \to \infty} \left[\sum_{i=1}^{n} f(x_i)\, \delta_i + \sum_{i=n+1}^{2n} f(x_i)\, \delta_i\right]$$

$$= \lim_{n \to \infty} \left[\sum_{i=1}^{n} -(x_i)\, \delta_i + \sum_{i=n+1}^{2n} x_i\, \delta_i\right]$$

$$= \lim_{n \to \infty} \left[\sum_{i=1}^{n} \left(1 - \frac{i}{n}\right) \frac{1}{n} + \sum_{i=n+1}^{2n} \left(\frac{i-n}{n}\right) \frac{1}{n}\right]$$

$$= \lim_{n \to \infty} \left[n \cdot \frac{1}{n} - \frac{1}{n^2} \sum_{i=1}^{n} i + \frac{1}{n^2} \sum_{i=1}^{n} i \right] = 1$$

$$(\because \; n + 1 \le i \le 2n \; \Rightarrow \; 1 \le i - n \le n)$$

$$\Rightarrow \qquad \int_{-1}^{1} f\, dx = \lim_{n \to \infty} \sum_{i=1}^{2n} f(x_i)\, \delta_i = 1.$$

Example 11. *Show that* $\int_0^a \sin x \, dx = 1 - \cos a$, *where a is a fixed real number.*

Sol. Since $\sin x$ is bounded and continuous, $\sin x$ is integrable on $[0, a]$.

Consider a partition $P = \left\{0, \frac{a}{n}, \frac{2a}{n}, \ldots\ldots, n\frac{a}{n}\right\}$ dividing $[0, a]$ into n equal sub-intervals each of length $\frac{a}{n}$ so that $\| P \| \to 0$ as $n \to \infty$.

Also
$$\left.\begin{aligned} x_i &= \frac{ia}{n} \\ \delta_i &= \frac{a}{n} \end{aligned}\right\}, (i = 1, 2, \ldots\ldots, n)$$
and

Now, $\lim_{n \to \infty} \sum_{i=1}^{n} f(x_i)\, \delta_i$ (Taking $\xi_i = x_i$)

$$= \lim_{n \to \infty} \sum_{i=1}^{n} \left(\sin \frac{ia}{n}\right) \cdot \frac{a}{n}$$

$$= \lim_{n \to \infty} \frac{a}{n} \left[\sin \frac{a}{n} + \sin \frac{2a}{n} + \ldots.. + \sin \frac{na}{n}\right]$$

$$= \lim_{n \to \infty} \frac{a}{n} \cdot \frac{\sin \dfrac{\left(\frac{a}{n} + \frac{na}{n}\right)}{2} \sin\left(n \cdot \dfrac{a}{2n}\right)}{\sin \dfrac{a}{2n}}$$

$$\left[\because \; \sin \alpha + \sin(\alpha + \beta) + \ldots\ldots + \sin(\alpha + (n-1)\beta) = \frac{\sin\left(\alpha + \frac{(n-1)}{2}\beta\right) \sin \frac{n\beta}{2}}{\sin \frac{\beta}{2}}\right]$$

$$= \lim_{n \to \infty} \frac{2}{\dfrac{\sin \frac{a}{2n}}{\frac{a}{2n}}} \sin\left(\frac{a}{2n} + \frac{a}{2}\right) \sin \frac{a}{2}$$

$$= 2 \sin^2 \frac{a}{2} = 1 - \cos a.$$

Example 12. *Evaluate by summation* $\int_a^b e^x\,dx$.

Sol. Consider a partition P : $a, a + h, a + 2h, ..., a + nh = b$ dividing $[a, b]$ into n equal parts so that $\| P \| = h = \dfrac{b-a}{n}$.

$$\therefore \qquad \| P \| \to 0 \text{ as } n \to \infty$$

$$\therefore \qquad \int_a^b e^x\,dx = \lim_{n\to\infty}\left(\sum_{i=1}^{n} f(\xi_i)\,.\,\delta_i\right), \text{ where } f(x) = e^x$$

$$= \lim_{n\to\infty}\left(\sum_{i=1}^{n} e^{a+(i-1)h}\,.\,h\right) \qquad \text{(Taking } \xi_i = a + (i-1)h\text{)}$$

$$= \lim_{n\to\infty} h\,[e^a + e^{a+h} + e^{a+2h} + \ldots\ldots + e^{a+(n-1)h}]$$

$$= \lim_{n\to\infty} h\,e^a\,[1 + e^h + e^{2h} + \ldots\ldots + e^{(n-1)h}]$$

$$= \lim_{n\to\infty} h\,e^a\,\frac{e^{nh}-1}{e^h-1}$$

$$= \lim_{n\to\infty}\left(\frac{h}{e^h-1}\right) e^a\,(e^{b-a}-1) \qquad [\text{Putting } nh = b - a]$$

$$= 1.e^a\,.\,(e^{b-a}-1) = e^b - e^a.$$

Example 13. *Evaluate by summation* $\int_a^b x^m\,dx$.

Sol. Consider a partition P :

$$a, ah, ah^2, ..., ah^n = b \text{ of } [a, b].$$

Then, $\| P \| \to 0$ as $n \to \infty$ and $h \to 1$, where $h^n = \dfrac{b}{a}$.

$$\therefore \qquad \int_a^b x^m\,dx = \lim_{n\to\infty}\left(\sum_{i=1}^{n} f(\xi_i)\,\delta_i\right), \text{ where } f(x) = x^m.$$

$$= \lim_{n\to\infty}\left(\sum_{i=1}^{n} f(ah^{i-1})\,.\,(ah^i - ah^{i-1})\right)$$

(Taking $\xi_i = ah^{i-1}$ and $\delta_i = ah^i - ah^{i-1}$)

$$= \lim_{n\to\infty}\left(\sum_{i=1}^{n} (ah^{i-1})^m\,.\,(ah^i - ah^{i-1})\right)$$

$$= \lim_{n\to\infty} (h-1)\,[a^{m+1} + (ah)^{m+1} + (ah^2)^{m+1} + \ldots\ldots + (ah^{n-1})^{m+1}]$$

$$= \lim_{n\to\infty} (h-1)\,.\,a^{m+1}\,.\left[\frac{(h^{m+1})^n - 1}{h^{m+1}-1}\right]$$

$$= \lim_{n \to \infty} \frac{h-1}{h^{m+1}-1} \cdot a^{m+1}\,[(h^n)^{m+1} - 1]$$

$$= \lim_{n \to \infty} \frac{h-1}{h^{m+1}-1} \cdot a^{m+1}\left[\left(\frac{b}{a}\right)^{m+1} - 1\right] \qquad \left[\text{Replacing } h^n \text{ by } \frac{b}{a}\right]$$

$$= \frac{1}{m+1} \cdot a^{m+1}\left[\frac{b^{m+1} - a^{m+1}}{a^{m+1}}\right]$$

$$\left[\because \text{ by L' Hospital's rule, } \lim_{h \to 1} \frac{h-1}{h^{m+1}-1} = \frac{1}{m+1}\right]$$

$$= \frac{b^{m+1} - a^{m+1}}{m+1}.$$

9.3.3. Summation of Infinite Series of a Special Type by Integration

We know that $\displaystyle\int_a^b f\,dx = \lim_{n \to \infty} \sum_{r=1}^{n} h\,f(a + rh)$, where $h = \dfrac{b-a}{n}$

Putting $a = 0$ and $b = 1$, we have

$$\int_0^1 f\,dx = \lim_{n \to \infty} \sum_{r=1}^{n} h\,f(rh) = \lim_{n \to \infty} \sum_{r=1}^{n} \frac{1}{n} f\left(\frac{r}{n}\right)$$

This result can be used in finding limit of some series.

Example 14. *Find the limit when $n \to \infty$, of the series :*

$$\frac{1}{n} + \frac{n^2}{(n+1)^3} + \frac{n^2}{(n+2)^3} + \ldots\ldots + \frac{1}{8n}.$$

Sol. $\displaystyle\lim_{n \to \infty}\left[\frac{n^2}{n^3} + \frac{n^2}{(n+1)^3} + \frac{n^2}{(n+2)^3} + \ldots\ldots + \frac{n^2}{8n^3}\right]$

$$= \lim_{n \to \infty} \sum_{r=1}^{n} \frac{n^2}{(n+r-1)^3} = \lim_{n \to \infty} \sum_{r=1}^{n} \frac{1}{n} \frac{1}{\left(1+\dfrac{r-1}{n}\right)^3}$$

$$= \lim_{n \to \infty} \sum_{r=0}^{n} \frac{1}{n} \frac{1}{\left(1+\dfrac{r}{n}\right)^3} = \int_0^1 \frac{dx}{(1+x)^3} \qquad \left[\text{Putting } \frac{1}{n} = dx \text{ and } \frac{r}{n} = x\right]$$

$$= \frac{3}{8} \text{ (on evaluation).}$$

EXERCISE 9 (*b*)

From definition, prove the following (Q. 1 to 3) :

1. $\displaystyle\int_1^2 x\,dx = \frac{3}{2}.$

2. $\displaystyle\int_0^{\pi/2} \sin x\,dx = 1.$

3. $\int_0^a \cos x \, dx = \sin a$, for a fixed number a.

4. Evaluate $\int_1^3 (x^2 + 2x + 3)\, dx$ by using as the limit of Riemann sums.

5. Use definition as limit of a sum to evaluate $\int_a^b \sin x \, dx$.

6. Prove by summation : $\int_a^b x^3 \, dx = \frac{1}{4}(b^4 - a^4)$.

7. Prove by summation : $\int_a^b \frac{1}{\sqrt{x}} \, dx = 2(\sqrt{b} - \sqrt{a})$.

8. Show that $\lim_{n \to \infty} \left(\frac{n}{n^2 + 1^2} + \frac{n}{n^2 + 2^2} + \frac{n}{n^2 + 3^2} + \ldots\ldots + \frac{1}{2n} \right) = \frac{\pi}{4}$.

9. If a bounded function f is Riemann integrable according to second definition, then prove that it is Riemann integrable according to definition first.

10. State and prove Darboux's Theorem.

9.4. PROPERTIES OF RIEMANN INTEGRAL

9.4.1. Theorem. (*i*) *If f and g are bounded on $[a, b]$ and $f(x) \le g(x)$, $x \in [a, b]$, then*

$$\underline{\int_a^b} f\,dx \le \underline{\int_a^b} g\,dx \quad \text{and} \quad \overline{\int_a^b} f\,dx \le \overline{\int_a^b} g\,dx.$$

(*ii*) *If f and g are integrable, then*

$$\int_a^b f\,dx \le \int_a^b g\,dx.$$

Proof. (*i*) Let m and M denote the g.l.b. and l.u.b. of f on $[a, b]$. Let m' and M$'$ denote the g.l.b. and l.u.b. of g on $[a, b]$.

Then, $\qquad m \le f(x) \le g(x) \le M', \ \forall\, x \in [a, b].$

$\Rightarrow \qquad m \le m'$ and $M \le M'$.

Let P : $\{a = x_0, x_1, \ldots, x_n = b\}$ be a partition of $[a, b]$.

Let m_i and M_i denote the bounds of f in $\delta_i = [x_{i-1}, x_i]$. Let m_i' and M_i' denote the corresponding bounds of g in δ_i. Then,

$$m_i \le m_i' \text{ and } M_i \le M_i', \qquad (i = 1, 2, 3, \ldots, n)$$

$$\Rightarrow \qquad L(f, P) \le L(g, P) \text{ and } U(f, P) \le U(g, P)$$

$$\Rightarrow \qquad L(f, P) \le \underline{\int_a^b} g\,dx \text{ and } \overline{\int_a^b} f\,dx \le U(g, P),$$

for all partitions P of $[a, b]$.

$$\Rightarrow \qquad \underline{\int_a^b} f\,dx \le \underline{\int_a^b} g\,dx \quad \text{and} \quad \overline{\int_a^b} f\,dx \le \overline{\int_a^b} g\,dx.$$

(*ii*) Proof follows from (*i*).

Cor. *If f is integrable on $[a, b]$ and $f(x) \ge 0 \ \forall\, x \in [a, b]$, then*

$$\int_a^b f\,dx \ge 0.$$

Notes 1. If f is continuous on $[a, b]$ and $f(x) \geq 0, x \in [a, b]$ but $f(c) \neq 0$ for some $c \in [a, b]$, then $\int_a^b f\,dx > 0$.

2. If f is continuous on $[a, b]$ and $f(x) \leq 0, x \in [a, b]$ but $f(c) \neq 0$ for some $c \in [a, b]$, then $\int_a^b f\,dx < 0$.

3. If f is continuous on $[a, b]$ and $\int_a^b f^2\,dx = 0$, then $f(x) = 0 \;\forall\; x \in [a, b]$.

9.4.2. Theorem. *If a function f is integrable on [a, b] and c is a real number, then cf is integrable on [a, b]. Moreover,*

$$\int_a^b cf\,dx = c\int_a^b f\,dx.$$

Proof. If $c = 0$, the result is obviously true.

So, let $c \neq 0$. Let $P = \{a = x_0, x_1, \ldots\ldots, x_n = b\}$ be a partition of $[a, b]$. Let m_i and M_i denote the bounds of f on $\delta_i = [x_{i-1}, x_i]$.

If m_i' and M_i' denote the corresponding bounds of cf on δ_i, then

$$m_i' = \begin{cases} cm_i & \text{if } c > 0 \\ cM_i & \text{if } c < 0 \end{cases} \quad \text{and} \quad M_i' = \begin{cases} cM_i, & \text{if } c > 0 \\ cm_i, & \text{if } c < 0 \end{cases}$$

$$\Rightarrow \qquad L(cf, P) = \begin{cases} cL(f, P), & \text{if } c > 0 \\ cU(f, P), & \text{if } c < 0 \end{cases}$$

and
$$U(cf, P) = \begin{cases} cU(f, P), & \text{if } c > 0 \\ cL(f, P), & \text{if } c < 0 \end{cases}$$

$\Rightarrow$ for any partition P of $[a, b]$,

$$U(cf, P) - L(cf, P) = |\,c\,|\,(U(f, P) - L(f, P))$$

Let $\varepsilon > 0$ be given.

Since f is integrable over $[a, b]$, there exists a partition P of $[a, b]$ such that

$$U(f, P) - L(f, P) < \frac{\varepsilon}{|c|}.$$

$$\Rightarrow \qquad |\,c\,|\,(U(f, P) - L(f, P)) < \varepsilon$$

$$\Rightarrow \qquad U(cf, P) - L(cf, P) < \varepsilon$$

$\Rightarrow$ cf is integrable on $[a, b]$.

Also,
$$\int_a^b cf\,dx = \int_a^{\bar{b}} cf\,dx = \underset{P \in \mathcal{P}}{\text{g.l.b.}}\; U(cf, P)$$

$$= \begin{cases} \underset{P \in \mathcal{P}}{\text{g.l.b.}}\; cU(f, P), & \text{if } c > 0 \\ \underset{P \in \mathcal{P}}{\text{l.u.b.}}\; cL(f, P), & \text{if } c < 0 \end{cases}$$

$$= \begin{cases} c\;\underset{P \in \mathcal{P}}{\text{g.l.b.}}\; U(f, P), & \text{if } c > 0 \\ c\;\underset{P \in \mathcal{P}}{\text{l.u.b.}}\; L(f, P), & \text{if } c < 0 \end{cases}$$

$$= \begin{cases} c\displaystyle\int_a^{\bar{b}} f\,dx, & \text{if } c > 0 \\ c\displaystyle\int_{\underline{a}}^{b} f\,dx, & \text{if } c < 0 \end{cases}$$

Since f is integrable,

$$\int_a^b f\,dx = \int_{\underline{a}}^b f\,dx = \int_a^{\bar{b}} f\,dx$$

Hence, $$\int_a^b cf\,dx = c\int_a^b f\,dx.$$

9.4.3. Theorem. *If f and g are two integrable functions on $[a, b]$, then $f+g$ is integrable on $[a, b]$ and*

$$\int_a^b (f+g)\,dx = \int_a^b f\,dx + \int_a^b g\,dx.$$

Proof. f and g, being integrable, are bounded on $[a, b]$.

$\Rightarrow$ $f+g$ is bounded on $[a, b]$.

Let $P = \{a = x_0, x_1, \ldots\ldots, x_n = b\}$ be a partition of $[a, b]$.

Let m_i and M_i be the bounds of f on $\delta_i = [x_{i-1}, x_i]$. Let m_i' and M_i' denote the corresponding bounds of g on δ_i.

Then $$m_i \le f(x) \le M_i \text{ and } m_i' \le g(x) \le M_i', \quad \forall\, x \in \delta_i, (i = 1, 2, \ldots\ldots, n)$$

$$\Rightarrow \quad m_i + m_i' \le f(x) + g(x) \le M_i + M_i', \quad \forall\, x \in \delta_i, \qquad (i = 1, 2, \ldots\ldots, n)$$

$\Rightarrow$ for any partition P of $[a, b]$,

$$L(f, P) + L(g, P) \le L(f+g, P) \le U(f+g, P)$$
$$\le U(f, P) + U(g, P) \qquad \ldots(1)$$

Let $\varepsilon > 0$ be given.

Since f and g are integrable on $[a, b]$, there exist partitions P_1 and P_2 of $[a, b]$ such that

$$U(f, P_1) - L(f, P_1) < \frac{\varepsilon}{2} \quad \text{and} \quad U(g, P_2) - L(g, P_2) < \frac{\varepsilon}{2}$$

Let $$P^* = P_1 \cup P_2.$$

Then, $$U(f, P^*) - L(f, P^*) \le U(f, P_1) - L(f, P_1) < \frac{\varepsilon}{2} \qquad \ldots(2)$$

and $$U(g, P^*) - L(g, P^*) \le U(g, P_2) - L(g, P_2) < \frac{\varepsilon}{2} \qquad \ldots(3)$$

Using (1), we have

$$U(f+g, P^*) - L(f+g, P^*)$$
$$\le [U(f, P^*) - L(f, P^*)] + [U(g, P^*) - (g, P^*)]$$
$$< \frac{\varepsilon}{2} + \frac{\varepsilon}{2} \qquad \text{[By (2) and (3)]}$$
$$= \varepsilon$$

$\Rightarrow$ $f+g$ is integrable on $[a, b]$.

Now, let $\varepsilon > 0$ be given. There exist partitions P_1 and P_2 of $[a, b]$ such that

$$\int_a^b f\,dx - \frac{\varepsilon}{2} = \int_{\underline{a}}^b f\,dx - \frac{\varepsilon}{2} < L(f, P_1)$$

and $$\int_a^b g\,dx - \frac{\varepsilon}{2} = \int_{\underline{a}}^b g\,dx - \frac{\varepsilon}{2} < L(g, P_2)$$

If $P = P_1 \cup P_2$, then $L(f, P_1) \le L(f, P)$ and $L(g, P_2) \le L(g, P)$.

$$\Rightarrow \quad \int_a^b f\,dx + \int_a^b g\,dx - \varepsilon < L(f, P_1) + L(g, P_2)$$

$$\leq L(f, P) + L(g, P) \leq L(f+g, P) \leq \int_a^b (f+g)\, dx$$

$$\Rightarrow \quad \int_a^b f\, dx + \int_a^b g\, dx - \varepsilon < \int_a^b (f+g)\, dx$$

Since $\varepsilon > 0$ is arbitrary,

$$\int_a^b f\, dx + \int_a^b g\, dx \leq \int_a^b (f+g)\, dx \qquad ...(4)$$

Replacing f by $-f$ and g by $-g$, we have from (4),

$$\int_a^b -f\, dx + \int_a^b -g\, dx \leq \int_a^b -(f+g)\, dx$$

$$\Rightarrow \quad \int_a^b f\, dx + \int_a^b g\, dx \geq \int_a^b (f+g)\, dx \qquad ...(5)$$

From (4) and (5), we get

$$\int_a^b f\, dx + \int_a^b g\, dx = \int_a^b (f+g)\, dx.$$

Cor. *If f and g are integrable on $[a, b]$ and α, β are two real numbers, then $\alpha f + \beta g$ is integrable on $[a, b]$*

and
$$\int_a^b (\alpha f + \beta g)\, dx = \alpha \int_a^b f\, dx + \beta \int_a^b g\, dx$$

9.4.4. Theorem. *If f and g are two integrable functions on $[a, b]$, then fg is integrable on $[a, b]$.* (K.U. BCA (II) 2003)

Proof. Since f and g are integrable on $[a, b]$, f and g are bounded on $[a, b]$.

$\Rightarrow$ there exist $k > 0$ such that

$$| f(x) | \leq k \quad \text{and} \quad | g(x) | \leq k, \quad \forall\, x \in [a, b]$$

$$\Rightarrow \quad | f(x) . g(x) | \leq k^2, \quad \forall\, x \in [a, b]$$

$$\Rightarrow \quad | (fg)(x) | \leq k^2, \quad \forall\, x \in [a, b]$$

$\Rightarrow$ fg is bounded on $[a, b]$.

Let $P : \{a = x_0, x_1, \ldots\ldots, x_n = k\}$ be a partition of $[a, b]$. Let m_i', M_i' ; m_i'', M_i'' ; and m_i, M_i be the bounds of f, g and fg respectively on $\delta_i = [x_{i-1}, x_i]$.

Now, $\quad \forall\, x', x'' \in \delta_i$,

$$\begin{aligned} | (fg)(x'') - (fg)\, x' | &= | f(x'')\, g(x'') - f(x')\, g(x') | \\ &= | g(x'')\, (f(x'') - f(x')) + f(x')\, [g(x'') - g(x')] | \\ &\leq | g(x'') || f(x'') - f(x') | + | f(x') || g(x'') - g(x') | \\ &\leq k\, (M_i' - m_i') + k(M_i'' - m_i'') \end{aligned}$$

$$\Rightarrow \quad M_i - m_i \leq k(M_i' - m_i') + k(M_i'' - m_i'')$$

$$\Rightarrow \quad \sum_{i=1}^{n} (M_i - m_i)\, \delta_i \leq k \sum_{i=1}^{n} (M_i' - m_i')\, \delta_i + k \sum_{i=1}^{n} (M_i'' - m_i'')\, \delta_1 \qquad ...(1)$$

Since f and g are integrable, there exists $\delta > 0$ such that for each partition P with $\| P \| < \delta$, the corresponding oscillatory sums of f and g are both $< \dfrac{\varepsilon}{2k}$.

$\therefore$ from (1),

$$U(fg, P) - L(fg, P) \le k \cdot \frac{\varepsilon}{2k} + k\,\frac{\varepsilon}{2k} = \varepsilon$$

$\Rightarrow$ fg is integrable on $[a, b]$.

Cor. If f is integrable on $[a, b]$, then f^2 is also integrable on $[a, b]$.

9.4.5. Theorem. *If f and g are two integrable functions on $[a, b]$, and there exists $t \geq 0$ such that $|\,g(x)\,| > t$, $\forall\, x \in [a, b]$, then $\dfrac{f}{g}$ is integrable on $[a, b]$.*

Proof. Since f and g are integrable, f and g are bounded on $[a, b]$. Therefore, there exists a positive real number k such that

$$|\,f(x)\,| \le k \quad \text{and} \quad |\,g(x)\,| \le k, \quad \forall\, x \in [a, b].$$

Now,
$$\left|\left(\frac{f}{g}\right)(x)\right| = \left|\frac{f(x)}{g(x)}\right| = \frac{|f(x)|}{|g(x)|} \le \frac{k}{t}, \ \forall\, x \in [a, b].$$

$\therefore$ $\dfrac{f}{g}$ is bounded on $[a, b]$.

Consider a partition $P : [a = x_0, x_1, x_2, \ldots\ldots, x_n = b]$ of $[a, b]$. Let m_i' and M_i' be the g.l.b. and l.u.b. of f in $[x_{i-1}, x_i]$; m_i'' and M_i'' the g.l.b. and l.u.b. of g in $[x_{i-1}, x_i]$; m_i and M_i the g.l.b. and l.u.b. of $\dfrac{f}{g}$ in $[x_{i-1}, x_i]$.

Now, $\forall\, x', x'' \in [x_{i-1}, x_i]$,

$$\left|\left(\frac{f}{g}\right)(x'') - \frac{f}{g}(x')\right| = \left|\frac{f(x'')}{g(x'')} - \frac{f(x')}{g(x')}\right| = \left|\frac{f(x'')g(x') - f(x')g(x'')}{g(x'')g(x')}\right|$$

$$= \left|\frac{g(x').[f(x'') - f(x')] + f(x')\,[g(x') - g(x'')]}{g(x'')g(x')}\right|$$

$$\le \left|\frac{|g(x')|\,|f(x'') - f(x')| + |f(x')|\,|g(x'') - g(x')|}{|g(x'')|\,|g(x')|}\right|$$

$$= \frac{|f(x'') - f(x')|}{|g(x'')|} + \frac{|f(x')|}{|g(x')|\,|g(x'')|}\,|\,g(x'') - g(x')\,|$$

$$\le \frac{1}{t}(M_i' - m_i') + \frac{k}{t^2}(M_i'' - m_i') \qquad [\because\ |\,g(x)\,| > t,\ \forall\, x \in [a, b]]$$

$$\Rightarrow \quad M_i - m_i \le \frac{1}{t}(M_i' - m_i') + \frac{k}{t^2}(M_i'' - m_i'')$$

$$\Rightarrow \quad \sum_{i=1}^{n} (M_i - m_i)\,\delta_i \le \frac{1}{t}\sum_{i=1}^{n} (M_i' - m_i')\,\delta_i + \frac{k}{t^2}\sum_{i=1}^{n} (M_i'' - m_i'')\,\delta_i$$

$$\Rightarrow \quad U\left(\frac{f}{g}, P\right) - L\left(\frac{f}{g}, P\right) \le \frac{1}{t}\,[U(f, P) - L(f, P)] + \frac{k}{t^2}\,[U(g, P) - L(g, P)]$$

for a partition P of $[a, b]$ ...(1)

Let $\varepsilon > 0$ be given.

Since f and g are integrable on $[a, b]$, therefore, there exist partitions P_1 and P_2 of $[a, b]$ such that

$$U(f, P_1) - L(f, P_1) < \frac{\varepsilon}{2} \cdot t \qquad ...(2)$$

$$U(g, P_2) - L(g, P_2) < \frac{\varepsilon \cdot t^2}{2 \cdot k} \qquad ...(3)$$

Let $P^* = P_1 \cup P_2$. Then P^* is the common refinement of P_1 and P_2.

$$\therefore \quad U(f, P^*) - L(f, P^*) \le U(f, P_1) - L(f, P_1) < \frac{\varepsilon}{2} \cdot t$$

and $$U(g, P^*) - L(g, P^*) \le U(g, P_2) - L(g, P_2) < \frac{\varepsilon}{2} \cdot \frac{t^2}{k} \qquad \text{[By (2) and (3)]}$$

Since (1) is true for any partition P of $[a, b]$,

$$\therefore \quad U\left(\frac{f}{g}, P^*\right) - L\left(\frac{f}{g}, P^*\right) < \frac{1}{t} \cdot \frac{\varepsilon}{2} \cdot t + \frac{k}{t^2} \cdot \frac{\varepsilon}{2} \cdot \frac{t^2}{k} = \varepsilon.$$

Hence, $\frac{f}{g}$ is integrable on $[a, b]$.

9.4.6. Theorem. *If f is bounded and integrable on $[a, b]$, then $|f|$ is also bounded and integrable on $[a, b]$. Moreover,*

$$\left| \int_a^b f\,dx \right| \le \int_a^b |f|\,dx.$$

Proof. Since the function f is bounded, there exists a positive number k such that

$$|f(x)| \le k, \ \forall \ x \in [a, b]$$

$\Rightarrow |f|$ is bounded.

Let $\varepsilon > 0$ be given.

Since f is integrable on $[a, b]$, there exists a partition

$$P : [a = x_0, x_1, ..., x_n = b] \text{ of } [a, b]$$

such that the corresponding oscillatory sum for f is $< \varepsilon$.

i.e., $$U(f, P) - L(f, P) < \varepsilon \qquad ...(1)$$

Let M_i', m_i' and M_i, m_i be the bounds of f and $|f|$ respectively on the sub-interval $[x_{i-1}, x_i]$, $(i = 1, 2, ..., n)$.

Now, $\forall\, x', x'' \in [x_{i-1}, x_i]$, we have

$$\left| |f|(x'') - |f|(x') \right| = \left| \left[|f(x'')| - |f(x')| \right] \right|$$
$$\le |f(x'') - f(x')| \le M_i' - m_i'$$

$$\Rightarrow \quad M_i - m_i \le M_i' - m_i', \qquad (i = 1, 2, ..., n)$$

$$\Rightarrow \quad \sum_{i=1}^{n} (M_i - m_i)\,\delta_i \le \sum_{i=1}^{n} (M_i' - m_i')\,\delta_i$$

i.e., $$U(|f|, P) - L(|f|, P) \le U(f, P) - L(f, P) < \varepsilon \qquad \text{[By (1)]}$$

$\Rightarrow |f|$ is integrable on $[a, b]$.

To show : $\left| \int_a^b f\,dx \right| \le \int_a^b |f|\,dx.$

Since $\quad -x \le |x|$ and $x \le |x|, \forall\, x$

$\therefore \qquad -f(x) \le |f(x)|$ and $f(x) \le |f(x)|$

$\Rightarrow \qquad -f(x) \le |f|(x)$ and $f(x) \le |f|(x), \forall x \in [a, b]$

$$\Rightarrow \qquad \int_a^b -f(x)\,dx \le \int_a^b |f|(x)\,dx \quad \text{and} \quad \int_a^b f(x)\,dx \le \int_a^b |f|(x)\,dx,$$

$$\Rightarrow \qquad -\int_a^b f(x)\,dx \le \int_a^b |f|(x)\,dx \quad \text{and} \quad \int_a^b f(x)\,dx \le \int_a^b |f|(x)\,dx$$

$$\Rightarrow \qquad \left|\int_a^b f(x)\,dx\right| \le \int_a^b |f|(x)\,dx.$$

Remark. *Converse of the above theorem is not true. For example, let f be a function defined on [0, 1] by :*

$$f(x) = \begin{cases} 1, & \text{if } x \text{ is rational} \\ -1, & \text{if } x \text{ is irrational} \end{cases}$$

Then, $\qquad \underline{\int_0^1} f\,dx = -1$ and $\overline{\int_0^1} f\,dx = 1$

$\therefore \quad f$ is not integrable.

But, $\qquad |f(x)| = 1 \quad \forall x \in [0, 1]$

$\Rightarrow \qquad |f|$ is integrable on $[0, 1]$.

9.4.7. Theorem. (*i*) *If a function is integrable on [a, b], then it is also integrable on [a, c] and [c, b], where c is a point of [a, b].* (K.U. BCA (II) 2003)

(*ii*) *Conversely, if f is integrable on [a, c] and [c, b], then it is also integrable on [a, b].*

(*iii*) *Moreover,* $\int_a^b f\,dx = \int_a^c f\,dx + \int_c^b f\,dx, a \le c \le b.$

Proof. Clearly, f is bounded on $[a, c]$ and $[c, b]$ if and only if f is bounded on $[a, b]$, $c \in [a, b]$.

(*i*) Let f be integrable on $[a, b]$.

For $\varepsilon > 0$, there exists a partition P of $[a, b]$ such that

$$U(f, P) - L(f, P) < \varepsilon$$

Let $\qquad P^* = P \cup \{c\}$

Then, $\qquad L(f, P) \le L(f, P^*) \le U(f, P^*) \le U(f, P)$

$\Rightarrow \qquad U(f, P^*) - L(f, P^*) \le U(f, P) - L(f, P)$

$$< \varepsilon \qquad \text{...(1)}$$

If P_1, P_2 denote the set of points of P^* on $[a, c]$, $[c, b]$ respectively, then P_1, P_2 are partitions of $[a, c]$ and $[c, b]$ respectively and $P^* = P_1 \cup P_2$.

$\therefore \qquad U(f, P^*) = U(f, P_1) + U(f, P_2)$

and $\qquad L(f, P^*) = L(f, P_1) + L(f, P_2)$

$\Rightarrow \qquad \{U(f, P_1) - L(f, P_1)\} + \{U(f, P_2) - L(f, P_2)\}$

$$= U(f, P^*) - L(f, P^*) < \varepsilon \qquad \text{[By (1)]}$$

Since each bracket on the left is non-negative, this implies that

$$U(f, P_1) - L(f, P_1) < \varepsilon$$

and $\qquad U(f, P_2) - L(f, P_2) < \varepsilon \qquad$ ...(2)

for partitions P_1, P_2 of $[a, c]$ and $[c, b]$.

Hence, f is integrable on $[a, c]$ and $[c, b]$.

(*ii*) Conversely, let f be integrable on $[a, c]$ and $[c, b]$.

Let $\varepsilon > 0$ be given. There exist partitions P_1 and P_2 of $[a, c]$ and $[c, b]$ respectively such that

$$U(f, P_1) - L(f, P_1) < \frac{\varepsilon}{2} \quad ...(3)$$

and

$$U(f, P_2) - L(f, P_2) < \frac{\varepsilon}{2} \quad ...(4)$$

If $P^* = P_1 \cup P_2$, then P^* is a partition of $[a, b]$.

$$\therefore \quad U(f, P^*) - L(f, P^*) = \{U(f, P_1) + U(f, P_2)\} - \{L(f, P_1) + L(f, P_2)\}$$

$$= \{U(f, P_1) - L(f, P_1)\} + \{U(f, P_2) - L(f, P_2)\}$$

$$< \frac{\varepsilon}{2} + \frac{\varepsilon}{2} = \varepsilon \quad \text{[By (3) and (4)]}$$

$\Rightarrow f$ is integrable on $[a, b]$.

(*iii*) If P_1 and P_2 are partitions of $[a, c]$ and $[c, b]$ respectively, then $P^* = P_1 \cup P_2$ is a partition of $[a, b]$.

Also, $\quad U(f, P^*) = U(f, P_1) + U(f, P_2)$.

Taking greatest lower bounds for all partitions,

$$\text{g.l.b. } U(f, P^*) \geq \text{g.l.b. } U(f, P_1) + \text{g.l.b. } U(f, P_2)$$

$$\Rightarrow \quad \int_a^{\bar{b}} f\,dx \geq \int_a^{\bar{c}} f\,dx + \int_c^{\bar{b}} f\,dx$$

Since f is integrable on $[a, b]$, $[a, c]$ and $[c, b]$.

$$\therefore \quad \int_a^b f\,dx \geq \int_a^c f\,dx + \int_c^b f\,dx \quad ...(5)$$

Proceeding with $-f$, we can show that

$$\int_a^b f\,dx \leq \int_a^c f\,dx + \int_c^b f\,dx \quad ...(6)$$

From (5) and (6), we have

$$\int_a^b f\,dx = \int_a^c f\,dx + \int_c^b f\,dx.$$

9.5. CONTINUITY AND DIFFERENTIABILITY OF INTEGRABLE FUNCTIONS

Theorem. *If a function f is integrable on $[a, b]$ and $F(t) = \int_a^t f\,dx$, where $a \leq t \leq b$, then F is continuous on $[a, b]$.*

Or

Show that the integral of an integrable function is continuous.

Proof. Since f is integrable on $[a, b]$,

f is bounded on $[a, b]$. Therefore, there exists a positive number k such that

$$|f(x)| \leq k, \quad \forall\, x \in [a, b]. \quad ...(1)$$

Let $x, y \in [a, b]$ such that $a \leq x < y \leq b$.

$$\therefore \quad F(x) = \int_a^x f\,dt \quad \text{and} \quad F(y) = \int_a^y f\,dt.$$

$$\Rightarrow \quad F(y) - F(x) = \int_a^y f\,dt - \int_a^x f\,dt = \int_a^y f\,dt + \int_x^a f\,dt = \int_x^y f\,dt$$

$$\therefore \qquad |F(y)-F(x)| = \left|\int_x^y f\,dx\right| \le \int_x^y |f|\,dt \le k(y-x) \qquad \text{(By (1))}$$

$$\Rightarrow \qquad |F(y)-F(x)| < \varepsilon \text{ if } |y-x| < \frac{\varepsilon}{k}$$

$\Rightarrow$ F is continuous on $[a, b]$.

Theorem. *If a function f is continuous on [a, b] and* $F(t) = \int_a^t f\,dx$, *then F is differentiable on [a, b] and* $F' = f$.

Proof. Let x_0 be an arbitrary point of $[a, b]$.

Then, $F(x_0) = \int_a^{x_0} f\,dx$ and $F(x_0 + h) = \int_a^{x_0+h} f\,dx$.

$$\therefore \qquad F(x_0+h) - F(x_0) = \int_a^{x_0+h} f\,dx - \int_a^{x_0} f\,dx$$

$$= \int_{x_0}^{x_0+h} f\,dx = h\,f(x_0 + \theta h),\ 0 \le \theta \le 1.$$

$$\Rightarrow \qquad \lim_{h\to 0}\left[\frac{F(x_0+h)-F(x_0)}{h}\right] = \lim_{h\to 0} f(x_0+\theta h)$$

$$= f(x_0) \qquad [\because f \text{ is continuous on } [a, b]]$$

$$\therefore \qquad F'(x_0) = f(x_0).$$

Since x_0 is arbitrary, F is differentiable on $[a, b]$ and $F' = f$.

9.5.1. Primitive. *If f and F are two functions defined on [a, b] and* $F'(x) = f(x)$, $x \in [a, b]$, *then F is called the primitive of f on [a, b].*

The First Fundamental theorem of Calculus states that we can always construct a primitive of a continuous function by integration.

Now, we prove the second Fundamental theorem of calculus which establishes integration as reverse process of differentiation.

9.5.2. Theorem. *(Fundamental Theorem of Integral Calculus)*

If f is integrable on $[a, b]$ and F is the primitive of f on $[a, b]$, then

$$\int_a^b f\,dx = F(b) - F(a).$$

Proof. By definition of a primitive, $F'(x) = f(x)$, $\forall\, x \in [a, b]$.

Consider a partition $P : \{a = x_0, x_1, \ldots\ldots, x_n = b\}$ of $[a, b]$.

Applying Lagrange's Mean Value Theorem for each sub-interval $[x_{i-1}, x_i]$, $(1 \le i \le n)$, we have

$$F(x_i) - F(x_{i-1}) = F'(\xi_i)\,\delta_i = f(\xi_i)\,\delta_i$$

where $x_{i-1} < \xi_i < x_i$, $(i = 1, 2, \ldots\ldots, n)$.

$$\Rightarrow \qquad \sum_{i=1}^{n} f(\xi_i)\,\delta_i = \sum_{i=1}^{n} (F(x_i) - F(x_{i-1})) = F(b) - F(a) \qquad \ldots(1)$$

$$\text{Also,} \qquad L(f, P) \le \sum_{i=1}^{n} f(\xi_i)\,\delta_i \le U(f, P)$$

$$\Rightarrow \quad L(f, P) \le F(b) - F(a) \le U(f, P) \qquad \text{[By (1)]}$$

$$\Rightarrow \quad \underline{\int_a^b} f\,dx \le F(b) - F(a) \le \overline{\int_a^b} f\,dx$$

But, integrability of f on $[a, b]$ implies

$$\underline{\int_a^b} f\,dx = \overline{\int_a^b} f\,dx = \int_a^b f\,dx.$$

$$\therefore \quad \int_a^b f\,dx = F(b) - F(a).$$

Remarks 1. *The above theorem does not state that if f is integrable, then f has a primitive on $[a, b]$. It only states that if f has a primitive on $[a, b]$, then this primitive can be used to evaluate $\int_a^b f\,dx$.*

2. *A function may have a primitive without being integrable.*

For example, consider the function F defined on $[-1, 1]$ as follows :

$$F(x) = \begin{cases} x^2 \sin\left(\dfrac{1}{x^2}\right), & \text{if } x \ne 0 \\ 0, & \text{if } x = 0 \end{cases}$$

Consider another function f defined on $[-1, 1]$ by :

$$f(x) = \begin{cases} -\left(\dfrac{2}{x}\right)\cos\left(\dfrac{1}{x^2}\right) + 2x \sin\left(\dfrac{1}{x^2}\right), & \text{if } x \ne 0 \\ 0, & \text{if } x = 0 \end{cases}$$

Clearly, $\quad F'(x) = f(x).$

Therefore, F is the primitive of f on $[-1, 1]$.

But, f is not integrable on $[-1, 1]$ because f is not bounded on $[-1, 1]$.

3. *A function may be integrable without having a primitive on $[a, b]$.*

For example, define a function f on $[0, 2]$ as follows :

$$f(x) = \begin{cases} -1, & \text{if } 0 \le x < 1 \\ 1, & \text{if } 1 \le x \le 2 \end{cases}$$

Since f is defined and monotonically increasing on $[0, 2]$, f is integrable on $[0, 2]$.

But, f has no primitive on $[0, 2]$ because otherwise f would have intermediate value property on $[0, 2]$, which is not the case here.

9.6. SOME MORE RESULTS ON INTEGRATION

1. We know that for an integrable function f,

$$m(b-a) \le \int_a^b f\,dx \le M(b-a) \text{ when } b \ge a \qquad ...(1)$$

If $b < a$, then $a > b$.

$\therefore$ interchanging a and b in (1), we have

$$m(a-b) \le \int_b^a f\,dx \le M(a-b)$$

$$\Rightarrow \quad -m(a-b) \ge -\int_b^a f\,dx \ge -M(a-b), \text{ when } a > b$$

$$\Rightarrow \qquad m(b-a) \geq \int_a^b f\,dx \geq M(b-a), \text{ when } b < a \qquad ...(2)$$

2. If f is integrable on $[a, b]$, then there exists a number λ lying between the bounds of f such that

$$\int_a^b f\,dx = \lambda(b-a).$$

[For, Since f is integrable on $[a, b]$,

$$\therefore \qquad m(b-a) \leq \int_a^b f\,dx \leq M(b-a) \text{ (show it)}$$

If $b \neq a$, then $m \leq \dfrac{1}{b-a}\displaystyle\int_a^b f\,dx \leq M.$

If $\lambda = \dfrac{1}{b-a}\displaystyle\int_a^b f\,dx$, then $\displaystyle\int_a^b f\,dx = \lambda(b-a),\ m \leq \lambda \leq M.$

If $a = b$, then the result is obviously true.]

3. If f is integrable on $[a, b]$ and k is a number such that $|f(x)| \leq k$ for all $x \in [a, b]$, then

$$\left|\int_a^b f\,dx\right| \leq k\,|b-a|$$

For, let M, m be the bounds of $f(x)$.

Now, $\qquad |f(x)| \leq k, \quad \forall\, x \in [a, b]$

$$\Rightarrow \qquad -k \leq f(x) \leq k \;\Rightarrow\; -k \leq m \leq f(x) \leq M \leq k$$

$$\Rightarrow \qquad -k(b-a) \leq m(b-a) \leq \int_a^b f\,dx \leq M(b-a) \leq k(b-a)$$

If $\qquad b > a \;\Rightarrow\; \left|\displaystyle\int_a^b f\,dx\right| \leq k(b-a)$

Similarly, $\left|\displaystyle\int_a^b f\,dx\right| \leq k\,|a-b|$ if $b < a$, *i.e.*, $a > b$

The result is obviously true if $a = b$.

Hence $\left|\displaystyle\int_a^b f\,dx\right| \leq k\,|a-b|.$

4. If f is integrable on $[a, b]$ and $f(x) \geq 0$ for $x \in [a, b]$, then

$$\int_a^b f\,dx \begin{cases} \geq 0, & \text{if } b \geq a \\ \leq 0, & \text{if } b \leq a \end{cases}$$

5. If f and g are two integrable functions on $[a, b]$,

and $\qquad f(x) \geq g(x) \quad \forall\, x \in [a, b]$, then

(*i*) $\displaystyle\int_a^b f\,dx \geq \int_a^b g\,dx$, if $b \geq a$. $\qquad$ (*ii*) $\displaystyle\int_a^b f\,dx \leq \int_a^b g\,dx$, if $b \leq a$.

[**Hint.** $f(x) \geq g(x) \Rightarrow (f-g)(x) \geq 0\ \forall\, x \in [a, b]$].

6. If a function f is continuous on $[a, b]$, then there exists a number c on $[a, b]$ such that

$$\int_a^b f\,dx = (b-a)\,f(c).$$

Proof. Since f is continuous on $[a, b]$, f is integrable on $[a, b]$.

If m, M be the g.l.b. and l.u.b. of f on $[a, b]$, then

$$m(b-a) \leq \int_a^b f\,dx \leq M(b-a).$$

$\Rightarrow$ $\exists$ a number $\lambda \in [m, M]$ such that

$$\int_a^b f\,dx = \lambda\,(b-a) \qquad \text{...(1)}$$

Since f is continuous on $[a, b]$, it attains every value between its bounds m, M.

$\therefore$ $\exists$ a number $c \in [a, b]$ such that $f(c) = \lambda$.

$\therefore$ from (1), $$\int_a^b f\,dx = (b-a)\,f(c).$$

7. **Mean Value Theorem.** *If f and g are integrable on $[a, b]$ and g keeps the same sign over $[a, b]$, then there exists a number λ between the bounds of f on $[a, b]$ such that*

$$\int_a^b fg\,dx = \lambda \int_a^b g\,dx.$$

Proof. Let $g(x) \geq 0\ \forall\, x \in [a, b]$

if m, M are the bounds of f on $[a, b]$, then $m \leq f(x) \leq M$

$$\Rightarrow \qquad m\,g(x) \leq f(x)\,g(x) \leq M\,g(x)$$

$$\Rightarrow \qquad m\int_a^b g\,dx \leq \int_a^b fg\,dx \leq M\int_a^b g\,dx, \text{ if } b \geq a$$

Let λ be a number lying between m and M, then

$$\int_a^b fg\,dx = \lambda \int_a^b g\,dx$$

Similarly, the result can be proved when either $b \leq a$ or $g(x) \leq 0$.

Cor. If in the above theorem, the function f is continuous on $[a, b]$, then $\exists$ a number $c \in [a, b]$ such that

$$\int_a^b fg\,dx = f(c) \int_a^b g\,dx.$$

Example 15. *If f is a function defined by*

$$f(x) = \begin{cases} \dfrac{1}{2^n}, & \dfrac{1}{2^{n+1}} < x \leq \dfrac{1}{2^n}, (n = 0, 1, 2, ...) \\ 0, & x = 0 \end{cases}$$

evaluate $\int_0^1 f\,dx$.

Sol. We have $f(0) = 0$

$$f(x) = 1, \quad \text{where } \frac{1}{2} < x \leq 1 \qquad (\text{Putting } n = 0)$$

$$f(x) = \frac{1}{2}, \quad \text{where } \frac{1}{2^2} < x \leq \frac{1}{2} \qquad (\text{Putting } n = 1)$$

....

$$f(x) = \frac{1}{2^{n-1}}, \text{ where } \frac{1}{2^n} < x \leq \frac{1}{2^{n-1}}$$

Since f is bounded and monotonically increasing, f is Riemann integrable

$$\therefore \quad \int_{1/2^n}^{1} f\,dx = \int_{1/2}^{1} f\,dx + \int_{1/2^2}^{1/2} f\,dx + \int_{1/2^3}^{1/2^2} f\,dx + \ldots\ldots + \int_{1/2^n}^{1/(2^{n-1})} f\,dx$$

$$= \int_{1/2}^{1} dx + \frac{1}{2}\int_{1/2^2}^{\frac{1}{2}} dx + \frac{1}{2^2}\int_{1/2^3}^{1/2^2} dx + \ldots\ldots + \frac{1}{2^{n-1}}\int_{1/2^n}^{1/(2^{n-1})} dx$$

$$= \frac{1}{2} + \frac{1}{2}\left(\frac{1}{2} - \frac{1}{2^2}\right) + \frac{1}{2^2}\left(\frac{1}{2^2} - \frac{1}{2^3}\right) + \ldots\ldots + \frac{1}{2^{n-1}}\left(\frac{1}{2^{n-1}} - \frac{1}{2^n}\right)$$

$$= \frac{1}{2}\left(1 + \frac{1}{2^2} + \left(\frac{1}{2^2}\right)^2 + \ldots\ldots + \left(\frac{1}{2^2}\right)^{n-1}\right)$$

$$= \frac{1}{2}\,\frac{1 - \left(\frac{1}{4}\right)^n}{1 - \frac{1}{4}} = \frac{2}{3}\left(1 - \frac{1}{4^n}\right)$$

Taking limits as $n \to \infty$,

$$\int_0^1 f\,dx = \frac{2}{3}.$$

Example 16. *Applying mean value theorem, show that*

$$x > \log(1 + x) > \frac{x}{1+x} \text{ for } x > 0.$$

Sol. Let $\quad f(x) = \dfrac{1}{1+x}$ and $g(x) = 1$

Then, f and g are both integrable on $[0, x]$.

If m and M are the bounds of f on $[0, x]$, then

$$m = \frac{1}{1+x} \text{ and } M = \frac{1}{1+0} = 1.$$

$\therefore$ by mean value theorem, $\exists\ \lambda$ between the bounds of f such that

$$\int_0^x \frac{1}{1+x}\,dx = \lambda \int_0^x 1\,dx$$

$$\Rightarrow \quad \log(1 + x) = \lambda x \;\Rightarrow\; \lambda = \frac{\log(1+x)}{x}.$$

Since λ lies between the bounds m and M,

$$\therefore \quad \frac{1}{1+x} < \frac{\log(1+x)}{x} < 1 \;\Rightarrow\; \frac{x}{1+x} < \log(1 + x) < x.$$

Example 17. *Prove that* $\dfrac{\pi^2}{9} \le \displaystyle\int_{\pi/6}^{\pi/2} \frac{x}{\sin x}\,dx \le \frac{2\pi^2}{9}.$ (K.U. BCA (II) 2003)

Sol. Let $f(x) = \dfrac{1}{\sin x}$, $g(x) = x$.

Then, f and g are both continuous on $\left[\dfrac{\pi}{6}, \dfrac{\pi}{2}\right]$ and hence integrable. If m and M are bounds of f on $[\pi/6, \pi/2]$, then $m = 1$, $M = 2$.

$\therefore$ by mean value theorem, $\exists\ \lambda$ between the bounds of f such that

$$\int_{\pi/6}^{\pi/2} fg\, dx = \lambda \int_{\pi/6}^{\pi/2} g\, dx$$

i.e.,

$$\int_{\pi/6}^{\pi/2} \frac{x}{\sin x}\, dx = \lambda \int_{\pi/6}^{\pi/2} x\, dx = \lambda \,.\left[\frac{x^2}{2}\right]_{\pi/6}^{\pi/2} = \lambda\, \frac{\pi^2}{9}.$$

$\Rightarrow$

$$\lambda = \frac{9}{\pi^2} \int_{\pi/6}^{\pi/2} \frac{x}{\sin x}\, dx.$$

But, $1 \le \lambda \le 2$ [$\because$ λ lies between the bounds of f]

$\therefore$

$$1 \le \frac{9}{\pi^2} \int_{\pi/6}^{\pi/2} \frac{x}{\sin x}\, dx \le 2$$

or

$$\frac{\pi^2}{9} \le \int_{\pi/6}^{\pi/2} \frac{x}{\sin x}\, dx \le \frac{2\pi^2}{9}.$$

EXERCISE 9 (*c*)

1. Examine whether true or false ?

(*i*) If f is Riemann integrable on $[a, b]$ and if $f(x) = g(x)$ except for a countable number of points $x \in [a, b]$, then g is Riemann integrable.

(*ii*) If f is Riemann integrable on $[a, b]$ and if $f(x) = g(x)$ except for a finite number of points $x \in [a, b]$, then g is Riemann integrable.

2. Using only the results, show that :

$$\int_0^1 (2x^2 - 3x + 5)\, dx = \frac{25}{6}.$$

3. If $0 \le x \le 1$, show that :

$$\frac{x^2}{\sqrt{2}} \le \frac{x^2}{\sqrt{1+x}} \le x^2$$

Hence, prove that :

$$\frac{1}{3\sqrt{2}} \le \int_0^1 \frac{x^2}{\sqrt{1+x^2}}\, dx \le \frac{1}{3}.$$

4. Prove that : $\dfrac{2\pi^2}{9} \le \displaystyle\int_{\pi/6}^{\pi/2} \frac{2x}{\sin x}\, dx \le \frac{4\pi^2}{9}.$ (*BCA (II) 2007*)

[**Hint.** For $\dfrac{\pi}{6} \le x \le \dfrac{\pi}{2}$, $\dfrac{1}{2} = \sin\dfrac{\pi}{6} \le \sin x \le \sin\dfrac{\pi}{2} = 1$

$\Rightarrow$ $1 \le \dfrac{1}{\sin x} \le 2\ \forall\, x \in \left[\dfrac{\pi}{6}, \dfrac{\pi}{2}\right]$

$$\Rightarrow \qquad x \le \frac{x}{\sin x} \le 2x \ \forall\, x \in \left[\frac{\pi}{6}, \frac{\pi}{2}\right].\Bigg]$$

5. Show that a function f defined on [0, 1] by

$$f(x) = \begin{cases} \dfrac{1}{n}, & \dfrac{1}{n+1} < x \le \dfrac{1}{n} \\ 0, & x = 0 \end{cases}$$

is integrable on [0, 1]. Also show that

$$\int_0^1 f\,dx = \frac{\pi^2}{6} - 1.$$

[**Hint.** The function is discontinuous at the points

$1, \frac{1}{2}, \frac{1}{3}$, ..., and the set of points of discontinuity has 0 as the only limit point. So the function is integrable.

Also, $$\int_0^1 f\,dx = \lim_{n\to\infty} \sum_{n=1}^{\infty} \int_{1/(n+1)}^{1/n} \frac{1}{n}\,dx$$

$$= \sum_{n=1}^{\infty} \frac{1}{n}\left(\frac{1}{n} - \frac{1}{n+1}\right) = 1\left(1-\frac{1}{2}\right) + \frac{1}{2}\left(\frac{1}{2}-\frac{1}{3}\right) + \frac{1}{3}\left(\frac{1}{3}-\frac{1}{4}\right) + \ldots\ldots$$

$$= \left(1 + \frac{1}{2^2} + \frac{1}{3^2} + \ldots\ldots\right) - \left(\frac{1}{2} + \frac{1}{2.3} + \frac{1}{3.4} + \ldots\ldots\right)$$

$$= \frac{\pi^2}{6} - \left(\left(1-\frac{1}{2}\right) + \left(\frac{1}{2}-\frac{1}{3}\right) + \left(\frac{1}{3}-\frac{1}{4}\right) + \ldots\ldots\right) \qquad \left(\because \frac{1}{1^2} + \frac{1}{2^2} + \frac{1}{3^2} + \ldots = \frac{\pi^2}{6}\right)$$

$$= \frac{\pi^2}{6} - 1.\Bigg]$$

6. Show that the greatest integer function $f(x) = [x]$ is integrable on [0, 4] and $\int_0^4 [x]\,dx = 6$.

[**Sol.** By definition, $$f(x) = [x] = \begin{cases} 0 \text{ when } 0 \le x < 1 \\ 1 \text{ when } 1 \le x < 2 \\ 2 \text{ when } 2 \le x < 3 \\ 3 \text{ when } 3 \le x < 4 \\ 4 \text{ when } x = 4 \end{cases}$$

Clearly, f is bounded on [0, 4] and has 4 points of discontinuity, namely x = 1, 2, 3, 4.

$\therefore$ f is integrable on [0, 4].

Also $$\int_0^4 [x]\,dx = \int_0^1 [x] + \int_1^2 [x]\,dx + \int_2^3 [x]\,dx + \int_3^4 [x]\,dx + \int_4^4 [x]\,dx$$

$$= \int_0^1 0\,dx + \int_1^2 1\,dx + \int_2^3 2\,dx + \int_3^4 4\,dx + 0$$

$$= 0 + 1 + 3 + 4 = 8]$$

7. Using mean value theorem, prove that :

$$\sin^{-1} x \le \int_0^x \frac{dx}{\sqrt{(1-x^2)(1-\lambda x^2)}} \le \frac{\sin^{-1} x}{\sqrt{1-\lambda x^2}}, \text{ where } 0 < \lambda < 1 \text{ and } 0 < x < 1.$$

$$\left[\textbf{Hint.}\ \text{Here, } f(x) = \frac{1}{\sqrt{1-\lambda x^2}},\ g(x) = \frac{1}{\sqrt{1-x^2}}.\right.$$

f is integrable.

Since g is bounded and monotonically increasing, therefore, g is integrable.

$$m = \text{g.l.b. of } f = 1 \text{ and } M = \text{l.u.b. of } f = \frac{1}{\sqrt{1-\lambda x^2}}.$$

$\therefore$ by mean value theorem,

$$1 . \int_0^x \frac{dx}{\sqrt{1-x^2}} \le \int_0^x \frac{1}{\sqrt{(1-x^2)(1-\lambda x^2)}}\, dx$$

$$\le \frac{1}{\sqrt{1-\lambda x^2}} \int_0^x \frac{dx}{\sqrt{1-x^2}}$$

$$\Rightarrow \qquad \sin^{-1} x \le \int_0^x \frac{dx}{\sqrt{(1-x^2)(1-\lambda x^2)}} \le \left.\frac{\sin^{-1} x}{\sqrt{1-\lambda x^2}}\right]$$

8. Show that : $\int_0^{\pi/2} \sin^{n+1} x\, dx < \int_0^{\pi/2} \sin^n x\, dx$, x being a non-negative integer.

[**Sol.** Since $\sin^n x$, $\sin^{n+1} x$ are both continuous functions on $\left[0, \frac{\pi}{2}\right]$.

$\therefore$ $\int_0^{\pi/2} \sin^n x\, dx$ and $\int_0^{\pi/2} \sin^{n+1} x\, dx$ exist.

Moreover, $\quad \sin^{n+1} x < \sin^n x, \ \forall\, x \in \left(0, \frac{\pi}{2}\right).$

$$\therefore \qquad \int_0^{\pi/2} \sin^{n+1} x\, dx < \int_0^{\pi/2} \sin^n x\, dx, \quad x \in \left(0, \frac{\pi}{2}\right)$$

$$\Rightarrow \qquad \int_0^{\pi/2} \sin^{n+1} x\, dx < \int_0^{\pi/2} \sin^n x\, dx, \quad x \in \left[0, \frac{\pi}{2}\right] \text{ also.} \qquad \left[\because \int_a^a f(x)\, dx = 0\right]\Bigg]$$

9. Show that the function f defined by

$$f(x) = \begin{cases} \frac{1}{2^n}, & \text{when } \frac{1}{2^{n+1}} < x \le \frac{1}{2^n} \\ 0, & \text{when } x = 0 \end{cases} \qquad (n = 0, 1, 2, \ldots\ldots)$$

is integrable on [0, 1] although it has an infinite number of points of discontinuity. Also, evaluate $\int_0^1 f\, dx$.

[**Sol.** Clearly the function f is bounded on [0, 1].

It is also continuous on [0, 1] except at the points $0. \frac{1}{2}, \frac{1}{2^2}, \frac{1}{2^3}, \ldots\ldots$

Since the set $\left[0, \frac{1}{2}, \frac{1}{2^2}, \frac{1}{2^3}, \ldots\ldots\right]$ of points of discontinuity of f on [0, 1] has 0 as the only limit point, the function f is integrable on [0, 1].

For set, see example 14.]

10

Improper Integrals

While defining the definite integral $\int_a^b f\,dx$, it is assumed that the function f is defined at every point of the finite integral $[a, b]$. However, it is possible to relax this restriction on f and on the interval to obtain extensions of the definite integral that are often useful in practical applications. For example, an important concept in electromagnetic field theory is that of the electric potential at a point due to an electric charge distribution. It is known that two like electric charges q_1 and q_2 repel each other with a force F given by

$$F = k\,\frac{q_1 q_2}{r^2}$$

where r is the distance between the charges and k is the constant of proportionality. The electric potential V_P at a point P in an electromagnetic field due to a charge q is defined by the physicist as the work done in bringing a unit charge from '*infinity*' to the point P. Using the definition of work, and letting r be the distance from the charge q, we may express the potential at P by the 'integral'

$$V_P = \int_{\infty}^{x} k\,\frac{q}{r^2}\,dr$$

where the symbol $\int_{\infty}^{x}$ indicates that the charge must 'start' at 'infinity' and be moved to the point x. In this example, the integral is not an ordinary one since we are concerned with an infinite interval. From the mathematician's point of view, this new concept is to be given a meaning which is consistent with that of the definite integral.

10.1. IMPROPER INTEGRAL

10.1.1. Improper Integral. If either a or b or both a and b are infinite or the integrand f becomes infinite on the interval $[a, b]$, then the integral $\int_a^b f\,dx$ is called an *improper integral** or infinite integral or generalized integral.

For example, $\int_1^{\infty} \frac{dx}{x}, \int_{-\infty}^{0} \frac{dx}{1+x^2}, \int_{-\infty}^{\infty} \frac{dx}{1+x^2}, \int_0^2 \frac{dx}{x(2-x)}, \int_{-1}^{\infty} \frac{dx}{x}$

are all improper integrals.

* For the sake of distinction, a definite integral is also called a proper integral.

(*i*) If in $\int_a^b f\,dx$, either a or b or both a and b become infinite, then $\int_a^b f\,dx$ is called an improper integral of first kind.

(*ii*) If f has a point of infinite discontinuity (*i.e.*, f becomes infinite) on $[a, b]$, then $\int_a^b f\,dx$ is called an improper integral of second kind.

If f has both the conditions (*i*) and (*ii*), then $\int_a^b f\,dx$ is also called an improper integral of third kind.

10.2. IMPROPER INTEGRAL AS LIMIT OF A PROPER INTEGRAL

10.2.1. When the improper integral is of first kind, *i.e.*, either a or b or both a and b are infinite, we define

(*i*) $\int_a^\infty f\,dx = \lim_{t\to\infty} \int_a^t f\,dx$, $(t > a)$,

provided the limit exists as a finite number.

If the limit exists as a finite number, we say that the improper integral $\int_a^\infty f\,dx$ is convergent, otherwise it is said to be divergent (or that it does not exist).

(*ii*) $\int_{-\infty}^b f\,dx = \lim_{t\to-\infty} \int_t^b f\,dx$, $(t < b)$, provided the limit exists as a finite number. If the limit exists as a finite number, we say that the improper integral $\int_{-\infty}^b f\,dx$ is convergent, otherwise it is said to be divergent.

(*iii*) $\int_{-\infty}^\infty f\,dx = \int_{-\infty}^c f\,dx + \int_c^\infty f\,dx$, for every c

$$= \lim_{t_1\to-\infty} \int_{t_1}^c f\,dx + \lim_{t_2\to\infty} \int_c^{t_2} f\,dx,$$

provided both the limits exist finitely and independent of each other. If the limits exist, we say that the improper integral $\int_{-\infty}^\infty f\,dx$ is convergent, otherwise it is said to be divergent.

Remark. The statement

$$\int_{-\infty}^\infty f\,dx = \lim_{t_1\to-\infty} \int_{t_1}^c f\,dx + \lim_{t_2\to\infty} \int_c^{t_2} f\,dx$$

is not equivalent to

$$\int_{-\infty}^\infty f\,dx = \lim_{t\to\infty} \left[\int_{-t}^c f\,dx + \int_c^t f\,dx\right].$$

The last limit may exist whereas the first may not.

10.2.2. When the improper integral is of second kind *i.e.*, when the integrand has an infinite discontinuity on $[a, b]$.

(*i*) If f becomes infinite at b only, we define

$$\int_a^b f\,dx = \lim_{\varepsilon\to0^+} \int_a^{b-\varepsilon} f\,dx,$$

provided the limit exists as a finite number. If the limit exists as a finite number, we say that the improper integral $\int_a^b f\,dx$ is convergent, otherwise it is said to be divergent.

(*ii*) If f becomes infinite at $x = a$ only, we define

$$\int_a^b f\,dx = \lim_{\varepsilon \to 0^+} \int_{a+\varepsilon}^b f\,dx,$$

provided the limit exists as a finite number.

(*iii*) If f becomes infinite at some point c only, $a < c < b$, we define

$$\int_a^b f\,dx = \lim_{\varepsilon_1 \to 0^+} \int_a^{c-\varepsilon_1} f\,dx + \lim_{\varepsilon_2 \to 0^+} \int_{c+\varepsilon_2}^b f\,dx,$$

provided both the limits exist finitely and independent of each other.

10.2.3. General case. Let $c_1, c_2, \ldots\ldots, c_n$ be any finite number of points of infinite discontinuity of f on $[a, b]$, where

$$a \le c_1 < c_2 < \ldots\ldots < c_{n-1} < c_n \le b.$$

If the improper integrals :

$$\int_a^{c_1} f\,dx, \int_{c_1}^{c_2} f\,dx, \ldots\ldots, \int_{c_{n-1}}^{c_n} f\,dx, \int_{c_n}^b f\,dx,$$

all exist as finite numbers in accordance with our definitions given above, regarding

$$\int_a^{c_1} f\,dx = 0 \quad \text{if} \quad c_1 = a$$

and
$$\int_{c_n}^b f\,dx = 0 \quad \text{if} \quad c_n = b,$$

then we say that $\int_a^b f\,dx$ exists and we write

$$\int_a^b f\,dx = \int_a^{c_1} f\,dx + \int_{c_1}^{c_2} f\,dx + \ldots\ldots + \int_{c_n}^b f\,dx.$$

Consideration of functions having an infinite number of points of discontinuity is beyond the scope of the book.

Remarks 1. If the discontinuity of f occurs at an end point of the interval of integration, then the point of discontinuity is approached from within the interval, as indicated by requiring that $\varepsilon \to 0^+$. If the discontinuity occurs at an interior point of the interval, then the two integrals on right hand side must converge independently before we say that $\int_a^b f\,dx$ converges.

2. A proper integral is always convergent.

3. If $\int_a^b f\,dx$ is convergent at a and k is a constant, then

(*i*) $\int_a^c f\,dx$, $(a < c < b)$ is convergent at a.

(*ii*) $\int_a^b k\,f dx$ is convergent at a.

Example 1. *Examine the convergence of the improper integrals*

(*i*) $\int_1^\infty xe^{-x}\, dx$ (*ii*) $\int_{-\infty}^\infty \frac{x}{1+x^2}\, dx$.

If convergent, find its value.

Sol. (*i*) By definition,

$$\int_1^\infty xe^{-x}\, dx = \lim_{t\to\infty} \int_1^t xe^{-x}\, dx \qquad \textbf{(Integrating by parts)}$$

$$= \lim_{t\to\infty} [-te^{-t} - e^{-t} + 2e^{-1}]$$

$$= \lim_{t\to\infty} \left(\frac{-t}{e^t}\right) - \lim_{t\to\infty} e^{-t} + \frac{2}{e}$$

(Applying L' Hospital's Rule to first limit)

$$= \lim_{t\to\infty} \left(\frac{-1}{e^t}\right) - \lim_{t\to\infty} \frac{1}{e^t} + \frac{2}{e}$$

$$= 0 - 0 + \frac{2}{e} = \frac{2}{e} \qquad \left[\because \frac{1}{e^t} \to 0 \text{ as } t\to\infty\right]$$

which exists as a finite number.

Hence, the given integral is convergent and $\frac{2}{e}$ is its value.

(*ii*) To examine the convergence, we must select a convenient number a (say) and examine the convergence of two improper integrals

$$\int_{-\infty}^a f\, dx \quad \text{and} \quad \int_a^\infty f\, dx.$$

Let us take $a = 0$.

Then,
$$\int_{-\infty}^\infty \frac{x\, dx}{x^2+1} = \int_{-\infty}^0 \frac{x}{x^2+1}\, dx + \int_0^\infty \frac{x\, dx}{x^2+1}$$

Now,
$$\int_0^\infty \frac{x}{x^2+1}\, dx = \lim_{t\to\infty} \int_0^t \frac{x}{x^2+1}\, dx = \lim_{t\to\infty} \frac{1}{2}\int_0^t \frac{2x}{x^2+1}\, dx$$

$$= \lim_{t\to\infty} \left[\frac{1}{2}\log(x^2+1)\right]_0^t = \lim_{t\to\infty} \frac{1}{2}\log(t^2+1) = \infty$$

Since the limit does not exist as a finite number,

$$\int_0^\infty \frac{x}{x^2+1}\, dx$$

is divergent and hence $\int_{-\infty}^\infty \frac{x}{x^2+1}\, dx$ is divergent.

Example 2. *Evaluate the improper integrals*

(*i*) $\int_1^2 \frac{dx}{\sqrt{x-1}}$ (*ii*) $\int_0^2 \frac{dx}{x(2-x)}$ (*iii*) $\int_a^{3a} \frac{dx}{(x-2a)^2}$,

if convergent.

Sol. (*i*) 1 is the only point of infinite discontinuity of the integrand on [1, 2].

$$\therefore \quad \int_1^2 \frac{dx}{\sqrt{x-1}} = \lim_{\varepsilon \to 0^+} \int_{1+\varepsilon}^2 \frac{dx}{\sqrt{x-1}}$$

$$= \lim_{\varepsilon \to 0^+} \left[2\sqrt{x-1}\right]_{1+\varepsilon}^2 = 2 - \lim_{\varepsilon \to 0^+} 2\sqrt{\varepsilon} = 2$$

Hence, the integral converges to the value 2.

(*ii*) Both the end points 0, 2 are points of infinite discontinuity of the integrand and are infact the only such points on [0, 2].

Take a point 1 (say) on [0, 2].

$$\text{Then,} \quad \int_0^2 \frac{dx}{x(2-x)} = \lim_{\varepsilon_1 \to 0^+} \int_{\varepsilon_1}^1 \frac{dx}{x(2-x)} + \lim_{\varepsilon_2 \to 0^+} \int_1^{2-\varepsilon_2} \frac{dx}{x(2-x)}$$

$$= \frac{1}{2} \lim_{\varepsilon_1 \to 0^+} \left[\log \frac{x}{2-x}\right]_{\varepsilon_1}^1 + \frac{1}{2} \lim_{\varepsilon_2 \to 0^+} \left[\log \frac{x}{2-x}\right]_1^{2-\varepsilon_2}$$

$$= -\frac{1}{2} \lim_{\varepsilon_1 \to 0^+} \log \frac{\varepsilon_1}{2-\varepsilon_1} + \frac{1}{2} \lim_{\varepsilon_2 \to 0^+} \log \frac{2-\varepsilon_2}{\varepsilon_2}$$

$$= \infty.$$

Thus, the given integral diverges.

(*iii*) Here, $x = 2a$, $(a < 2a < 3a)$ is a point of infinite discontinuity of the integrand.

$$\therefore \quad \int_a^{3a} \frac{dx}{(x-2a)^2} = \lim_{\varepsilon_1 \to 0^+} \int_a^{2a-\varepsilon_1} \frac{dx}{(x-2a)^2} + \lim_{\varepsilon_2 \to 0^+} \int_{2a+\varepsilon_2}^{3a} \frac{dx}{(x-2a)^2}$$

$$= \lim_{\varepsilon_1 \to 0^+} \left[-\frac{1}{x-2a}\right]_a^{2a-\varepsilon_1} + \lim_{\varepsilon_2 \to 0^+} \left[-\frac{1}{x-2a}\right]_{2a+\varepsilon_2}^{3a}$$

$$= \lim_{\varepsilon_1 \to 0^+} \left[\frac{1}{\varepsilon_1} - \frac{1}{a}\right] + \lim_{\varepsilon_2 \to 0^+} \left[-\frac{1}{a} + \frac{1}{\varepsilon_2}\right]$$

Since the limits do not exist finitely, the integral is not convergent.

EXERCISE 10 (*a*)

1. Examine the convergence of the following improper integrals :

(*i*) $\int_1^\infty \frac{1}{x}\, dx$ (*ii*) $\int_0^\infty e^{-mx}\, dx \; (m > 0)$

(*iii*) $\int_0^\infty x e^{-x^2}\, dx$ (*iv*) $\int_0^\infty \frac{dx}{x^2 + 4a^2}$

(*v*) $\int_0^\infty \frac{dx}{(1+x)^3}\, dx$ (*vi*) $\int_{-\infty}^0 e^{2x}\, dx$

(*vii*) $\int_{-\infty}^0 \frac{dx}{p^2 + q^2 x^2}$ (*viii*) $\int_{-\infty}^\infty \frac{dx}{1+x^2}$

(ix) $\int_{-\infty}^{\infty} \frac{dx}{e^x + e^{-x}}$ (x) $\int_0^{\infty} e^{-x} \cos x \, dx$.

2. Test the improper integral for convergence and evaluate the integral, if it converges :

(i) $\int_0^1 \frac{1}{x^2} dx$ (ii) $\int_0^1 \frac{dx}{\sqrt{1-x}}$

(iii) $\int_{-a}^{a} \frac{x \, dx}{a\sqrt{a^2 - x^2}}$ (iv) $\int_0^{\pi/2} \frac{\cos x}{\sqrt{1 - \sin x}} dx$

(v) $\int_0^1 \frac{dx}{\sqrt{x}}$ (vi) $\int_0^1 \log x \, dx$

(vii) $\int_{-1}^1 \frac{1}{x^2} dx$ (viii) $\int_0^{1/e} \frac{dx}{x(\log x)^2}$.

3. Show that for every value of a, $\int_a^{\infty} \frac{x}{x^2+1} dx$ diverges.

4. Verify that $\lim_{t\to\infty} \left\{ \int_{-t}^{a} \frac{x}{x^2+1} dx + \int_a^t \frac{x}{x^2+1} dx \right\} = 0$ but $\int_{-\infty}^{\infty} \frac{x}{x^2+1} dx$ diverges.

In the examples considered so far, the improper integrals are such that the integrands admit of primitives in terms of elementary functions. In such cases the examination of the convergence of integrals is generally easy. But, it is not uncommon to encounter an improper integral of a function that does not possess a primitive in terms of elementary functions. For example, in Statistics, an integral of the form

$$\int_0^{\infty} e^{-x^2} dx$$

often occurs. This integral cannot be examined for convergence by the procedure discussed so far. Consequently, some more advanced methods are necessary for testing the convergence of such integrals.

10.3. TEST FOR CONVERGENCE AT a OF $\int_a^b f \, dx$

Let a be the only point of infinite discontinuity of f on $[a, b]$.
If the integrand f is positive in (a, c) for some $c > 0$, then

$$\int_{a+\varepsilon}^{b} f \, dx = \int_{a+\varepsilon}^{c} f \, dx + \int_c^b f \, dx$$

This implies that $\int_a^c f \, dx$ and $\int_a^b f \, dx$ are either both convergent at a or both non-convergent. Without any loss of generality, we assume that f is positive on $[a, b]$.

In case f is negative, we can replace it by $-f$, for testing the convergence of $\int_a^b f \, dx$.
The case $f = 0$ is trivial.

10.3.1. Theorem. *A necessary and sufficient condition for the convergence of the improper integral* $\int_a^b f\,dx$ *at a, where f is positive on (a, b], is that there exists a positive number k, independent of* $\varepsilon > 0$ *such that*

$$\int_{a+\varepsilon}^{b} f\,dx < k,\ \forall\ \varepsilon \in (0, b-a).$$

Proof. Since f is positive on $(a, b]$, $\int_{a+\varepsilon}^{b} f\,dx$ is a positive function of ε and is monotonically increasing as ε decreases and will therefore, tends to a finite limit if and only if it is bounded above, *i.e.*, there exists a positive number k independent of ε such that

$$\int_{a+\varepsilon}^{b} f\,dx < k.$$

Note. *If no such k exists, the monotonically increasing function*

$$\int_{a+\varepsilon}^{b} f\,dx$$

is not bounded above and hence tends to ∞ *as* $\varepsilon \to 0^+$, *i.e., the improper integral diverges to* ∞.

10.3.2. Theorem (*Comparison Test I*)

If f and g are two functions integrable on

$$[a+\varepsilon, b],\ (0 < \varepsilon \le b-a),$$

a being the only point of infinite discontinuity, such that

$$0 \le f(x) \le g(x),\quad \forall\ x \in (a, b],\ \text{then}$$

(*i*) $\int_a^b g\,dx$ *is convergent* $\Rightarrow$ $\int_a^b f\,dx$ *is convergent*

(*ii*) $\int_a^b f\,dx$ *is divergent* $\Rightarrow$ $\int_a^b g\,dx$ *is divergent.*

Proof. Since $0 \le f(x) \le g(x),\ \forall\ x \in (a, b]$

$$\therefore \quad \int_{a+\varepsilon}^{b} f\,dx \le \int_{a+\varepsilon}^{b} g(dx),\ \forall\ \varepsilon \in (0, b-a] \qquad ...(1)$$

(*i*) Let $\int_a^b g\,dx$ be convergent at a.

Then, there exists a positive number k such that

$$\int_{a+\varepsilon}^{b} g\,dx < k,\quad \forall\ \varepsilon \in (0, b-a] \qquad ...(2)$$

From (1) and (2),

$$\int_{a+\varepsilon}^{b} f\,dx < k,\quad \forall\ \varepsilon \in (0, b-a]$$

$$\Rightarrow \quad \int_a^b f\,dx \text{ is convergent at } a.$$

(*ii*) Let $\int_a^b f\,dx$ be divergent at a. Then, $\int_{a+\varepsilon}^{b} f\,dx$ is unbounded above and hence from (1) above, $\int_{a+\varepsilon}^{b} g\,dx$ is unbounded above. Hence, $\int_a^b g\,dx$ is divergent at a.

10.3.3. Theorem (*Comparison Test II—Practical form*). *If f and g are two positive functions on (a, b], a being the only point of infinite discontinuity, such that* $\lim_{x \to a^+} \frac{f(x)}{g(x)} = l \; (\neq 0, \infty)$, *then the two integrals*

$$\int_a^b f\,dx \quad and \quad \int_a^b g\,dx$$

converge or diverge together at a.

Proof. Since $\frac{f(x)}{g(x)} > 0 \; \forall \; x, l$ cannot be negative.

$\therefore \qquad l > 0.$

Choose a positive number ε such that $l - \varepsilon > 0$.

Since $\lim_{x \to a^+} \frac{f(x)}{g(x)} = l$, therefore, there exists a neighbourhood (a, c), $(a < c < b)$, such that

$$\left| \frac{f(x)}{g(x)} - l \right| < \varepsilon, \quad \forall \; x \in (a, c)$$

$$\Rightarrow \qquad (l - \varepsilon)\, g(x) < f(x) < (l + \varepsilon)\, g(x), \quad \forall \; x \in (a, c) \qquad \text{...(1)}$$

Now, $\int_a^b f\,dx$ converges at a.

$\Rightarrow \quad \int_a^c f\,dx$ converges at a.

$\Rightarrow \quad (l - \varepsilon) \int_a^c g\,dx$ converges at a. (Using (1) and the comparison test I)

$\Rightarrow \quad \int_a^b g\,dx$ converges at a

Again, from second part of the inequality in (1),

$$f(x) < (l + \varepsilon)\, g(x), \quad \forall \; x \in (a, c)$$

$\Rightarrow \quad \int_a^b f\,dx$ diverges at $a \Rightarrow \int_a^c f\,dx$ diverges at a

$\Rightarrow \quad (l + \varepsilon) \int_a^c g\,dx$ diverges at a (comparison test)

$\Rightarrow \quad \int_a^b g\,dx$ diverges at a.

It may similarly be proved that

$$\int_a^b g\,dx \text{ converges at } a \;\Rightarrow\; \int_a^b f\,dx \text{ converges at } a$$

and

$$\int_a^b g\,dx \text{ diverges at } a \;\Rightarrow\; \int_a^b f\,dx \text{ diverges at } a.$$

Remark. One can easily show that

(*i*) if $\lim_{x \to a^+} \frac{f(x)}{g(x)} = 0$ and $\int_a^b g\,dx$ converges, then $\int_a^b f\,dx$ converges and (*ii*) if $\lim_{x \to a^+} \frac{f(x)}{g(x)} = \infty$ and $\int_a^b g\,dx$ diverges, then

$\int_a^b f\,dx$ also diverges.

10.4. USEFUL COMPARISON INTEGRALS

10.4.1. (*a*) *The improper integral* $\int_a^b \frac{dx}{(x-a)^n}$ *is convergent if and only if* $n < 1$.

(*b*) *The improper integral* $\int_a^b \frac{dx}{(b-x)^n}$ *is convergent if and only if* $n < 1$.

Proof. (*a*) It is a proper integral if $n \le 0$.

If $n > 0$, it is an improper integral having a as the only point of infinite discontinuity of the integrand.

Case (*i*) n ≠ 1

$$\int_a^b \frac{dx}{(x-a)^n} = \lim_{\varepsilon \to 0^+} \int_{a+\varepsilon}^b \frac{dx}{(x-a)^n},\ \varepsilon \in (0, b-a)$$

$$= \lim_{\varepsilon \to 0^+} \left[\frac{(x-a)^{-n+1}}{-n+1}\right]_{a+\varepsilon}^b = \lim_{\varepsilon \to 0^+} \frac{1}{1-n}\left[\frac{1}{(b-a)^{n-1}} - \frac{1}{\varepsilon^{n-1}}\right]$$

$$= \begin{cases} \dfrac{1}{(1-n)(b-a)^{n-1}}, & \text{if } n < 1 \\ \infty, & \text{if } n > 1 \end{cases}.$$

Case (*ii*) n = 1

$$\int_a^b \frac{dx}{(x-a)^n} = \int_a^b \frac{dx}{x-a} = \lim_{\varepsilon \to 0^+} \int_{a+\varepsilon}^b \frac{dx}{x-a}$$

$$= \lim_{\varepsilon \to 0^+} [\log(x-a)]_{a+\varepsilon}^b = \lim_{\varepsilon \to 0^+} [\log(b-a) - \log \varepsilon] = \infty$$

Thus, $\int_a^b \frac{dx}{(x-a)^n}$ is convergent only if $n < 1$.

(*b*) The proof is similar to that of part (*a*) and is left as an exercise for the reader.

10.4.2. Some more tests for testing the convergence of improper integrals. Combining the above integrals with the comparison tests, we can have the following forms of comparison tests which are more useful in solving questions :

1. If f is positive in a neighbourhood of a, then the integral $\int_a^b f\,dx$ converges at a if there exists a number n, $(0 < n < 1)$ and a fixed positive number k such that

$$f(x) \le \frac{k}{(x-a)^n},\quad \forall\, x \in (a, b].$$

Also, $\int_a^b f\,dx$ diverges if there exists a number $n \ge 1$ and a fixed positive number k such that

$$f(x) \ge \frac{k}{(x-a)^n},\quad \forall\, x \in (a, b].$$

2. *If 'a' is the only point of infinite discontinuity of f on [a, b] and* $\lim_{x \to a} [(x-a)^n f(x)]$ *exists and is non-zero finite, then* $\int_a^b f\,dx$ *converges if and only if n < 1.*

[**Proof.** Take $g(x) = \dfrac{1}{(x-a)^n}$.

$$\therefore \quad \lim_{x \to a^+} \frac{f(x)}{g(x)} = \lim_{x \to a^+} \frac{f(x)}{\dfrac{1}{(x-a)^n}} = \lim_{x \to a^+} [(x-a)^n f(x)],$$

which is finite non-zero.

$\therefore$ by comparison test $\int_a^b f\,dx$ and $\int_a^b g\,dx$ converge or diverge together. But, $\int_a^b g\,dx = \int_a^b \dfrac{1}{(x-a)^n}\,dx$ converges if and only if $n < 1$.

Hence $\int_a^b f\,dx$ converges if $n < 1$.]

3. *If b is the only point of infinite discontinuity of f on [a, b] and* $\lim_{x \to b^-} (b-x)^n f(x)$ *exists and is non-zero finite, then* $\int_a^b f\,dx$ *converges if and only if n < 1.*

Example 3. *Test the convergence of the improper integrals*

(*i*) $\displaystyle\int_0^1 \frac{dx}{\sqrt{1-x^3}}$ (*ii*) $\displaystyle\int_0^{\pi/2} \frac{\sin x}{x^p}\,dx$.

Sol. (*i*) Here, $$f(x) = \frac{1}{\sqrt{1-x^3}} = \frac{1}{\sqrt{(1-x)(1+x+x^2)}}$$

$\dfrac{1}{\sqrt{1+x+x^2}}$ being bounded above by k (say),

$$\therefore \quad f(x) \le \frac{k}{(1-x)^{1/2}}$$

But, $\displaystyle\int_0^1 \frac{dx}{(1-x)^{1/2}}$ is convergent $\quad (\because \; n = \frac{1}{2} < 1)$

$\therefore$ by comparison test, $\displaystyle\int_0^1 \frac{dx}{\sqrt{1-x^3}}$ is convergent.

(*ii*) $\displaystyle\int_0^{\pi/2} \frac{\sin x}{x^p}\,dx$

If $p \le 0$, then the given integral is a proper integral and hence convergent.

So, let $p > 0$.

Here, 0 is the only point of infinite discontinuity.

Also, $$\frac{\sin x}{x^p} = \frac{1}{x^{p-1}} \cdot \frac{\sin x}{x}$$

Since $$\frac{\sin x}{x} \le 1,$$

$$\therefore \quad \frac{\sin x}{x^p} \le \frac{1}{x^{p-1}} \qquad (p > 1)$$

But, $\int_0^{\pi/2} \frac{dx}{x^{p-1}}$ is convergent if $p - 1 < 1$ or $p < 2$.

$\therefore$ by comparison test, $\int_0^{\pi/2} \frac{\sin x}{x^p}\, dx$ is convergent if $p < 2$ and divergent if $p \ge 2$.

Example 4. *Examine the convergence of*

(*i*) $\int_0^1 \frac{dx}{x^{1/2}(1+x^2)}$ (*ii*) $\int_0^1 \frac{dx}{x^3(2+x^2)^5}$

(*iii*) $\int_0^1 \frac{dx}{x^{1/3}(1-x)^{1/2}}$.

Sol. All integrands are positive on (0, 1].

(*i*) Here, 0 is the only point of infinite discontinuity

$$f(x) = \frac{1}{x^{1/2}(1+x^2)}$$

Take, $$g(x) = \frac{1}{x^{1/2}}$$

Now, $$\lim_{x \to 0} \frac{f(x)}{g(x)} = \lim_{x \to 0} \frac{1}{1+x^2} = 1 \ (\ne 0, \infty)$$

$\therefore$ by comparison test,

$\int_0^1 f\, dx$ and $\int_0^1 g\, dx$ converge or diverge together.

But, $\int_0^1 g\, dx$ is convergent $(\because \ n = \frac{1}{2} < 1)$

$\therefore \quad \int_0^1 f\, dx$ is convergent.

(*ii*) Here, 0 is the only point of infinite discontinuity.

$$f(x) = \frac{1}{x^3(2+x^2)^5}$$

Take, $$g(x) = \frac{1}{x^3}.$$

Now, $$\lim_{x \to 0} \frac{f(x)}{g(x)} = \lim_{x \to 0} \frac{1}{(2+x^2)^5} = \frac{1}{2^5} \ (\ne 0, \infty)$$

$\therefore$ by comparison test,

$\int_0^1 f\, dx$ and $\int_0^1 g\, dx$ converge or diverge together.

But, $\int_0^1 g\,dx$ is divergent. $(\because \quad n = 3 > 1)$

$\therefore \quad \int_0^1 f\,dx$ is divergent.

(*iii*) Here, 0 and 1 are the only points of infinite discontinuity.

$$f(x) = \frac{1}{x^{1/3}(1-x)^{1/2}}.$$

Choose a number $\frac{1}{2}$ (say) between 0 and 1, and examine the convergence behaviour of the improper integrals

$$\int_0^{1/2} f\,dx \text{ and } \int_{1/2}^1 f\,dx \text{ at 0 and 1 respectively.}$$

To examine the convergence of $\int_0^{1/2} f\,dx$

Take
$$g(x) = \frac{1}{x^{1/3}}$$

$$\lim_{x\to 0} \frac{f(x)}{g(x)} = \lim_{x\to 0} \frac{1}{(1-x)^{1/2}} = 1 \qquad (\neq 0, \infty)$$

$\therefore$ by comparison test, the integrals $\int_0^{1/2} f\,dx$ and $\int_0^{1/2} g\,dx$ converge or diverge together.

But, $\int_0^{1/2} g\,dx$ is convergent $(\because \quad n = \frac{1}{3} < 1)$

$\therefore \quad \int_0^{1/2} f\,dx$ is convergent.

To examine the convergence of $\int_{1/2}^1 f\,dx$

Take
$$g(x) = \frac{1}{(1-x)^{1/2}}$$

$$\therefore \quad \lim_{x\to 1} \frac{f(x)}{g(x)} = \lim_{x\to 1} \frac{1}{x^{1/3}} = 1 \qquad (\neq 0, \infty)$$

$\therefore$ by comparison test, $\int_{1/2}^1 f\,dx$

and $\int_{1/2}^1 g\,dx$ converge or diverge together.

But, $\int_{1/2}^1 g\,dx = \int_{1/2}^1 \frac{dx}{(1-x)^{1/2}}$ is convergent. $(\because \quad n = \frac{1}{2} < 1)$

$\Rightarrow \quad \int_{1/2}^1 f\,dx$ is convergent.

Hence, $\int_0^1 f\,dx$ is convergent.

Example 5. *Show that*

(i) $\int_0^1 \frac{\log x}{\sqrt{x}}\, dx$ *is convergent* *but* *(ii)* $\int_0^2 \frac{\sqrt{x}}{\log x}\, dx$ *is divergent.*

Sol. (*i*) Since $\frac{\log x}{\sqrt{x}}$ is negative on (0, 1), we take

$$f(x) = -\frac{\log x}{\sqrt{x}} = \frac{\log \frac{1}{x}}{\sqrt{x}}$$

Here, 0 is the only point of infinite discontinuity.

Take $$g(x) = \frac{1}{x^{3/4}}.$$

$$\therefore \quad \lim_{x \to 0^+} \frac{f(x)}{g(x)} = \lim_{x \to 0^+} \frac{\log \frac{1}{x}}{x^{-1/4}} \left(\frac{\infty}{\infty}\right) \quad \text{(Applying L' Hospital's Rule)}$$

$$= \lim_{x \to 0^+} 4x^{1/4} = 0$$

The integral $\int_0^1 g\, dx = \int_0^1 \frac{dx}{x^{3/4}}$ is convergent at $x = 0$ $(\because \ n < 1)$

$$\therefore \quad \int_0^1 f\, dx = \int_0^1 \frac{\log \frac{1}{x}}{\sqrt{x}}\, dx$$

and hence $\int_0^1 \frac{\log x}{\sqrt{x}}\, dx$ is convergent.

(*ii*) Here, 1 is the only point of infinite discontinuity

$$f(x) = \frac{\sqrt{x}}{\log x}.$$

Take $$g(x) = \frac{1}{x-1} \text{ so that}$$

$$\lim_{x \to 1^+} \frac{f(x)}{g(x)} = \lim_{x \to 1^+} \frac{(x-1)\sqrt{x}}{\log x} \left(\frac{0}{0}\right)$$

$$= \lim_{x \to 1^+} \frac{\frac{3}{2}x^{1/2} - \frac{1}{2}x^{-1/2}}{\frac{1}{x}} = 1 \quad (\neq 0, \infty)$$

But, $\int_1^2 \frac{dx}{x-1}$ is divergent. $(\because \ n = 1)$

$\therefore$ by comparison test, $\int_1^2 \frac{\sqrt{x}}{\log x}\, dx$ is divergent.

Example 6. *Show that* $\int_0^{\pi/2} x^m \operatorname{cosec}^n x\, dx$ *exists if and only if* $n < (m + 1)$.

Sol. $$f(x) = x^m \operatorname{cosec}^n x = \frac{x^m}{\sin^n x}$$

$$= \left(\frac{x}{\sin x}\right)^n . x^{m-n} = \left(\frac{x}{\sin x}\right)^n \frac{1}{x^{n-m}}$$

Now,
$$\lim_{x \to 0} f(x) = \begin{cases} 0 & \text{if } m - n > 0 \\ 1 & \text{if } m = n \\ \infty & \text{if } m - n < 0 \end{cases}$$

$\therefore$ $\int_0^{\pi/2} x^m \operatorname{cosec}^n x \, dx$ is a proper integral if $m - n \geq 0$ and an improper integral if $m - n < 0$; 0 being the only point of infinite discontinuity of the integrand in this case.

Let $n - m > 0$.

Take
$$g(x) = \frac{1}{x^{n-m}}.$$

Then,
$$\lim_{x \to 0} \frac{f(x)}{g(x)} = \lim_{x \to 0} \left(\frac{x}{\sin x}\right)^n = 1.$$

Also,
$$\int_0^{\pi/2} g \, dx = \int_0^{\pi/2} \frac{1}{x^{n-m}} \, dx$$

is convergent if and only if $n - m < 1$, *i.e.*, $n < m + 1$.

$\therefore$ by comparison test, the given integral is convergent if and only if $n < m + 1$, which also includes the case $n \leq m$ when the integral is proper.

Example 7. (Beta Function). *Show that*

$$\int_0^1 x^{m-1} (1 - x)^{n-1} \, dx$$

exists if and only if m and n are both positive.

Sol. The integral is proper if $m \geq 1$ and $n \geq 1$.

0 is the only point of infinite discontinuity if $m < 1$

and 1 is the only point of infinite discontinuity if $n < 1$.

For m < 1 and n < 1.

Take a number $\frac{1}{2}$ (say) between 0 and 1 and examine the convergence of the improper integrals

$$\int_0^{1/2} x^{m-1} (1 - x)^{n-1} \, dx,$$

$$\int_{1/2}^1 x^{m-1} (1 - x)^{n-1} \, dx$$

at 0 and 1 respectively.

Convergence at 0, when m < 1

$$f(x) = x^{m-1} (1 - x)^{n-1} = \frac{(1 - x)^{n-1}}{x^{1-m}}$$

Take,
$$g(x) = \frac{1}{x^{1-m}}$$

Then,
$$\lim_{x \to 0} \frac{f(x)}{g(x)} = \lim_{x \to 0} (1 - x)^{n-1} = 1 \qquad (\neq 0, \infty)$$

Also, $$\int_0^{1/2} g\,dx = \int_0^{1/2} \frac{dx}{x^{1-m}} = \int_0^{1/2} \frac{dx}{(x-0)^{1-m}}$$

is convergent if and only if $1 - m < 1$, *i.e.*, $m > 0$

$\therefore$ by comparison test,

$$\int_0^{1/2} f\,dx = \int_0^{1/2} x^{m-1}(1-x)^{n-1}\,dx$$

is convergent at $x = 0$ if $m > 0$.

Convergence at 1, when n < 1

$$f(x) = x^{m-1}(1-x)^{n-1} = \frac{x^{m-1}}{(1-x)^{1-n}}$$

Take $$g(x) = \frac{1}{(1-x)^{1-n}}$$

Then, $$\lim_{x\to 1} \frac{f(x)}{g(x)} = \lim_{x\to 1} x^{m-1} = 1.$$

But, $$\int_{1/2}^{1} g\,dx = \int_{1/2}^{1} \frac{1}{(1-x)^{1-n}}\,dx$$

is convergent if and only if $1 - n < 1$ *i.e.*, $n > 0$.

$\therefore$ by comparison test,

$$\int_{1/2}^{1} f\,dx = \int_{1/2}^{1} x^{m-1}(1-x)^{n-1}\,dx$$

is convergent at 1 if $n > 0$.

Hence, $\int_0^1 x^{m-1}(1-x)^{n-1}\,dx$ exists for positive values of m and n only.

Example 8. *Examine the convergence of*

$$\int_0^1 x^{n-1} \log x\,dx.$$

Sol. Since $x^{n-1}\log x$ is negative in (0, 1),

let $$f(x) = -x^{n-1}\log x = x^{n-1}\log\left(\frac{1}{x}\right)$$

The integral $\int_0^1 x^{n-1}\log\frac{1}{x}\,dx$ is proper if $n - 1 > 0$ *i.e.*, $n > 1$, because the integrand tends to 0 as $x \to 0$ and hence 0 is not a point of infinite discontinuity.

Let n ≤ 1

Case (*i*) n = 1

$$\int_0^1 x^{n-1}\log x\,dx = \int_0^1 \log x\,dx$$

$$= \lim_{\varepsilon\to 0^+} \int_\varepsilon^1 \log x\,dx \qquad \text{(Integrating by parts)}$$

$$= \lim_{\varepsilon \to 0^+} \Big[\log x \,.\, x - x \Big]_{\varepsilon}^{1}$$

$$= \lim_{\varepsilon \to 0^+} [0 - 1 - \varepsilon \log \varepsilon + \varepsilon] = -1 \qquad \left(\because \lim_{\varepsilon \to 0^+} \varepsilon \log \varepsilon = 0\right).$$

Hence, the given integral is convergent when $n = 1$.

Case (*ii*) n < 1

$$f(x) = -x^{n-1} \log x.$$

Take $$g(x) = \frac{1}{x^{\lambda}}$$

$$\therefore \quad \lim_{x \to 0^+} \frac{f(x)}{g(x)} = \lim_{x \to 0^+} (-x^{\lambda + n - 1} \log x)$$

$$= \begin{cases} 0 & \text{if } \lambda + n - 1 > 0, \text{ i.e., if } \lambda > 1 - n \\ \infty & \text{if } \lambda + n - 1 < 0, \text{ i.e., if } \lambda < 1 - n \end{cases}.$$

But, $\int_0^1 g \, dx = \int_0^1 \frac{1}{x^{\lambda}}$ is convergent if $\lambda < 1$.

$\therefore \quad \int_0^1 f \, dx$ is convergent if $\lambda < 1$, *i.e.*, $1 - n < \lambda < 1$ *i.e.*, $n > 0$.

$\therefore \quad$ for $0 < n < 1$, $\int_0^1 f \, dx$ is convergent.

Consider n < 0.

Taking $\lambda = 1$, $$\int_0^1 g \, dx = \int_0^1 \frac{1}{x} dx = \lim_{\varepsilon \to 0^+} \int_{\varepsilon}^1 \frac{1}{x} dx$$

$$= \lim_{\varepsilon \to 0} \Big[\log x \Big]_{\varepsilon}^{1} = \lim_{\varepsilon \to 0} (-\log \varepsilon) = \infty.$$

$\therefore \quad \int_0^1 g \, dx$ diverges.

But, $$\lim_{x \to 0} \frac{f(x)}{g(x)} = \infty \quad \text{if } \lambda < 1 - n$$

i.e., $\quad$ if $1 < 1 - n$

i.e., $\quad$ if $n < 0$

$\therefore \quad \int_0^1 f \, dx$ diverges.

EXERCISE 10 (*b*)

1. Examine the convergence of

(*i*) $\int_0^1 \frac{x^n}{1 - x} dx$ (*ii*) $\int_0^1 \frac{x^n}{1 + x} dx$

(*iii*) $\int_0^1 \frac{dx}{x^{1/3}(1 + x^2)}$ (*iv*) $\int_0^1 \frac{dx}{x^2(2 + x^2)}$ (*BCA (II) 2007*)

(v) $\int_0^1 \frac{dx}{x^{1/2}(1-x)^{1/3}}$ (vi) $\int_0^1 \frac{dx}{\sqrt{x(1-x)}}$.

2. Find the values of m and n for which the following integrals converge :

$$\int_0^1 e^{-mx} x^n \, dx.$$

[**Hint.** For each m, the integral is proper for $n \geq 0$.
If $n < 0$, the integral is improper and 0 is the only point of infinite discontinuity.

Let $$f(x) = e^{-mx} x^n \text{ and } g(x) = x^n = \frac{1}{x^{-n}}$$

$$\therefore \quad \lim_{x \to 0^+} \frac{f(x)}{g(x)} = \lim_{x \to 0} e^{-mx} = 1.$$

Hence $\int_0^1 f\, dx$ and $\int_0^1 g\, dx$ converge and diverge together.

But, $\int_0^1 g\, dx$ is convergent if $-n < 1$ *i.e.*, $n > -1$.

Hence, the given integral is convergent for $n > -1$ irrespective of values of m.]

3. Show that $\int_0^{\pi/2} \frac{\sin^m x}{x^n}$ exists if and only if $n < m + 1$.

[**Hint.** $f(x) = \frac{\sin^m x}{x^n} = \left(\frac{\sin x}{x}\right)^m \cdot \frac{1}{x^{n-m}}$.

$$\therefore \quad \lim_{x \to 0^+} f(x) = \lim_{x \to 0^+} \left(\frac{\sin x}{x}\right)^m \cdot \frac{1}{x^{n-m}}$$

$$= \lim_{x \to 0^+} \frac{1}{x^{n-m}} = \begin{cases} 0 & \text{if } n - m < 0 \\ 1 & \text{if } n - m = 0 \\ \infty & \text{if } n - m > 0 \end{cases}$$

So, 0 is the point of infinite discontinuity of f when $n > m$.

Consider n > m

Take $$g(x) = \frac{1}{x^{n-m}}$$

Then, $\lim_{x \to 0^+} \frac{f(x)}{g(x)} = 1$ (finite and non-zero) (Show it)

$\therefore \int_0^{\pi/2} f\, dx$ and $\int_0^{\pi/2} g\, dx$ converge or diverge together but $\int_0^{\pi/2} g\, dx$ is convergent if $n - m < 1$.

Hence, $\int_0^{\pi/2} f\, dx$ is convergent if $n - m < 1$.]

4. Show that $\int_0^{\pi/2} (\sin x) \log \sin x\, dx$ is convergent with value $\log\left(\frac{2}{e}\right)$.

[**Hint.** Integrate by parts]

5. Show that $\int_0^{\pi/2} \log \sin x\, dx$ is convergent and hence evaluate it.

[**Hint.** Here, f is negative on $\left(0, \frac{\pi}{2}\right]$ and 0 is the only point of infinite discontinuity.

Take $$g(x) = \frac{1}{x^m}, \; 0 < m < 1.$$

$$\therefore \quad \lim_{x \to 0^+} \frac{-f(x)}{g(x)} = \lim_{x \to 0^+} \frac{-\log \sin x}{\frac{1}{x^m}} \qquad \left(\frac{\infty}{\infty}\right)$$

$$= -\lim_{x \to 0^+} \frac{\log \sin x}{x^{-m}} = -\lim_{x \to 0^+} \frac{\frac{1}{\sin x} \cos x}{-mx^{-m-1}}$$

$$= \lim_{x \to 0^+} \left(\frac{\cos x}{1} \cdot \frac{x}{\sin x} \cdot x^m \right) = 0$$

Hence, the integral is convergent. $\left(\because \int_0^{\pi/2} g\, dx \text{ is convergent}\right)\Bigg]$

6. Show that $\int_0^1 \frac{\operatorname{cosec} x}{x^n}\, dx$ is divergent if $n \geq 1$.

[**Hint.** $|\operatorname{cosec} x| \geq 1$

$\therefore \quad \left| \frac{\operatorname{cosec} x}{x^n} \right| \geq \frac{1}{x^n}.$]

7. Examine the convergence of $\int_a^b \frac{dx}{(x-a)\sqrt{b-x}}$.

[**Hint.** Here, a, b are the only points of discontinuity. Write the given integral as

$$\int_a^c \frac{dx}{(x-a)\sqrt{b-x}} + \int_c^b \frac{dx}{(x-a)\sqrt{b-x}}, \; a < c < b.$$

To test the convergence of $\int_a^c \frac{dx}{(x-a)\sqrt{b-x}}$ at $x = a$

Take $$g(x) = \frac{1}{x-a}.$$

$$\therefore \quad \lim_{x \to a^+} \frac{f(x)}{g(x)} = \lim_{x \to a^+} \frac{1}{\sqrt{b-x}} = \frac{1}{b-a}. \qquad \text{(finite and non-zero)}$$

$\therefore$ by comparison test, $\int_a^c f\, dx$ and $\int_a^c g\, dx$ converge or diverge together.

But, $\int_a^c g\, dx$ is divergent. $(\because \; n = 1)$

$\therefore \quad \int_a^c f\, dx$ is divergent.

To test the convergence of $\int_c^b \frac{dx}{(x-a)\sqrt{b-x}}$ at b, take $g(x) = \frac{1}{\sqrt{b-x}}$.]

8. Examine the convergence of $\int_0^2 \frac{\log x}{\sqrt{2-x}}\, dx$. *(BCA (II) 2006)*

[**Hint.** 0 and 2 are both points of infinite discontinuity of f on [0, 2].

Write the given integral as $\int_0^1 \frac{\log x}{\sqrt{2-x}}\, dx + \int_1^2 \frac{\log x}{\sqrt{2-x}}\, dx$

To test the convergence of $\int_0^1 \frac{\log x}{\sqrt{2-x}}\,dx$ at $x = 0$

Here, integrand is negative on (0, 1).

Let $$f(x) = -\frac{\log x}{\sqrt{2-x}}.$$

Take $$g(x) = \frac{1}{x^n}.$$

$$\therefore \quad \lim_{x\to 0^+} \frac{-f(x)}{g(x)} = \lim_{x\to 0^+} \frac{-x^n \log x}{\sqrt{2-x}} = 0$$

$$\left(\because \lim_{x\to 0^+} x^n \log x = \lim_{x\to 0} \frac{\log x}{x^{-n}} = 0 \text{ by L'Hospital's Rule}\right)$$

It we choose n, $0 < n < 1$, then $\int_0^1 g\,dx$ is convergent

$\therefore \int_0^1 -f\,dx$ is convergent and hence $\int_0^1 f\,dx$ is convergent.

To test the convergence of $\int_1^2 \frac{\log x}{\sqrt{2-x}}\,dx$ at $x = 2$

Here, integrand is positive in (1, 2).

Take $g(x) = \frac{1}{\sqrt{2-x}}$.

Then, $\lim_{x\to 2^-} \frac{f(x)}{g(x)} = \lim_{x\to 2^-} \log x = \log 2$ (finite and non-zero)

Since $\int_1^2 g\,dx$ is convergent, hence $\int_1^2 f(x)\,dx$ is convergent.]

10.5. GENERAL TEST FOR CONVERGENCE AT a (f NEED NOT BE POSITIVE)

10.5.1. Theorem. *The necessary and sufficient condition for the convergence of the improper integral* $\int_a^b f\,dx$ *at a, a being the only point of infinite discontinuity, is that to each* $k > 0$, *there corresponds a* $\delta > 0$ *such that*

$$\left| \int_{a+\varepsilon_1}^{a+\varepsilon_2} f\,dx \right| < k$$

for all positive numbers ε_1, ε_2 *less than* δ.

Proof. Let $\phi(\varepsilon) = \int_{a+\varepsilon}^{b} f\,dx,\ \varepsilon > 0.$

Now, the necessary and sufficient condition for $\lim_{\varepsilon\to 0^+} \phi(\varepsilon)$ to exist finitely is that to each $k > 0$, there corresponds a $\delta > 0$ such that

$$|\ \phi(\varepsilon_1) - \phi(\varepsilon_2)\ | < k,\ \forall\ 0 < \varepsilon_1, \varepsilon_2 < \delta$$

$$\Leftrightarrow \quad \left| \int_{a+\varepsilon_1}^{b} f\,dx - \int_{a+\varepsilon_2}^{b} f\,dx \right| < k,\ \forall\ 0 < \varepsilon_1, \varepsilon_2 < \delta$$

$$\Leftrightarrow \qquad \left| \int_{a+\varepsilon_1}^{a+\varepsilon_2} f\, dx \right| < k, \forall\, 0 < \varepsilon_1, \varepsilon_2 < \delta.$$

Cor. $\int_a^b |f|\, dx$ exists $\Rightarrow$ $\int_a^b f\, dx$ exists, provided $\int_{a+\varepsilon}^b f\, dx$ exists.

10.5.2. Absolute Convergence. The improper integral $\int_a^b f\, dx$ is said to be absolutely convergent if $\int_a^b |f|\, dx$ is convergent.

It follows from the above corollary that every absolutely convergent integral is convergent. Converse is however false.

Example 9. *Test the convergence of* $\int_0^1 \frac{\sin \frac{1}{x}}{\sqrt{x}}\, dx.$

Sol. Let $f(x) = \frac{\sin \frac{1}{x}}{\sqrt{x}}$.

Clearly, in a neighbourhood of '0', f does not keep the same sign.

Now, $$\left| \frac{\sin \frac{1}{x}}{\sqrt{x}} \right| = \frac{\left| \sin \frac{1}{x} \right|}{|\sqrt{x}|} \le \frac{1}{\sqrt{x}}, \forall\, x \in (0, 1]$$

But, $\int_0^1 \frac{1}{\sqrt{x}}\, dx = 2$, is convergent at $x = 0$

$$\Rightarrow \int_0^1 \left| \frac{\sin \frac{1}{x}}{\sqrt{x}} \right| dx \text{ is convergent at } 0$$

$$\Rightarrow \int_0^1 \frac{\sin \frac{1}{x}}{\sqrt{x}} \text{ is absolutely convergent and hence convergent at } x = 0.$$

Example 10. *Show that* $\int_0^\infty f\, dx$*, where*

$$f(x) = \begin{cases} 1, & 0 \le x \le 1 \\ 0, & n-1 < x < n - \frac{1}{n}, \\ (-1)^{n+1}, & n - \frac{1}{n} \le x \le n \end{cases}$$

$n = 2, 3, 4, \ldots\ldots$ *is convergent but not absolutely.*

Sol. $$\int_0^\infty f\, dx = \int_0^1 f\, dx + \int_1^{3/2} f\, dx + \int_{3/2}^2 f\, dx + \ldots\ldots$$

$$= \int_0^1 dx + \int_1^{3/2} 0\, dx - \int_{3/2}^2 1\, dx + \ldots\ldots$$

$$= 1 - \frac{1}{2} + \frac{1}{3} - \frac{1}{4} + \ldots\ldots$$

Since the series $1 - \frac{1}{2} + \frac{1}{3} - \frac{1}{4} + \ldots\ldots$ is convergent to log 2, the given integral is convergent to log 2.

But, $$\int_0^\infty |f|\, dx = \int_0^1 |f|\, dx + \int_1^{3/2} |f|\, dx + \int_{3/2}^2 f\, dx + \ldots\ldots$$

$$= 1 + \frac{1}{2} + \frac{1}{3} + \frac{1}{4} + \ldots\ldots$$

Since the series $1 + \frac{1}{2} + \frac{1}{3} + \frac{1}{4} + \ldots\ldots$ is divergent, the integral $\int_0^\infty |f|\, dx$ is divergent and hence $\int_0^\infty f\, dx$ is not absolutely convergent.

10.6. CONVERGENCE AT ∞, THE INTEGRAND f BEING POSITIVE

10.6.1. Theorem. *Let $f(x) > 0$, $\forall\, x \in [a, t]$. The necessary and sufficient condition for $\int_a^\infty f\, dx$ to be convergent is that there exists a positive number k such that*

$$\int_a^t f\, dx \le k, \quad \forall\, t \ge a.$$

Proof. Proof follows from the fact that f being positive for each x, the integral $\int_a^t f\, dx$ monotonically increases with t and hence tends to a finite limit or to ∞ according as it is bounded above or not.

Cor. 1. *If f and g are two functions such that*

$$0 < g(x) \le f(x), \quad \forall\, x \in [a, \infty),$$

then convergence of $\int_a^\infty f\, dx$ at ∞

$\Rightarrow$ convergence of $\int_a^\infty g\, dx$ at ∞.

Cor. 2. If f and g are positive functions of x with

$$\lim_{x \to \infty} \frac{f(x)}{g(x)} = l$$

$(\ne 0, \infty)$, then $\int_a^\infty f\, dx$ and $\int_a^\infty g\, dx$ converge or diverge together.

10.6.2. Theorem. (*i*) *If $\lim_{x\to\infty} \frac{f(x)}{g(x)} = 0$ and $\int_a^\infty g\, dx$ is convergent, then $\int_a^\infty f\, dx$ is convergent.*

(*ii*) *If $\lim_{x\to\infty} \frac{f(x)}{g(x)} = \infty$ and $\int_a^\infty g\, dx$ is divergent, then $\int_a^\infty f\, dx$ is divergent.*

Proof. (*i*) If $\lim_{x\to\infty} \frac{f(x)}{g(x)} = 0$, we can find a number k such that

$$\frac{f(x)}{g(x)} < \varepsilon, \quad \forall\, x \ge k$$

i.e., $$f(x) < \varepsilon\, g(x), \quad \forall\, x > k$$

$\therefore$ convergence of $\int_a^\infty g(x)\,dx$

$\Rightarrow$ the convergence of $\int_a^\infty f\,dx$.

(*ii*) If $\lim_{x\to\infty} \frac{f(x)}{g(x)} = \infty$, we can find numbers k and K such that

$$\frac{f(x)}{g(x)} > \text{K} \quad \text{or} \quad f(x) > \text{K}\,g(x), \quad \forall\, x \ge k$$

$\therefore$ divergence of $\int_a^\infty g\,dx$

$\Rightarrow$ divergence of $\int_a^\infty f\,dx$.

10.7. A USEFUL COMPARISON INTEGRAL

10.7.1. *The improper integral* $\int_a^\infty \frac{dx}{x^n}$, *(a > 0) is convergent if and only if n > 1.*

Proof. $$\int_a^\infty \frac{dx}{x^n} = \lim_{t\to\infty} \int_a^t \frac{dx}{x^n} = \lim_{t\to\infty} \left[\frac{x^{-n+1}}{-n+1}\right]_a^t, \text{ if } n \ne 1$$

$$= \lim_{t\to\infty} \frac{1}{1-n}\,[t^{1-n} - a^{1-n}] = \lim_{t\to\infty} \frac{t^{1-n}}{1-n} - \frac{a^{1-n}}{1-n}$$

$$= \begin{cases} -\dfrac{a^{1-n}}{1-n} & \text{which is finite, if } n > 1 \\ \infty, & \text{if } n < 1 \end{cases}$$

If $n = 1$, $$\int_a^\infty \frac{1}{x^n}\,dx = \int_a^\infty \frac{1}{x}\,dx$$

$$= \lim_{t\to\infty} \int_a^t \frac{1}{x}\,dx = \lim_{t\to\infty}\,[\log t - \log a] = \infty.$$

Hence, $\int_a^\infty \frac{1}{x^n}\,dx$ is convergent if and only if $n > 1$.

10.7.2. Using the above integral, we have the following two comparison tests which are useful in solving the questions.

I. If f is positive in $[a, \infty)$, then the integral $\int_a^\infty f\,dx$ converges if there exists a positive number $n > 1$ and a fixed positive number K such that

$$f(x) \le \text{K}/x^n, \quad \forall\, x \ge a.$$

Also, the integral diverges if there exists a positive number K such that

$$f(x) \geq K/x^n, \quad \forall\, x \geq a.$$

II. If $\lim_{x\to\infty} x^n f(x)$ exists and is non-zero finite, then the integral $\int_a^\infty f\,dx$ converges if and only if $n > 1$.

Example 11. *Examine the convergence of*

(i) $\int_1^\infty \frac{x^3}{(1+x)^5}\,dx$ (BCA (II) 2006) (ii) $\int_1^\infty \frac{dx}{(2+x)\sqrt{x}}$

(iii) $\int_1^\infty \frac{dx}{x^{1/2}(5+x)^{1/3}}$ (iv) $\int_0^\infty \frac{\sin^2 x}{x^2}\,dx$

(v) $\int_1^\infty \frac{\log x}{x^2}\,dx$ (vi) $\int_1^\infty x^n e^{-x}\,dx$

(vii) $\int_0^\infty \frac{x^{2m}}{1+x^{2n}}\,dx$, *where m, n are positive integers.*

Sol. (i) Let $\qquad f(x) = \frac{x^3}{(1+x)^5} \qquad \left(\text{which behaves like } \frac{x^3}{x^5} = \frac{1}{x^2} \text{ as } x \to \infty\right)$

Take $\qquad g(x) = \frac{1}{x^2}.$

$$\therefore \quad \lim_{x\to\infty} \frac{f(x)}{g(x)} = \lim_{x\to\infty} \frac{x^5}{(1+x)^5} = 1 \; (\neq 0, \infty)$$

$\therefore$ the integrals $\int_1^\infty f\,dx$ and $\int_1^\infty g\,dx$ converge or diverge together.

But, $\int_1^\infty g\,dx = \int_1^\infty \frac{1}{x^2}\,dx$ is convergent $\qquad (\because \; n = 2 > 1)$

$$\therefore \quad \int_1^\infty f\,dx = \int_1^\infty \frac{x^3}{(1+x)^5}\,dx \text{ is convergent.}$$

(ii) $\qquad f(x) = \frac{1}{(2+x)\sqrt{x}}$

Take $\qquad g(x) = \frac{1}{x\sqrt{x}} = \frac{1}{x^{3/2}}$

$$\lim_{x\to\infty} \frac{f(x)}{g(x)} = \lim_{x\to\infty} \frac{x}{2+x} = 1 \quad (\neq 0, \infty)$$

Also, $\int_1^\infty g\,dx = \int_1^\infty \frac{1}{x^{3/2}}\,dx$ is convergent.

$$\therefore \quad \int_1^\infty f\,dx = \int_1^\infty \frac{1}{(1+x)\sqrt{x}}\,dx \text{ is convergent.}$$

(iii) $\qquad f(x) = \frac{1}{x^{1/2}(5+x)^{1/3}}$

Take $\qquad g(x) = \frac{1}{x^{1/2}\,.\,x^{1/3}} = \frac{1}{x^{5/6}}$

$$\lim_{x\to\infty}\frac{f(x)}{g(x)}=\lim_{x\to\infty}\frac{x^{5/6}}{x^{1/2}(5+x)^{1/3}}=1(\neq 0,\infty)$$

Also, $\int_1^\infty g\,dx=\int_1^\infty \frac{1}{x^{5/6}},\left(n=\frac{5}{6}<1\right)$ is divergent.

$\therefore\quad \int_1^\infty f\,dx$ is divergent.

(*iv*) $$f(x)=\frac{\sin^2 x}{x^2}.$$

Since $\lim_{x\to 0}\frac{\sin^2 x}{x^2}=1$, 0 is not a point of infinite discontinuity.

If we write $\int_0^\infty \frac{\sin^2 x}{x^2}dx=\int_0^1\frac{\sin^2 x}{x^2}dx+\int_1^\infty\frac{\sin^2 x}{x^2}dx$, the first integral on the right is proper and so we test the convergence of the second integral.

Since $\frac{\sin^2 x}{x^2}\le\frac{1}{x^2}$ and $\int_1^\infty\frac{1}{x^2}dx$ is convergent,

$\therefore\quad \int_1^\infty\frac{\sin^2 x}{x^2}dx$ is convergent at ∞.

Hence, $\int_1^\infty\frac{\sin^2 x}{x^2}dx$ is convergent.

(*v*) Here, $x^{3/2}\frac{\log x}{x^2}=\frac{\log x}{x^{1/2}}\to 0$ as $x\to\infty$

and $\int_1^\infty\frac{dx}{x^{3/2}}$ is convergent. $\left[\text{Here } g(x)=\frac{1}{x^{3/2}}\right]$

$\therefore$ By comparison test, the integral $\int_1^\infty\frac{\log x}{x^2}dx$ is convergent.

(*vi*) Now, $x^2\,.\,x^n\,e^{-x}=x^{n+2}/e^x\to 0$ as $x\to\infty,\quad\forall\, n$

and $\int_1^\infty\frac{dx}{x^2}$ is convergent. $\left[\text{Here, } g(x)=\frac{1}{x^2}\right]$

$\therefore$ by comparison test, $\int_1^\infty x^n\,e^{-x}\,dx$ is convergent.

(*vii*) $\int_0^\infty\frac{x^{2m}}{1+x^{2n}}dx=\int_0^a\frac{x^{2m}}{1+x^{2n}}dx+\int_a^\infty\frac{x^{2m}}{1+x^{2n}}dx$, for some $a>0$.

Here $\int_0^a\frac{x^{2m}}{1+x^{2n}}dx$, being a proper integral is convergent.

Consider $\int_a^\infty\frac{x^{2m}}{1+x^{2n}}dx$.

Now, $$\lim_{x\to\infty} x^{\lambda} f(x) = \lim_{x\to\infty} x^{\lambda} \frac{x^{2m}}{1+x^{2n}}$$

[We have used λ instead of n as n is used in the statement of the question]

$$= \lim_{x\to\infty} \frac{x^{\lambda+2m}}{1+x^{2n}} = 1 \text{ if } \lambda + 2m = 2n$$

i.e., if $$\lambda = 2(n - m).$$

Since n and m are positive integers, we have $\lambda > 1$ if $n > m$.

Hence $\int_a^{\infty} \frac{x^{2m}}{1+x^{2n}}\, dx$ is convergent iff $\lambda > 1$ *i.e.*, $n > m$.

It follows that the given integral is convergent iff $n > m$.

Example 12. (Gamma function). *Show that the integral* $\int_0^{\infty} x^{n-1} e^{-x}\, dx$ *is convergent if and only if* $n > 0$. (K.U. BCA (II) 2003)

Sol. Let $$f(x) = x^{n-1} e^{-x} = \frac{e^{-x}}{x^{1-n}}.$$

The integrand f has infinite discontinuity at $x = 0$ if $n < 1$. Thus, we have to examine the convergence at 0 and ∞ both. Consider any positive number > 0, say 1 and examine the convergence of

$$\int_0^1 f\, dx \text{ and } \int_1^{\infty} f\, dx$$

at 0 and ∞ respectively.

Convergence at 0, n < 1.

Take $$g(x) = \frac{1}{x^{1-n}}.$$

Then $$\lim_{x\to 0} \frac{f(x)}{g(x)} = \lim_{x\to 0} e^{-x} = 1$$

Also, $\int_0^1 \frac{1}{x^{1-n}}\, dx$ is convergent, if and only if, $1 - n < 1$, *i.e.*, $n > 0$.

Convergence at ∞

For a given n, we have $e^x > x^{n+1}$ for sufficiently large values of x.

$\therefore$ $$e^x > x^{n-1} . x^2$$

$\therefore$ $$x^{n-1} . e^{-x} < \frac{1}{x^2}, \text{ for sufficiently large } x.$$

Since $\int_1^{\infty} \frac{1}{x^2}\, dx$ is convergent,

$\therefore$ by comparison test, $\int_1^{\infty} x^{n-1} e^{-x}\, dx$ is convergent for all n.

Thus, $\int_0^{\infty} x^{n-1} e^{-x}\, dx$ converges, if and only if, $n > 0$.

Remark. For evaluating improper integrals, one can apply all the methods, namely, change of variables, integration by parts, etc., for evaluating proper integrals provided all the integrals entering into them are convergent.

For example, the integral $\int_2^\infty \frac{dx}{x^2+x-2}$, as can be seen by comparing it with $\int_2^\infty \frac{dx}{x^2}$, is convergent. But if we resolve the integrand into partial fractions, then we have

$$\frac{1}{x^2+x-2} = \frac{1}{3(x-1)} - \frac{1}{3(x+2)}$$

and the integrals $\int_2^\infty \frac{dx}{x+2}$ and $\int_2^\infty \frac{dx}{x-1}$ are both divergent.

Therefore, the equality

$$\int_2^\infty \frac{dx}{x^2+x-2} = \frac{1}{3}\int_2^\infty \frac{dx}{x-1} - \frac{1}{3}\int_2^\infty \frac{dx}{x+2}$$

is not correct.

EXERCISE 10 (*c*)

1. Examine the convergence of the following improper integrals :

(*i*) $\int_0^\infty \frac{x}{1+x^3}\,dx$ (*ii*) $\int_1^\infty \frac{x^2+2}{x^4}\,dx$

(*iii*) $\int_1^\infty \frac{x}{\sqrt{x^5+1}}\,dx$ (*iv*) $\int_1^\infty \frac{dx}{x\sqrt{x^2+1}}$.

2. Examine the convergence of $\int_0^\infty \frac{x\tan^{-1}x}{(1+x^4)^{1/3}}\,dx$.

[**Hint.** $f(x) = \frac{x\tan^{-1}x}{(1+x^4)^{1/3}} = \frac{\tan^{-1}x}{x^{1/3}(1+x^{-4})^{1/3}} \simeq x^{-1/3}$ at ∞

Take $g(x) = x^{-1/3}$

$$\lim_{x\to\infty}\frac{f(x)}{g(x)} = \frac{\pi}{2}, \text{ etc.}]$$

3. Show that : $\int_0^\infty \left(\frac{1}{1+x} - e^{-x}\right)\frac{dx}{x}$ is convergent.

[**Hint.** $f(x) = \frac{e^x-1-x}{x(1+x)e^x} > 0,\ \forall\, x>0,\ f(x)\to 0$ as $x\to 0$

$\therefore$ 0 is not a point of infinite discontinuity.

For convergence at ∞, $f(x)$ can be written as

$$\frac{e^x-(1+x)}{e^x}\cdot\frac{1}{1+x}\cdot\frac{1}{x} \quad\text{and}\quad \frac{e^x-(1+x)}{e^x}\to 1 \text{ as } x\to\infty$$

Take $g(x) = \frac{1}{x^2}$, etc.]

10.8. GENERAL TEST FOR CONVERGENCE AT ∞, THE INTEGRAND MAY NOT NECESSARY BE POSITIVE

10.8.1. Cauchy's Test. *The necessary and sufficient condition for the convergence of the improper integral* $\int_a^{\infty} f\,dx$ *at* ∞ *is that to each* $\varepsilon > 0$, *there corresponds a positive number k such that*

$$\left|\int_{t_1}^{t_2} f\,dx\right| < \varepsilon, \quad \forall\, t_1, t_2 > k.$$

Proof. Let $\qquad \phi(t) = \int_a^t f\,dx$

Now, the necessary and sufficient condition for $\lim_{t\to\infty} \phi(t)$ to exist finitely is that to each $\varepsilon > 0$, there corresponds a number $k > 0$ such that

$$|\,\phi(t_2) - \phi(t_1)\,| < \varepsilon, \quad \forall\, t_1, t_2 > k$$

$$\Leftrightarrow \qquad \left|\int_a^{t_2} f\,dx - \int_a^{t_1} f\,dx\right| < \varepsilon, \quad \forall\, t_1, t_2 > k$$

$$\Leftrightarrow \qquad \left|\int_{t_1}^{t_2} f\,dx\right| < \varepsilon, \quad \forall\, t_1, t_2 > k.$$

Example 13. *Show that* $\int_0^{\infty} \frac{\sin x}{x}\,dx$ *is convergent.*

Sol. As $\frac{\sin x}{x} \to 1$ as $x \to 0$, 0 is not a point of infinite discontinuity.

Writing,

$$\int_0^{\infty} \frac{\sin x}{x}\,dx = \int_0^{1} \frac{\sin x}{x}\,dx + \int_1^{\infty} \frac{\sin x}{x}\,dx \qquad \text{...(1)}$$

We discuss the convergence of $\int_1^{\infty} \frac{\sin x}{x}\,dx$ at ∞.

Let $\varepsilon > 0$ be given.

Now, $$\int_{t_1}^{t_2} \frac{\sin x}{x}\,dx = \left[-\frac{\cos x}{x}\right]_{t_1}^{t_2} - \int_{t_1}^{t_2} \frac{\cos x}{x^2}\,dx \qquad \text{(Integrating by parts)}$$

$$= \frac{\cos t_1}{t_1} - \frac{\cos t_2}{t_2} - \int_{t_1}^{t_2} \frac{\cos x}{x^2}\,dx$$

$$\Rightarrow \qquad \left|\int_{t_1}^{t_2} \frac{\sin x}{x}\,dx\right| \le \left|\frac{\cos t_1}{t_1} - \frac{\cos t_2}{t_2}\right| + \left|\int_{t_1}^{t_2} \frac{\cos x}{x^2}\,dx\right|$$

$$\le \frac{1}{t_1} + \frac{1}{t_2} + \left|\int_{t_1}^{t_2} \frac{dx}{x^2}\right| \qquad (\because \ |\cos x| \le 1)$$

$$\le \frac{1}{t_1} + \frac{1}{t_2} + \frac{1}{t_1} + \frac{1}{t_2} \le 4 \cdot \frac{\varepsilon}{4}, \text{ if } t_1, t_2 \text{ are both} > \frac{4}{\varepsilon}$$

$$= \varepsilon$$

$\therefore$ by Cauchy's test, $\int_1^\infty \frac{\sin x}{x}\,dx$ is convergent and hence $\int_0^\infty \frac{\sin x}{x}\,dx$ is convergent.

10.9. TESTS FOR CONVERGENCE OF THE INTEGRAL OF A PRODUCT OF TWO FUNCTIONS

10.9.1. Abel's test. *If* $\int_a^\infty f\,dx$ *is convergent and if g is a bounded and monotonic function for* $x > a$, *then* $\int_a^\infty fg\,dx$ *is convergent.*

Proof. Since g is a monotonic function for $x > a$, g is integrable on $[a, t]$, $\forall\, t \ge a$.

Also, f is integrable on $[a, t]$, $\forall\, t \ge a$,

therefore by **second mean value theorem***, we have

$$\int_{t_1}^{t_2} fg\,dx = g(t_1)\int_{t_1}^{\xi} f\,dx + g(t_2)\int_{\xi}^{t_2} f\,dx \qquad \text{...(1)}$$

where $a < t_1 \le \xi \le t_2$.

As g is bounded on $[a, \infty)$, therefore there exists a positive number k such that

$$|\,g(x)\,| \le k,\ \forall\, x \ge a$$

In particular, $|\,g(t_1)\,| \le k$ and $|\,g(t_2)\,| \le k$ ...(2)

Since $\int_a^\infty f\,dx$ is convergent, there exists a +ve number t_0 such that

$$\left|\int_{t_1}^{t_2} f\,dx\right| < \frac{\varepsilon}{2k},\ \forall\, t_1, t_2 \ge t_0 \qquad \text{...(3)}$$

Choose the numbers t_1 and t_2 in (1) such that both are $\ge t_0$ and hence $\xi \ge t_0$ also.

$$\therefore \text{ from (3), } \left|\int_{t_1}^{\xi} f\,dx\right| < \frac{\varepsilon}{2k} \text{ and } \left|\int_{\xi}^{t_2} f\,dx\right| < \frac{\varepsilon}{2k} \qquad \text{...(4)}$$

$$\therefore \text{ from (1), } \left|\int_{t_1}^{t_2} fg\,dx\right| \le |g(t_1)|\left|\int_{t_1}^{\xi} f\,dx\right| + |g(t_2)|\left|\int_{\xi}^{t_2} f\,dx\right|$$

$$< k\,.\,\frac{\varepsilon}{2k} + k\,.\,\frac{\varepsilon}{2k} = \varepsilon \qquad \text{[Using (2) and (4)]}$$

Hence, by Cauchy's test $\int_a^\infty fg\,dx$ is convergent at ∞.

Example 14. *Test the convergence of* $\int_0^\infty e^{-ax}\frac{\sin x}{x}\,dx$.

Sol. Here e^{-ax} is monotonically decreasing function of x for $x > 0$ and $\int_0^\infty \frac{\sin x}{x}\,dx$ is convergent (refer to Example 13).

---

* **Second Mean Value Theorem.** If $\int_a^b f\,dx$ and $\int_a^b g\,dx$ both exist and g is monotonic on $[a, b]$, then there exists $\xi \in [a, b]$ such that

$$\int_a^b fg\,dx = g(a)\int_a^{\xi} f\,dx + g(b)\int_{\xi}^{b} f\,dx.$$

Therefore, by Abel's test $\int_0^{\infty} e^{-ax} \dfrac{\sin x}{x} dx$ is convergent.

10.9.2. Dirichlet's test. *If g is a bounded and monotonic function of x tending to 0 as $x \to \infty$, and $\int_a^t f\, dx$ is bounded for all $t \geq a$, then $\int_a^{\infty} fg\, dx$ is convergent at ∞.*

Or

An infinite integral which oscillates finitely becomes convergent after the insertion of a monotonic factor which tends to zero as a limit.

Proof. The function g, being monotonic on $[a, \infty)$, is integrable on $[a, t]$, $\forall\, t \geq a$.

Also, since f is integrable on $[a, t]$, therefore, applying second mean value theorem, we have

$$\int_{t_1}^{t_2} fg\, dx = g(t_1) \int_{t_1}^{\xi} f\, dx + g(t_2) \int_{\xi}^{t_2} f dx \qquad \text{...(1)}$$

where $a < t_1 \leq \xi \leq t_2$.

Since $\int_a^t f\, dx$ is bounded $\forall\, t \geq a$, there exists a positive number k such that

$$\left| \int_a^t f\, dx \right| \leq k \;\forall\, t \geq a \qquad \text{...(2)}$$

$$\therefore \qquad \left| \int_{t_1}^{\xi} f\, dx \right| = \left| \int_a^{\xi} f\, dx - \int_a^{t_1} f\, dx \right|$$

$$\leq \left| \int_a^{\xi} f\, dx \right| + \left| \int_a^{t_1} f\, dx \right|$$

$$\leq k + k = 2k, \quad \forall\, t_1, \xi \geq a \qquad \text{...(3) [By (2)]}$$

Similarly, $$\left| \int_{\xi}^{t_2} f\, dx \right| \leq 2k, \quad \forall\, t_2, \xi \geq a \qquad \text{...(4)}$$

Let $\varepsilon > 0$ be given.

Since $g(x) \to 0$ as $x \to \infty$, there exists a number t_0 such that

$$|\, g(x)\, | < \frac{\varepsilon}{4k}, \quad \forall\, x \geq t_0.$$

If t_1, t_2 are any two numbers $\geq t_0$, then

$$|\, g(t_1)\, | < \frac{\varepsilon}{4k} \quad \text{and} \quad |\, g(t_2)\, | < \frac{\varepsilon}{4k} \qquad \text{...(5)}$$

From (1), (3), (4) and (5), we deduce that there exists a number t_0 such that $\forall\, t_1, t_2 \geq t_0$

$$\left| \int_{t_1}^{t_2} fg\, dx \right| \leq |\, g(t_1)| \left| \int_{t_1}^{\xi} f\, dx \right| + |\, g(t_2)| \left| \int_{\xi}^{t_2} f\, dx \right|$$

$$< \frac{\varepsilon}{4k}\, 2k + \frac{\varepsilon}{4k} \,.\, 2k = \varepsilon$$

Hence, by Cauchy's test, $\int_a^{\infty} fg\, dx$ is convergent at ∞.

Example 15. *Show that* $\int_1^\infty \frac{\sin x}{x^p}\, dx$ *is convergent for* $p > 0$.

Sol. Let $f(x) = \sin x$ and $g(x) = \frac{1}{x^p}$

Now, $$\left| \int_1^t f\, dx \right| = \left| \int_1^t \sin x\, dx \right|$$

$$= \left| \Big[-\cos x \Big]_1^t \right| = | \cos 1 - \cos t |$$

$$\le | \cos 1 | + | \cos t | \le 2, \quad \forall\, t \in [1, \infty). \qquad (\because \ | \cos x | \le 1)$$

Also, $g(x) = \frac{1}{x^p}$ is a monotonically decreasing function of x and tends to 0 as $x \to \infty$ for $p > 0$.

$\therefore$ by Dirichlet's test,

$$\int_1^\infty fg\, dx = \int_1^\infty \frac{\sin x}{x^p}\, dx$$

is convergent for $p > 0$.

Example 16. Using Dirichlet's test, *show that* $\int_0^\infty \frac{\sin x}{x}\, dx$ *is convergent.*

Sol. Since $\frac{\sin x}{x} \to 1$ as $x \to 0$, 0 is not a point of infinite discontinuity.

Writing $$\int_0^\infty \frac{\sin x}{x}\, dx = \int_0^1 \frac{\sin x}{x}\, dx + \int_1^\infty \frac{\sin x}{x}\, dx,$$

we see that $\int_0^1 \frac{\sin x}{x}\, dx$ is a proper integral.

So, let us examine the convergence of $\int_1^\infty \frac{\sin x}{x}\, dx$.

Since $$\left| \int_1^t \sin x\, dx \right| = | \cos 1 - \cos t | \le | \cos 1 | + | \cos t | \le 2,$$

$$\left| \int_1^t \sin x\, dx \right| \text{ is bounded above for all } t \ge 1.$$

Also, $\frac{1}{x}$ is monotonically decreasing function of x and tends to 0 as $x \to \infty$.

$\therefore$ by Dirichlet's test, $\int_1^\infty \frac{\sin x}{x}\, dx$ is convergent.

Hence, $\int_0^\infty \frac{\sin x}{x}\, dx$ is convergent.

Example 17. *Show that* $\int_0^\infty \sin x^2\, dx$ *is convergent.*

Sol. Writing $\sin x^2 = \frac{1}{2x} \cdot 2x \sin x^2$, we consider the improper integrals $\int_1^\infty \sin x^2\, dx$, *i.e.*, $\int_1^\infty \frac{1}{2x} \cdot 2x \sin x^2\, dx$.

$\dfrac{1}{2x}$ is a monotonically decreasing function of x and tends to 0 as $x \to \infty$.

Also, $\left| \int_1^t 2x \sin x^2 \right| = | -\cos t^2 + \cos 1 |$

$\leq | \cos t^2 | + | \cos 1 | \leq 2.$

$\Rightarrow \int_1^t 2x \sin x^2\, dx$ is bounded $\forall\, t \geq 1$.

$\therefore$ by Dirichlet's test, $\int_1^\infty \dfrac{1}{2x} \cdot 2x \sin x^2\, dx$, *i.e.,*

$$\int_1^\infty \sin x^2\, dx \text{ is convergent.}$$

Also, $\int_0^1 \sin x^2\, dx$ is a proper integral. Hence, the given integral is convergent.

Example 18. *Examine the convergence of* $\int_1^\infty \dfrac{\sin x^m}{x^n}\, dx$.

Sol. Case (*i*) $m = 0$.

The given integral reduces to $\sin 1. \int_1^\infty \dfrac{dx}{x^n}$, which is convergent at ∞ for $n > 1$.

Case (*ii*) $m \neq 0$.

Substituting $x^m = t$, we get

$$\int_1^\infty \frac{\sin x^m}{x^n}\, dx = \frac{1}{m} \int_1^\infty \frac{\sin t}{t^{\frac{n-1}{m}+1}}\, dt$$

Now, $\left| \int_1^\infty \sin t\, dt \right| \leq 2$, bounded and $\dfrac{1}{t^{\frac{n-1}{m}+1}}$ is monotonically decreasing function, tending to 0 as $t \to \infty$ when $\dfrac{n-1}{m} + 1 > 0$ *i.e.,* $\dfrac{n-1}{m} > -1$.

$\therefore$ by Dirichlet's test, $\int_1^\infty \dfrac{\sin t}{t^{\frac{n-1}{m}+1}}\, dt$ is convergent at ∞, when $\dfrac{n-1}{m} > -1$, *i.e.,* $n > 1 - m$, $\forall\, m \neq 0$.

EXERCISE 10 (*d*)

1. Establish the convergence of :

(*i*) $\int_0^\infty e^{-px} \dfrac{\sin x}{x}\, dx,\ p \geq 0$ (*ii*) $\int_e^\infty \dfrac{\log x \cdot \sin x}{x}\, dx$.

[**Hint.** (*ii*) $\left| \int_e^\infty \sin x\, dx \right|$ is bounded above by 2 and $\dfrac{\log x}{x}$ is a monotonically decreasing function and tends to 0 as $x \to \infty$.]

(iii) $\int_0^{\infty} \frac{\sin kx}{x}\, dx$ (iv) $\int_0^{\infty} \frac{\sin x}{\sqrt{x}}\, dx$

(v) $\int_0^{\infty} e^{-ax^2} \cos bx\, dx, (a > 0)$ (vi) $\int_0^{\infty} \frac{\cos x}{\sqrt{x+x^2}}\, dx$

[**Hint.** (*vi*) 0 and ∞ are both points of infinite discontinuity. The given integral can be written as $\int_0^{a} \frac{\cos x}{\sqrt{x+x^2}}\, dx + \int_a^{\infty} \frac{\cos x}{\sqrt{x+x^2}}\, dx$. For $\int_0^{a} \frac{\cos x}{\sqrt{x+x^2}}\, dx$, take $g(x) = \frac{1}{\sqrt{x}}$ and apply comparison test. For $\int_0^{\infty} \frac{\cos x}{\sqrt{x+x^2}}\, dx$, take $g(x) = \frac{1}{\sqrt{x+x^2}}$ and apply Dirichlet's test.]

(vii) $\int_0^{\infty} \frac{x}{1+x^2} \sin x\, dx$ (viii) $\int_0^{\infty} \frac{\cos ax - \cos bx}{x}\, dx$

(ix) $\int_a^{\infty} \frac{\sin x}{x^m}\, dx$, where a and m are both positive.

(x) $\int_0^{\infty} \cos x^2\, dx$ (xi) $\int_0^{\infty} \frac{\cos x}{1+x^2}\, dx$.

11

Beta and Gamma Functions

In this chapter, we are going to study Beta and Gamma functions, their properties and relationship

11.1. BETA FUNCTION

We have seen in example 7 of chapter 10 that the integral

$$\int_0^1 x^{m-1}(1-x)^{n-1}\,dx$$

is convergent for all positive values of m and n. *This integral, obviously a function of m and n, is called a* **Beta function** *and is denoted by B (m, n), i.e.,*

$$B(m, n) = \int_0^1 x^{m-1}(1-x)^{n-1}\,dx, \quad \forall\ m > 0, n > 0 \qquad \text{(K.U. BCA II 2007)}$$

Beta function is also called the **First Eulerian Integral***.

For example,

(*i*) $\int_0^1 x^4(1-x)^5\,dx$ is a Beta function B(5, 6).

The integral can be easily evaluated by expanding $(1-x)^5$.

(*ii*) $\int_0^1 x^{2/3}(1-x)^6\,dx$ is a Beta function $B\left(\frac{5}{3}, 7\right)$

The integral can be easily evaluated by expanding $(1-x)^6$.

(*iii*) $\int_0^1 x^3(1-x)^{4/3}\,dx = B\left(4, \frac{7}{3}\right)$

This integral can be evaluated using the symmetric property (given below) of the beta function.

Note that B (m, n) can be easily evaluated if at least one of m and n is a positive integer.

11.1.1. Symmetry of Beta Function. $B(m, n) = B(n, m)$. (K.U. BCA II 2007)

Proof. By definition,

$$B(m, n) = \int_0^1 x^{m-1}(1-x)^{n-1}\,dx, \ m > 0, n > 0$$

---

*After the name of a great mathematician and physicist L. Euler (1707 – 1783), who was a member of St. Petersburg Academy of Sciences.

Changing x to $1 - x$, we have

$$B(m, n) = \int_0^1 (1-x)^{m-1} [1-(1-x)]^{n-1}\, dx \qquad \left[\because \int_0^a f(x)\, dx = \int_0^a f(a-x)\, dx\right]$$

$$= \int_0^1 (1-x)^{m-1} x^{n-1}\, dx$$

$$= \int_0^1 x^{n-1} (1-x)^{m-1}\, dx = B(n, m).$$

Note. Integral of example (*iii*) above, can be written as

$$B\left(4, \frac{7}{3}\right) = B\left(\frac{7}{3}, 4\right) = \int_0^1 x^{4/3} (1-x)^3\, dx$$

and can be easily integrated.

Example 1. *Express* $\int_0^2 (8 - x^3)^{-1/3}\, dx$ *in terms of a Beta function.*

Sol. Let $\quad I = \int_0^2 (8 - x^3)^{-1/3}\, dx \quad$ ...(1)

Put $\quad x^3 = 8z$

$\therefore \quad x = 2z^{1/3}$ and $dx = \frac{2}{3} z^{-2/3}\, dz$

Also, when $x = 0, z = 0$; when $x = 2, z = 1$

$\therefore$ (1) becomes

$$I = \int_0^1 (8 - 8z)^{-1/3} \cdot \frac{2}{3} z^{-2/3}\, dz = 8^{-1/3} \cdot \frac{2}{3} \int_0^1 (1-z)^{-1/3} z^{-2/3}\, dz$$

$$= \frac{1}{3} \int_0^1 (1-x)^{-1/3} x^{-2/3}\, dx \qquad \left[\because \int_a^b f(z)\, dz = \int_a^b f(x)\, dx\right]$$

$$= \frac{1}{3} \int_0^1 x^{-2/3} (1-x)^{-1/3}\, dx$$

$$= \frac{1}{3} B\left(\frac{-2}{3} + 1, \frac{-1}{3} + 1\right) = \frac{1}{3} B\left(\frac{1}{3}, \frac{2}{3}\right).$$

Example 2. *Evaluate* $\int_a^b (x-a)^{m-1} (b-x)^{n-1}\, dx,\ m > 0, n > 0.$

Sol. Put $x = a + (b-a)z$ so that $dx = (b-a)\, dz$

$$\therefore \quad \int_a^b (x-a)^{m-1} (b-x)^{n-1}\, dx$$

$$= \int_0^1 [(b-a)z]^{m-1} [b - a - (b-a) z]^{n-1} (b-a)\, dz$$

$$= (b-a)^{m+n-1} \int_0^1 z^{m-1} (1-z)^{n-1}\, dz$$

$$= (b-a)^{m+n-1} \int_0^1 x^{m-1} (1-x)^{n-1}\, dx$$

$$= (b-a)^{m+n-1} \cdot B(m, a).$$

Example 3. *Show that*

$$\int_0^1 \frac{x^{m-1}(1-x)^{n-1}}{(a+bx)^{m+n}}\,dx = \frac{1}{(a+b)^m a^n} B(m, n).$$

Sol. Put $\qquad \dfrac{x}{a+bx} = \dfrac{z}{a+b\,.\,1}$

Differentiating w.r.t. x, we have

$$\frac{(a+bx)\,.\,1 - x\,.\,b}{(a+bx)^2}\,dx = \frac{dz}{a+b} \quad \Rightarrow \quad \frac{dx}{(a+bx)^2} = \frac{dz}{a(a+b)}$$

Also, when $x = 0, z = 0$; when $x = 1, z = 1$.

$$\therefore \quad \int_0^1 \frac{x^{m-1}(1-x)^{n-1}}{(a+bx)^{m+n}}\,dx = \int_0^1 \left(\frac{x}{a+bx}\right)^{m-1}\left(\frac{1-x}{a+bx}\right)^{n-1} . \frac{1}{(a+bx)^2}dx$$

$$= \int_0^1 \left(\frac{z}{a+b}\right)^{m-1}\left(\frac{1-z}{a}\right)^{n-1} . \frac{dz}{a(a+b)}$$

$$\left[\because \frac{x}{a+bx} = \frac{z}{a+b} \quad \Rightarrow \quad (a+b)x = az + bxz\right.$$

$$\Rightarrow \quad (a+b-bz)x = az \quad \Rightarrow \quad x = \frac{az}{a+b-bz}$$

$$\Rightarrow \quad \frac{1-x}{a+bx} = \frac{1 - \dfrac{az}{a+b-bz}}{a + \dfrac{abz}{a+b-bz}} = \frac{a+b-bz-az}{a^2+ab-abz+abz}$$

$$\left. = \frac{(a+b)(1-z)}{a(a+b)} = \frac{1-z}{a}\right]$$

$$= \frac{1}{a^n(a+b)^m}\int_0^1 z^{m-1}(1-z)^{n-1}\,dz = \frac{B(m, n)}{(a+b)^m\,.\,a^n}.$$

Example 4. *Show that*

$$\int_0^\infty \frac{x^{m-1}}{(1+x)^{m+n}}\,dx = B(m, n),\ m > 0,\ n > 0.$$

Sol. Put $\qquad \dfrac{x}{1+x} = z$

$$\Rightarrow \quad x = \frac{z}{1-z}$$

$$\therefore \quad dx = \frac{(1-z)\,.\,1 - z(-1)}{(1-z)^2}\,dz = \frac{dz}{(1-z)^2}$$

Also, $\qquad 1 + x = 1 + \dfrac{z}{1-z} = \dfrac{1}{1-z}$

Further, when $x = 0, z = 0$;

when $x \to \infty$, $z = \lim\limits_{x\to\infty} \dfrac{x}{1+x} = \lim\limits_{x\to\infty} \dfrac{1}{1+\dfrac{1}{x}} = 1$

$$\therefore \quad \int_0^\infty \frac{x^{m-1}}{(1+x)^{m+n}}\,dx = \int_0^1 \frac{\left(\dfrac{z}{1-z}\right)^{m-1}}{\left(\dfrac{1}{1-z}\right)^{m+n}} \cdot \frac{dz}{(1-z)^2}$$

$$= \int_0^1 z^{m-1}(1-z)^{n-1}\,dz = \mathrm{B}(m, n).$$

Remark. Result of the above example is also used in solving questions.

Example 5. *Show that*

$$\int_0^1 \frac{x^{m-1}+x^{n-1}}{(1+x)^{m+n}}\,dx = B(m, n),\ m > 0,\ n > 0. \qquad \text{(BCA (II) 2006)}$$

Sol.

$$\mathrm{B}(m, n) = \int_0^\infty \frac{x^{m-1}}{(1+x)^{m+n}}\,dx$$

$$= \int_0^1 \frac{x^{m-1}}{(1+x)^{m+n}}\,dx + \int_1^\infty \frac{x^{m-1}\,dx}{(1+x)^{m+n}} \qquad \text{...(1)}$$

Consider $\displaystyle\int_1^\infty \frac{x^{m-1}}{(1+x)^{m+n}}\,dx$

Put $\displaystyle x = \frac{1}{t}$

$\therefore \quad \displaystyle dx = -\frac{1}{t^2}\,dt$

Also, when $x = 1$, $t = 1$; when $x \to \infty$, $t \to 0$

$$\therefore \quad \int_1^\infty \frac{x^{m-1}}{(1+x)^{m+n}}\,dx = \int_1^0 \frac{\left(\dfrac{1}{t}\right)^{m-1}}{\left(1+\dfrac{1}{t}\right)^{m+n}}\left(-\frac{1}{t^2}\right)dt$$

$$= -\int_1^0 \frac{t^{n-1}}{(1+t)^{m+n}}\,dt = \int_0^1 \frac{t^{n-1}}{(1+t)^{m+n}}\,dt$$

$$= \int_0^1 \frac{x^{n-1}}{(1+x)^{m+n}}\,dx \qquad \text{...(2)}$$

Using (2), (1) becomes

$$\mathrm{B}(m, n) = \int_0^1 \frac{x^{m-1}}{(1+x)^{m+n}}\,dx + \int_0^1 \frac{x^{n-1}}{(1+x)^{m+n}}\,dx = \int_0^1 \frac{x^{m-1}+x^{n-1}}{(1+x)^{m+n}}\,dx.$$

Example 6. *If p, q are positive, show that*

(i) $\displaystyle \frac{B(p, q+1)}{q} = \frac{B(p+1, q)}{p}$ *(ii)* $B(p, q) = B(p + 1, q) + B(p, q + 1)$.

Sol. (*i*) $$B(p, q+1)^* = \int_0^1 x^{p-1}(1-x)^q\, dx$$

Integrating by parts taking $(1-x)^q$ as first function, we have

[$\because$ on comparing the two integrals $B(p, q+1)$ and $B(p+1, q)$, we find that power of $1-x$ is to be decreased by 1]

$$B(p, q+1) = \left[(1-x)^q \frac{x^p}{p}\right]_0^1 - \int_0^1 -q(1-x)^{q-1} \frac{x^p}{p}\, dx$$

$$= 0 + \frac{q}{p}\int_0^1 x^p (1-x)^{q-1}\, dx = \frac{q}{p} B(p+1, q)$$

$$\Rightarrow \qquad \frac{B(p, q+1)}{q} = \frac{B(p+1, q)}{p}$$

(*ii*) $$\text{R.H.S.} = B(p+1, q) + B(p, q+1)$$

$$= \int_0^1 x^p(1-x)^{q-1}\, dx + \int_0^1 x^{p-1}(1-x)^q\, dx$$

$$= \int_0^1 [x^p (1-x)^{q-1} + x^{p-1}(1-x)^q\, dx$$

$$= \int_0^1 x^{p-1} (1-x)^{q-1} (x + 1 - x)\, dx$$

$$= \int_0^1 x^{p-1}(1-x)^{q-1}\, dx = B(p, q) = \text{L.H.S.}$$

11.1.2. *If m, n are positive integers, prove that*

$$B(m, n) = \frac{(m-1)!\,(n-1)!}{(m+n-1)!}. \qquad \text{(K.U. BCA (II) 2001)}$$

Proof. $$B(m, n) = \int_0^1 x^{m-1} (1-x)^{n-1}\, dx.$$

Integrating by parts (taking x^{m-1} as first function), we have

$$B(m, n) = \left[x^{m-1} \frac{(1-x)^n}{-n}\right]_0^1 - \int_0^1 (m-1)\, x^{m-2} \frac{(1-x)^n}{-n}\, dx$$

$$= \frac{m-1}{n}\int_0^1 x^{m-2} (1-x)^n\, dx$$

$$\Rightarrow \qquad B(m, n) = \frac{m-1}{n} B(m-1, n+1)$$

Changing m to $m-1$ and n to $n+1$, we have

$$B(m-1\,; n+1) = \frac{m-2}{n+1} B(m-2, n+2)$$

Similarly, $B(m-2, n+2) = \dfrac{m-3}{n+2} B(m-3, n+3)$

---

*To change a Beta function to integral decrease each part by 1, *i.e.*, $B(p, q) = \int_0^1 x^{p-1} (1-x)^{q-1}\, dx$.

Repeating the above process $m - 2$ times, we have

$$B(2, n + m - 2) = \frac{1}{n+m-2} B(1, n + m - 1)$$

Multiplying the above $m - 1$ equations, we have

$$B(m, n) = \frac{(m-1)(m-2)\ldots\ldots 1}{n(n+1)\ldots\ldots(n+m-2)} B(1, n + m - 1) \qquad \ldots(1)$$

But, $\quad B(1, n + m - 1) = \int_0^1 x^{1-1} (1 - x)^{n+m-1-1}\, dx$

$$= \int_0^1 (1 - x)^{n+m-2}\, dx$$

$$= \left[\frac{(1-x)^{n+m-1}}{-(n+m-1)}\right]_0^1 = \frac{1}{n+m-1}$$

Substituting this value in (1), we have

$$B(m, n) = \frac{(m-1)(m-2)\ldots\ldots 1}{n(n+1)\ldots\ldots(n+m-2)(n+m-1)} \qquad \ldots(2)$$

$$= \frac{(m-1)!}{(n+m-1)(n+m-2)\ldots\ldots(n+1)n}$$

$$= \frac{(m-1)!\,(n-1)!}{(n+m-1)(n+m-2)\ldots\ldots(n+1)n\,.\,(n-1)!}$$

$$= \frac{(m-1)!\,(n-1)!}{(n+m-1)!}.$$

Cor. Putting $m = 1$, in the above result, we have

$$B(1, n) = \frac{1}{n}.$$

EXERCISE 11 (*a*)

1. Express in terms of Beta functions :

(*i*) $\int_0^1 \frac{x^2}{\sqrt{1-x^5}}\, dx$ (*ii*) $\int_0^2 x^3(8 - x^3)^{-1/3}\, dx$

(*iii*) $\int_0^1 \frac{x^2}{1-x^5}\, dx$ (*iv*) $\int_0^1 x^{l-1}(1 - x^2)^{m-1}\, dx$

(*v*) $\int_0^p x^m (p^a - x^a)^n\, dx.$ *(K.U. BCA (II) 2001)*

[**Hint.** (*i*) Put $x^5 = z$ (*ii*) Put $x^3 = 8z$ (*v*) Put $x^a = p^a z$.]

2. Prove that $\int_0^a (a - x)^{m-1} x^{n-1}\, dx = a^{m+n-1} B(m, n)$.

3. Using the property $B(m, n) = B(n, m)$, evaluate

$$\int^1 x^3(1 - x)^{4/3}\, dx.$$

4. Show that $B(m, n) = 2 \int_0^{\pi/2} \sin^{2m-1}\theta \cos^{2n-1}\theta \, d\theta$, m, n being positive integers.

[**Hint.** $$B(m, n) = \int_0^1 x^{m-1}(1-x)^{n-1}\,dx$$

Put $$\sqrt{x} = \sin\theta, \text{ etc.}]$$

5. Show that $\displaystyle\int_0^1 \frac{x^{m-1}(1-x)^{n-1}}{(a+x)^{m+n}}\,dx = \frac{B(m, n)}{a^n(1+a)^m}$.

[**Hint.** See example 3.]

6. Show that

(*i*) $\displaystyle\int_0^\infty \frac{x^{m-1}+x^{n-1}}{(1+x)^{m+n}}\,dx = 2B(m, n), m > 0, n > 0$

(*ii*) $\displaystyle\int_0^\infty \frac{x^{m-1}-x^{n-1}}{(1+x)^{m+n}}\,dx = 0, m > 0, n > 0.$

11.2. GAMMA FUNCTION

We have seen in Example 11 of chapter 10, that the integral $\int_0^\infty x^{n-1}e^{-x}\,dx$ is convergent for each positive n. *The integral $\int_0^\infty x^{n-1}e^{-x}\,dx$ for $n > 0$, is obviously a function of n and is called a Gamma function of n. It is denoted by $\Gamma(n)$, i.e.,*

$$\Gamma(n) = \int_0^\infty x^{n-1}e^{-x}\,dx, \quad \forall \quad n > 0.$$

Gamma function is also called the Second Eulerian Integral.

For example,

(*i*) $\int_0^\infty x^5 e^{-x}\,dx$ is a Gamma function $\Gamma(6)$

(*ii*) $\int_0^\infty x^{5/3} e^{-x}\,dx$ is a Gamma function $\Gamma\left(\frac{8}{3}\right)$.

11.2.1. Recurrence Formula for Gamma Function $\Gamma(n)$. *To prove that $\Gamma(n) = (n-1)\Gamma(n-1)$, when $n > 1$.*

Proof. By definition,

$$\Gamma(n) = \int_0^\infty x^{n-1}e^{-x}\,dx.$$

Integrating by parts,

$$= \left[x^{n-1}\cdot\frac{e^{-x}}{-1}\right]_0^\infty - \int_0^\infty (n-1)x^{n-2}\frac{e^{-x}}{-1}\,dx$$

$$= -\left[\lim_{x\to\infty}\frac{x^{n-1}}{e^x} - 0\right] + (n-1)\int_0^\infty x^{n-2}e^{-x}\,dx$$

$$= (n-1)\int_0^\infty x^{n-2} e^{-x}\, dx \qquad \left[\because \lim_{x\to\infty} \frac{x^{n-1}}{e^x} = 0 \text{ for } n > 0\right]$$

$$= (n-1)\Gamma(n-1)$$

Thus, $\Gamma(n) = (n-1)\Gamma(n-1)$.

Note. $\Gamma(n) > 0$ always.

Cor. *If n is a positive integer, then* $\Gamma(n) = (n-1)\,!$.

Proof. $\Gamma(n) = (n-1)\,\Gamma(n-1)$

Changing n to $n-1, n-2, \ldots\ldots$, successively $n-1$ times.

$$\Gamma(n-1) = (n-2)\,\Gamma(n-2)$$

$$\Gamma(n-2) = (n-3)\,\Gamma(n-3)$$

$$\ldots\ldots\ldots\ldots\ldots\ldots\ldots\ldots\ldots\ldots$$

$$\Gamma(2) = 1.\,\Gamma(1)$$

Multiplying, we have

$$\Gamma(n) = (n-1)(n-2)\ldots\ldots 1\,.\,\Gamma(1) \qquad \ldots(1)$$

But,

$$\Gamma(1) = \int_0^\infty x^{1-1} e^{-x}\, dx = \int_0^\infty e^{-x}\, dx$$

$$= \lim_{t\to\infty} \int_0^t e^{-x}\, dx = \lim_{t\to\infty} -[e^{-t} - 1] = 1$$

Hence from (1), $\Gamma(n) = (n-1)\,!$

For example, $\Gamma(4) = 3\,! = 6$.

11.2.2. Relation between Beta and Gamma Functions. *To prove that*

$$B(m, n) = \frac{\Gamma(m)\,\Gamma(n)}{\Gamma(m+n)}, \text{ where } m > 0,\ n > 0.$$

Proof. To prove the given relation, we first prove

$$\Gamma(n) = z^n \int_0^\infty x^{n-1} e^{-zx}\, dx$$

Putting $x = az$ in $\Gamma(n) = \int_0^\infty x^{n-1} e^{-x}\, dx$, we have

$$\Gamma(n) = \int_0^\infty (az)^{n-1} e^{-az}\, a\, dz = a^n \int_0^\infty z^{n-1} e^{-az}\, dz$$

Replacing z by x,

$$= a^n \int_0^\infty x^{n-1} e^{-ax}\, dx$$

Replacing a by z, we have

$$\Gamma(n) = z^n \int_0^\infty x^{n-1} e^{-zx}\, dx \qquad \ldots(1)$$

$$\Rightarrow \qquad \Gamma(n) = \int_0^\infty x^{n-1} z^n e^{-zx}\, dx$$

Multiplying both sides by $e^{-z} z^{m-1}$, we have

$$\Gamma(n)\,.\,e^{-z} z^{m-1} = \int_0^\infty x^{n-1} z^{n+m-1} e^{-z(1+x)}\, dx$$

Integrating both sides w.r.t. z between the limits 0 to ∞, we have

$$\int_0^\infty \Gamma(n) \,.\, e^{-z} z^{m-1}\, dz = \int_0^\infty \left[\int_0^\infty x^{n-1}\, z^{n+m-1}\, e^{-z(1+x)}\, dx\right] dz$$

$$\Gamma(n) \int_0^\infty e^{-z} z^{m-1}\, dz = \int_0^\infty x^{n-1} \left[\int_0^\infty z^{n+m-1}\, e^{-z(1+x)}\, dz\right] dx.$$

[Put $z(1 + x) = y$, etc. on R.H.S.]

$$\Rightarrow \qquad \Gamma(n)\Gamma(m) = \int_0^\infty x^{n-1} \left[\int_0^\infty \frac{y^{m+n-1}\, e^{-y}\, dy}{(1+x)^{m+n}}\right] dx$$

$$= \int_0^\infty \frac{x^{n-1}}{(1+x)^{m+n}} \left[\int_0^\infty y^{m+n-1}\, e^{-y}\, dy\right] dx$$

$$= \int_0^\infty \frac{x^{n-1}}{(1+x)^{m+n}} [\Gamma(m+n)]dx$$

$$= \Gamma(m+n) \int_0^\infty \frac{x^{n-1}}{(1+x)^{m+n}}\, dx$$

$$= \Gamma(m+n)\, \mathrm{B}(m, n)$$

$$\left(\because \int_0^\infty \frac{x^{n-1}}{(1+x)^{m+n}} = \mathrm{B}(m,n) \text{ Example 4 after 11.1.1}\right)$$

Hence $\qquad \mathrm{B}(m, n) = \dfrac{\Gamma(m)\Gamma(n)}{\Gamma(m+n)}.$

Cor. 1. $\Gamma\left(\frac{1}{2}\right) = \sqrt{\pi}$.

Proof. Substituting $m = \frac{1}{2}$ and $n = \frac{1}{2}$ in the relation

$$\mathrm{B}(m, n) = \frac{\Gamma(m)\Gamma(n)}{\Gamma(m+n)}, \text{ we have}$$

$$\mathrm{B}\left(\tfrac{1}{2}, \tfrac{1}{2}\right) = \frac{\Gamma(\frac{1}{2})\Gamma(\frac{1}{2})}{\Gamma(1)}$$

$$= \left[\Gamma\left(\tfrac{1}{2}\right)\right]^2 \qquad [\because \ \Gamma(1) = 1]$$

$$\Rightarrow \qquad \left[\Gamma\left(\tfrac{1}{2}\right)\right]^2 = \mathrm{B}\left(\tfrac{1}{2}, \tfrac{1}{2}\right)$$

$$= \int_0^1 x^{\frac{1}{2}-1} (1-x)^{\frac{1}{2}-1}\, dx$$

$$= \int_0^1 \frac{1}{\sqrt{x}\,\sqrt{1-x}}\, dx \qquad \ldots(1)$$

Let $\qquad x = \sin^2\theta$

$\therefore \qquad dx = 2 \sin\theta \cos\theta\, d\theta$

Also, when $x = 0$, $\theta = 0$; when $x = 1$, $\theta = \dfrac{\pi}{2}$

$\therefore$ from (1),

$$\left[\Gamma\left(\tfrac{1}{2}\right)\right]^2 = \int_0^{\pi/2} \frac{1}{\sin\theta\sqrt{1-\sin^2\theta}} \,.\, 2\sin\theta\cos\theta\, d\theta$$

$$= \int_0^{\pi/2} 2 \, . \, d\theta = 2\left(\frac{\pi}{2}\right) = \pi$$

$\Rightarrow$ $\qquad \Gamma(\tfrac{1}{2}) = \sqrt{\pi}$.

Cor. 2. $\qquad \displaystyle\int_0^{\infty} e^{-x^2}\, dx = \frac{\sqrt{\pi}}{2}$.

Proof. Let $\qquad \displaystyle I = \int_0^{\infty} e^{-x^2}\, dx$

Put $\qquad x^2 = z$

$\Rightarrow$ $\qquad \displaystyle 2x\, dx = dz \;\Rightarrow\; dx = \frac{dz}{2x} = \frac{1}{2\sqrt{z}}\, dz$

Also, when $x = 0, z = 0$; when $x \to \infty, z \to \infty$

$$\therefore \qquad I = \int_0^{\infty} e^{-z} \frac{dz}{2\sqrt{z}} = \frac{1}{2}\int_0^{\infty} \frac{e^{-x}}{\sqrt{z}}\, dz$$

$$= \frac{1}{2}\int_0^{\infty} z^{-\frac{1}{2}} e^{-z}\, dz$$

$$= \tfrac{1}{2}\Gamma(-\tfrac{1}{2}+1) = \tfrac{1}{2}\Gamma(\tfrac{1}{2}) = \tfrac{1}{2}\sqrt{\pi} \qquad \text{(By Cor. 1)}$$

Cor. 3. $\qquad \displaystyle\int_{-\infty}^{0} e^{-x^2}\, dx = \frac{\sqrt{\pi}}{2}$

One can prove the result by substituting $x = -z$.

Cor. 4. $\qquad \displaystyle\int_{-\infty}^{\infty} e^{-x^2}\, dx = \sqrt{\pi}$.

For, $\qquad \displaystyle\int_{-\infty}^{\infty} e^{-x^2}\, dx = 2\int_0^{\infty} e^{-x^2}\, dx = 2 \, . \frac{\sqrt{\pi}}{2} = \sqrt{\pi}$

$$\left[\because \text{ if } (f)\, x \text{ is an even function of } x, \text{ then } \int_{-a}^{a} f(x)\, dx = 2\int_0^{a} f(x)\, dx\right]$$

Example 7. *Show that* $\displaystyle\int_0^{\infty} \sqrt{x}\, e^{-x^3}\, dx = \frac{1}{3}\sqrt{\pi}$.

Sol. Put $x^3 = z$* so that $\displaystyle dx = \frac{dz}{3x^2}$

Also, when $x = 0, z = 0$; when $x \to \infty, z \to \infty$.

$$\therefore \qquad \int_0^{\infty} \sqrt{x}\, e^{-x^3}\, dx = \int_0^{\infty} \sqrt{x}\, e^{-z} \frac{dz}{3x^2}$$

$$= \frac{1}{3}\int_0^{\infty} x^{-3/2} e^{-z}\, dz = \frac{1}{3}\int_0^{\infty} (z^{1/3})^{-3/2} e^{-z}\, dz$$

$$= \frac{1}{3}\int_0^{\infty} z^{-1/2} e^{-z} = \frac{1}{3}\Gamma(\tfrac{1}{2}) = \frac{1}{3}\sqrt{\pi} \qquad (\because \; \Gamma(\tfrac{1}{2}) = \sqrt{\pi})$$

---

*In an integral of the type $\int x^n e^{-f(x)}\, dx$, put $f(x) = z$.

Example 8. *Show that* $\dfrac{B(m+2, n-2)}{B(m, n)} = \dfrac{m(m+1)}{(n-1)(n-2)}$.

Sol. $$\frac{B(m+2, n-2)}{B(m, n)} = \frac{\dfrac{\Gamma(m+2)\Gamma(n-2)}{\Gamma(m+2+n-2)}}{\dfrac{\Gamma(m)\Gamma(n)}{\Gamma(m+n)}}$$

$$= \frac{\Gamma(m+2)\Gamma(n-2)}{\Gamma(m)\Gamma(n)} = \frac{(m+1)m\Gamma(m)\Gamma(n-2)}{\Gamma(m)(n-1)(n-2)\Gamma(n-2)}$$

$[\because \ \Gamma(n) = (n-1)\,\Gamma(n-1)]$

$$= \frac{m(m+1)}{(n-1)(n-2)}.$$

Example 9. *Evaluate*

(*i*) $\displaystyle\int_0^\infty e^{-4x}\, x^{3/2}\, dx$ (*ii*) $\displaystyle\int_0^1 x^4 (1-x)^3\, dx$ (*iii*) $\displaystyle\int_0^2 \frac{x^2}{\sqrt{2-x}}\, dx.$

Sol. (*i*) Putting $4x = y$, we have $dx = \frac{1}{4}\, dy$.

Also, when $x = 0, y = 0$; when $x \to \infty, y \to \infty$

$$\therefore \quad \int_0^\infty e^{-4x}\, x^{3/2}\, dx = \frac{1}{32} \int_0^\infty e^{-y}\, y^{5/2-1}\, dy$$

$$= \frac{1}{32} \cdot \Gamma\left(\frac{5}{2}\right) = \frac{1}{32} \cdot \frac{3}{2} \cdot \frac{1}{2} \cdot \Gamma\left(\frac{1}{2}\right) = \frac{3}{128} \sqrt{\pi}.$$

(*ii*) $$\int_0^1 x^4 (1-x)^3\, dx = B(5, 4)$$

$$= \frac{\Gamma(5)\Gamma(4)}{\Gamma(9)} = \frac{4!\,3!}{8!} = \frac{1}{280}.$$

(*iii*) Putting $x = 2y$ so that $dx = 2dy$

Also, when $x = 0, y = 0$; when $x = 2, y = 1$

$$\therefore \quad \int_0^2 \frac{x^2 \,.\, dx}{\sqrt{2-x}} = 4\sqrt{2} \int_0^1 y^2 (1-y)^{-1/2}\, dy$$

$$= 4\sqrt{2}\ B(3, \tfrac{1}{2}) = 4\sqrt{2}\ \frac{\Gamma(3)\Gamma(\frac{1}{2})}{\Gamma\left(\frac{7}{2}\right)} = \frac{4\sqrt{2}\,.\,2!\,\Gamma(\frac{1}{2})}{\frac{5}{2} \cdot \frac{3}{2} \cdot \frac{1}{2} \cdot \Gamma\left(\frac{1}{2}\right)} = \frac{64\sqrt{2}}{15}.$$

Example 10. *Evaluate*

(*i*) $\displaystyle\int_0^\infty \frac{x^8 (1-x^6)}{(1+x)^{24}}\, dx$. (*ii*) $\displaystyle\int_0^\infty \frac{x^4 (1+x^5)}{(1+x)^{15}}\, dx.$

Sol. (*i*) $$I = \int_0^\infty \frac{x^6 (1-x^6)}{(1+x)^{24}}\, dx$$

$$= \int_0^\infty \frac{x^8}{(1+x)^{24}}\, dx - \int_0^\infty \frac{x^{14}}{(1+x)^{24}}\, dx$$

$$= \int_0^\infty \frac{x^{9-1}}{(1+x)^{9+15}}\,dx - \int_0^\infty \frac{x^{15-1}}{(1+x)^{15+9}}\,dx$$

$$= B(9, 15) - B(15, 9) = 0 \qquad [\because \; B(m, n) = B(n, m)]$$

(*ii*)

$$I = \int_0^\infty \frac{x^4(1+x^5)}{(1+x)^{15}}\,dx$$

$$= \int_0^\infty \frac{x^4}{(1+x)^{15}}\,dx + \int_0^\infty \frac{x^9}{(1+x)^{15}}\,dx$$

$$= \int_0^\infty \frac{x^{5-1}}{(1+x)^{5+10}}\,dx + \int_0^\infty \frac{x^{10-1}}{(1+x)^{10+5}}\,dx$$

$$= B(5, 10) + B(10, 5) = 2B(5, 10)$$

$$= \frac{2\Gamma(5)\,.\,\Gamma(10)}{\Gamma(5+10)} = \frac{2\,.\,4!\,.\,9!}{14!} = \frac{1}{5005}.$$

11.2.3. *To evaluate* $\int_0^{\pi/2} \sin^p x \cos^q x\,dx,\ p > -1,\ q > -1.$

Proof. Put $\sin^2 x = y$ so that $2 \sin x \cos x\,dx = dy$.

Also, $\cos x = \sqrt{1-\sin^2 x} = \sqrt{1-y}$

$\therefore$ $dx = \dfrac{dy}{2 \sin x \cos x} = \dfrac{dy}{2\sqrt{y}\,\sqrt{1-y}}$

Also, when $x = 0$, $y = 0$; when $x = \dfrac{\pi}{2}$, $y = 1$

Therefore,

$$\int_0^{\pi/2} \sin^p x \cos^q x\,dx = \int_0^1 (\sqrt{y})^p (\sqrt{1-y})^q \frac{2}{2\sqrt{y}\,\sqrt{1-y}}\,dy$$

$$= \frac{1}{2}\int_0^1 y^{\frac{p-1}{2}} (1-y)^{\frac{q-1}{2}}\,dy$$

$$= \frac{1}{2} B\left(\frac{p-1}{2}+1, \frac{q-1}{2}+1\right)$$

$$= \frac{1}{2} B\left(\frac{p+1}{2}, \frac{q+1}{2}\right) = \frac{1}{2} \frac{\Gamma\left(\frac{p+1}{2}\right)\Gamma\left(\frac{q+1}{2}\right)}{\Gamma\left(\frac{p+1}{2}+\frac{q+1}{2}\right)}$$

Thus,

$$\int_0^{\pi/2} \sin^p x \cos^q x\,dx = \frac{\Gamma\left(\frac{p+1}{2}\right)\Gamma\left(\frac{q+1}{2}\right)}{2\Gamma\left(\frac{p+1}{2}+\frac{q+1}{2}\right)},\ p > -1, q > -1.$$

Example 11. *Evaluate*

(*i*) $\int_0^{\pi/2} \sin^3 x \cos^{5/2} dx$ (*ii*) $\int_0^{\pi/2} \sin^5 x\,dx.$

Sol. (*i*) $\int_0^{\pi/2} \sin^3 x \cos^{5/2} dx$ $\left(\text{Here } p = 3, q = \frac{5}{2}\right)$

$$= \frac{\Gamma\left(\frac{3+1}{2}\right)\Gamma\left(\frac{\frac{5}{2}+1}{2}\right)}{2\Gamma\left(\frac{3+1}{2}+\frac{\frac{5}{2}+1}{2}\right)} = \frac{\Gamma(2)\,\Gamma\left(\frac{7}{4}\right)}{2\Gamma\left(\frac{15}{4}\right)} = \frac{1.\,\Gamma\left(\frac{7}{4}\right)}{2.\frac{11}{4}\frac{7}{4}\Gamma\left(\frac{7}{4}\right)} = \frac{8}{77}.$$

(*ii*) $\int_0^{\pi/2} \sin^5 x\, dx = \int_0^{\pi/2} \sin^5 x \cos^0 x\, dx$ (Here $p = 5, q = 0$)

$$= \frac{\Gamma\left(\frac{5+1}{2}\right)\Gamma\left(\frac{0+1}{2}\right)}{2\Gamma\left(\frac{5+1}{2}+\frac{0+1}{2}\right)} = \frac{\Gamma(3)\,\Gamma(\frac{1}{2})}{2\Gamma\left(\frac{7}{2}\right)}$$

$$= \frac{2\,!\,\Gamma(\frac{1}{2})}{2.\frac{5}{2}.\frac{3}{2}.\frac{1}{2}\Gamma\left(\frac{1}{2}\right)} = \frac{8}{15}.$$

11.2.4. Duplication Formula. *To prove that*

$$\Gamma(m)\,\Gamma\left(m+\frac{1}{2}\right) = \frac{\sqrt{\pi}}{2^{2m-1}}\,\Gamma(2m).$$ (K.U. BCA (II) 2003, 7)

Proof. We know

$$B(m, m) = \int_0^1 x^{m-1}(1-x)^{m-1}\,dx \qquad ...(1)$$

Putting $x = \sin^2\theta$, we have

$$dx = 2\sin\theta\cos\theta\,d\theta$$

Also, when $x = 0, \theta = 0$; when $x = 1, \theta = \frac{\pi}{2}$

$\therefore$ (1) becomes

$$B(m, m) = \int_0^{\pi/2} (\sin^2\theta)^{m-1}(1-\sin^2\theta)^{m-1}\,2\sin\theta\cos\theta\,d\theta$$

$$= 2\int_0^{\pi/2} (\sin\theta\cos\theta)^{2m-1}\,d\theta = 2\int_0^{\pi/2}\left(\frac{\sin 2\theta}{2}\right)^{2m-1} d\theta$$

$$= \frac{2}{2^{2m-1}}\int_0^{\pi/2} \sin^{2m-1} 2\theta\,d\theta \qquad ...(2)$$

Putting $2\theta = \phi$, we have $d\theta = \frac{1}{2}\,d\phi$.

Also, when $\theta = 0, \phi = 0$; when $\theta = \frac{\pi}{2}, \phi = \pi$

$\therefore$ from (2),

$$B(m, m) = \frac{1}{2^{2m-1}} \int_0^{\pi} \sin^{2m-1} \phi \, d\phi$$

$$= \frac{2}{2^{2m-1}} \int_0^{\pi/2} \sin^{2m-1} \phi \, d\phi \qquad (\because \sin^{2m-1}(\pi - \phi) = \sin^{2m-1} \phi)$$

$$= \frac{2}{2^{2m-1}} \int_0^{\pi/2} \sin^{2m-1} \phi \,.\, \cos^0 \phi \, d\phi$$

$$= \frac{2}{2^{2m-1}} \frac{\Gamma\left(\frac{2m-1+1}{2}\right)\Gamma\left(\frac{0+1}{2}\right)}{2\Gamma\left(\frac{2m-1+1}{2}+\frac{0+1}{2}\right)} = \frac{1}{2^{2m-1}} \frac{\Gamma(m)\,\Gamma(\frac{1}{2})}{\Gamma(m+\frac{1}{2})}$$

$$\Rightarrow \qquad \frac{\Gamma(m)\,\Gamma(m)}{\Gamma(m+m)} = \frac{1}{2^{2m-1}} \frac{\Gamma(m)\,\Gamma(\frac{1}{2})}{\Gamma(m+\frac{1}{2})}$$

Cancelling $\Gamma(m)$ and using $\Gamma(\frac{1}{2}) = \sqrt{\pi}$, we have

$$\frac{\Gamma(m)}{\Gamma(2m)} = \frac{\sqrt{\pi}}{2^{2m-1}\,\Gamma(m+\frac{1}{2})}$$

$$\Rightarrow \qquad \Gamma(m)\,\Gamma(m+\tfrac{1}{2}) = \frac{\sqrt{\pi}\,\Gamma(2m)}{2^{2m-1}}$$

Example 12. *Prove that*

$$B(m, m)\,.\,B\left(m+\frac{1}{2}, m+\frac{1}{2}\right) = \frac{\pi\, m^{-1}}{2^{4m-1}}.$$

Sol. $\quad$ L.H.S. $= B(m, m).\, B(m+\frac{1}{2}, m+\frac{1}{2})$

$$= \frac{\Gamma(m)\Gamma(m)}{\Gamma(2m)} \cdot \frac{\Gamma(m+\frac{1}{2})\,\Gamma(m+\frac{1}{2})}{\Gamma(m+\frac{1}{2}+m+\frac{1}{2})}$$

$$= \frac{[\Gamma(m)\Gamma(m+\frac{1}{2})]^2}{\Gamma(2m)\,.\,\Gamma(2m+1)}$$

$$= \frac{[\Gamma(m)\Gamma(m+\frac{1}{2})]^2}{\Gamma(2m)\,.\,2m\,\Gamma(2m)} \qquad [\because \Gamma(n) = (n-1)\Gamma(n-1)]$$

$$= \left(\frac{\sqrt{\pi}\,\Gamma(2m)}{2^{2m-1}}\right)^2 \cdot \frac{1}{2m\,[\Gamma(2m)]^2} \qquad \text{[by Duplication Formula]}$$

$$= \frac{\pi}{2^{4m-2}\,.\,2m} = \frac{\pi\, m^{-1}}{2^{4m-1}}.$$

EXERCISE 11 (*b*)

1. Show that

(*i*) $\int_0^\infty x^3 e^{-x} \, dx = 6$

(*ii*) $\int_0^\infty \sqrt{x} \, e^{-x^3} \, dx = \frac{1}{3}\sqrt{\pi}$

(*iii*) $\int_0^\infty 4x^4 e^{-x^4} \, dx = \Gamma\left(\frac{5}{4}\right)$

(*iv*) $\int_0^\infty x^6 e^{-2x} \, dx = \frac{45}{8}$

(*v*) $\int_0^1 x^2 (1-x)^3 \, dx = \frac{1}{60}$

(*vi*) $\int_0^3 \frac{dx}{\sqrt{3x - x^2}} = \pi$

(*vii*) $\int_0^1 \frac{dx}{\sqrt{1-x^3}} = \frac{\sqrt{\pi}\,\Gamma\left(\frac{1}{3}\right)}{3\Gamma\left(\frac{5}{6}\right)}$.

2. Express $\int_0^1 x^m (1 - x^n)^p \, dx$ in terms of the Beta function and hence evaluate $\int_0^1 x^5 (1 - x^3)^{10} \, dx$.

3. Prove that $\Gamma(n + 1) = n\,\Gamma(n)$, where $n > 0$.

4. Prove that $\int_0^\infty e^{-ax} x^{n-1} \, dx = \frac{\Gamma(n)}{a^n}$, where $a > 0, n > 0$.

[**Hint.** Put $ax = z$.]

5. (*i*) Prove that $\int_0^\infty x^n e^{-a^2x^2} \, dx = \frac{1}{2a^{n+1}} \Gamma\left(\frac{n+1}{2}\right)$

Hence show that $\int_{-\infty}^\infty e^{-a^2x^2} \, dx = \frac{\sqrt{\pi}}{a}$.

[**Hint.** Put $a^2x^2 = z$.]

(*ii*) Prove that $\int_0^\infty e^{-a^2x^2} \, dx = \frac{\sqrt{\pi}}{2a}$, $a > 0$.

6. Show that $\int_0^1 \frac{x^{m-1} (1-x)^{n-1}}{(a+x)^{m+n}} \, dx = \frac{\Gamma(m)\Gamma(n)}{a^n (1+a)^m \, \Gamma(m+n)}$.

[**Hint.** See Example 3.]

7. Show that $\int_0^1 \frac{dx}{\sqrt{1-x^n}} = \frac{\sqrt{\pi}\,\Gamma\left(\frac{1}{n}\right)}{n\,\Gamma\left(\frac{1}{n} + \frac{1}{2}\right)}$.

$\left[\textbf{Hint}. \text{ Put } x^n = z \text{ and use } B(m, n) = \frac{\Gamma(m)\,\Gamma(n)}{\Gamma(m+n)}.\right]$

8. (*i*) Prove that $\frac{B(p, q+1)}{q} = \frac{B(p+1, q)}{p} = \frac{B(p, q)}{p+q}$, $p > 0, q > 0$.

(*ii*) Prove that $\frac{B(m+1, n)}{B(m, n)} = \frac{m}{m+n}$, $m > 0, n > 0$.

9. (*i*) Show that

$$\int_0^{\pi/2} \sin^n \theta \, d\theta = \frac{\sqrt{\pi}}{2} \frac{\Gamma\left(\frac{n+1}{2}\right)}{\Gamma\left(\frac{n+2}{2}\right)}, \; n > -1.$$

(*ii*) Show that $\int_0^{\pi/2} \sqrt{\sin \theta} \, d\theta \times \int_0^{\pi/2} \frac{d\theta}{\sqrt{\sin \theta}} = \pi.$

10. Show that $\int_0^{\infty} \frac{x^a}{a^x} \, dx = \frac{\Gamma(a+1)}{(\log a)^{n+1}}$, if $a > 1$.

[**Hint.** $a^x = e^{x \log a}$. Put $x \log a = z$.]

11. Prove that $B(m, m) = 2^{1-2m} B(m, \frac{1}{2})$. [See solution of Duplication formula]

12

Differentiation under the Integral Sign

In this chapter, we are going to discuss the technique of differentiation under the integral sign due to Leibnitz, in evaluating some definite integrals which are difficult otherwise.

12.1. LEIBNITZ'S RULE FOR DIFFERENTIATION UNDER INTEGRAL SIGN

If $f(x, \alpha)$ is a function of two variables x and α, α being a parameter, then $\int_a^b f(x, \alpha)\, dx$ is a function of α.

For example,

$$\int_1^2 (x + \alpha)^2\, dx = \left[\frac{(x + \alpha)^3}{3}\right]_1^2 = \frac{1}{3}\, [3\alpha^2 + 9\alpha + 7]$$

which is a function of α.

12.1.1. Leibnitz's Rule. *If $f(x, \alpha)$ and $\frac{\partial}{\partial \alpha} f(x, \alpha)$ are continuous functions of x and α for $a \le x \le b$, $c \le \alpha \le d$, a, b being independent of α, then*

$$\frac{d}{d\alpha} \int_a^b f(x, \alpha)\, dx = \int_a^b \frac{\partial}{\partial \alpha} f(x, \alpha)\, dx.$$

Proof. Let $$\phi(\alpha) = \int_a^b f(x, \alpha)\, dx \qquad ...(1)$$

Let α change to $\alpha + \delta\alpha$, (α and $\alpha + \delta\alpha$ both in $[c, d]$) then a, b, x being independent of α, remain unaltered and $\phi(\alpha)$ changes to $\phi(\alpha + \delta\alpha)$.

$$\therefore \quad \phi\,(\alpha + \delta\alpha) = \int_a^b f(x, \alpha + \delta\alpha)\, dx. \qquad ...(2)$$

(2) – (1) gives,

$$\phi(\alpha + \delta\alpha) - \phi(\alpha) = \int_a^b (f(x, \alpha + \delta\alpha) - f(x, \alpha))\, dx$$

But, by mean value theorem of differential calculus,

$$f(x, \alpha + \delta\alpha) - f(x, \alpha) = \delta\alpha.\, \frac{\partial}{\partial \alpha} f(x, \alpha + \theta\delta\alpha),\ 0 < \theta < 1$$

$$\therefore \quad \phi(\alpha + \delta\alpha) - \phi(\alpha) = \int_a^b \delta\alpha\, .\, \frac{\partial}{\partial \alpha} f(x, \alpha + \theta\delta\alpha)\, dx$$

$$\Rightarrow \quad \frac{\phi(\alpha + \delta\alpha) - \phi(\alpha)}{\delta\alpha} = \int_a^b \frac{\partial}{\partial \alpha} f(x, \alpha + \theta\delta\alpha)\, dx$$

Taking limits as $\delta\alpha \to 0$, we have

$$\frac{d\phi}{d\alpha} = \lim_{\delta\alpha \to 0} \int_a^b \frac{\partial}{\partial \alpha} f(x, \alpha + \theta\delta\alpha)\, dx$$

$$= \int_a^b \left(\lim_{\delta\alpha \to 0} \frac{\partial}{\partial \alpha} f(x, \alpha + \theta\delta\alpha) \right) dx$$

$$= \int_a^b \frac{\partial}{\partial \alpha} f(x, \alpha)\, dx \qquad \left[\because \frac{\partial}{\partial \alpha} f(x,\alpha) \text{ is continuous}\right]$$

Remark. Leibnitz's rule shows that the operations of differentiation w.r.t. the parameter and integration w.r.t. the variable are interchangeable, if limits of integration are independent of the parameter.

We state below Leibnitz's rule (without proof) when limits of integration involve the parameter.

12.1.2. Leibnitz's rule when limits are functions of parameter. If a and b are not constants but as continuous and differentiable functions of α, then

$$\frac{d}{d\alpha} \int_a^b f(x, d)\, dx = \int_a^b \frac{\partial}{\partial \alpha} f(x, \alpha)\, dx + f(b, \alpha) \frac{db}{d\alpha} - f(a, \alpha) \frac{da}{d\alpha}.$$

Remark. The above results may not be applicable in the case of improper integrals, as the question of validity of the results of improper integrals, requires more investigations. This is, however, beyond the scope of this book. Whenever we deal with any improper integral, it is assumed that the necessary conditions for the validity of the results are satisfied.

12.1.3. Procedure for using the rule of differentiation under the integral sign

(*i*) Denote the given definite integral by $\phi(\alpha)$, if α is the parameter.

(*ii*) Differentiate both sides w.r.t. α, and interchange the sign of differentiation and integration on the right hand side (Leibnitz's rule).

(*iii*) Integrate right hand side w.r.t. the variable within the given limits.

(*iv*) Integrate both sides w.r.t. α.

(*v*) Evaluate the constant of integration by giving suitable value to the parameter α.

Following examples will illustrate the above procedure.

Example 1. *If* $|\ a\ | < 1$, *evaluate*

$$\int_0^\pi \frac{\log(1 + a\cos x)}{\cos x}\, dx.$$

Sol. Here a is a parameter.

Let
$$\phi(a) = \int_0^\pi \frac{\log(1 + a\cos x)}{\cos x}\, dx \qquad \text{...(1)}$$

Differentiating both sides w.r.t. the parameter a, we have

$$\frac{d\phi(a)}{da} = \int_0^\pi \frac{\partial}{\partial a}\left[\frac{\log(1 + a\cos x)}{\cos x}\right] dx \qquad \text{(By Leibnitz's Rule)}$$

$$= \int_0^\pi \frac{1}{\cos x} \cdot \frac{\partial}{\partial a} \log(1 + a\cos x)\, dx$$

$$= \int_0^\pi \frac{1}{\cos x} \cdot \frac{1}{1 + a\cos x} \cdot \cos x\, dx$$

$$= \int_0^{\pi} \frac{1}{1+a\cos x}\,dx$$

$$= \frac{1}{\sqrt{1-a^2}}\left[\cos^{-1}\frac{a+\cos x}{1+a\cos x}\right]_0^{\pi}$$

$\left[\because \text{ if } a^2 > b^2, \text{ then } \int \frac{1}{a+b\cos x}\,dx = \frac{1}{\sqrt{a^2-b^2}}\cos^{-1}\frac{b+a\cos x}{a+b\cos x}\right.$

Here, $|\ a\ | < 1$ so that $a^2 < 1$]

$$= \frac{1}{\sqrt{1-a^2}}\left[\cos^{-1}\frac{a+\cos\pi}{1+a\cos\pi} - \cos^{-1}\frac{a+\cos 0}{1+a\cos 0}\right]$$

$$= \frac{1}{\sqrt{1-a^2}}\left[\cos^{-1}\frac{a-1}{1-a} - \cos^{-1}\frac{a+1}{1+a}\right]$$

$$= \frac{1}{\sqrt{1-a^2}}(\cos^{-1}(-1) - \cos^{-1}1)$$

$$\Rightarrow \qquad \frac{d}{da}\phi(a) = \frac{1}{\sqrt{1-a^2}}(\pi - 0) = \frac{\pi}{\sqrt{1-a^2}}$$

Integrating both sides w.r.t. a, we have

$$\phi(a) = \pi\sin^{-1}a + C \qquad \text{...(2)}$$

where C is constant of integration.

To find C, put $a = 0$ in (1) and (2).

From (1), $\qquad \phi(0) = \int_0^{\pi}\frac{\log 1}{\cos x}\,dx = 0.$

From (2), $\qquad \phi(0) = \pi\sin^{-1}0 + C = 0 + C = C$

Thus, C = 0.

$\therefore$ from(2), $\qquad \phi(a) = \pi\sin^{-1}a$

or $\qquad \int_0^{\pi}\frac{\log(1+a\cos x)}{\cos x}\,dx = \pi\sin^{-1}a.$

Example 2. *Prove that if* $a > b$,

$$\int_0^{\pi/2} \log\left(\frac{a+b\sin\theta}{a-b\sin\theta}\right)\frac{d\theta}{\sin\theta} = \pi\sin^{-1}\frac{b}{a}. \qquad \text{(K.U. BCA (II) 2003)}$$

Sol. Here, the integrand contains two parameters a and b. Because of the presence of $\frac{1}{\sin\theta}$ as a factor in the integrand, we consider only b as parameter.

Let $$\phi(b) = \int_0^{\pi/2}\log\left(\frac{a+b\sin\theta}{a-b\sin\theta}\right)\frac{d\theta}{\sin\theta} \qquad \text{...(1)}$$

$$\Rightarrow \qquad \frac{d\phi(b)}{db} = \int_0^{\pi/2}\frac{\partial}{\partial b}\log\left(\frac{a+b\sin\theta}{a-b\sin\theta}\right)\frac{d\theta}{\sin\theta} \qquad \text{(By Leibnitz's Rule)}$$

$$= \int_0^{\pi/2} \frac{\partial}{\partial b} [\log (a + b \sin \theta) - \log (a - b \sin \theta)] \frac{d\theta}{\sin \theta}$$

$$= \int_0^{\pi/2} \left[\frac{\sin \theta}{a + b \sin \theta} - \frac{-\sin \theta}{a - b \sin \theta} \right] \frac{d\theta}{\sin \theta}$$

$$= \int_0^{\pi/2} \left[\frac{2a}{a^2 - b^2 \sin^2 \theta} \right] d\theta$$

$$= 2a \int_0^{\pi/2} \frac{d\theta}{a^2 - b^2 \sin^2 \left(\frac{\pi}{2} - \theta \right)} \qquad \left(\because \int_0^a f(x)\, dx = \int_0^a f(a - x)\, dx \right)$$

$$= 2a \int_0^{\pi/2} \frac{d\theta}{a^2 - b^2 \cos^2 \theta}$$

Multiplying the numerator and denominator of the integrand by $\sec^2 \theta$, we have

$$\frac{d\phi(b)}{db} = 2a \int_0^{\pi/2} \frac{\sec^2 \theta}{a^2 \sec^2 \theta - b^2} d\theta$$

Put $\tan \theta = t$ so that $\sec^2 \theta \, d\theta = dt$

Also, when $\theta = 0$; $t = 0$; when $\theta = \frac{\pi}{2}$, $t = \infty$

$$\therefore \qquad \frac{d\phi(b)}{db} = 2a \int_0^{\infty} \frac{dt}{(a^2 - b^2) + a^2 t^2}$$

$$= \frac{2a}{a^2} \int_0^{\infty} \frac{dt}{(t)^2 + \left(\frac{\sqrt{a^2 - b^2}}{a} \right)^2}$$

$$= \frac{2}{a} \cdot \frac{1}{\frac{\sqrt{a^2 - b^2}}{a}} \left[\tan^{-1} \left(\frac{at}{\sqrt{a^2 - b^2}} \right) \right]_0^{\infty} = \frac{2}{\sqrt{a^2 - b^2}} \cdot \frac{\pi}{2}$$

$$\Rightarrow \qquad \frac{d\phi(b)}{db} = \frac{\pi}{\sqrt{a^2 - b^2}}$$

Integrating both sides w.r.t. the parameter b, we have

$$\phi(b) = \pi \sin^{-1} \frac{b}{a} + C \qquad \ldots(2)$$

where C is the constant of integration.

To find C, put $b = 0$ in (1) and (2).

From (1), $\qquad \phi(0) = \int_0^{\pi/2} \log (1) \frac{d\theta}{\sin \theta} = 0$

From (2), $\qquad \phi(0) = \pi \sin^{-1} 0 + C = C$

$\therefore \qquad C = 0$

$\therefore$ from (2), $\qquad \phi(b) = \pi \sin^{-1} \dfrac{b}{a}$

i.e.,
$$\int_0^{\pi/2} \log \left(\frac{a + b\sin\theta}{a - b\sin\theta} \right) \frac{d\theta}{\sin\theta} = \pi \sin^{-1} \frac{b}{a}.$$

Example 3. *If $\alpha > 0$, $\beta > 0$, prove that*
$$\int_0^{\pi/2} \log(\alpha\cos^2\theta + \beta\sin^2\theta)\, d\theta = \pi \log\left[\tfrac{1}{2}(\sqrt{\alpha} + \sqrt{\beta})\right].$$

Sol. Here, the integrand involves two arbitrary constants α and β.

Let
$$\phi(\alpha, \beta) = \int_0^{\pi/2} \log(\alpha\cos^2\theta + \beta\sin^2\theta)\, d\theta \qquad ...(1)$$

Differentiating w.r.t. α,
$$\frac{\partial\phi(\alpha,\beta)}{\partial\alpha} = \int_0^{\pi/2} \frac{\partial}{\partial\alpha} \log(\alpha\cos^2\theta + \beta\sin^2\theta)\, d\theta$$
$$= \int_0^{\pi/2} \frac{1}{\alpha\cos^2\theta + \beta\sin^2\theta} \cdot \cos^2\theta\, d\theta$$

Dividing the numerator and denominator of the integrand by $\cos^2\theta$, we have
$$\frac{\partial\phi(\alpha,\beta)}{\partial\alpha} = \int_0^{\pi/2} \frac{d\theta}{\alpha + \beta\tan^2\theta}$$

Put $\tan\theta = t$ so that $\sec^2\theta\, d\theta = dt$

Also, when $\theta = 0$, $t = 0$; when $\theta = \dfrac{\pi}{2}$, $t = \infty$

$$\therefore \quad \frac{\partial\phi(\alpha,\beta)}{\partial\alpha} = \int_0^{\infty} \frac{dt}{(\alpha + \beta t^2)(1 + t^2)} \qquad ...(2)$$

To resolve $\dfrac{1}{(\alpha + \beta t^2)(1 + t^2)}$ into partial fractions, let $t^2 = y$

$$\therefore \quad \frac{1}{(\alpha + \beta t^2)(1 + t^2)} = \frac{1}{(a + \beta y)(1 + y)}$$
$$= \frac{\beta}{\beta - \alpha} \cdot \frac{1}{\alpha + \beta y} + \frac{1}{\alpha - \beta} \cdot \frac{1}{1 + y} \qquad \text{(Partial fractions)}$$
$$= \frac{\beta}{\beta - \alpha} \cdot \frac{1}{\alpha + \beta t^2} + \frac{1}{\alpha - \beta} \cdot \frac{1}{1 + t^2} \qquad \text{(Replacing } y \text{ by } t^2\text{)}$$

$\therefore$ (2) becomes
$$\frac{\partial\phi(\alpha,\beta)}{\partial\alpha} = \frac{1}{\beta - \alpha} \int_0^{\infty} \frac{\beta}{\alpha + \beta t^2} - \frac{1}{t^2 + 1}\, dt$$
$$= \frac{1}{\beta - \alpha} \left[\int_0^{\infty} \frac{dt}{\frac{\alpha}{\beta} + t^2} - \int_0^{\infty} \frac{dt}{t^2 + 1} \right]$$

$$= \frac{1}{\beta - \alpha}\left[\left[\frac{1}{\sqrt{\frac{\alpha}{\beta}}}\tan^{-1}\frac{t}{\sqrt{\frac{\alpha}{\beta}}}\right]_0^{\infty} - \left[\tan^{-1} t\right]_0^{\infty}\right]$$

$$= \frac{1}{\beta - \alpha}\left[\sqrt{\frac{\beta}{\alpha}}(\tan^{-1}\infty - \tan^{-1} 0) - (\tan^{-1}\infty - \tan^{-1} 0)\right]$$

$$= \frac{1}{\beta - \alpha}\left[\sqrt{\frac{\beta}{\alpha}}\frac{\pi}{2} - \frac{\pi}{2}\right]$$

$$= \frac{\sqrt{\beta} - \sqrt{\alpha}}{\sqrt{\alpha}\,(\beta - \alpha)} \cdot \frac{\pi}{2} = \frac{\sqrt{\beta} - \sqrt{\alpha}}{\sqrt{\alpha}\,(\sqrt{\beta} + \sqrt{\alpha})(\sqrt{\beta} - \sqrt{\alpha})}\,\frac{\pi}{2}$$

$$\Rightarrow \qquad \frac{\partial \phi(\alpha, \beta)}{\partial \alpha} = \frac{\pi}{2}\,\frac{1}{\sqrt{\alpha}\,(\sqrt{\beta} + \sqrt{\alpha})}$$

Integrating both sides w.r.t. α,

$$\phi(\alpha, \beta) = \pi \int \frac{\frac{1}{2\sqrt{\alpha}}}{\sqrt{\alpha} + \sqrt{\beta}}\,d\alpha + C$$

where C is constant of integration.

$$\Rightarrow \qquad \phi(\alpha, \beta) = \pi \log(\sqrt{\alpha} + \sqrt{\beta}) + C \qquad \ldots(3) \qquad \left(\because \int \frac{f'(x)}{f(x)}\,dx = \log f(x)\right)$$

To find C, put $\alpha = 1, \beta = 1$, in (1) and (3).

From (1), $$\phi(1, 1) = \int_0^{\pi/2} \log(\cos^2\theta + \sin^2\theta)\,d\theta$$

$$= \int_0^{\pi/2} \log 1\,d\theta = 0.$$

From (3), $\phi(1, 1) = \pi \log(1 + 1) + C$

$\therefore \qquad \pi \log 2 + C = 0 \Rightarrow C = -\pi \log 2$

$\therefore$ from (3), $$\phi(\alpha, \beta) = \pi \log(\sqrt{\alpha} + \sqrt{\beta}) - \pi \log 2$$

$$= \pi \log \frac{\sqrt{\alpha} + \sqrt{\beta}}{2}$$

i.e. $$\int_0^{\pi/2} \log(\alpha\cos^2\theta + \alpha\sin^2\theta)\,d\theta = \pi \log \frac{\sqrt{\alpha} + \sqrt{\beta}}{2}.$$

Example 4. *Show that* $\int_0^{\infty} \frac{\tan^{-1} ax}{x(1 + x^2)}\,dx = \frac{\pi}{2} \log(1 + a),\ a \geq 0.$

Also, find the value of the integral if $a < 0$. (K.U. BCA (II) 2001)

Sol. Let $$\phi(a) = \int_0^{\infty} \frac{\tan^{-1} ax}{x(1 + x^2)}\,dx \qquad \ldots(1)$$

$$\Rightarrow \qquad \frac{d\phi(a)}{da} = \int_0^\infty \frac{\partial}{\partial a}\left(\frac{\tan^{-1} ax}{x(1+x^2)}\right) dx$$

$$= \int_0^\infty \frac{1}{(1+x^2)(1+a^2x^2)}\, dx$$

(Resolving the integrand into partial fractions)

$$= \frac{1}{1-a^2}\int_0^\infty \left(\frac{1}{1+x^2} - \frac{a^2}{1+a^2x^2}\right) dx$$

$$= \frac{1}{1-a^2}\left[\tan^{-1} x - a\tan^{-1}(ax)\right]_0^\infty \qquad ...(2)$$

Case (*i*) **a ≥ 0**

$$\frac{d\phi(a)}{da} = \frac{1}{1-a^2}\left[\frac{\pi}{2} - a\,.\frac{\pi}{2}\right] = \frac{\pi}{2(1+a)}$$

Integrating w.r.t. a, we get

$$\phi(a) = \frac{\pi}{2}\log(1+a) + C \qquad ...(3)$$

where C is the constant of integration.

To find C, let $a = 0$ in (1) as well as in (3).

From (1), $\phi(0) = 0$

and from (3), $\phi(0) = 0 + C$

$\therefore$ C = 0

Hence, $\phi(a) = \dfrac{\pi}{2}\log(1+a)$.

Case (*ii*) **a < 0.**

From (2), $$\frac{d\phi(a)}{da} = \frac{1}{1-a^2}\left[\frac{\pi}{2} - a\left(-\frac{\pi}{2}\right)\right] = \frac{\pi}{2(1-a)}$$

$$\left(\because \text{ when } a < 0,\ \lim_{x\to\infty} \tan^{-1}(ax) = -\frac{\pi}{2}\right)$$

Integrating, we have

$$\phi(a) = -\frac{\pi}{2}\log(1-a) + C' \qquad ...(4)$$

where C′ is constant of integration.

Using the initial conditions as in case (*i*), we find that C′ = 0.

Hence, from (4),

$$\phi(a) = -\frac{\pi}{2}\log(1-a).$$

Example 5. *Evaluate* $\displaystyle\int_0^a \frac{\log(1+ax)}{1+x^2}\, dx,\ a > 0.$

Sol. Here, the limits of integration involve the parameter a.

Let $$\phi(a) = \int_0^a \frac{\log(1+ax)}{1+x^2}\,dx \qquad ...(1)$$

Differentiating w.r.t. a,

$$\frac{d\phi(a)}{da} = \int_0^a \frac{\partial}{\partial a}\frac{\log(1+ax)}{1+x^2}\,dx + \frac{\log(1+a^2)}{1+a^2}\frac{da}{da} - \frac{\log(1+a.0)}{1+0}\cdot\frac{d(0)}{da}$$

$$= \int_0^a \frac{1}{1+x^2}\,\frac{1}{1+ax}\,.\,x\,dx\;\frac{\log(1+a^2)}{1+a^2} \qquad ...(2)$$

To resolve the integrand into partial fractions

Let $$\frac{x}{(1+ax)(1+x^2)} = \frac{A}{1+ax} + \frac{Bx+C}{1+x^2} \qquad ...(3)$$

$\Rightarrow$ $$x = A(1+x^2) + (Bx+C)(1+ax)$$

Putting $1 + ax = 0$, *i.e.*, $x = -\dfrac{1}{a}$, we get

$$A = \frac{-a}{1+a^2}.$$

Comparing coefficients of x^2, we get

$$0 = A + aB \qquad \Rightarrow \quad B = -\frac{A}{a} = \frac{1}{1+a^2}.$$

Comparing constant terms, we have

$$0 = A + C \qquad \Rightarrow \quad C = -A = \frac{a}{1+a^2}$$

$\therefore$ from (3),

$$\frac{x}{(1+ax)(1+x^2)} = \frac{1}{1+a^2}\left[\frac{-a}{1+ax} + \frac{x+a}{1+x^2}\right]$$

$$\Rightarrow \int_0^a \frac{x}{(1+x^2)(1+ax)}\,dx$$

$$= \frac{1}{1+a^2}\left[-\log(1+ax) + \frac{1}{2}\log(1+x^2) + a\tan^{-1}x\right]_0^a$$

$$= \frac{1}{1+a^2}\left[-\log(1+a^2) + \frac{1}{2}\log(1+a^2) + a\tan^{-1}a\right]$$

$$= \frac{1}{1+a^2}\left[-\frac{1}{2}\log(1+a^2) + a\tan^{-1}a\right]$$

Using this value of the integral in (2), we have

$$\frac{d\phi(a)}{da} = \frac{1}{1+a^2}\left[-\frac{1}{2}\log(1+a^2) + a\tan^{-1}a\right] + \frac{\log(1+a^2)}{1+a^2}$$

$$= \frac{1}{1+a^2}\left[\frac{1}{2}\log(1+a^2)+a\tan^{-1}a\right]$$

Integrating w.r.t. a,

$$\phi(a) = \frac{1}{2}\int \frac{\log(1+a^2)}{1+a^2}\,da + \int \frac{a\tan^{-1}a}{1+a^2}\,da + C_1$$

where C_1 is constant of integration.

Integrating by parts the first integral taking log $(1 + a^2)$ as first function, we have

$$\phi(a) = \frac{1}{2}\left[\log(1+a^2)\,.\tan^{-1}a - \int \frac{2a}{1+a^2}\,.\tan^{-1}a\,da\right] + \int \frac{a\tan^{-1}a}{1+a^2}\,da + C_1$$

$\Rightarrow$ $$\phi(a) = \frac{1}{2}\log(1+a^2)\,.\tan^{-1}a + C_1 \quad \ldots(4)$$

To find C_1. Put $a = 0$ in (1) and (4).

From (1), $\phi(0) = 0$ and from (4),

$$\phi(0) = 0 + C_1 = C_1.$$

$\therefore$ $$C_1 = 0$$

Hence, from (4),

$$\int_0^a \frac{\log(1+ax)}{1+x^2}\,dx = \frac{1}{2}\log(1+a^2)\tan^{-1}a.$$

12.1.4. Sometimes an integral is given with its value. Then, differentiating both the sides w.r.t. a parameter, we get a new integral on one side and its value on the other.

For example,

(*i*) Consider the known integral

$$\int_0^\infty e^{-ax}\,dx = \frac{1}{a},\ a > 0.$$

Differentiating both sides n times w.r.t. a, we get

$$\int_0^\infty x^n e^{-ax}\,dx = \frac{1.2.3\ldots\ldots n}{a^{n+1}} = \frac{n!}{a^{n+1}}.$$

(*ii*) Consider the integral

$$\int_0^\infty \frac{dx}{x^2+a} = \frac{\pi}{2a^{1/2}}$$

Differentiating n times w.r.t. a,

$$\int_0^\infty \frac{(-1)(-2)\ldots\ldots(-n)\,dx}{(x^2+a)^{n+1}} = \frac{\pi}{2}\left(-\frac{1}{2}\right)\left(-\frac{3}{2}\right)\ldots\ldots\left(-\frac{2n-1}{2}\right)\frac{1}{a^{n+1/2}}$$

$\Rightarrow$ $$\int_0^\infty \frac{dx}{(x^2+a)^{n+1}} = \frac{\pi}{2}\cdot\frac{1.3.5\ldots\ldots(2n-1)}{2.4.6\ldots..2n}\,\frac{1}{a^{n+1/2}}$$

(*iii*) Consider the integral

$$\int_0^1 x^m\,dx = \frac{1}{m+1},\ m > -1,$$

Differentiating both sides w.r.t. m, n times, we get

$$\int_0^1 x^m (\log x)^n \, dx = \frac{(-1)(-2)\ldots\ldots(-n)}{(m+1)^{n+1}}.$$

$$= \frac{(-1)^n \, n!}{(m+1)^{n+1}} \qquad \left(\because \frac{d}{dx}(a^x) = a^x \log a\right)$$

Example 6. *Evaluate* $\int_0^\pi \frac{dx}{a + b\cos x}$, $a > 0$, $|\, b \,| < a$ *and deduce that*

(*i*) $\int_0^\pi \frac{dx}{(a + b\cos x)^2} = \frac{\pi a}{(a^2 - b^2)^{3/2}}$ (*ii*) $\int_0^\pi \frac{\cos x \, dx}{(a + b\cos x)^2} = -\frac{\pi b}{(a^2 - b^2)^{3/2}}$.

Sol. Let

$$\text{I} = \int_0^\pi \frac{dx}{a + b\cos x}$$

$$= \int_0^\pi \frac{dx}{a\left(\cos^2\frac{x}{2} + \sin^2\frac{x}{2}\right) + b\left(\cos^2\frac{x}{2} - \sin^2\frac{x}{2}\right)}$$

$$= \frac{1}{a-b}\int_0^\pi \frac{\sec^2\frac{x}{2}\,dx}{\frac{a+b}{a-b} + \tan^2\frac{x}{2}}$$

$$= \frac{2}{a-b}\sqrt{\frac{a-b}{a+b}}\left[\tan^{-1}\left(\tan\frac{x}{2}\sqrt{\frac{a-b}{a+b}}\right)\right]_0^\pi$$

$$= \frac{2}{\sqrt{a^2 - b^2}}\,[\tan^{-1}\infty - \tan^{-1} 0] = \frac{\pi}{\sqrt{a^2 - b^2}}$$

Thus,

$$\int_0^\pi \frac{dx}{a + b\cos x} = \frac{\pi}{\sqrt{a^2 - b^2}} \qquad \ldots(1)$$

Differentiating both sides w.r.t. a, we have

$$\int_0^\pi \frac{-dx}{(a + b\cos x)^2} = \frac{\pi\left(-\frac{1}{2}\right)2a}{(a^2 - b^2)^{3/2}}$$

or

$$\int_0^\pi \frac{dx}{(a + b\cos x)^2} = \frac{\pi a}{(a^2 - b^2)^{3/2}}$$

which proves the first result.

Again, differentiating both sides of (1) w.r.t. b, we have

$$\int_0^\pi \frac{-\cos x}{(a + b\cos x)^2}\,dx = \frac{\pi\left(-\frac{1}{2}\right)(-2b)}{(a^2 - b^2)^{3/2}}$$

or

$$\int_0^\pi \frac{\cos x}{(a + b\cos x)^2}\,dx = -\frac{\pi b}{(a^2 - b^2)^{3/2}}.$$

EXERCISE 12

1. Evaluate the integral $\int_0^1 \frac{x^\alpha - 1}{\log x}\, dx$, $(\alpha > -1)$, by applying differentiation under the integral sign.

(BCA (II) 2006)

2. Prove that $\int_0^{\pi/2} \frac{\log(1+\cos\alpha\cos x)}{\cos x}\, dx = \frac{1}{2}\left(\frac{\pi^2}{4} - \alpha^2\right)$

[**Hint.** If $\phi(\alpha) = \int_0^{\pi/2} \frac{\log(1+\cos\alpha\cos x)}{\cos x}\, dx$, then

$$\phi'(\alpha) = -\sin\alpha \int_0^{\pi/2} \frac{dx}{1+\cos\alpha\cos x}$$

$$= -\sin\alpha \frac{1}{\sqrt{1-\cos^2\alpha}} \left| \cos^{-1} \frac{\cos\alpha + \cos x}{1+\cos\alpha\cos x} \right|_0^{\pi/2}$$

$$= -\alpha$$

Integrating both sides w.r.t. α,

$$\phi(\alpha) = -\frac{\alpha^2}{2} + C.$$

To find C, take $\alpha = \frac{\pi}{2}$.]

3. Prove that

$$\int_0^{\pi/2} \log(\alpha^2\cos^2\theta + \beta^2\sin^2\theta)\, d\theta = \pi \log\left(\frac{\alpha+\beta}{2}\right), \alpha > 0, \beta > 0.$$

4. Prove that

$$\int_{-\pi/2}^{\pi/2} \frac{\log(1+a\sin x)}{\sin x}\, dx = \pi \sin^{-1} a, \text{ where } |a| < 1.$$

[**Hint.** To evaluate $\int_{-\pi/2}^{\pi/2} \frac{1}{1+a\sin x}\, dx$, put $x = \frac{\pi}{2} - y$ etc.]

5. Evaluate $\int_0^\infty \frac{\log(1+a^2x^2)}{1+b^2x^2}\, dx$.

6. Prove that

$$\int_0^{\pi/2} \frac{\log(1+y\sin^2 x)}{\sin^2 x}\, dx = \pi[\sqrt{1+y} - 1], y > -1.$$ (BCA (II) 2007)

[**Hint.** If $\phi(y) = \int_0^{\pi/2} \frac{\log(1+y\sin^2 x)}{\sin^2 x}\, dx$, then

$$\phi'(y) = \int_0^{\pi/2} \frac{dx}{1+y\sin^2 x}$$

To integrate, divide Num. and Den. by $\cos^2 x$, etc.]

7. Prove that for $a > 0$, $\int_0^{\infty} \frac{e^{-ax} \sin bx}{x} dx = \tan^{-1} \frac{b}{a}$ and deduce that $\int_0^{\infty} \frac{\sin bx}{x} dx = \frac{\pi}{2}$ or $-\frac{\pi}{2}$ according as b is positive or negative.
[**Hint.** Differentiate partially w.r.t. b]

8. Using $\int_0^x \frac{dx}{x^2 + a^2} = \frac{1}{a} \tan^{-1} \frac{x}{a}$, deduce that

$$\int_0^x \frac{dx}{(x^2 + a^2)^2} = \frac{1}{2a^3} \tan^{-1} \frac{x}{a} + \frac{x}{2a^2(x^2 + a^2)}.$$

9. Using $\int_0^{\infty} \frac{dx}{x^2 + a^2} = \frac{\pi}{2a}$, $a > 0$ deduce that

(i) $\int_0^{\infty} \frac{dx}{(x^2 + a^2)^n} = \frac{\pi}{2 . a^{2n-1}} \cdot \frac{1 . 3 . 5 \ldots\ldots (2n-3)}{2 . 4 . 6 \ldots\ldots (2n-2)}$, where n is a positive integer.

(ii) $\int_0^{\pi/2} \cos^{2n} \theta \, d\theta = \frac{1 . 3 . 5 \ldots\ldots (2n-1)}{2 . 4 . 6 \ldots\ldots 2n} \cdot \frac{\pi}{2}$.

[Put $x = a \tan \theta$ in (i).]

10. Show that if x^2 is < or = 1,

$$\int_0^{\pi/2} \log (1 - x^2 \cos^2 \theta) \, d\theta = \pi \left[\log (1 + \sqrt{1 - x^2}) - \log 2.\right]$$

[Note that here x is the parameter]

11. Prove that $\int_0^{\pi/2} \frac{dx}{a^2 \cos^2 x + b^2 \sin^2 x} = \frac{\pi}{2ab}$

Hence deduce that

$$\int_0^{\pi/2} \frac{dx}{(a^2 \cos^2 x + b^2 \sin^2 x)^2} = \frac{\pi}{4a^3 b^3} (a^2 + b^2).$$

13

Sequences and Series of Functions

In chapters 4 and 5, we discussed convergence of sequences and series of real numbers. In this chapter, we discuss the convergence of sequences and series of functions. The properties of '*pointwise convergence*' and '*uniform convergence*' are discussed. As before, we restrict our discussion to real valued functions of a real variable.

13.1. POINTWISE CONVERGENCE OF SEQUENCES OF FUNCTIONS

Let $<f_n>$, $(n = 1, 2, 3, \ldots\ldots)$ be a sequence of functions, defined on a set D (subset of the set R of real numbers). To each point $x \in$ D, there corresponds a sequence of real numbers $<f_n(x)>$ with terms $f_1(x), f_2(x), f_3(x), \ldots\ldots$

13.1.1. Definitions. *The sequence of functions $<f_n>$ is said to be convergent at $x (x \in D)$ if the sequence of numbers $<f_n(x)>$ is convergent.*

The sequence of functions $<f_n>$ is said to be pointwise convergent on $S \subseteq D$, if the sequence of numbers $<f_n(x)>$ converges for each $x \in S$.

The limit of $<f_n(x)>$ is unique for each $x \in S$ and S is then called the domain of convergence.

If $\lim\limits_{n \to \infty} f_n(x) = f(x)$, $x \in S$, then the function f, (defined on S) is called the pointwise limit or just limit of the sequence of functions $<f_n>$ on S. We also say that the sequence of functions $<f_n>$ is pointwise convergent to the function f on S.

Let us consider a few examples :

(*i*) Let $f_n(x) = x^n$, $x \in [0, 1]$

Then the sequence of functions $<f_n>$ converges to a function f on $[0, 1]$, where

$$f(x) = \begin{cases} 0, & \text{if } 0 \le x < 1 \\ 1 & \text{if } x = 1. \end{cases}$$

(*ii*) Let $f_n(x) = \dfrac{x}{1+nx}$, $0 \le x < \infty$

Then, for $x > 0$, $0 < f_n(x) \le \dfrac{x}{nx} = \dfrac{1}{n}$

$\therefore$ $$\lim_{n \to \infty} f_n(x) = 0$$

Also, $f_n(0) = 0$, $\forall\, n \in$ N.

Thus, the sequence of functions $<f_n>$ converges to 0 (the function identically 0) on $[0, \infty)$.

(*iii*) Let $f_n(x) = \dfrac{nx}{1+n^2x^2}$, $-\infty < x < \infty$

For $\quad x > 0, f_n(x) = \dfrac{\dfrac{1}{nx}}{\dfrac{1}{n^2x^2}+1} \to 0$ as $n \to \infty$

Also, $\quad f_n(0) = 0, \quad \forall\, n \in \mathrm{N}.$

Thus, the sequence of functions $<f_n>$ converges to 0 on $(-\infty, \infty)$.

Now, the important question is : If each function of a sequence $<f_n>$ has a certain property, such as continuity, differentiability or integrability, to what extent is this property transferred to the limit function ? For example, if each function f_n is continuous at a point x_0, is the limit function also continuous at x_0 ? We shall see that, in general, it is not true. In fact, we shall find that *pointwise convergence* is not strong enough to transfer any of the properties mentioned above from the terms f_n of $<f_n>$ to the limit function. To illustrate, we consider the following examples :

(*i*) *A sequence of continuous functions with a discontinuous limit function.*

Let $\quad f_n(x) = \dfrac{x^{2n}}{1+x^{2n}}, x \in \mathrm{R}.$

Then, $\lim\limits_{n\to\infty} f_n(x)$ exists for each $x \in \mathrm{R}$ and the limit function f is given by

$$f(x) = \begin{cases} 0, & \text{if } |x|<1 \\ \frac{1}{2}, & \text{if } |x|=1 \\ 1, & \text{if } |x|>1 \end{cases}$$

Here, each f_n is continuous on R but f is discontinuous at $x = \pm 1$.

(*ii*) *A sequence of differentiable functions in which the limit of the derivatives is not equal to the derivative of the limit.*

Let $\quad f_n(x) = \dfrac{\sin nx}{\sqrt{n}}, x \in \mathrm{R}$

Then, $\quad f(x) = \lim\limits_{n\to\infty} f_n(x) = \lim\limits_{n\to\infty} \dfrac{\sin nx}{\sqrt{n}} = 0, \quad \forall\, x \in \mathrm{R}$

$\therefore \quad f'(x) = 0 \;\Rightarrow\; f'(0) = 0$

But $\quad f_n'(x) = \sqrt{n}\, \cos nx$

$\Rightarrow \quad f_n'(0) = \sqrt{n} \to \infty$ as $n \to \infty$

Thus, at the origin, the sequence of derivatives $<f_n'>$ diverges, whereas the limit function $f'(x)$ is zero.

(*iii*) *A sequence of functions $<f_n>$ for which the limit of integrals is not equal to the integral of the limit function.*

Let $\quad f_n(x) = nx\,(1-x^2)^n, x \in [0, 1]$

Then, $\quad f(x) = \lim\limits_{n\to\infty} f_n(x) = 0, \quad$ for $x \in [0, 1]$

$\therefore \quad \displaystyle\int_0^1 f\,dx = 0$

But, $$\int_0^1 f_n\, dx = \int_0^1 nx\,(1-x^2)^n\, dx = \frac{n}{2\,(n+1)}$$

$$\Rightarrow \quad \lim_{n\to\infty}\left(\int_0^1 f_n dx\right) = \lim_{n\to\infty}\frac{n}{2\,(n+1)} = \frac{1}{2} \neq 0$$

The above examples show that we need to study stronger methods of convergence that do preserve these properties. This leads us to study the concept of *uniform convergence.*

13.2. UNIFORM CONVERGENCE OF SEQUENCES OF FUNCTIONS

A sequence of functions $<f_n>$ converges pointwise to a function f on S if for each $x \in$ S and for each $\varepsilon > 0$, there exists a positive integer m (depending upon both x and ε) such that $| f_n(x) - f(x) | < \varepsilon\,;\ \forall\, n \geq m.$

The number m depends upon both x and ε. It is not always possible to find an m which works for each x ∈ S.

For example, if $f_n(x) = x^n$, $x \in [0, 1]$, then $<f_n>$ converges to the function f on [0, 1] where

$$f(x) = \begin{cases} 0, & \text{if } 0 \leq x < 1 \\ 1, & \text{if } x = 1 \end{cases}$$

Take $\varepsilon = \frac{1}{2}$.

Then, for each $x \in [0, 1]$, there exists a positive integer m such that

$$| f_n(x) - f(x) | < \tfrac{1}{2}, \quad \forall\, n \geq m \qquad ...(1)$$

If $x = 0$ or 1, then (1) is true for $m = 1$. If $x = \frac{3}{4}$, then the smallest value of m for which (1) in true is 3. [For, if $x = \frac{3}{4}$, then $f_n(x) = \left(\frac{3}{4}\right)^n$ and $f(x) = 0$. Thus, $| f_n(x) - f(x) | = \left(\frac{3}{4}\right)^n < \frac{1}{2}$ iff $n \geq 3$].

If $x = \dfrac{9}{10}$, then the smallest value of m for which (1) is true is 7.

In fact, there is no single m for which (1) holds for all $x \in [0, 1]$.

[For, if such an m exists, then

$$x^n < \tfrac{1}{2},\ n \geq m \quad \forall\, x \in [0, 1]$$

This implies $x^m < \frac{1}{2}$, $(0 \leq x < 1)$.

Let $x \to 1$, then $1 < \frac{1}{2}$, a contradiction.]

Consider another example.

Let $$f(x) = \frac{x}{1+nx},\ 0 \leq x < \infty$$

Then, $$0 \leq f_n(x) \leq \frac{x}{nx} = \frac{1}{n} \text{ and } f(x) = \lim_{n\to\infty} f_n(x) = 0$$

Hence, for any $\varepsilon > 0$, the statement

$$| f_n(x) - f(x) | < \varepsilon,\ \forall\, n \geq m \qquad ...(2)$$

is true for all $x \in [0, \infty)$ simultaneously provided $m > \dfrac{1}{\varepsilon}$.

$$\left[\because \; |f_n(x)-f(x)|=|f_n(x)-0|\leq \frac{1}{n}\leq \frac{1}{m}<\varepsilon, \forall\, x \in [0,\infty)\right]$$

Thus, in this example, we can find an m such that (2) is true for each $x \in [0, \infty)$.

In this case we say that the sequence $<f_n>$ is uniformly convergent to f on $[0, \infty)$.

13.2.1. Definition. *A sequence of functions $<f_n>$ is said to be uniformly convergent to a function f on a set S if for each $\varepsilon > 0$, there exists a natural number m (depending only on ε) such that*

$$| f_n(x) - f(x) | < \varepsilon, \; \forall \; n \geq m \text{ and } x \in S.$$

and we write $f_n \to f$ uniformly on S.

13.2.2. Geometrical Interpretation of uniform convergence. If a sequence of functions $<f_n>$ is uniformly convergent to a function f on S, then the inequality $| f_n(x) - f(x) | < \varepsilon$ is equivalent to

$$f(x) - \varepsilon < f_n(x) < f(x) + \varepsilon$$

Since it holds for each $n \geq m$ (some natural number) and for each $x \in$ S, it means that the entire graph of the function f_n (*i.e.*, the set $\{(x\, f_n(x)) \mid x \in S\}$) lies within a band of height 2ε situated symmetrically about the graph of f, as shown below :

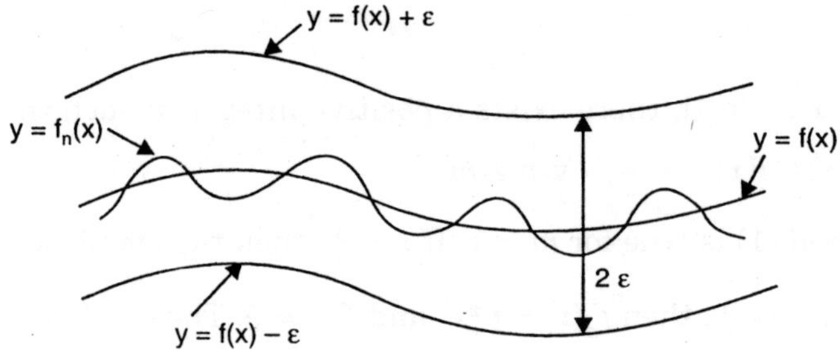

Remarks 1. *It is clear from the definition of uniform convergence that if a sequence of functions $<f_n>$ converges uniformly to a function f on a set S, then the sequence $<f_n>$ converges pointwise to f also. However, converse is not true.* For example, if $f_n(x) = x^n$, $x \in [0, 1]$, then the sequence of functions $<f_n>$ converges pointwise to the function f on [0, 1], where $f(x) = \begin{cases} 0, & \text{if } 0 \leq x < 1 \\ 1, & \text{if } x = 1 \end{cases}$ but not uniformly on [0, 1].

2. *The sequence of functions $<f_n>$ does not converge uniformly to f on S if and only if there exists some $\varepsilon > 0$ such that there is no natural number m for which the statement '$| f_n(x) - f(x) | < \varepsilon, \forall\, n \geq m, x \in S$' holds.*

3. *If $<f_n>$ converges to 0 uniformly on S, then to each $\varepsilon > 0$, $\exists$ a natural number m such that $| f_n(x) | < \varepsilon, \forall\, n \geq m$ and $x \in S$.*

$$\Rightarrow \qquad \underset{x \in S}{l.u.b.}\; | f_n(x) | \leq \varepsilon, \quad \forall\, n \geq m$$

Hence, $<f_n>$ converges uniformly to 0 on S if and only if

$$\lim_{n \to \infty} \left(\underset{x \in S}{l.u.b.} | f_n(x) | \right) = 0.$$

In general, a sequence $<f_n>$ converges uniformly to f on S if and only if

$$\underset{x \in S}{l.u.b.}\; | f_n(x) - f(x) | \to 0 \text{ as } n \to \infty.$$

4. *We have defined uniform convergence with reference to any set S ($\subseteq$ R). S may or may not be an interval, but S must be an infinite set. We never talk about uniform convergence at a single point.*

13.2.3. Uniformly Bounded. *A sequence of functions $<f_n>$ is said to be uniformly bounded on S if there exists a positive number K such that $| f_n(x) | < K, \quad \forall x \in S$ and $\forall n \in N$. The number K is called a uniform bound for $<f_n>$.*

If each individual function f_n of $<f_n>$ is bounded and if $<f_n>$ converges uniformly to f on S, then $<f_n>$ is uniformly bounded on S.

This fact often enables us to conclude that a sequence of functions is not uniformly convergent. For instance, the sequence of functions $<f_n>$, where $f_n(x) = n^2x (1-x)^n$, $(x \in R)$, is not uniformly convergent on any subset containing a neighbourhood of origin, because the sequence $<f_n>$ is not uniformly bounded in the neighbourhood of origin as is clear from its graph.

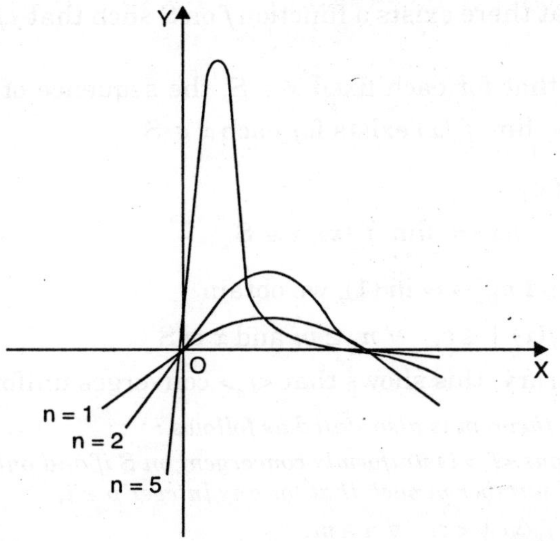

13.2.4. Point of Non-uniform Convergence. *A point which is such that the sequence is not uniformly convergent on any neighbourhood (however small) of the point, is called a point of non-uniform convergence.*

For example, the point $x = 0$ in the above example, is a point on non-uniform convergence.

Analogous to the Cauchy's criterion for the convergence of a sequence of real numbers, we have Cauchy's criterion for uniform convergence and is given in the following theorem.

13.2.5. Theorem. *(Cauchy's criterion for uniform convergence). A sequence of functions $<f_n>$ is uniformly convergent on a set S if and only if for each $\varepsilon > 0$ and for all $x \in S$, there exists a natural number m such that*

$$|f_{n_1}(x) - f_{n_2}(x)| < \varepsilon, \quad \forall\, n_1, n_2 \geq m. \qquad ...(1)$$

Proof. Necessary. Let a sequence of functions $<f_n>$ be uniformly convergent on S.

Let $<f_n>$ converge to f on S.

Let $\varepsilon > 0$ be given. Then, there exists a natural number m such that

$$| f_n(x) - f(x) | < \frac{\varepsilon}{2}, \quad \forall\, n \geq m \text{ and } x \in S$$

If $n_1, n_2 \geq m$, then

$$| f_{n_1}(x) - f(x) | < \frac{\varepsilon}{2}, \quad \forall\, x \in S \qquad ...(i)$$

and $$| f_{n_2}(x) - f(x) | < \frac{\varepsilon}{2}, \quad \forall\, x \in S \qquad ...(ii)$$

$$\therefore \quad |f_{n_1}(x) - f_{n_2}(x)| = | f_{n_1}(x) - f(x) + f(x) - f_{n_2}(x) |$$
$$\leq | f_{n_1}(x) - f(x) | + | f_{n_2}(x) - f(x) |$$
$$< \frac{\varepsilon}{2} + \frac{\varepsilon}{2} = \varepsilon, \quad \forall\, n_1, n_2 \geq m \text{ and } x \in S \qquad [\text{By } (i) \text{ and } (ii)]$$

Sufficient. Let $<f_n>$ be a sequence of functions on S such that for each $\varepsilon > 0$, there exists a natural number m such that (1) holds.

We are to show that there exists a function f on S such that $<f_n>$ converges uniformly to f on S.

From (1), we find that for each fixed $x \in S$, the sequence of real numbers $<f_n(x)>$ is a Cauchy sequence. Hence, $\lim\limits_{n \to \infty} f_n(x)$ exists for each $x \in S$.

Define a function f by

$$f(x) = \lim_{n \to \infty} f_n(x),\ x \in S.$$

Keeping n_1 fixed and $n_2 \to \infty$ in (1), we obtain

$$| f_{n_1}(x) - f(x) | \leq \varepsilon, \quad \forall\, n_1 \geq m \text{ and } x \in S$$

Since $\varepsilon > 0$ is arbitrary, this shows that $<f_n>$ converges uniformly to f on S.

Remark. *The above theorem is also stated as follows :*

'A sequence of functions $<f_n>$ is uniformly convergent on S if and only if for each $\varepsilon > 0$, and for all $x \in S$, there exists a natural number m such that for any integer $p \geq 1$,

$$| f_{n+p}(x) - f_n(x) | < \varepsilon, \quad \forall\, n \geq m.$$

Example 1. *Show that the sequence of functions $<f_n>$, where $f_n(x) = x^n$ is uniformly convergent on [0, k], k < 1 but only pointwise convergent on [0, 1].*

Sol. Here, $f(x) = \lim\limits_{n \to \infty} f_n(x) = \begin{cases} 0, & \text{if } 0 \leq x < 1 \\ 1, & \text{if } x = 1 \end{cases}$

Thus, the sequence $<f_n>$ converges pointwise to a discontinuous function f on [0, 1].

To see whether the sequence $<f_n>$ is uniformly convergent, let $\varepsilon > 0$ be given.

For 0 < x < 1,

$$| f_n(x) - f(x) | = | x^n - 0 | = x^n < \varepsilon$$

if $$\frac{1}{x^n} > \frac{1}{\varepsilon}$$

i.e., if $$n > \frac{\log\left(\frac{1}{\varepsilon}\right)}{\log\left(\frac{1}{x}\right)}$$

The number $\dfrac{\log\left(\frac{1}{\varepsilon}\right)}{\log\left(\frac{1}{x}\right)}$ increases with x with maximum value $\dfrac{\log\left(\frac{1}{\varepsilon}\right)}{\log\left(\frac{1}{k}\right)}$ on $[0, k]$, $k > 0$.

Choose an integer $m \geq \dfrac{\log\left(\frac{1}{\varepsilon}\right)}{\log\left(\frac{1}{k}\right)}$.

Then, $\quad | f_n(x) - f(x) | < \varepsilon, \quad \forall\, n \geq m$ and $0 < x < 1$.

At $x = 0$, $| f_n(x) - f(x) | = 0 < \varepsilon\ \forall\, n \geq 1$.

Thus, there exists a positive integer m such that

$| f_n(x) - f(x) | < \varepsilon$, $\forall\, n \geq m$ and $x \in [0, k]$.

$\Rightarrow$ $<f_n>$ is uniformly convergent in $[0, k]$, $k < 1$.

Since the number $\dfrac{\log\left(\frac{1}{\varepsilon}\right)}{\log\left(\frac{1}{x}\right)} \to \infty$ as $x \to 1$, it is not possible to find an integer m such that

$$| f_n(x) - f(x) | < \varepsilon, \ \forall\, n \geq m \text{ and } x \in [0, 1].$$

Hence, the sequence $<f_n>$ is not uniformly convergent on any interval containing 1 and in particular on [0, 1].

Example 2. *Show that $x = 0$ is a point of non-uniform convergence of the sequence of functions $<f_n>$ where $f_n(x) = \dfrac{nx}{1+n^2x^2}$.*

Sol. $f(x) = \lim\limits_{n \to \infty} f_n(x) = \lim\limits_{n \to \infty} \dfrac{nx}{1+n^2x^2}$

$$= \lim_{n \to \infty} \frac{\frac{x}{n}}{\frac{1}{n^2} + x^2} = 0, \ \forall\, x \in \mathbf{R}$$

Let $\varepsilon > 0$ be given. Then,

$$| f_n(x) - f(x) | < \varepsilon$$

if $$\left| \frac{nx}{1+n^2x^2} - 0 \right| < \varepsilon$$

i.e., if $$\frac{n\,|x|}{1+n^2x^2} < \varepsilon$$

i.e., if $$n\ |x| < \varepsilon + n^2x^2\,\varepsilon$$

i.e., if $$n^2x^2\,\varepsilon - n\ |x| + \varepsilon > 0$$

i.e., if $$n = \frac{|x| + \sqrt{x^2 - 4x^2\varepsilon^2}}{2x^2\varepsilon}$$

Thus, for an integer $m \geq \dfrac{1+\sqrt{1-4\varepsilon^2}}{2x\varepsilon}$, $(x \neq 0)$

$| f_n(x) - f(x) | < \varepsilon$, $\forall\, n \geq m$ and $x \neq 0$.

This implies that the sequence $<f_n>$ is uniformly convergent in every interval which does not contain 0.

But, when $x = 0$, $m \to \infty$ and therefore, it is not possible to choose an integer m such that

$$| f_n(x) - f(x) | < \varepsilon, \ \forall \ n \geq m \text{ and } x \in \mathbf{R}.$$

Hence, $x = 0$ is a point of non-uniform convergence.

Example 3. *Test for uniform convergence, the sequence* $<e^{-nx}>$, *for* $x \geq 0$.

Sol. Let $f_n(x) = e^{-nx}$.

$$\therefore \quad f(x) = \lim_{n \to \infty} f_n(x) = \lim_{n \to \infty} e^{-nx} = \begin{cases} 1, & \text{if } x = 0 \\ 0, & \text{if } x > 0 \end{cases}$$

Consider $x > 0$. Let $\varepsilon > 0$ be given.

Now, $\quad | f_n(x) - f(x) | = e^{-nx} < \varepsilon,$

if $\quad e^{nx} > \dfrac{1}{\varepsilon}$, *i.e.*, if $nx > \log\left(\dfrac{1}{\varepsilon}\right)$

i.e., if $\quad n > \dfrac{\log\left(\frac{1}{\varepsilon}\right)}{x}$

Choose an integer m, just $\geq \dfrac{\log\left(\frac{1}{\varepsilon}\right)}{x}$.

Thus, $| f_n(x) - f(x) | < \varepsilon, \ \forall \ n \geq m$ and $x > 0$

$\Rightarrow$ $<f_n>$ is uniformly convergent for $x > 0$.

However, when $x \to 0$, $n \to \infty$. So, the sequence $<f_n>$ is not uniformly convergent for $x \geq 0$.

Example 4. *Show that the sequence* $<\tan^{-1} nx>$, $x \geq 0$ *is uniformly convergent on any interval* $[a, b]$, $a > 0$ *but is only pointwise convergent on* $[0, b]$.

Sol. $f_n(x) = \tan^{-1} nx$, $x \geq 0$

$$f(x) = \lim_{n \to \infty} f_n(x) = \lim_{n \to \infty} \tan^{-1} nx = \begin{cases} \dfrac{\pi}{2}, & \text{if } x > 0 \\ 0, & \text{if } x = 0 \end{cases}$$

Let $\varepsilon > 0$ be given.

For $x > 0$,

$$| f_n(x) - f(x) | = \left| \tan^{-1} nx - \frac{\pi}{2} \right| < \varepsilon$$

if $\quad \cot^{-1} nx < \varepsilon \qquad \left(\because \ \cot^{-1}\theta = \dfrac{\pi}{2} - \tan^{-1}\theta\right)$

i.e., if $\quad n > \dfrac{1}{x} \cot \varepsilon$

The number $\frac{1}{x}$ cot ε decreases as x increases, the maximum value of $\frac{1}{x}$ cot ε, being $\frac{1}{a}$ cot ε in $[a, b]$, $a > 0$. Choose an integer $m \geq \frac{1}{x}$ cot ε.

Then, $| f_n(x) - f(x) | < \varepsilon, \quad \forall\, n \geq m$ and $x > 0$

$\Rightarrow$ the sequence $<f_n>$ is uniformly convergent on $[a, b]$, $a > 0$

But, as $x \to 0$, $\frac{1}{x} \cot \varepsilon \to \infty$.

Therefore, it is not possible to choose an integer m such that

$$| f_n(x) - f(x) | < \varepsilon, \forall\, n \geq m.$$

Hence, $<f_n>$ is not uniformly convergent on $[0, b]$, but is only pointwise convergent on $[0, b]$.

EXERCISE 13 (*a*)

1. Show that the sequence of functions $<f_n>$, where

$$f_n(x) = \frac{n^2 x}{1 + n^2 x^2} \text{ is non-uniformly convergent on } [0, 1].$$

2. Show that the sequence of functions $<f_n>$, where

$$f_n(x) = \frac{1}{n + x}, \text{ is uniformly convergent in any interval } [0, k],\ k > 0.$$

3. Show that the sequence $<e^{-nx}>$ is uniformly convergent on any interval $[a, b]$, where a, b are positive numbers, but only pointwise on $[0, b]$.
4. Show that the following functions are not uniformly convergent on the indicated interval :
   (*i*) $<x^n>$ on $[0, 1]$
   (*ii*) $<e^{-nx}>$ on $[0, k]$
   (*iii*) $<\frac{x^2}{n}>$ on $(-\infty, \infty)$.

To examine whether a sequence of functions $<f_n>$ is uniformly convergent or not, we have to find a positive integer m, which is independent of x and finding such an integer in most of the cases is not easy. We give below a test which does not involve the computation of integer m and is more convenient in practice.

13.2.6. Theorem (M_n Test). *Let $<f_n>$ be a sequence of functions on S such that*

$$\lim_{n \to \infty} f_n(x) = f(x), \forall\, x \in S$$

and let M_n be the maximum value of $| f_n(x) - f(x) |$ for fixed n and varying x, $x \in S$.

Then, $<f_n>$ is uniformly convergent on S if and only if $M_n \to 0$ as $n \to \infty$.

Proof. Necessary. Let $<f_n>$ be uniformly convergent to f on S.

Then, for each $\varepsilon > 0$, there exists an integer m such that

$$| f_n(x) - f(x) | < \varepsilon, \forall\, n \geq m, x \in S.$$

$$\Rightarrow \qquad M_n = \max_{x \in S} | f_n(x) - f(x) | \leq \varepsilon, \forall\, n \geq m$$

Since ε is arbitrary, $M_n \to 0$ as $n \to \infty$.

Sufficient. Let $M_n \to 0$ as $n \to \infty$.

Then, for each $\varepsilon > 0$, there exists an integer m such that $M_n < \varepsilon, \quad \forall\, n \geq m, x \in S$.

Since M_n is the maximum value of

$$| f_n(x) - f(x) |, \forall\, x \in S$$

$$\therefore \qquad | f_n(x) - f(x) | < \varepsilon, \quad \forall\, n \geq m \text{ and } x \in S.$$

$\Rightarrow$ the sequence $<f_n>$ is uniformly convergent to f on S.

Remark. *If M_n does not tend to 0, the sequence $<f_n>$ is not uniformly convergent.*

Example 5. *Show that the sequence of functions $<f_n>$, where $f_n(x) = \dfrac{x}{1+nx^2}$, $x \in R$ converges uniformly on any closed interval $[a, b]$.*

Sol. $f(x) = \lim\limits_{n \to \infty} f_n(x) = \lim\limits_{n \to \infty} \dfrac{x}{1+nx^2} = 0, \forall\, x \in [a, b]$

Also,
$$M_n = \max_{x \in [a,b]} | f_n(x) - f(x) | = \max_{x \in [a,b]} \left| \frac{x}{1+nx^2} \right| = \frac{1}{2\sqrt{n}} \to 0, \text{ as } n \to \infty$$

For, let
$$y = \frac{x}{1+nx^2}.$$

Differentiating w.r.t. x,

$$\frac{dy}{dx} = \frac{(1+nx^2).1 - x\,.\,2nx}{(1+nx^2)^2}$$

$$= \frac{1-nx^2}{(1+nx^2)^2} = 0 \qquad \text{(for maxima and minima)}$$

$$\Rightarrow \qquad x = \frac{1}{\sqrt{n}}.$$

Also,
$$\frac{d^2y}{dx^2} = \frac{(1+nx^2)^2(-2nx) - 2(1+nx^2)(2nx)(1-nx^2)}{(1+nx^2)^4}$$

$$\Rightarrow \qquad \left.\frac{d^2y}{dx^2}\right]_{x=\frac{1}{\sqrt{n}}} = -\frac{\sqrt{n}}{2} < 0$$

$\Rightarrow$ y is maximum when $x = \dfrac{1}{\sqrt{n}}$.

$$\Rightarrow \qquad M_n = \text{Maximum value of } | f_n(x) - f(x) |$$

$$= \qquad \frac{1}{2\sqrt{n}} \to 0 \text{ as } n \to \infty.$$

Hence, $<f_n>$ converges uniformly to f on $[a, b]$.

Example 6. *Show that the sequence of functions $<f_n>$, where $f_n(x) = nx(1-x)^n$, $0 \le x \le 1$ is not uniformly convergent on [0, 1].*

Proof. For $0 < x < 1$, $f(x) = \lim\limits_{n \to \infty} f_n(x) = \lim\limits_{n \to \infty} nx\,(1-x)^n$

$$= \lim_{n \to \infty} \frac{nx}{(1-x)^{-n}} \left(\frac{\infty}{\infty}\right)$$

(Differentiating numerator and denominator w.r.t. n)

$$= \lim_{n \to \infty} \frac{x}{-(1-x)^{-n} \log(1-x)}$$

$$= \lim_{n \to \infty} \frac{-x(1-x)}{\log(1-x)}$$

$$= 0, \text{ as } (1-x)^n \to 0 \text{ as } n \to \infty$$

But, $f(x) = 0$ for $x = 0, 1$

$\therefore$ $f(x) = 0 \quad \forall\, x \in [0, 1]$.

Let $y = |\, f_n(x) - f(x)\,| = nx\,(1-x)^n$

$\therefore$ $\dfrac{dy}{dx} = n(1-x)^n - n^2x\,(1-x)^{n-1} = 0$ (for maxima and minima)

$\Rightarrow$ $x = \dfrac{1}{n+1}$

Also, $\dfrac{d^2y}{dx^2} = -(n+1)\,n(1-x)^{n-1} - n\,(n-1)\,(1-x)^{n-2}\,(1-x-nx)$

$$= -\left(\frac{n}{n+1}\right)^{n-1} < 0 \text{ at } x = \frac{1}{n+1}.$$

$\Rightarrow$ M_n = maximum value of y

$$= \frac{n}{n+1}\left(1 - \frac{1}{n+1}\right)^n \to \frac{1}{e} \text{ as } n \to \infty$$

Since does not tend to 0 as $n \to \infty$, the convergence of $<f_n>$ is not uniform on [0, 1].

EXERCISE 13 (*b*)

1. Show that the sequence $<\dfrac{1}{n+x}>$ is uniformly convergent for all real $x \ge 0$.

(K.U. B.C.A. (II). 2001)

2. Show that the sequence $< x^{n-1}\,(1-x) >$ is uniformly convergent on [0, 1].

$\left[\textbf{Hint. } M_n = \dfrac{x^{n-1}}{n} \to 0 \text{ as } n \to \infty.\right]$

3. Show that the sequence $<f_n>$, where $f_n(x) = \dfrac{nx}{1+n^2x^2}$, is not uniformly convergent on any interval containing zero.

[**Hint.** $M_n = \frac{1}{2}$.]

4. Show that the sequence $<f_n>$, where $f_n(x) = nxe^{-nx^2}$, $x \geq 0$ is not uniformly convergent on $[0, b]$, $b > 0$.

$\left[\textbf{Hint. } M_n = \sqrt{\dfrac{n}{2e}}.\right]$

13.3. POINTWISE AND UNIFORM CONVERGENCE OF SERIES

By the convergence of a series of real numbers we mean the convergence of the sequence of partial sums of the series. Likewise, by the convergence of a series of functions $\sum_{n=1}^{\infty} f_n$, defined on a set S, we mean the convergence of the sequence $<S_n>$ of partial sums of the series, where $S_n(x) = f_1(x) + f_2(x) + \ldots\ldots + f_n(x)$, $x \in S$.

The limit function f of the sequence $<S_n>$ is called the sum function or pointwise sum of the series $\sum_{n=1}^{\infty} f_n$. We write

$$\sum_{n=1}^{\infty} f_n(x) = f(x),\ x \in S$$

or

$$\sum_{n=1}^{\infty} f_n = f.$$

For example, if $f_n(x) = x^n$, $(-1 < x < 1)$, then

$$S_n(x) = f_1(x) + f_2(x) + \ldots\ldots + f_n(x)$$
$$= x + x^2 + \ldots\ldots + x^n$$
$$= \frac{x(1-x^n)}{1-x} \to \frac{x}{1-x} \text{ as } n \to \infty$$

Since the sequence of partial sums $<S_n>$ converges to $\dfrac{x}{1-x}$, we say that the series $\sum_{n=1}^{\infty} f_n$ is (pointwise) convergent to f, where $f(x) = \dfrac{x}{1-x}$. We write

$$\sum_{n=1}^{\infty} f_n(x) = \frac{x}{1-x} = f(x),\ (-1 < x < 1).$$

13.3.1. Definition. *A series of functions $\sum_{n=1}^{\infty} f_n$ is said to be uniformly convergent on a set S if the sequence $<S_n>$ of partial sums, where $S_n(x) = \sum_{i=1}^{n} f_i(x) = f_1(x) + f_2(x) + \ldots\ldots + f_n(x)$, is uniformly convergent on S.*

Thus, *a series of function* $\sum_{n=1}^{\infty} f_n$ *converges uniformly on a set S to a function f, if for each* $\varepsilon > 0$ *and for each* $x \in S$, *there exists a natural number m (dependent on ε but not on x) such that*

$$| S_n(x) - f(x) | < \varepsilon, \quad \forall\, n \geq m$$

i.e.,
$$\left| \sum_{i=1}^{n} f_i(x) - f(x) \right| < \varepsilon, \; \forall\, n \geq m.$$

13.3.2. Theorem. (*Cauchy's Criterion for uniform convergence*)

A series of functions $\sum_{n=1}^{\infty} f_n$, *defined on a set S is uniformly convergent on S if and only if for each* $\varepsilon > 0$ *and for all* $x \in S$, *there exists a natural number m (depending only on ε) such that*

$$| f_{n+1}(x) + f_{n+2}(x) + \ldots\ldots + f_{n+p}(x) | < \varepsilon, \quad \forall\, n \leq m,\; p \in N.$$

Proof. Proof is left as an exercise to the reader.

Example 7. *Show that the series* $\sum_{n=1}^{\infty} \frac{x}{n(n+1)}$ *is uniformly convergent in (0, b) b > 0, but is not so in (0, ∞).*

Sol. Here, $S_n(x) = \sum_{1}^{n} \frac{x}{n(n+1)} = \sum_{1}^{n} x \left[\frac{1}{n} - \frac{1}{n+1} \right]$ (Partial fractions)

$$= x \left[\left(1 - \frac{1}{2}\right) + \left(\frac{1}{2} - \frac{1}{3}\right) + \left(\frac{1}{3} - \frac{1}{4}\right) + \ldots\ldots + \left(\frac{1}{n} - \frac{1}{n+1}\right) \right]$$

$$= x \left(1 - \frac{1}{n+1}\right) = \frac{nx}{n+1}$$

$$\therefore \qquad S(x) = \lim_{n \to \infty} S_n(x) = \lim_{n \to \infty} \frac{nx}{n+1} = \begin{cases} 0, & \text{if } x = 0 \\ x, & \text{if } x > 0 \end{cases}$$

If $x > 0$, then

$$| S_n(x) - S(x) | < \varepsilon$$

if
$$\left| \frac{nx}{n+1} - x \right| < \varepsilon$$

i.e., if
$$\left| \frac{-x}{n+1} \right| < \varepsilon$$

if
$$\frac{x}{n+1} < \varepsilon \qquad (\because \; x > 0)$$

i.e., if
$$n > \frac{x}{\varepsilon} - 1$$

Choose an integer just $\geq \dfrac{x}{\varepsilon} - 1$.
If $x = 0$, then $m = 1$ works.
But, when $x \to \infty$, $n \to \infty$.
Thus, the sequence $<S_n>$ converges uniformly in $(0, b)$ but not in $(0, \infty)$.

Hence, the series $\sum\limits_{n=1}^{\infty} \dfrac{x}{n(n+1)}$ is uniformly convergent in $(0, b)$, $b > 0$ but not so in $(0, \infty)$.

Example 8. *Show that $x = 0$ is a point of non-uniform convergence of the series*

$$x^2 + \frac{x^2}{1+x^2} + \frac{x^2}{(1+x^2)^2} + \ldots\ldots$$

Sol. Here,
$$S_n(x) = \frac{x^2\left[1 - \dfrac{1}{(1+x^2)^n}\right]}{1 - \dfrac{1}{1+x^2}}$$

$$= (1 + x^2) - \frac{1}{(1+x^2)^{n-1}}$$

$$\therefore \qquad S(x) = \lim_{n \to \infty} S_n(x) = \begin{cases} 1+x^2, & \text{if } x \gtrless 0 \\ 0 \quad , & \text{if } x = 0 \end{cases}$$

Consider $x \gtrless 0$. Then

$$| S_n(x) - S(x) | < \varepsilon$$

if
$$\left| (1+x^2) - \frac{1}{(1+x^2)^{n-1}} - (1+x^2) \right| < \varepsilon$$

i.e., if
$$\frac{1}{(1+x^2)^{n-1}} < \varepsilon$$

i.e., if
$$(1 + x^2)^{n-1} > \frac{1}{\varepsilon}$$

i.e., if
$$n - 1 > \frac{\log\left(\dfrac{1}{\varepsilon}\right)}{\log(1+x^2)} \qquad \ldots(1)$$

From (1), it is clear that as $x \to 0$, $n \to \infty$.
Thus, 0 is a point of non-uniform convergence and hence of the given series.
We give below a very useful test for uniform convergence of series.

13.3.3. Theorem. *(Weierstrass's M-test). A series of functions* $\sum\limits_{n=1}^{\infty} f_n$ *converges uniformly (and absolutely) on a set S if there exists a convergent series* $\sum\limits_{n=1}^{\infty} M_n$ *of non-negative terms* M_n *such that*

$$| f_n(x) | \le M_n. \quad \forall x \in S \text{ and } n \in N.$$

Proof. Let $\varepsilon > 0$ be given.

If $\sum_{n=1}^{\infty} M_n$ is convergent, then $\exists$ a positive integer m such that

$$| M_{n+1} + M_{n+2} + \ldots\ldots + M_{n+p} | < \varepsilon, \quad \forall\, n \ge m \text{ and } p \ge 1 \qquad \ldots(1)$$

Therefore, $\forall\, x \in S$,

$$\begin{aligned} &| f_{n+1}(x) + f_{n+2}(x) + \ldots\ldots + f_{n+p}(x) | \\ &\le | f_{n+1}(x) | + | f_{n+2}(x) | + \ldots\ldots + | f_{n+p}(x) | \\ &\le M_{n+1} + M_{n+2} + \ldots\ldots + M_{n+p} \\ &< \varepsilon, \quad \forall\, n \ge m, p \ge 1 \text{ and } n \in S \end{aligned} \qquad \text{[By (1)]}$$

$\Rightarrow$ the series $\sum_{n=1}^{\infty} f_n$ is uniformly convergent on S.

Since $\quad | f_{n+1}(x) | + | f_{n+2}'(x) | + \ldots\ldots + | f_{n+p}(x) | < \varepsilon, \quad \forall\, n \ge m, p \ge 1$ and $x \in S$,

the series $\sum_{n=1}^{\infty} f_n$ is also absolutely convergent on S.

For example,

(*i*) Each of the series $\sum_{n=1}^{\infty} r^n \cos nx, \ \sum_{n=1}^{\infty} r^n \sin nx, \ \sum_{n=1}^{\infty} r^n \cos n^2x, \ \sum_{n=1}^{\infty} r^n \sin a^n x, (0 < r < 1)$ is uniformly convergent for all real values of x.

The result follows by applying Weierstrass's M-test, taking $M_n = r^n$.

(*ii*) $\sum_{n=1}^{\infty} \dfrac{\sin (x^2 + n^2 x)}{n\,(n+2)}$ and $\sum_{n=1}^{\infty} \dfrac{\cos (x^2 + n^2 x)}{n\,(n+2)}$ are both uniformly convergent for all real x.

$$\left[\text{Take } M_n = \frac{1}{n\,(n+2)}\right]$$

(*iii*) $\sum_{n=1}^{\infty} \dfrac{\sin nx}{n^p}$ and $\sum_{n=1}^{\infty} \dfrac{\cos nx}{n^p}$ are both uniformly and absolutely convergent for all real values of x and $p > 1$.

$$\left[\text{Here, } \left|\frac{\sin nx}{n^p}\right| \le \frac{1}{n^p} \text{ and } \sum_{n=1}^{\infty} \frac{1}{n^p} \text{ is convergent for } p > 1\right]$$

Example 9. *Test for uniform convergence of the series*

$$\sum_{n=1}^{\infty} \frac{x}{(n+x^2)^2}.$$

Sol. Here, $\qquad f_n(x) = \dfrac{x}{(n+x^2)^2}$

$$\Rightarrow \qquad \frac{d}{dx} f_n(x) = \frac{(n+x^2)^2 \cdot 1 - x \cdot 2(n+x^2) \cdot 2x}{(n+x^2)^4}$$

$$\therefore \quad \frac{d}{dx} f_n(x) = 0 \;\Rightarrow\; 3x^4 + 2nx^3 - n^2 = 0$$

$$\Rightarrow \quad x^2 = \frac{n}{3} \qquad \text{(leaving negative value)}$$

$$\Rightarrow \quad x = \sqrt{\frac{n}{3}}\,.$$

One can see that $\frac{d^2 f_n(x)}{dx^2}$ is negative at $x = \sqrt{\frac{n}{3}}$.

$\therefore$ $f_n(x)$ is maximum at $x = \sqrt{\frac{n}{3}}$ and maximum value is

$$\frac{\sqrt{\frac{n}{3}}}{\left(n + \frac{n}{3}\right)^2} = \frac{3\sqrt{3}}{16n^{3/2}} = M_n.$$

But $\sum_{n=1}^{\infty} M_n$ is convergent.

$\therefore$ by Weierstrass's M-test, the given series is uniformly convergent.

Example 10. *Apply Weierstrass's M-test to show that the series $\sum_{n=1}^{\infty} \frac{a_n x^n}{1+x^{2n}}$ is uniformly convergent for all real x if $\sum_{n=1}^{\infty} a_n$ is absolutely convergent.*

Sol. One can see that $\frac{x^n}{1+x^{2n}}$ is maximum when $x = 1$. (The reader should verify !) and its maximum value is $\frac{1}{2} < 1$.

$$\therefore \quad \left| \frac{a_n x^n}{1+x^{2n}} \right| < | a_n |, \text{ for all real values of } x.$$

If $\sum_{n=1}^{\infty} a_n$ is absolutely convergent, then $\sum_{n=1}^{\infty} | a_n |$ is convergent and hence by Weierstrass's M-test, the given series is uniformly convergent for all real x.

EXERCISE 13 (c)

1. Show that the series :

$$\frac{x}{x+1} + \frac{x}{(x+1)(2x+1)} + \frac{x}{(2x+1)(3x+1)} + \ldots\ldots$$

is uniformly convergent on $[a, \infty]$, $a > 0$. But, it is not uniformly convergent near $x = 0$.

$$\left[\textbf{Hint. } S_n(x) = \frac{x}{x+1} + \frac{x}{(x+1)(2x+1)} + \ldots\ldots + \frac{x}{[(n-1)x+1][nx+1]} = 1 - \frac{1}{nx+1} = \frac{nx}{nx+1}.\right]$$

2. Show that the series $\sum_{n=1}^{\infty} \frac{x}{1+n^2x^2}$ is uniformly convergent on $[a, 1]$; $0 < a < 1$.

3. Show that $\sum_{n=1}^{\infty} \frac{(-1)^n}{n^p} \frac{x^{2n}}{1+x^{2n}}$ is absolutely and uniformly convergent for all real x if $p > 1$.

$\left[\textbf{Hint.} \text{ Take } M_n = \frac{1}{n^p}.\right]$

4. Show that $\sum_{n=1}^{\infty} \frac{x}{n(1+nx^2)}$ is uniformly convergent for all real x.

$\left[\textbf{Hint.} \text{ Max. } f_n(x) = \frac{1}{2n^{3/2}} \text{ and apply M-test.}\right]$

5. Apply Weierstrass's M-test to show that $\sum_{n=1}^{\infty} \frac{a_n x^n}{1+x^{2n}}$ is uniformly convergent for all real x if $\sum_{n=1}^{\infty} a_n$ is absolutely convergent.

6. The sum to n terms of a series is $S_n(x) = \frac{n^2x}{1+n^4x}$. Show that the series is uniformly convergent on $[0, 1]$.

7. Discuss the uniform convergence of

$$\sum_{n=1}^{\infty} x^n(1-x) \text{ on } [0, 1].$$

[**Hint.** Show that 0 is a point of non-uniform convergence.]

8. Show that $\sum_{n=1}^{\infty} \frac{x}{n(1+nx^2)}$ is uniformly convergent for all real x.

9. (*i*) Show that (*i*) $\sum_{n=1}^{\infty} \frac{1}{n^p + n^p x^2}$ is uniformly convergent for all real x and $p > 1$.

$\left[\textbf{Hint.} \frac{1}{n^p + n^q x^2} < \frac{1}{n^p}, \forall x \in \mathbb{R}.\right]$

(*ii*) $\sum_{n=1}^{\infty} \frac{x}{n^p + n^q x^2}$ is uniformly convergent for all real x if $p + q > 2$.

$\left[\textbf{Hint.} \text{ Maximum value of } \frac{x}{n^p + n^q x^2} \text{ is } \frac{1}{2n^{\frac{p+q}{2}}}.\right]$

10. Show that the series

$$1 + \frac{e^{-2x}}{2^2-1} - \frac{e^{-4x}}{4^2-1} + \frac{e^{-6x}}{6^2-1} - \ldots\ldots$$

is uniformly convergent for $x \geq 0$.

$\left[\text{Here, } |f_n(x)| = \frac{e^{-nx}}{4n^2-1} < \frac{1}{n}, \forall x \geq 0.\right]$

13.4. UNIFORM CONVERGENCE AND CONTINUITY

13.4.1. Theorem. *If a sequence of continuous functions $<f_n>$ is uniformly convergent to a function f on [a, b], then f is continuous on [a, b].*

Proof. Let $\varepsilon > 0$ be given.

If $<f_n>$ is uniformly convergent to f on $[a, b]$, then there exists a natural number m such that

$$| f_n(x) - f(x) | < \frac{\varepsilon}{3}, \quad \forall\, n \geq m \text{ and } x \in [a, b] \qquad ...(1)$$

Let x_0 be any point of $[a, b]$.

Then from (1),

$$| f_n(x) - f(x_0) | < \frac{\varepsilon}{3}, \quad \forall\, n \geq m \qquad ...(2)$$

Since f_n is continuous at x_0, there exists $\delta > 0$ such that

$$| f_n(x) - f_n(x_0) | < \frac{\varepsilon}{3}, \text{ whenever } | x - x_0 | < \delta \text{ and } x \in [a, b] \qquad ...(3)$$

Now, $$| f(x) - f(x_0) | = | f(x) - f_n(x) + f_n(x) - f_n(x_0) + f_n(x_0) - f_n(x_0) |$$
$$\leq | f(x) - f_n(x) | + | f_n(x) - f_n(x_0) | + | f_n(x_0) - f(x_0) |$$
$$< \frac{\varepsilon}{3} + \frac{\varepsilon}{3} + \frac{\varepsilon}{3} = \varepsilon \quad \forall\, n \geq m \text{ and } | x - x_0 | < \delta,\ x \in [a, b]$$

[By (1) to (3)]

Hence, f is continuous at $x_0 \in [a, b]$.

Since x_0 is arbitrary, f is continuous on $[a, b]$.

Cor. *If a series $\sum_{n=1}^{\infty} f_n$ of continuous functions is uniformly convergent to a function f on [a, b], then the sum function f is also continuous on [a, b].*

Remark. *Uniform convergence of the sequence $<f_n>$ (series $\sum_{n=1}^{\infty} f_n$) is sufficient but not a necessary condition for the limit function (sum function) to be continuous. The theorem (corollary) shows that if the sequence (series) of continuous functions has the limit function (sum function) discontinuous, the sequence (series) cannot be continuous. Thus, the theorem (corollary) provides a good negative test for uniform convergence of a sequence (series).*

For example,

(*i*) The sequence $<x^n>$ has a discontinuous limit function f given by

$$f(x) = \begin{cases} 0, & \text{if } 0 \leq x < 1 \\ 1, & \text{if } x = 1 \end{cases}$$

Therefore, the sequence cannot converge uniformly on [0, 1].

(*ii*) The sequence $<\dfrac{nx}{1+n^2x^2}>$ of continuous functions has the limit function

$$f(x) = 0, \quad \forall\, x \in \mathrm{R}.$$

Thus, the limit function is continuous but the given sequence is not uniformly convergent (See Example 2).

(*iii*) The sum function of the series $\sum_{n=1}^{\infty} (1-x)\, x^n$ of continuous function is

$$f(x) = \begin{cases} 1, & \text{if } x \neq 1 \\ 0, & \text{if } x = 1 \end{cases}$$

which is discontinuous on [0, 1]. Therefore, the given series is not uniformly convergent on [0, 1].

(*iv*) Consider the series $\sum_{n=1}^{\infty} f_n$, where

$$f_n(x) = \frac{nx}{1+n^2x^2} - \frac{(n-1)x}{1+(n-1)^2x^2}, \quad \forall\, x \in \mathrm{R}$$

The series $\sum_{n=1}^{\infty} f_n$ is the series of continuous functions on R and converges to 0 but the convergence is not uniform as the sequence of its partial sums $<\frac{nx}{1+n^2x^2}>$ does not converge uniformly.

Example 11. *Show that the series* $x^4 + \frac{x^4}{1+x^4} + \frac{x^4}{(1+x^4)^2} + \ldots\ldots$ *is not uniformly convergent on [0, 1].*

Sol. Clearly, the terms of the series are continuous functions and converge to the function f given by

$$f(x) = \begin{cases} 1+x^4, & \text{if } x \neq 0 \\ 0, & \text{if } x = 0 \end{cases}$$

Since the limit function $f(x)$ is discontinuous at $x = 0 \in [0, 1]$, the series cannot be uniformly convergent on [0, 1].

13.5. UNIFORM CONVERGENCE AND INTEGRATION

13.5.1. Theorem. *If* $<f_n>$ *is a sequence of integrable functions on [a, b] and if* $<f_n>$ *is uniformly convergent to a function f on [a, b], then f is integrable on [a, b] and the sequence* $<\int_a^b f_n\, dx>$ *converges uniformly to* $\int_a^b f\, dx$.

i.e.
$$\lim_{n \to \infty} \int_a^b f_n\, dx = \int_a^b f\, dx.$$

Proof. Let $\varepsilon > 0$ be given.

If $<f_n>$ is uniformly convergent to a function f on $[a, b]$, then there exists a natural number m such that

$$|\, f_n(x) - f(x)\, | < \frac{\varepsilon.}{2(b-a)}, \ \forall\, n \geq m \text{ and } x \in [a, b],\ a \neq b.$$

$$\Rightarrow \quad f(x) - \frac{\varepsilon}{2(b-a)} < f_n(x) < f(x) + \frac{\varepsilon}{2(b-a)}, \quad \forall\, n \geq m,\ x \in [a, b]$$

$$\Rightarrow \qquad f(x) < f_n(x) + \frac{\varepsilon}{2(b-a)}$$

and
$$f(x) > f_n(x) - \frac{\varepsilon}{2(b-a)}, \quad \forall\, n \geq m, x \in [a, b]$$

$$\Rightarrow \qquad \overline{\int_a^b} f\,dx < \overline{\int_a^b} f_n\,dx + \frac{\varepsilon}{2} = \int_a^b f_n\,dx + \frac{\varepsilon}{2} \qquad ...(1)$$

and
$$\underline{\int_a^b} f\,dx > \underline{\int_a^b} f_n(x)\,dx - \frac{\varepsilon}{2} = \int_a^b f_n\,dx - \frac{\varepsilon}{2} \qquad ...(2)$$

($\because$ f_n is integrable on $[a, b]$)

$$\Rightarrow \qquad 0 \leq \overline{\int_a^b} f\,dx - \underline{\int_a^b} f\,dx < \int_a^b f_n\,dx + \frac{\varepsilon}{2} - \int_a^b f_n\,dx + \frac{\varepsilon}{2} = \varepsilon$$

$\Rightarrow$ f is integrable on $[a, b]$.

Again, from (1) and (2)

$$\int_a^b f_n\,dx - \frac{\varepsilon}{2} < \int_a^b f\,dx < \int_a^b f_n\,dx + \frac{\varepsilon}{2}$$

$$\Rightarrow \qquad \lim_{n \to \infty} \int_a^b f_n\,dx = \int_a^b f\,dx.$$

Cor. *If a series of functions $\sum_{n=1}^{\infty} f_n$ is uniformly convergent to a function f on [a, b] and each term f_n is integrable on [a, b], then f is integrable on [a, b] and $\sum_{n=1}^{\infty} \int_a^b f_n\,dx$ is uniformly convergent to $\int_a^b f\,dx$.*

i.e.,
$$\sum_{n=1}^{\infty} \int_a^b f_n\,dx = \int_a^b f\,dx.$$

(We also say that the series is term by term integrable).

Proof. Let $\sum_{n=1}^{\infty} f_n$ be uniformly convergent to f on $[a, b]$.

$\Rightarrow$ The sequence $<S_n>$ of partial sums of $\sum_{n=2}^{\infty} f_n$ is uniformly convergent to f an $[a, b]$.

Also, S_n, being the sum of first n integrable functions is integrable for each n.

$\therefore$ f is integrable on $[a, b]$.

and
$$\lim_{n \to \infty} \int_a^b S_n\,dx = \int_a^b f\,dx$$

$$\Rightarrow \qquad \sum_{n=1}^{\infty} \int_a^b f_n\,dx = \int_a^b f\,dx.$$

Remark. *Uniform convergence of the sequence $<f_n>$ (series $\sum_{n=1}^{\infty} f_n$) is sufficient but not a necessary condition for the limit function (sum function) to be integrable.*

Example 12. *Show that the sequence $<f_n>$, where*

$$f_n(x) = nex^{-nx^2}, \; n \in N$$

is not uniformly convergent on [0, 1].

Sol. The sequence $<f_n>$ converges positive to 0 on [0, 1].

For, $$f(x) = \lim_{n\to\infty} f_n(x) = \lim_{n\to\infty} \frac{nx}{e^{nx^2}}$$

$$= \lim_{n\to\infty} \frac{nx}{1+\frac{nx^2}{1!}+\frac{n^2x^4}{2!}+\ldots\ldots} = 0 \text{ for } x \in [0, 1]$$

Also, $$\int_0^1 f\,dx = 0$$

and $$\int_0^1 f_n\,dx = \frac{1}{2}\left[-e^{-nx^2}\right]_0^1 = \frac{1}{2}(1-e^{-n})$$

$$\Rightarrow \quad \lim_{n\to\infty}\int_0^1 f_n\,dx = \lim_{n\to\infty}\frac{1}{2}(1-e^{-n}) = \frac{1}{2} \neq \int_0^1 f\,dx$$

$\Rightarrow$ the sequence is not uniformly convergent on [0, 1].

In fact, $x = 0$ is a point of non-uniform convergence.

Example 13. *Prove that*

$$\int_0^1\left(\sum_{n=1}^{\infty}\frac{x^n}{n^2}\right)dx = \sum_{n=1}^{\infty}\frac{1}{n^2(n+1)}.$$

Sol. Applying Weierestrass's M-test, we can see that $\sum_{n=1}^{\infty}\frac{x^2}{n^2}$ is uniformly convergent for $0 \le x \le 1$.

$\therefore$ it can be integrated term by term

$$\therefore \quad \int_0^1\left(\sum_{n-1}^{\infty}\frac{x^n}{n^2}\right)dx = \sum_{n=1}^{\infty}\left(\int_0^1\frac{x^n}{n^2}dx\right)$$

$$= \sum_{n=1}^{\infty}\left[\frac{x^{n+1}}{n^2(n+1)}\right]_0^1$$

$$= \sum_{n=1}^{\infty}\frac{1}{n^2(n+1)}.$$

Example 14. *Show that the series :*

$$1 - x + x^2 - \ldots\ldots = \frac{1}{1+x}, \; (0 \le x \le 1)$$

can be integrated term by term, though the series is not uniformly convergent on [0, 1].

Sol. Since the series oscillates when $x = 1$ and hence is not convergent.

But,
$$\int_0^1 dx - \int_0^1 x\,dx + \int_0^1 x^2\,dx - \int_0^1 x^2\,dx + \ldots\ldots$$
$$= 1 - \frac{1}{2} + \frac{1}{3} - \frac{1}{4} + \ldots\ldots = \log 2$$

and
$$\int_0^1 \frac{1}{1+x}\,dx = \left[\log(1+x)\right]_0^1 = \log 2$$

Thus, the series can be integrated term by term even though the series is not uniformly convergent near $x = 1$.

EXERCISE 13 (*d*)

1. Examine for term by term integration for the series $\sum_{n=1}^{\infty} f_n$, where $S(n) = \sum_{i=1}^{n} f_i = nx\,e^{-nx^2}$, over the intervals
   (*i*) [0, 1] (*ii*) $[a, 1]$, $0 < a < 1$.
   [**Ans.** (*i*) Term by term integration is not justified. (*ii*) Justified.]
2. Examine for term by term integration the series the sum of whose first n terms is $n^2 x(1-x)^n$, $0 \le x \le 1$.
   [**Ans.** Not justified]
3. Show that the series for which
   (*i*) $S_n(x) = \dfrac{1}{1+nx}$, (*ii*) $S_n(x) = nx\,(1-x)^n$
   can be integrated term by term on [0, 1], though they are not uniformly convergent on [0, 1].

13.6. UNIFORM CONVERGENCE AND DIFFERENTIATION

We now take up the following question :

If $<f_n>$ be a sequence of functions on $[a, b]$ such that, for each $n \in \mathbb{N}$, $f_n'(x)$ exists $\forall\, x \in [a, b]$. Let also $<f_n>$ converge to f on $[a, b]$. We ask (*i*) does $f'(x)$ exist for all x ? and (*ii*) if so, does $<f_n'>$ converge to f' ?

The answer to first question is 'no' because it is possible to have a sequence of differentiable functions which converges uniformly to a function that does not have a derivative at any point.

Even if $<f_n>$ converges uniformly to f, f_n' and f' exist for all $x \in [a, b]$, it may happen that $<f_n'>$ does not converge to f' at some x. For example, if $f_n(x) = \dfrac{x^n}{n}$, $(0 \le x \le 1)$, then $<f_n>$ converges uniformly to $f = 0$ but $<f_n'(1)>$ does not converge to $f'(1)$.

Thus, in this example,
$$\lim_{n\to\infty} f_n'(x) \ne (\lim_{n\to\infty} f_n)'(x) \text{ at } x = 1$$

In this connection, we prove the following theorem :

13.6.1. Theorem. *If a sequence of functions* $< f_n >$ *is such that*
(*i*) *each* f_n' *is differentiable on [a, b]*
(*ii*) *each* f_n' *is continuous on [a, b]*

(iii) $<f_n>$ *is convergent to the function f on [a, b]*

(iv) $<f_n'>$ *is uniformly convergent to the function g on [a, b], then f is differentiable and* $f'(x) = g(x), (a \le x \le b)$.

Proof. Since each f_n' in continuous on $[a, b]$ and $<f_n'>$ is uniformly convergent to g on $[a, b]$, therefore g is continuous and hence integrable on $[a, b]$.

Moreover, since $<f_n'>$ is uniformly convergent to g on $[a, y]$, where y is any point of $[a, b]$.

$$\lim_{n \to \infty} \int_a^y f_n' \, dx = \int_a^y g \, dx$$

By Fundamental Theorem of Integral Calculus, this implies

$$\lim_{n \to \infty} [f_n(y) - f_n(a)] = \int_a^y g \, dx$$

But, $\lim_{n \to \infty} f_n(y) = f(y)$ and $\lim_{n \to \infty} f_n(a) = f(a)$

$$\therefore \qquad f(y) - f(a) = \int_a^y g \, dx, \ (a \le y \le b)$$

$$\Rightarrow \qquad f'(y) = g(y), \ (a \le y \le b)$$

Changing y to x,

$$f'(x) = g(x), \ (a \le x \le b).$$

Cor. If a series of functions $\sum_{n=1}^{\infty} f_n$ is such that

(i) each f_n is differentiable on $[a, b]$

(ii) each f_n' is continuous on $[a, b]$

(iii) $\sum_{n=1}^{\infty} f_n$ is convergent to the function f on $[a, b]$

(iv) $\sum_{n=1}^{\infty} f_n'$ is uniformly convergent to the function g on $[a, b]$, then f is differentiable and $f'(x) = g(x), (a \le x \le b)$.

Proof. Let $S_n = f_1 + f_2 + \ldots\ldots + f_n$ so that $<S_n>$ is the sequence of partial sums of the series $\sum_{n=1}^{\infty} f_n$.

Since $\sum_{n=1}^{\infty} f_n$ is convergent to f on $[a, b]$, the sequence $<S_n>$ converges to f on $[a, b]$.

Since $\qquad S_n' = f_1' + f_2' + \ldots\ldots + f_n',$

the sequence $<S_n'>$ of partial sums of the series $\sum_{n=1}^{\infty} f_n'$ is uniformly convergent to g on $[a, b]$, where $g = \sum_{n=1}^{\infty} f_n'$.

Therefore, by above theorem,

$$f'(x) = g(x), \quad (a \le x \le b).$$

Remarks 1. *In the above theorem (the corollary), the derived sequence (series) must be uniformly convergent and the original sequence (series) need only be simply convergent.*

2. *The result of the corollary is written as :*

$$f' = \sum_{n=1}^{\infty} f_n' = f_1' + f_2' + f_3' + ...$$

i.e., the series is differentiable term by term.

Example 15. *Show that the sequence $<f_n>$, where*

$$f_n(x) = \frac{nx}{1+n^2x^2},\ 0 \le x \le 1$$

cannot be differentiated term by term at $x = 0$.

Sol. Here, $f(x) = 0,\quad \forall\, x \in [0, 1]$

$$\therefore \qquad f'(a) = 0$$

$$f_n'(0) = \lim_{h \to 0} \frac{f_n(0+h) - f_n(0)}{h}$$

$$= \lim_{h \to 0} \frac{\dfrac{nh}{1+n^2h^2} - 0}{h}$$

$$= \lim_{h \to 0} \frac{n}{1+n^2h^2} = n$$

$$\therefore \qquad f_n'(0) \to \infty \text{ as } n \to \infty$$

$$\therefore \qquad f'(0) \ne \lim_{n \to \infty} f_n'(0)$$

Hence, $<f_n>$ cannot be differentiated term by term at $x = 0$.

Example 16. *Show that the function represented by $\sum_{n=1}^{\infty} \frac{\sin nx}{n^3}$ is differentiable for every x and its derivative is $\sum_{n=1}^{\infty} \frac{\cos nx}{n^2}$.*

Sol. Here,

$$f_n(x) = \frac{\sin nx}{n^3}$$

$$\therefore \qquad f_n'(x) = \frac{\cos nx}{n^2}$$

$$\therefore \qquad \sum_{n=1}^{\infty} f_n'(x) = \sum_{n=1}^{\infty} \frac{\cos nx}{n^2}$$

Since $\left| \frac{\cos nx}{n^2} \right| \le \frac{1}{n^2}$, $\forall\, x$ and $\sum_{n=1}^{\infty} \frac{1}{n^2}$ is convergent, therefore by Weierstrass's M-test the series $\sum_{n=1}^{\infty} f_n$ is uniformly convergent for all x and hence $\sum_{n=1}^{\infty} f_n$ can be differentiated term by term.

$$\therefore \qquad \left(\sum_{n=1}^{\infty} \frac{\sin nx}{n^3}\right)' = \sum_{n=1}^{\infty} \left(\frac{\sin nx}{n^3}\right)'$$

$$= \sum_{n=1}^{\infty} \frac{\cos nx}{n^2}.$$

Example 17. *Show that the differential coefficient of*

$$\sum_{n=1}^{\infty} \frac{1}{n^3 + n^4 x^2} \text{ is } -2x \sum_{n=1}^{\infty} \frac{1}{n^2(1+nx^2)^2} \text{ for all real } x.$$

Sol. Here, $$f_n(x) = \frac{1}{n^3 + n^4 x^2}$$

$$\Rightarrow \qquad f_n'(x) = -\frac{2x}{n^2(1+nx^2)^2}$$

$f_n'(x)$ is max. when $\frac{d}{dx} f_n'(x) = 0$

i.e., when $(1 + nx^2)^2 - 4nx^2(1 + nx^2) = 0$

$$\Rightarrow \qquad 1 - 3nx^2 = 0$$

$$\Rightarrow \qquad x = \frac{1}{\sqrt{3n}}, \text{ (taking positive sign)}$$

$$\therefore \qquad \text{Max. } |f_n'(x)| = \frac{2}{\sqrt{3}\,.\,n^{5/2}\left(1+\frac{1}{3}\right)^2}$$

$$= \frac{3\sqrt{3}}{8n^{5/2}}$$

$$\Rightarrow \qquad |f_n'(x)| < \frac{1}{n^{5/2}}, \quad \forall\, x$$

Since $\sum_{n=1}^{\infty} \frac{1}{n^{5/2}}$ is convergent, by Weierstrass's M-test, the series $\sum_{n=1}^{\infty} f_n'$ is uniformly convergent for all real x. Therefore, term by term differentiation is justified.

Hence, $$\left(\sum_{n=1}^{\infty} \frac{1}{n^3 + n^4 x^2}\right)'$$

$$= \sum_{n=1}^{\infty} \left(\frac{1}{n^3 + n^4 x^2}\right)'$$

$$= -2x \sum_{n=1}^{\infty} \frac{1}{n^2(1+nx^2)^2}.$$

EXERCISE 13 (e)

1. Show that the series for which

$$S_n(x) = \frac{nx}{1+n^2x^2}, \; 0 \le x \le 1,$$

cannot be differentiated term by term at $x = 0$.

2. Given the series $\sum_{n=1}^{\infty} f_n$ for which

$$S_n(x) = \frac{1}{2n^2} \log (1 + n^4x^2), \; 0 \le x \le 1$$

Show that the series $\sum_{n=1}^{\infty} f_n'$ does not converge uniformly, but that the given series can be differentiated term by term.

[**Hint.** Show that 0 is *a* point of non-uniform convergence of $\sum_{n=1}^{\infty} f_n'$.]

3. Show that for the sequence $\langle f_n \rangle$, where $f_n(x) = \dfrac{x}{1+nx^2}$, the formula : $\lim_{n \to \infty} f_n'(x) = f'(x)$ is true if $x \ne 0$ and false if $x = 0$.

Answers

EXERCISE 2 (b)

2. (*i*) $x = b - a$, (*ii*) $x = \frac{b}{a}$ (if $a \neq 0$), (*iii*) $x = -\frac{b}{a}$ (if $a \neq 0$).

EXERCISE 2 (d)

1. (*a*) (*i*) Greatest element 20, least element 1, l.u.b. 20 and g.l.b. 1
   (*ii*) Greatest element 1, l.u.b. 1, g.l.b. 0
   (*iii*) Smallest element – 2, g.l.b. – 2
   (*iv*) Greatest element $\frac{1}{2}$, l.u.b. $\frac{1}{2}$
   (*v*) Smallest element 0, g.l.b. = 0, l.u.b. 1
   (*vi*) g.l.b. 0, l.u.b. 1, greatest element 1.
   (*b*) (*i*) g.l.b. = $\frac{-1}{5}$, l.u.b. = $\frac{1}{5}$
   (*ii*) g.l.b. = $\frac{3}{2}$, l.u.b. = $\frac{5}{3}$
2. (*i*) {– 1, – 2, – 3, ...} (*ii*) N,
   (*iii*) Z, Q or R (say), (*iv*) Any finite set, {2, 3, 5} (say).
3. No.
10. Singleton, {5} (say).

EXERCISE 3 (a)

1. (*i*) Any open interval (0, 1) (say)
   (*ii*) Any finite set {1, 2, 3} (say)
   (*iii*) A set which is not a neighbourhood of one point, two points n points : (0, 1], [0, 1], (0, 1) ∪ {1, 2, 3, ..., n}. Yes, if n = 1 or 2
   (*iv*) (0, 1], [0, 1) Q, etc. (*v*) φ and R.
   (*vi*) (0, 1] (*vii*) (2, 3) ∪ (4, 5).
5. (*i*) False, (*ii*) False, (*iii*) True,
   (*iv*) False, (*v*) False
   (For example, Q is an infinite set and is not closed.)

(*vi*) True, (*vii*) False as it is not a neighbourhood of 0,
(*viii*) True.

9. (*i*) ϕ (*ii*) (0, 1) (*iii*) $(0, 1) \cup (3, 5)$
(*iv*) ϕ (*v*) ϕ (*vi*) ϕ
(*vii*) R.

10. A = [0, 1], B = [1, 2].

EXERCISE 3 (*b*)

1. (*i*) N, (*ii*) $\left\{\frac{1}{n} : n \in N\right\}$,

(*iii*) $\left\{-1 + \frac{1}{n} : n \in N\right\} \cup \left\{1 - \frac{1}{n} ; n \in N\right\}$

(*iv*) Q, R, (*v*) [1, 2] (*vi*) $\left\{\sqrt{2} + \frac{1}{n} \mid n \in N\right\}$.

2. (*i*) {0} (*ii*) $\left\{\frac{1}{p} : p \in N\right\} \cup \{0\}$ (*iii*) {1}
(*iv*) $[1, \infty)$ (*v*) $(-\infty, -1]$ (*vi*) {1}
(*vii*) {– 1, 1} (*viii*) {0} (*ix*) R.

5. Let
$$A = [0, 1),\ B = (1, 2]$$
$$A \cap B = \phi,\ (A \cap B)' = \phi$$
$$A' = [0, 1)' = [0, 1],\ B' = (1, 2]' = [1, 2]$$
$\therefore$
$$A' \cap B' = \{1\}.$$

EXERCISE 4 (*a*)

1. (*i*), (*ii*), (*iii*), (*iv*), (*v*), (*vii*), (*x*) bounded ; (*vi*), (*viii*) and (*ix*) unbounded.

2. (*i*), (*iii*), (*iv*), (*v*), (*vii*) converge ; (*vi*), (*viii*) diverge to ∞ ;
(*ix*) diverges to $-\infty$; (*ii*) and (*x*) oscillate.

3. (*a*) All the four statements are false.
Example in (*ii*) is $a_n = n^2 + (-1)^n n$
Example in (*iii*) is $a_n = (-1)^n . n$
Example in (*iv*) is $< (-1)^n >$.

(*b*) $a_n = \sqrt{n}$.

8. (*i*) $a_n = n$, (*ii*) $a_n = \frac{2n+1}{n}$, (*iii*) $a_n = \frac{1}{n}$.

EXERCISE 4 (*b*)

2. Take $a_n = 2 - \frac{1}{n}$ and $b_n = 2 + \frac{1}{n}$. Then, $a_n < b_n \quad \forall\, n$ but,
$$\lim_{n \to \infty}\left(2 - \frac{1}{n}\right) = \lim_{n \to \infty}\left(2 + \frac{1}{n}\right) = 2.$$

6. False. For example, take $a_n = (-1)^n$.
7. (*i*) Take $a_n = n$, $b_n = -n$, (*ii*) Take $a_n = n$, $b_n = n$
(*iii*) Take $a_n = b_n = (-1)^n$ (*iv*) Take $a_n = b_n = n$.
9. (*i*) $a_n = n$, $b_n = -n$ (*ii*) $a_n = n^2$, $b_n = -n$
(*iii*) $a_n = n$, $a_n = -n^2$

(*iv*) $a_n = \begin{cases} n^2, & \text{if } n \text{ is odd} \\ n+5, & \text{if } n \text{ is even} \end{cases}$

$b_n = -n$ for all n.

EXERCISE 4 (*c*)

1. Read (*viii*) as $(-1) \cdot \dfrac{1}{n}$

(*i*), (*ii*), (*iii*), (*iv*), (*vi*) and (*viii*) are monotonically increasing
(*v*) and (*vii*) are monotonically decreasing.
2. (*i*), (*ii*), (*iii*), (*iv*), (*v*), (*vii*), (*viii*) converge (*vi*) diverges to ∞.

EXERCISE 4 (*d*)

1. (*i*) $< n >$ (*ii*) $< \dfrac{1}{n} >$ (*iii*) $< (-1)^n >$

(*iv*) the sequence $< 1, 1 + \frac{1}{2}, 2, 1 + \frac{1}{3}, 2 + \frac{1}{2}, 3, 1 + \frac{1}{4}, 2 + \frac{1}{3}, 3 + \frac{1}{2}, 4, \ldots.. >$.
2. (*i*) $\{-1, 1\}$, (*ii*) $\{5\}$, (*iii*) $\{0\}$ (*iv*) no cluster point (*v*) $\{-1, 1\}$
(*vi*) e.
3. (*i*) 19, 1 (*ii*) $\infty, -\infty$ (*iii*) $\dfrac{\sqrt{3}}{2}, \dfrac{-\sqrt{3}}{2}$,

(*iv*) $\infty, -\infty$
7. See question 7 of Exercise 4 (*b*).

EXERCISE 5 (*c*)

1. convergent. 2. convergent.
3. convergent. 4. divergent.
5. convergent 6. convergent.
7. divergent. 8. divergent.
9. divergent. 10. convergent.
12. convergent. 13. convergent.
14. divergent.
16. convergent $0 \le a < 1$ and divergent $a \ge 1$.
17. divergent. 18. divergent.
19. divergent. 20. convergent.

21. convergent.
22. convergent.
23. convergent.
24. convergent.
25. convergent.
26. convergent if $p > 2$ and divergent if $p \le 2$.
27. divergent.

31. (*i*) False. For example, take $a_n = \frac{1}{n^2}$,

(*ii*) False. For example, take $a_n = \frac{1}{n^2}$.

32. convergent for $q > 1$.
33. convergent for $q > 1$ and p arbitrary.

EXERCISE 5 (*d*)

1. divergent
2. convergent
3. convergent
4. divergent
5. divergent
6. convergent
7. convergent
8. convergent if $x \le 1$ and divergent if $x > 1$
9. convergent if $x \le 1$ and divergent if $x > 1$
10. convergent
11. convergent
12. convergent if $x < 1$ and divergent if $x \ge 1$
13. convergent if $x < 1$ and divergent if $x \ge 1$
14. convergent if $\beta > \alpha$ and divergent if $\alpha \ge \beta$.
15. [**Hint.** $\lim_{n \to \infty} \frac{a_n}{a_{n+1}}$ exists and > 1

$$\therefore \quad \lim_{n \to \infty} \frac{n\, a_n}{(n+1)\, a_{n+1}} = \lim_{n \to \infty} \left(\frac{n}{n+1}\right) \cdot \left(\frac{a_n}{a_{n+1}}\right)$$

$$= \lim_{n \to \infty} \left(\frac{1}{1 + \frac{1}{n}}\right) \cdot \left(\lim_{n \to \infty} \frac{a_n}{a_{n+1}}\right) > 1.\ ]$$

16. convergent for $0 \le a \le 1$ and divergent for $a > 1$.

EXERCISE 5 (*e*)

1. convergent
2. convergent if $x < 1$ and divergent if $x \ge 1$
3. convergent
4. convergent
5. convergent
6. convergent if $x < 1$ and divergent if $x \ge 1$
7. convergent.

EXERCISE 5 (*f*)

1. convergent if $x \le 1$ and divergent if $x > 1$
2. convergent if $x \le 1$ and divergent if $x > 1$

3. convergent if $x \le 1$ and divergent if $x > 1$
4. convergent if $x < 1$ and divergent if $x \ge 1$
5. (*i*) convergent if $b > a + 1$ and divergent if $b \le a + 1$
   (*ii*) convergent for each x.
6. convergent if $x^2 < 4$ and divergent if $x^2 \ge 4$.

EXERCISE 5 (*i*)

1. divergent
2. convergent if $\alpha > 2$ and divergent if $\alpha \le 2$
3. convergent if $x \le 1$ and divergent if $x > 1$
4. convergent if $x \le 1$ and divergent if $x > 1$
5. convergent if $x < 1$ and divergent if $x \ge 1$
6. divergent
7. convergent if $x < 4$ and divergent if $x \ge 4$
8. convergent if $x > 2$ and divergent if $x \le 2$.

EXERCISE 5 (*j*)

1. divergent
2. convergent
3. divergent
4. divergent
5. convergent

EXERCISE 5 (*k*)

1. divergent
2. convergent
3. divergent
4. convergent
5. convergent
6. convergent if $p > 1$ and divergent if $p \le 1$.

EXERCISE 5 (*l*)

1. Conditionally convergent
2. Absolutely convergent
3. Absolutely convergent
4. Conditionally convergent
5. Conditionally convergent
6. Conditionally convergent
7. Conditionally convergent
8. Absolutely convergent if $p > 1$, conditionally convergent if $0 < p \le 1$, oscillating if $p \le 0$
9. conditionally convergent
10. Absolutely convergent if $p > 1$, conditionally convergent if $0 \le p \le 1$ and oscillating if $p < 0$
11. Absolutely convergent
12. conditionally convergent
13. Absolutely convergent
14. conditionally convergent
15. Absolutely convergent
16. Absolutely convergent
17. Absolutely convergent
18. divergent
19. Absolutely convergent
20. Absolutely convergent
21. Absolutely convergent for each x
22. Absolutely convergent if $|x| < 1$ and convergent if $|x| \le 1$.

EXERCISE 6 (*a*)

2. (*i*) $\frac{1}{2}$, (*ii*) $\frac{36}{\pi}$, (*iii*) 1.

4. (*i*) 0, (*ii*) ∞, (*iii*) $\frac{a_0}{b_0}$.

5. (*i*) True. For example, let $f(x) = \sin\frac{1}{x}$, $g(x) = -\sin\frac{1}{x}$,

$(x \neq 0)$. Then $\lim_{x \to 0} f(x)$ and $\lim_{x \to 0} g(x)$ do not exist but

$\lim_{x \to \infty} (f + g)(x)$ exists.

Again, define $f(x) = g(x) = \begin{cases} 1 & \text{if } x \in Q \\ -1 & \text{if } x \in R - Q \end{cases}$

Then, $\lim_{x \to 0} f(x)$ and $\lim_{x \to 0} g(x)$ do not exist but $\lim_{x \to 0} (fg)(x) = 1$.

(In fact, we can take limits at any real number instead of 0).

(*ii*) True

(*iii*) False. For example, let $f(x) = x \quad \forall\, x \in R$

and $g(x) = \begin{cases} \sin\frac{1}{x} & \text{if } x \neq 0 \\ 0 & \text{if } x = 0 \end{cases}$

Then, $\lim_{x \to 0} (fg)(x)$ and $\lim_{x \to 0} f(x)$ both exists but $\lim_{x \to 0} g(x)$ does not exist.

6. Let $f(x) = 2x$, $g(x) = x$. Then, $\lim_{x \to 0} f(x) = \lim_{x \to 0} g(x)$.

But $f(x) \neq g(x)$.

7. Same as example of 6.

8. False. For example, let $f(x) = \begin{cases} 1, & x < a \\ -1, & x \geq a \end{cases}$ and $g(x) = 1 \;\forall\, x$.

Then, $f(x) \leq g(x)$ and $\lim_{x \to a} g(x)$ exist but $\lim_{x \to a} f(x)$ does not exist.

EXERCISE 6 (*b*)

1. (*i*) Let $f(x) = \begin{cases} x, & 0 < x < 1 \\ 1, & x = 0 \\ 2, & x = 1 \end{cases}$

(*ii*) $f(x) = x$ if $x \in [1, 2] - \left\{\frac{3}{2}\right\}$ and $f(x) = 5$ at $x = \frac{3}{2}$

(*iii*) $f(x) = \frac{1}{x}$

(*iv*) and (*v*) $f(x) = 1$ if $x \in Q$ and $f(x) = -1$ if $x \in R - Q$

2. (*i*), (*ii*) Removable discontinuity (*iii*) Continuous
(*iv*) Infinite discontinuity (*v*) Continuous
(*vi*) Removable discontinuity at $x = 0$ (*vii*) Discontinuity of first kind
(*viii*) Discontinuity of first kind (*ix*) Discontinuous
(*x*) Removable discontinuity (*xi*) Discontinuity of first kind.

3. Continuous at 1, discontinuous at 2, 3.

4. Removable discontinuity at $x = 0, 1$ and discontinuity of first kind at $x = \frac{1}{2}$.

10. (*i*) $f(x) = \dfrac{1}{x^2+1}$, (*ii*) $f(x) = \dfrac{1}{x^2+1}$, (*iii*) $f(x) = \dfrac{1}{x}$.

18. $a = 3, b = -8$.

EXERCISE 7 (*c*)

1. (*i*) Not valid (*ii*) valid, $c = \dfrac{6-\sqrt{21}}{6}$

(*iii*) Hypothesis not valid but conclusion is (*iv*) Valid, $c = \dfrac{7}{2}$.

EXERCISE 8 (*e*)

1. (*i*) convergent (*ii*) convergent (*iii*) divergent
(*iv*) divergent

4. convergent.

EXERCISE 9 (*a*)

1. $L(f, P) = 3, U(f, P) = 6$. 5. Example 1.

EXERCISE 9 (*c*)

1. (*i*) F (*ii*) F.

EXERCISE 10 (*a*)

1. (*ii*), (*iii*), (*iv*), (*v*), (*vi*), (*vii*), (*viii*), (*ix*) and (*x*) are convergent
(*i*), divergent.

2. (*i*) divergent (*ii*) convergent, 2
(*iii*) convergent, 0 (*iv*) convergent

(*v*) convergent
(*vi*) convergent (*vii*) divergent (*viii*) convergent, I.

EXERCISE 10 (*b*)

1. (*i*) divergent (*ii*) convergent if $n > -1$
(*iii*) convergent (*iv*) divergent
(*v*) convergent
2. (*i*) $n > -1$ and for all m (*ii*) $-1 < m < 0$
7. divergent.

EXERCISE 10 (*c*)

1. (*i*) to (*iv*) all convergent 2. Divergent.

EXERCISE 11 (*a*)

1. (*i*) $\frac{1}{5}\,\mathrm{B}\left(\frac{3}{5}, \frac{1}{2}\right)$ (*ii*) $\frac{8}{3}\,\mathrm{B}\left(\frac{4}{3}, \frac{2}{3}\right)$

(*iii*) $\frac{1}{5}\,\mathrm{B}\left(\frac{3}{5}, 0\right)$ (*iv*) $\frac{1}{2}\,\mathrm{B}\left(\frac{l}{2}, m\right)$

(*v*) $\frac{p^{na+m+1}}{a}\,\mathrm{B}\left(n+1, \frac{m+1}{a}\right)$ 3. $\frac{243}{7280}$

EXERCISE 11 (*b*)

2. $\frac{1}{n}\,\mathrm{B}\left(\frac{m+1}{n}, p+1\right)$; $\frac{1}{396}$

EXERCISE 12

1. $\log(1+\alpha)$ 5. $\frac{\pi}{b}\log\frac{a+b}{b}$.

EXAMINATION PAPER

K.U. KURUKSHETRA

BCA-II Examination, 2008

REAL ANALYSIS

Paper–BCA–206

Time : 3 Hours *Maximum Marks : 100*

Note : Attempt *five* questions in all, selecting at least *one* question from each section.

SECTION-I

1. (*a*) Define Finite set, Denumerable set and Bounded set. Give suitable examples. 6

(*b*) Prove that the union of two open sets is an open set. 7

(*c*) If x is a positive real number, y is any real number then prove that there exists a positive integer n such that $nx > y$. 7

2. (*a*) Show that the sequence $\left\{\frac{1}{3^n}\right\}$ converges to zero. 10

(*b*) Prove that the sequence $\left\{\frac{n+1}{n}\right\}$ is monotonically decreasing and bounded below. 10

3. (*a*) Test the convergence of the infinite series

$$\sum_{n=1}^{\infty} \frac{x^{n-1}}{1+x^n},\ x > 0.$$ 10

(*b*) Show that the infinite series $\sum_{n=1}^{\infty} \frac{1}{n^p}$ converges if $p > 1$ and diverges if $p \leq 1$. 10

4. (*a*) If $\{a_n\}$ is a sequence of positive terms, then prove that

$$\underset{n \to \infty}{\text{Lt}}\ (a_n)^{1/n} = \underset{n \to \infty}{\text{Lt}}\ \frac{a_{n+1}}{a_n}$$

provided that the limit on right hand side exists, whether finite or infinite. 10

(*b*) Show that every bounded sequence has the greatest and the least limit points. 10

SECTION-II

5. (*a*) Verify Rolle's theorem for function

$$f(x) = (x - a)^m (x - b)^n$$

where m, n are positive integers in the interval $[a, b]$. 10

(*b*) Show that the function

$$f(x) = \begin{cases} e^{i/x}; & \text{when } x \neq 0 \\ 0; & \text{when } x = 0 \end{cases}$$

is discontinuous of second kind at $x = 0$. 10

6. (*a*) Show that the infinite product

$$\left(1-\frac{1}{2^2}\right)\left(1-\frac{1}{3^2}\right)\left(1-\frac{1}{4^2}\right)\ldots\ldots \text{ converges to } \frac{1}{2}.$$ 10

(*b*) If the series $\sum_{n=1}^{\infty} a_n$ is convergent and the sequence $\{b_n\}$ is monotonic and bounded, then $\sum_{n=1}^{\infty} a_n b_n$ is convergent.

7. (*a*) Using Taylor's theorem, prove that

$$x-\frac{x^3}{3!} \le \sin x \le x-\frac{x^3}{3!}+\frac{x^5}{5!};\, x > 0.$$ 10

(*b*) Expand $\sin x$ as an infinite series. 10

SECTION-III

8. (*a*) If $f : [a, b] \to \mathrm{R}$ is a bounded function and $p \in p[a, b]$, prove that

$$m(b-a) \le \mathrm{L}(p, f) \le \mathrm{U}(p, f) \le \mathrm{M}(b-a)$$

where m, M are the infirmum and supermum of f on $[a, b]$ and L and U are lower and upper Darboux sum of f.

(*b*) If $f : [a, b] \to \mathrm{R}$ is a bounded function, then for $\in > 0$ there exists a $\delta > 0$ such that Prove that

$$\mathrm{U}(\mathrm{P}, f) < \int_a^b f(x)\, dx + \in \quad \text{for each} \quad p \in p\,[a, b] \text{ with } \|p\| < \delta.$$

9. (*a*) Examine the convergence of Improper Integral

$$\int_0^{2a} \frac{dx}{(x-a)^2}.$$ 10

(*b*) If m and n are positive integers, then

$$\mathrm{B}(m, n) = \frac{\overline{m-1}!.\,\overline{n-1}!}{\overline{m+n-1}!}.$$ 10

10. (*a*) Using differentiation under integral sign for $y > 0$, prove that

$$\int_0^{\infty} e^{-xy}\frac{\sin x}{x}\, dx = \cot^{-1} y.\Gamma$$ 10

(*b*) Prove that $\Gamma\left(\frac{1}{2}\right) = \sqrt{\pi}$. 10

Bibliography

1. Apostol, T.M., 'Mathematical Analysis', Addison-Wesley Publishing Company (1974).
2. Bartle, RG., 'The Elements of Real Analysis', John Wiley and Sons Inc. (2nd edition 1982).
3. Burkil J.C., 'A First Course in Mathematical Analysis', Cambridge University Press (1970).
4. Benjamin Williamson 'An Elementary Treatise on Integral Calculus' Scientific Book Company, Patna (India) (1955).
5. Gibson, G.A., 'Advanced Calculus', MacMillan and Company Ltd., London (1958).
6. Goldberg, RR, 'Methods of Real Analysis', (Indian edition), Oxford and IBH Publishing Company, Calcutta-Bombay-New Delhi (1970).
7. Hardy, G.H., 'A Course of Pure Mathematics' (10th edition), Cambridge University Press (1952).
8. Knopp, K., 'Infinite Sequences and Series', Dover Publications, Inc, New York (1956).
9. Kolmogorov, A.N. and Fomin, S.V., 'An Introductory Real Analysis', (Revised English edition by Richard A. Silverman), Printice Hall Inc., Englewood Chiffs, New Jersey (1970).
10. Malik, S.C., 'Mathematical Analysis' Wiley Eastern Limited, New Delhi, (1984).
11. Malik, S.C., 'Introduction to Pointwise Convergence', Wiley Eastern Limited, New Delhi (1984).
12. Nikolsky, S.M., 'A Course of Mathematical Analysis', Vol. I., MIR Publications, Moscow (1977).
13. Phillips, E.G., 'A Course of Analysis', Cambridge University Press (1960).
14. Piskunov, N., 'Differential and Integral Calculus', MIR Publishers, Moscow (1969) (English translation).
15. 'Protter, M.H. and Movey, C.B., 'Moderen Mathematical Analysis', Addison Wesley Publishing Company (4th edition) (1972).
16. Royden, H.L., 'Real Analysis' (2nd edition), The MacMillan Company, New York (1968).
17. Rudin, W., 'Principles of Mathematical Analysis', Mc Graw Hill book Company, New York (1954).
18. Saxena, S.C. and Shah, S.M., 'Introduction to Real Variable Theory' Prentice-Hall of India, New Delhi (1987).
19. Shanti Narayan, 'Real Analysis' S. Chand & Co. New Delhi, (1968).
20. Titchmarsh E.C., 'The Theory of Functions' (2nd edition), English Language Book Society, Oxford University Press (1939).
21. Wason, S.K. and Prakash, R., Real Analysis', Tata McGraw-Hill Publishing Co., New Delhi (1985).
22. Widder, D.V., 'Advanced Calculus', (2nd edition) Printice-Hall of India, New Delhi (1964).

